Kinematic Analysis of Parallel Manipulators by Algebraic Screw Theory

Jaime Gallardo-Alvarado

Kinematic Analysis of Parallel Manipulators by Algebraic Screw Theory

Springer

Jaime Gallardo-Alvarado
Department of Mechanical Engineering
Instituto Tecnologico de Celaya
Celaya, Mexico

ISBN 978-3-319-31124-1 ISBN 978-3-319-31126-5 (eBook)
DOI 10.1007/978-3-319-31126-5

Library of Congress Control Number: 2016939238

Printed on acid-free paper

This Springer imprint is published by Springer Nature
The registered company is Springer International Publishing AG Switzerland

A la entrañable memoria de mis padres,
Josefina y José.
A mi esposa Alma Delia e hijos Delia
Josefina,
Juan Pablo y José.
To my family and friends!

Preface

Robotics is a prolific and intensive research field whose applications have had a highly positive impact on both academia and industry. While old and modern theories are applied to solve challenging problems and improve existing solutions, mathematical models are currently used in the design, simulation, and control of industrial robots. Certainly, the possibilities offered by robot manipulators are immense, causing the variety of robot designs to continue to grow, which demands efficient and confident mathematical methods of analysis. It is noticeable that despite the diversity of journals and books concerned with the study of robot manipulators, there are particular issues that deserve to be deeply investigated or in some cases elucidated in the light of modern kinematics; these include singularities and higher-order kinematic analyses of robot manipulators. In this area, contributions approaching the jerk and hyper-jerk analyses of robot manipulators via screw theory are rather scarce, perhaps because of the lack of practical applications or the lack of credibility about the validity of the equations in screw form of the acceleration analysis of kinematic chains introduced almost two decades ago. In this book, relevant fundamentals of screw theory concerning the so-called first-order kinematic analysis of a rigid body are carefully reviewed through the introduction of propositions that are confirmed through detailed and explicit proofs. In addition, the acceleration, jerk, and hyper-jerk analyses of a rigid body are exhaustively investigated by taking advantage of the method outlined for the first-order kinematic analysis. Because of the recent successful introduction of parallel manipulators in the industry, those mechanical devices are selected as the means to exemplify the application of screw theory exposed in this book. With the material provided in this book, readers can extend screw theory into the kinematics of optional order of a rigid body and apply it to the analysis of parallel manipulators. Illustrative examples and exercises to reinforce learning are provided. Of particular note, the kinematics of emblematic parallel manipulators, such as the Delta robot as well as the original Gough and Stewart platforms, are revisited. In addition to screw theory, we apply new methods devoted to simplifying the corresponding forward displacement analysis, a challenging task for most parallel manipulators.

The material presented in this book is based on research published by the author over the last 20 years, including his doctoral dissertation developed at Instituto Tecnológico de la Laguna, México, under the supervision of Prof. José María Rico-Martínez. To the author's best knowledge, this is one of the first books devoted to the higher-order kinematic analyses of a rigid body by screw theory and its application to the study of mechanisms possessing parallel kinematic structures. The author is confident about the usefulness of the material provided in this book, which represents an excellent argument to ensure that screw theory is not confined to the so-called first-order kinematic analysis. The material presented here provides tools for engineers, researchers, and students dealing with the kinematics of parallel manipulators. As a consideration for readers unfamiliar with screw theory, the book contains a reasonable quantity of examples. Furthermore, as an unusual outcome, full answers to selected exercises are provided at the end of the volume.

Celaya, GTO, Mexico
May 2016

Jaime Gallardo-Alvarado

Acknowledgments

The author wishes to thank the support of two Mexican institutions: the Consejo Nacional de Ciencia y Tecnología (CONACYT) and the Tecnológico Nacional de México (TecNM). The author is especially indebted to Michael Luby, Editor, Mechanical Engineering, Springer, for his advice and guidance in the preparation of the final text. Also, Nicole Lowary and Brian Halm are to be thanked for the assistance they provided during the publication of this book.

Contents

Part VI Solved Exercises

List of Tables

List of Figures

Part I
General Introduction

Chapter 1
An Overview of the Theory of Screws

1.1 The Theory of Screws: Historical Contributions

The geometry of lines plays a central role in three-dimensional mechanics. In fact, lines can be used to represent the screw axes of spatial motions and also to represent the action of forces. With this assumption in mind, a screw may be understood as two concatenated vectors: The first vector, namely the primal part of the screw, is a unit vector along the axis of the screw, while the second vector, namely the dual part, is the moment produced by the primal part about a reference point named the reference pole.

The kinematics and dynamics of a rigid body may be elegantly addressed through the perception of screw theory. On that topic, the principle of transference is a noticeable result of screw theory (McCarthy and Soh 2010). One benefit of this principle is the analogy observed between geometric calculations for points using vectors with geometric calculations for lines generated by the substitution of vectors by screws. Screw theory allows us to express displacements, velocities, forces, and torques in three-dimensional space based on the idea that any rigid body motion can be represented as the inseparable union of a rotation about an axis and a translation along the same axis, where the axis is coincident with the object or particle undergoing displacement. The following results are among the highlights of the representation of rigid body motion.

Every rigid body motion can be realized by a rotation about an axis combined with a translation parallel to the same axis.
The most general rigid body displacement can be produced by a translation along a line followed (or preceded) by a rotation about that line.
The displacement in the Euclidean space of a rigid body can be represented by a rotation along a unique axis and a translation parallel to the axis.

Although Mozzi and Cauchy are credited with similar earlier results, this well-known theorem is attributed to Chasles (1830). It is not surprising that the terms

J. Gallardo-Alvarado, *Kinematic Analysis of Parallel Manipulators by Algebraic Screw Theory*, DOI 10.1007/978-3-319-31126-5_1

DISCORSO
MATEMATICO
SOPRA
IL ROTAMENTO MOMENTANEO DEI CORPI
DEL
CAVALIERE GIULIO MOZZI
Patrizio Fiorentino.

IN NAPOLI MDCCLXIII.
Nella Stamperia di Donato Campo
Con licenza de' Superiori.

Fig. 1.1 The portrait of the work *Discorso matematico sopra il rotamento momentaneo dei corpi* (Mozzi 1763)

"rotation" and "translation" may be related to other physical quantities such as force and couple. In that way, Poinsot's theorem establishes that

> *Rigid body action is equivalent to a wrench on a screw, that is, a force along a unique line and a couple parallel to the line.*

This is an old theorem that was formulated almost two centuries ago. Chasles' and Poinsot's theorems were decisive for the development of screw theory. It is worth noting that prior to the contributions of Chasles and Poinsot, Mozzi published in 1763 a work entitled *Discorso matematico sopra il rotamento momentaneo dei corpi* (see Fig. 1.1) approaching the same issue.

In his contribution Mozzi certainly describes the general motion of a rigid body based on the concept of screw axis. Hence, the instantaneous screw axis is frequently referred to as Mozzi's axis. Several decades later, Giorgini proved that although the results reported by Mozzi were in general accurate, the procedures Mozzi used had been partially incorrect. In 1866, Plñucker coined the term "dynamo" to describe the simultaneous effect of force and couple; this concept was widely studied by Routh in 1892. However, despite the pioneering contribution of Mozzi, today the framework of the theory of screws is credited to Ball (1900). To clarify this point, the fundamentals of screw theory were revisited by Lipkin and Duffy (2002). To provide a proper accreditation of the subject, note that part of the monumental work of Ball (1900) is based on Plücker's investigations concerning with the geometry of lines.

The concept of motors, or in other words the combined action of moments and rotors, was introduced by Clifford (1873) in his biquaternion algebra. However, the application of motors to the kinematics of a rigid body, which appears to be the immediate task, was curiously overlooked by Clifford. A half-century later, von Mises published, in two parts, his *Motorrechnung, ein neues Hilfsmittel in Mechanik.* In the first part, von Mises (1924a) introduced the concept of dual motor product and its role in the measure of the instantaneous change of a motor associated to a rigid body under the action of a second motor. In the second part, von Mises (1924b) successfully applied his motor algebra in several problems, including the derivation of the general form of the equations of motion of rigid bodies. Without a doubt, von Mises's seminal contributions were decisive for the beginning of a productive research field dealing with the higher-order kinematic analyses of a rigid body. However, although by the time scientists recognized the importance of the motor algebra, its practical applications were overlooked by most engineers, consider, for instance, that the pioneering work of von Mises was reprinted until 1998 (von Mises 1996).

In 1965, Dimentberg published his work entitled *The Screw Calculus and Its Applications in Mechanics*. In that contribution Dimentberg introduced the calculus of screws based on the framework of vector algebra and its applications in mechanics. Three years later, Dimentberg's work was translated into English (the corresponding portrait is shown in Fig. 1.2) by the Foreign Technology Division of U.S.A.F. (available through DTIC). According to Dimentberg (1965), after Clifford (1873) Study was the first mathematician to investigate the analogy between the screw and the twist of instantaneous kinematics. Furthermore, the analogy between the screw and the wrench of statics was also deeply investigated by Study through his algebra of complex numbers. Based on the work of von Mises, Brand (1947) in his book *Vector and Tensor Analysis*, which contains a chapter on motor algebra,

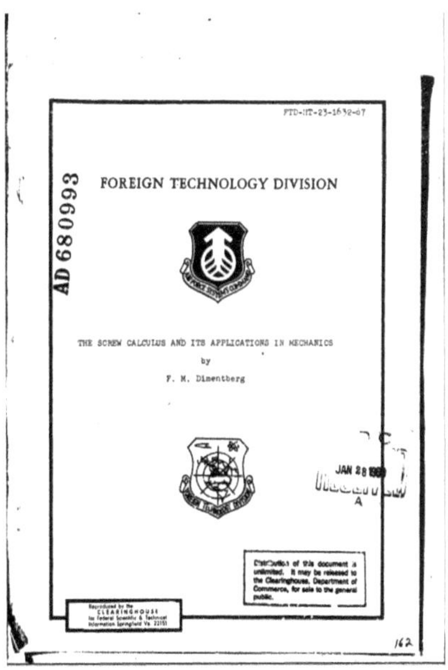

Fig. 1.2 Book cover of the work *The Screw Calculus and Its Applications in Mechanics* (Dimentberg 1965)

introduced a kinematic entity termed "accelerator" that allowed us to apply screw theory to the acceleration analysis of a rigid body, opening the door to glimpse the concept of the time derivative of a screw. However, applications of Brand's accelerator to the acceleration analysis of kinematic chains began five decades later because of the difficulty in expressing the accelerator in screw form. In fact, the acceleration analysis of kinematic chains by expressing the accelerator in screw form dates back to 1996 when Rico and Duffy introduced a kinematic entity termed "the reduced acceleration state" of a rigid body.

In the mid-1900s, the theory of screws was revitalized mainly because of the contributions of Hunt and Phillips. Phillips and Hunt (1964) reported a complete derivation of the theorem of three axes with application to the spatial motion of a rigid body. Roth (1967) and Keller (1979) approached the problem of finite screw displacements of a rigid body. At the beginning of the 1960s at Monash University, Hunt wrote his *Screw Systems in Spatial Kinematics*, an internal report that was later formally published as one of the most remarkable contributions concerned with modern screw theory (Hunt 1978). Certainly, the book entitled *Kinematic Geometry of Mechanisms* is considered the renaissance of screw theory. Other prominent works contributed by the Australian school include Phillips (1984, 1990). Using the cylindroid, Waldron (1966) showed how to determine the instantaneous screws in mechanisms having one degree of freedom. Woo and Freudenstein (1970) investigated the linear dependency of screw systems by using a variant of the representation of lines in Plücker coordinates. According to Sugimoto and Duffy (1982), Davis and Primrose were the first to apply linear algebra to the screw systems in order to represent the joint rates of mechanisms. Later, Keler (1970, 1971) introduced the dual vector algebra to approach the acceleration analysis of mechanisms. By this time, other prominent contributors included Yang and Freudenstein (1964) and Yang (1971a,b, 1974). Dai (2006) made a comprehensive and deep review about the contributions of Chasles, Cayley, Hamilton, Study, and Clifford concerned with the mathematical model of the first-order kinematics of a rigid body. Sugimoto and Duffy (1982) addressed the velocity analysis of mechanisms by using linear algebra and screw theory. A few years later, screw theory was applied not only to solve the velocity analysis of robots (Sugimoto 1984) but also to solve the corresponding dynamic analysis; see, for instance, Sugimoto (1987, 1989). The determination of overconstrained mechanisms was also approached by Sugimoto (1990) as an extension of the so-called motor algebra. Samuel et al. (1991) used invariant properties of orthogonal matrices to demonstrate the equivalence of screw geometry with the matrix representation of the Euclidean group. Chevallier (1991) proved that a screw may be considered the geometric representation of a helicoidal vector field. The concept of differential of a screw was introduced by Bokelberg et al. (1991) in an attempt to redefine the spatial infinitesimal motion. The acceleration analysis of a rigid body was the immediate benefit of such a contribution (Ridley et al. 1992).

Finally, the theory of screw systems and its potential applications to the finite kinematics of a rigid body have recursively attracted the attention of researchers; see, for instance, Gibson and Hunt (1990a,b), Parkin (1992), Rico-Martínez and

Duffy (1992a,b,c), Huang and Roth (1994), Dai and Jones (2001, 2002, 2003), Huang et al. (2008), Zlatanov (2012), Carricato and Zlatanov (2014), and Wu et al. (2016).

1.2 Notable Scientists in Screw Theory

Science is the means by which the fog of ignorance can be scattered. With each scientific advance, we have an opportunity to better understand natural laws as well as ourselves. However, knowledge cannot progress by itself; it needs the support, and talent, of extraordinary people whose efforts and dedication allow us to live in a better world, in harmony with nature. With that in mind, this section presents short biographies of scientists whose contributions were decisive for the development of screw theory.

Sir Robert Stawell Ball (1840–1913) An astronomer and mathematician, Ball was born in Dublin. He is formally credited as the founder of the theory of screws. Curiously, screw theory was his avocation but not his principal research. In fact, Ball was a prominent Royal Astronomer of Ireland. In 1876, Ball published a work entitled *The Theory of Screws: A study in the Dynamics of a Rigid Body* (Ball, 1876). That work earned him the Cunningham Medal of the Royal Irish Academy in 1879. In 1892, he was appointed Lowndean Professor of Astronomy and Geometry at Cambridge University, at the same time becoming director of the Cambridge Observatory. Almost one decade later, Cambridge University Press published a book entitled *A Treatise on the Theory of Screws*, a seminal emblematic contribution that today is considered the most comprehensive text on screw theory. It is worthwhile noting the importance modern applications like robot mechanics, mechanical design, computational geometry, and multibody dynamics have gained through Ball's monumental work published more than one century ago.

Giuseppe Giulio Mozzi del Garbo (1730–1813) Mozzi was an Italian mathematician, politician, and poet. In 1763, Mozzi published a work entitled *Discorso matematico sopra il rotamento momentaneo dei corpi*. However, that contribution, which was prior to Chasles' work, concerned the representation of rigid body motion and passed almost unnoticed. Today Mozzi's pioneering contribution is gaining in reputation because of relevant concepts applied in rational mechanics like the novelty of the theorem that would bear his name as well as the concept of force and couple specified by Poinsot.

Louis Poinsot (1777–1859) A mathematician and physicist, Poinsot was born in Paris. He proved that any system of forces acting on a rigid body can be reduced into a single force and a couple in a plane normal to the force, yielding a new research field called geometrical mechanics. Previous work done on the motion of a rigid body had been purely analytical without visualization of the motion. Poinsot was the first to describe the motion of the angular velocity vector of a rigid body with one point fixed, usually its center of mass, and proved that the endpoint of the angular velocity vector moves in a plane perpendicular to the angular momentum of the rigid body. From 1804 until 1809, Poinsot was a mathematics professor at the Lycée Bonaparte in Paris. Then he was appointed as inspector general of the Imperial University. On the death of Joseph-Louis Lagrange in 1813, Poinsot was elected to fill his place at the Académie des Sciences. Poinsot is remembered for his high principles. In fact, Poinsot was determined to publish only fully developed results and to present them with clarity and elegance.

Michel Chasles (1793–1880) Born in Épernon, France, Chasles introduced the concept of twist in kinematics in 1830. His description of the spatial motion of a rigid body as screw displacement was decisive for the development of the theories of rigid body dynamics. With the publication of one of his major works entitled *Historical View of the Origin and Development of Methods in Geometry*, which is a study of the method of reciprocal polars in projective geometry, Chasles gained considerable respect as a mathematician as well as a historian of mathematics. Then Chasles was appointed Professor at the École Polytechnique in 1841; a few years later he was awarded a chair at the Sorbonne. Chasles' name is one of the 72 inscribed on the Eiffel Tower.

Sir William Rowan Hamilton (1805–1865) Hamilton was an Irish physicist, astronomer, and mathematician who made important contributions to classical mechanics, optics, and algebra. His major contribution was the reformulation of Newtonian mechanics, now called Hamiltonian mechanics, enhancing the development of quantum mechanics. In pure mathematics, he is recognized as the inventor of quaternion algebra. A peculiar story is behind this, and the Brougham Bridge is witness. In 1830, Hamilton discovered the cylindroid, a surface that describes a two-system of screws. In 1833, Hamilton read a paper concerned with complex numbers expressed as algebraic couples, or ordered pairs of real numbers, to the Royal Irish Academy. A year later he applied it to rigid body dynamics in a work entitled *On a General Method in Dynamics*. Hamilton was honored with several awards throughout his prolific life; for instance, he was twice awarded the Cunningham Medal of the Royal Irish Academy.

Arthur Cayley (1821–1895) Cayley was a British mathematician who helped found the modern British school of pure mathematics. Cayley and Christian Klein noted that projective geometry is equally powerful in its application to non-Euclidean geometries. In pure mathematics, the Cayley–Hamilton theorem is one of the most celebrated concepts of linear algebra. Prior to the work of Plücker, in 1860 Cayley devised the system of six homogeneous coordinates of a line. Curiously, Klein deeply exploited the subject, influenced by Plücker and not by Cayley. The majority of Cayley's contributions, an amazing quantity of about 966 papers, are in *The Collected Mathematical Papers of Arthur Cayley*, 13 volumes, indexed in a 14th (Cambridge, 1889–1898). The printing of the first seven volumes was supervised by Cayley himself.

Julius Plücker (1801–1868) A German mathematician and physicist, in 1836 Plücker was appointed professor of physics at the University of Bonn. He is considered the inventor of what is now known as line geometry. In projective geometry, "Plücker coordinates" refer to a set of homogeneous coordinates originally introduced to embed the set of lines in three dimensions as a quadric in five dimensions; today that is part of the theory of Grassmannians. Two years before his death, Plücker was honored with the Copley Medal from the Royal Society in 1866.

Hermann Günther Grassmann (1809–1877) A German mathematician, physicist, and neohumanist, Grassmann is warmly remembered for the development of a general calculus of vectors exposed in his masterpiece *The Theory of Linear Extension, a New Branch of Mathematics*. In that contribution Grassmann proposed a new foundation for all of mathematics, showing that once geometry is put into the algebraic context he advocated, the number of possible dimensions in space is unbounded. In fact, Grassmann developed a groundbreaking n-dimensional algebra of space. In his 1862 *Ausdehnungslehre*, Grassmann reported a study about the reduction and decomposition of geometric structures into two components with applications to screw theory.

Christian Felix Klein (1849–1925) A German mathematician, Klein is remembered for his work comprising group theory, complex analysis, non-Euclidean geometry, and the connections between geometry and group theory. Klein finished the second part of Plücker's work *Neue Geometrie des Raumes*, after Plücker's death in 1868. When Klein was only 23, he was appointed professor at the University of Erlangen–Nuremberg. Klein defined geometry in terms of groups and invariants and completed the unification of projective and metrical geometries proposed by Cayley. Klein extended Ball's theory of screws to the study of the finite motions of a rigid body.

William Kingdon Clifford (1845–1879) Clifford was an English mathematician and philosopher. Following the work of Grassmann, he introduced what today is called geometric algebra, a special case of the Clifford algebra named in his honor. As part of the Clifford algebra, multivectors represent not only line segments but also plane segments in a comprehensive geometrical style. In that area, quantities like screws and wrenches may be represented by combinations of multivectors. In 1871, Clifford was appointed professor of mathematics and mechanics at University College London. His report entitled *On the Space-Theory of Matter*, published by Cambridge in 1876, anticipated Einstein's theory of general relativity by four decades. In 1878, Clifford published a seminal work entitled *Applications of Grassmann's Extensive Algebra.* Then he succeeded in unifying Grassmann's outer product with the quaternions, developed by William Rowan Hamilton. The importance of general Clifford algebras has continuously grown over time. Meanwhile, their isomorphism classes as real algebras have been identified within other mathematical resources as well as for the elusive quaternions. The journal *Advances in Applied Clifford Algebras* continues Clifford's legacy in kinematics and abstract algebra.

Marius Sophus Lie (1842–1899) A Norwegian mathematician, Lie created the theory of continuous symmetry, with applications to the study of geometry and differential equations. One of Lie's greatest achievements was the discovery that continuous transformation groups may be better understood by linearizing them and studying the corresponding generated vector fields. Lie proved that the infinitesimal generators are subject to a linearized version of the group law. In 1872, Lie showed the isomorphism between Plücker coordinates and sphere coordinates. Today the Lie bracket is widely used in the acceleration analysis of a rigid body, with applications to robot kinematics.

Christian Hugo Eduard Study (1862–1930) A German mathematician, Study is remembered for his contributions on the invariant theory of ternary forms and spherical trigonometry. Study was called in 1904 by the University of Bonn because the position held by Rudolf Lipschitz was vacant. He applied the principle of transference between spherical and spatial motion, using biquaternions and dual numbers.

Aleksandr Petrovich Kotelnikov (1865–1944) A Russian mathematician specializing in geometry and kinematics, in 1895 Kotelnikov developed screw theory using biquaternions. Kotelnikov obtained his doctorate in 1899 for a thesis entitled *The Projective Theory of Vectors*, a generalization of his master thesis, *The Cross-Product Calculus and Certain of Its Applications in Geometry and Mechanics.*

Then he was appointed professor and head of the Department of Pure Mathematics at the University of Kiev. In 1927, he published one of his major works,*The Principle of Relativity and Lobachevsky's Geometry*. He also worked on quaternions and applied them to mechanics and geometry.

Richard von Mises (1883–1953) Born in Lemberg, then part of Austria-Hungary, von Mises was a prolific German mathematician who worked in several areas, including solid mechanics, fluid mechanics, aerodynamics, aeronautics, statistics, and probability theory. In 1919, he was appointed director of the Institute of Applied Mathematics of the University of Berlin. Although von Mises is popularly known as the scientist who developed the distortion energy theory of stress, which is one of the most important concepts used by engineers in material strength calculations, his *motor algebra* exposed in two contributions entitled *Motorrechnung, ein neues hilfsmittel der mechanik* and *Anwendungen der motorrenchung* represents a remarkable advance in the modeling of higher-order kinematics of a rigid body through the so-called motor product, perhaps an unexpected result for von Mises himself.

F. M. Dimentberg (1905–1999) Dimentberg was a Russian mathematician who investigated the kinematics of spatial mechanisms using motor algebra and Plücker coordinates, as well as the screw properties of inertia and stiffness. His major contribution was the book entitled *The Screw Calculus and Its Applications in Mechanics*, which was translated into English by the Foreign Technology Division of U.S.A.F. in 1968. In his transcendental work, Dimentberg set forth material covering a wide variety of topics ranging from sliding vectors to screw algebra.

Ferdinand Freudenstein (1926–2006) An American physicist and engineer, Freudenstein is considered the father of modern kinematics in the United States. Screw theory was revitalized after a persistent dissemination and popularization of the contributions of Ball, Klein, Brand, and Dimentberg, among others, by Freudenstein.

Kenneth Henderson Hunt (1920–2002) Because of his major works, *Mechanisms and Motion* and *Kinematic Geometry of Mechanisms*, two books published in 1949 and 1978, respectively, Hunt is considered one of the rediscoverers of screw theory. He applied screw theory to the kinematics of spatial mechanisms, with special emphasis on the analysis of singularities. He was also the author (with J. Davidson) of the book *Robots and Screw Theory* that was published in 2004.

Jack Raymond Phillips (1923–2009) Phillips and Hunt are gratefully remembered as the scientists who rediscovered many of the results of Ball's instantaneous screw system calculus and applied them to early mechanism problems. In 1984, the major work of Phillips was published by Cambridge University Press in two volumes. The first volume, a book entitled *Freedom in Machinery, Volume 1, Introducing Screw Theory*, deals with problems of freedom and constraint in machinery that are solved by investigating the corresponding screw system. Meanwhile, in the second volume, *Freedom in Machinery, Volume 2, Screw Theory Exemplified*, Phillips applies the theory of screws in the kinematic analysis and synthesis of mechanisms. The works of Hunt and Phillips contributed splendidly in renewing the framework of screw theory with fundamental applications in robot analysis.

References

Ball, R. S. (1876). *The theory of screws: a study in the dynamics of a rigid body*. Dublin: Hodges, Foster and Company.

Ball, R. S. (1900) (reprinted 1998) *A treatise on the theory of screws*. Cambridge: Cambridge University Press.

Bokelberg, E. H., Hunt, K. H., & Ridley, P. R. (1991). Spatial motion-I: points of inflection and the differential geometry of screws. *Mechanism and Machine Theory, 27*(1), 1–15.

Brand, L. (1947). *Vector and tensor analysis*. New York: Wiley.

Carricato, M., & Zlatanov, D. (2014). Persistent screw systems. *Mechanism and Machine Theory, 73*, 296–313.

Chasles, M. (1830). Notes on the general properties of a system of 2 identical bodies randomly located in space; and on the finite or infinitesimal motion of a free solid body. *Bulletin des Sciences Mathematiques, Astronomiques, Physiques et Chimiques, 14*, 321–326.

Chevallier, D. P. (1991). Lie algebras, modules, dual quaternions and algebraic methods in kinematics. *Mechanism and Machine Theory, 26*(6), 613–627.

Clifford, W. K. (1873). Preliminary sketch of bi-quaternions. *Proceedings London Mathematical Society, 4*(64/65), 381–395.

Dai, J. S. (2006). An historical review of the theoretical development of rigid body displacements from Rodrigues parameters to the finite twist. *Mechanism and Machine Theory, 41*(1), 41–52.

Dai, J. S., & Jones, J. R. (2001). Interrelationship between screw systems and corresponding reciprocal systems and applications. *Mechanism and Machine Theory, 36*(5), 633–651.

Dai, J. S., & Jones, J. R. (2002). Null space construction using cofactors from a screw algebra context. *Royal Society of London Series A: Mathematical, Physical and Engineering Sciences, 458*(2024), 1845–1866.

Dai, J. S., & Jones, J. R. (2003). A linear algebraic procedure in obtaining reciprocal screw systems. *Journal of Robotic Systems, 20*(7), 401–412.

Dimentberg, F. M. (1965). *Screw calculus and its applications to mechanics*, Moscow (in Russian), English translation: AD680993 (Clearinghouse for Federal and Scientific for Technical Information).

Gibson, C. G., & Hunt, K. H. (1990a). Geometry of screw systems-1. *Mechanism and Machine Theory, 25*(1), 1–10.

Gibson, C. G., & Hunt, K. H. (1990b). Geometry of screw systems-2, classification of screw systems. *Mechanism and Machine Theory, 25*(1), 11–27.

Huang, C., & Roth, B. (1994). Analytical expressions for the finite screw systems. *Mechanism and Machine Theory, 29*(2), 207–222.

Huang, C., Sugimoto, K., & Parkin, I. (2008). The correspondence between finite screw systems and projective spaces. *Mechanism and Machine Theory, 43*(1):50–56.

Hunt, K. H. (1978). *Kinematic geometry of mechanisms*. Oxford: Oxford University Press.

Keler, M. L. (1970). Analyse und Synthese räumlicher, sphärischer und ebner Getriebe in dual-komplexer Darstellung. *Feinwerktechnik, 74*, 341–351.

Keler, M. L. (1971). Analyse und Synthese räumlicher, sphärischer und ebner Getriebe in dual-komplexer Darstellung. In *Proceedings Third World Congress for the Theory of Machines and Mechanisms* (Vol. B, pp. 49–62). Kupari, Yugoslavia.

Keler, M. (1979). Dual-vector half-tangents for the representation of the finite motion of rigid bodies. *Environment and Planning B, 6*(4), 403–412.

Lipkin, H., & Duffy, J. (2002). Sir Robert Stawell Ball and methodologies of modern screw theory. *Proceedings Institution of Mechanical Engineers, Part C: Journal of Mechanical Engineering Science, 216*(1), 1–11.

McCarthy, J. M., & Soh, G. S. (2010). *Geometric design of linkages* (2nd ed.). New York: Springer.

Mozzi, G. (1763). *Discorso Matematico Sopra il Rotamento Momentaneo dei Corpi* [Mathematical treatise on the temporally revolving of bodies]. Napoli: Stamperia di Donato, Campo.

Parkin, I. A. (1992). A third conformation with the screw systems: finite twist displacements of a directed line and point. *Mechanism and Machine Theory, 27*(2), 177–188.

Phillips, J. (1984). *Freedom in machinery. Introducing screw theory* (Vol. 1). Cambridge: Cambridge University Press.

Phillips, J. (1990). *Freedom in machinery. Screw theory exemplified* (Vol. 2). Cambridge: Cambridge University Press.

Phillips, J., & Hunt, K. H. (1964). On the theorem of the three axes in the spatial motion of three bodies. *Australian Journal of Applied Sciences, 15*, 267–287.

Rico, J. M., & Duffy, J. (1996). An application of screw algebra to the acceleration analysis of serial chains. *Mechanism and Machine Theory, 31*(4), 445–457.

Rico-Martínez, J. M., & Duffy, J. (1992a). Orthogonal spaces and screw systems. *Mechanism and Machine Theory, 27*(4), 451–458.

Rico-Martínez, J. M., & Duffy, J. (1992b). Classification of screw systems-I. One- and two-systems. *Mechanism and Machine Theory, 27*(4), 459–470.

Rico-Martínez, J. M., & Duffy, J. (1992c). Classification of screw systems-II. Three-systems. *Mechanism and Machine Theory, 27*(4), 471–490.

Ridley, P. R., Bokelberg, E. H., & Hunt, K. H. (1992). Spatial motion II. Acceleration and the differential geometry of screws. *Mechanism and Machine Theory, 27*(1), 17–35.

Roth, B. (1967). On the screw axes and other special lines associated with spatial displacements of a rigid body. *Journal of Engineering for Industry, 89*, 102–112.

Samuel, A. E., McAree, R. R., & Hunt, K. H. (1991). Unifying screw geometry and matrix transformations. *International Journal of Robotics Research, 10*(5), 454–472.

Sugimoto, K. (1984). Determination of joint velocities of robots by using screw theory. *ASME Journal of Mechanisms, Transmissions and Automation in Design, 106*(2), 222–227.

Sugimoto, K. (1987). Kinematic and dynamic analysis of parallel manipulators by means of motor algebra. *ASME Journal of Mechanisms, Transmissions and Automation in Design, 109*(1), 3–7.
Sugimoto, K. (1989). Computational scheme for dynamic analysis of parallel manipulators. *ASME Journal of Mechanisms, Transmissions and Automation in Design, 111*(1), 29–33.
Sugimoto, K. (1990). Existence criteria for overconstrained mechanisms. An extension of motor algebra. *ASME Journal of Mechanical Design, 112*(3), 295–298.
Sugimoto, K., & Duffy, J. (1982). Application of linear algebra to screw systems. *Mechanism and Machine Theory, 17*(1), 73–83.
von Mises, R. (1924a). Motorrechnung, ein neues hilfsmittel der mechanik. *Zeitschrift für Angewandte Mathematik und Mechanik, 4*(2), 55–181.
von Mises, R. (1924b). Anwendungen der motorrenchung. *Zeitschrift für Angewandte Mathematik und Mechanik, 4*(3), 193–213.
von Mises, R. (1996). In E. J. Baker, K. Wohlhart (Eds.), *Motor calculus: a new theoretical device for mechanics*. Graz: Institute for Mechanics, University of Technology Graz.
Waldron, K. J. (1966). The constraint analysis of mechanisms. *Journal of Mechanisms, 1*(2), 101–115.
Woo, L. S., & Freudenstein, F. (1970). Application of line geometry to theoretical kinematics and the kinematic analysis of mechanical systems. *Journal of Mechanisms, 5*(3), 417–460.
Wu, Y., Löwe, H., Carricato, M., & Li, Z. (2016). Inversion symmetry of the Euclidean group: theory and application in robot kinematics. *IEEE Transactions on Robotics, 32*(2), 312–326.
Yang, A. T. (1971a). Acceleration analysis of spatial four-link mechanisms. *Journal of Engineering for Industry, 88*(3), 296–300.
Yang, A. T. (1971b). Inertia force analysis of spatial mechanisms. *Journal of Engineering for Industry, 93*(1), 27–32.
Yang, A. T. (1974). Calculus of screws. In W. R. Spillers (Ed.), *Basic questions of design theory* (pp. 266–281). New York: Elsevier.
Yang, A. T., & Freudenstein, F. (1964). Application of dual-number quaternion algebra to the analysis of spatial mechanisms. *ASME Journal of Applied Mechanics, 31*(2), 300–308.
Zlatanov, D. (2012). The representation of the general three-system of screws by a sphere. *Mechanism and Machine Theory, 49*(3), 315–331.

Chapter 2
An Overview of Parallel Manipulators

2.1 Typical Parallel Manipulators

The word *robota* means "hard work" or "serf labor" in Czech and was introduced in 1920 by Karel Čapek, one of the most influential Czech writers of the twentieth century, in his *Rossum's Universal Robots*. Čapek was widely known as a science fiction author, and perhaps his renown most people believe that a robot is a human-like artificial machine capable of having and expressing human feelings. This romantic perception of robots has been adapted as technologies have emerged over time. For our purposes in this book, a robot is a manipulator used to move materials without direct human contact in order to perform tasks in places that are inaccessible or dangerous for humans. Also, jobs demanding higher manufacturing standards, requiring repetitive tasks, or involving handling dangerous materials are good opportunities to use robots instead of humans. Although these uses do not constitute scintillating material to inspire Hollywood movies, robot manipulators have successfully been applied in applications such as simple pick-and-place operations, robotically assisted surgery, multiaxis machine tools, assembly, and arc welding. Robot manipulators can be roughly classified in three main types: serial, parallel, or hybrid. A serial manipulator consists of an end-effector connected to the base link via a single kinematic chain, while a parallel manipulator is a mechanism composed of a moving platform connected to the fixed platform by at least two kinematic chains. A hybrid manipulator is a combination of parallel and serial manipulators or is a mechanism realized by several parallel manipulators assembled in series connection, frequently referred to as hyper-redundant manipulators. Interest in parallel manipulators has been stimulated by some well-known proven advantages

---

Electronic supplementary material: The online version of this chapter (doi: 10.1007/978-3-319-31126-5_2) contains supplementary material, which is available to authorized users.

J. Gallardo-Alvarado, *Kinematic Analysis of Parallel Manipulators by Algebraic Screw Theory*, DOI 10.1007/978-3-319-31126-5_2

over their serial counterparts. In fact, in terms of kinematics and dynamics, parallel manipulators have better characteristics in areas where serial manipulators exhibit poor performance, and vice versa.

Part of this work is devoted to the kinematic analysis of parallel manipulators. According to the available mobility (degrees of freedom, or DOF) and the dimension of the task space, which are fundamental issues in mechanisms and robotics, it is possible to distinguish three classes of parallel manipulators: redundant, nonredundant, and lower-mobility mechanisms. In redundant parallel manipulators, the DOF are greater than the dimension of the task space, while in nonredundant parallel manipulators the DOF are equal to the dimension of the task space. Lower-mobility mechanisms, also known as limited-DOF or defective parallel manipulators, are mechanisms with fewer DOF than the dimension of the task space; translational and spherical parallel manipulators belong to this class. At this point it is worth noting that the current interest in the study of the hexapod is not as new as it appears. As noted by Huxley (1960), the study of complex mechanisms was first realized by Wren in 1645. In the 1800s, Augustin-Louis Cauchy, a brilliant French mathematician, investigated the stiffness of an articulated octahedron. It seems that the hexapod geometry was first specially studied by Bricard (1897) during the nineteenth century.

One of the first well-documented parallel manipulators was precisely a lower-mobility mechanism. After World War I, the cinematography industry became one of the most lucrative businesses in America. Hence, Gwinnet (1931) proposed a spherical parallel manipulator as a cinema motion simulator, *the Oxymoron*, and was granted the corresponding patent (Fig. 2.1). However, at that time the industry was not ready for Gwinnett's invention. Several years later, a more practical application of a parallel manipulator was introduced by Pollard (1940): a spray painting machine (Fig. 2.2).

Pollard's creative invention, a simple five-bar parallel robot, for which he was granted the corresponding patent, is considered the first industrial application of

Fig. 2.1 Possibly the first parallel robot, an invention credited to Gwinnett

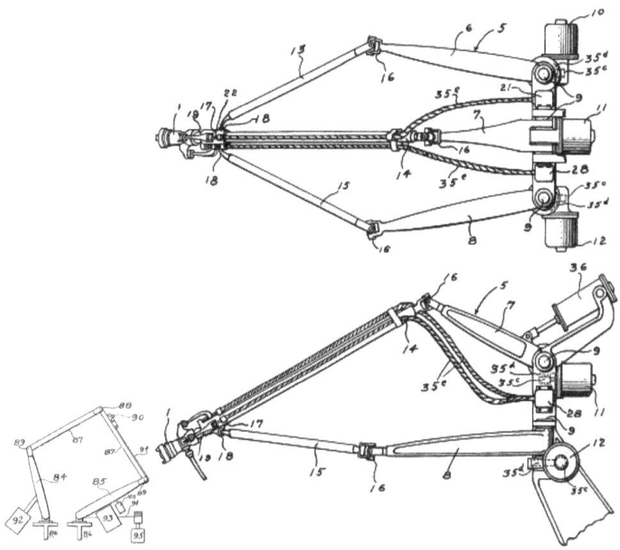

Fig. 2.2 Pollard's spray painting machine (U.S. Patent No. 2,286,571)

a parallel robot. Although Gwinnett and Pollard had pioneering contributions, undoubtedly the most celebrated parallel manipulator is the *universal tyre testing machine* credited to Dr. Eric Gough, an automotive engineer at Dunlop Rubber Company. As Gough noted, this mechanical device had been glimpsed in 1947 and its construction was finished in 1955 to respond to problems of aero-landing loads and to determine the properties of tyre under combined loads. The necessity to measure the deflections caused by forces applied in any direction to the tyre was one of the main reasons to develop Gough's hexapod. Evidently, during World War II aircraft manufacturers coveted such investigations. A full explanation of the *rig*, as Gough named his invention, was published by Gough and Whitehall (1962) a few years later. Stewart (1965–1966) published a theoretical paper approaching the design of an eight-bar linkage for purposes ranging from space applications to flight simulators. Furthermore, in an accurate prediction of the future, Stewart emphasized that his invention could be used to develop a new class of machine tools. Stewart's theoretical paper, which is one of the most cited contributions concerning parallel manipulators, included mainly the mobility and workspace analysis of the proposed parallel manipulator. The originality of the architecture Stewart introduced has sometimes been unjustly questioned, even by the same Dr. Gough. This is an opportune moment to mention that while Gough's hexapod consists of a moving platform and a fixed platform connected to each other by six extendable limbs, in the manipulator proposed by Stewart the moving and fixed platforms are connected by three extendable limbs, where the orientation of each connecting limb is constrained by a second extendable leg. Hence, one can assume that Stewart's platform is a partially decoupled robot with simpler forward kinematics and an improved workspace compared to Gough's machine. With these considerations, it is not completely true that... *the universal joint systems attaching the jacks to the platform are identical to those attaching the jacks to the foundation....* Today, paradoxically, mechanisms based on the topology of the octahedral hexapod are universally known as Gough–Stewart platforms. Curiously, neither Gough nor Stewart showed sufficient interest in obtaining a patent for his invention. Thus, the first patent concerning the octahedral hexapod and its applications was granted in 1967 to Klaus Cappel, who, years after Gough and Stewart, independently designed very nearly the hexapod. Because of Cappel's contribution, it was possible to achieve the first transcendental commercially available octahedral hexapod motion simulators. From the contributions of Gough, Stewart, and Cappel (Fig. 2.3), one can learn that a practical problem can be successfully addressed regardless of the spheres of activity. After these seminal works, in the subsequent two decades most significant efforts of the kinematician community devoted to the study of parallel manipulators focused on elucidating the forward kinematics of the Gough–Stewart platform as well as introducing lower-mobility parallel manipulators.

In the 1980s, at École Polytechnique Fédérale de Lausanne, EPFL, Professor Reymond Clavel revealed the original idea to design a lower-parallel manipulator to realize three translations and one rotation between the moving and fixed platforms (Fig. 2.4).

Clavel's invention, named the Delta robot, was based on the use of parallelograms to constrain the rotations of the moving platform and culminated in his doctoral dissertation (Clavel 1991); Dr. Clavel had been granted the associated patent one year before (Clavel 1990). As the Delta robot was well appreciated by engineers and scientists, Clavel was honored with the prestigious Golden Robot Award, an event sponsored by ABB Flexible Automation, for his innovative work. The Delta robot is one of the most successful parallel robot designs because the original design is capable of achieving accelerations of up to 50 G and 12 G in experimental and industrial environments, respectively, in a wide workspace. There are several hundred active robots of this class worldwide performing tasks ranging from simple pick-and-place operations to medical/surgical-assistance devices. In that trend, a family of 4-DOF parallel manipulators called H4 for high-speed pick-and-place applications was introduced by Pierrot and Company (1999). Three rotations plus one rotation, commonly known as Sch́onflies motions (Bottema and Roth 1979; Hervé 1999; Lee and Hervé 2005), are the available DOF of an H4 robot. Serial manipulators called SCARA robots, an acronym for selective-compliance assembly robot arm, were the first manipulators designed to realize Schönflies motions. Angeles et al. (2006) introduced a parallel manipulator composed of two identical 4-DOF serial chains in a parallel arrangement called the Schönflies motion generator robot. The successful introduction of translational parallel manipulators in the industry is indisputable. Consider, for instance, the Quattro robot released by Adept Technology©; also based on parallelograms, Quattro, with its 240 cycles per minute, is considered the industry's fastest pick-and-place robot today (Bonev et al. 2006; Pierrot et al. 2009). Not only have parallelograms been used to develop translational parallel manipulators, but spatial mechanisms built with serial chains play the role

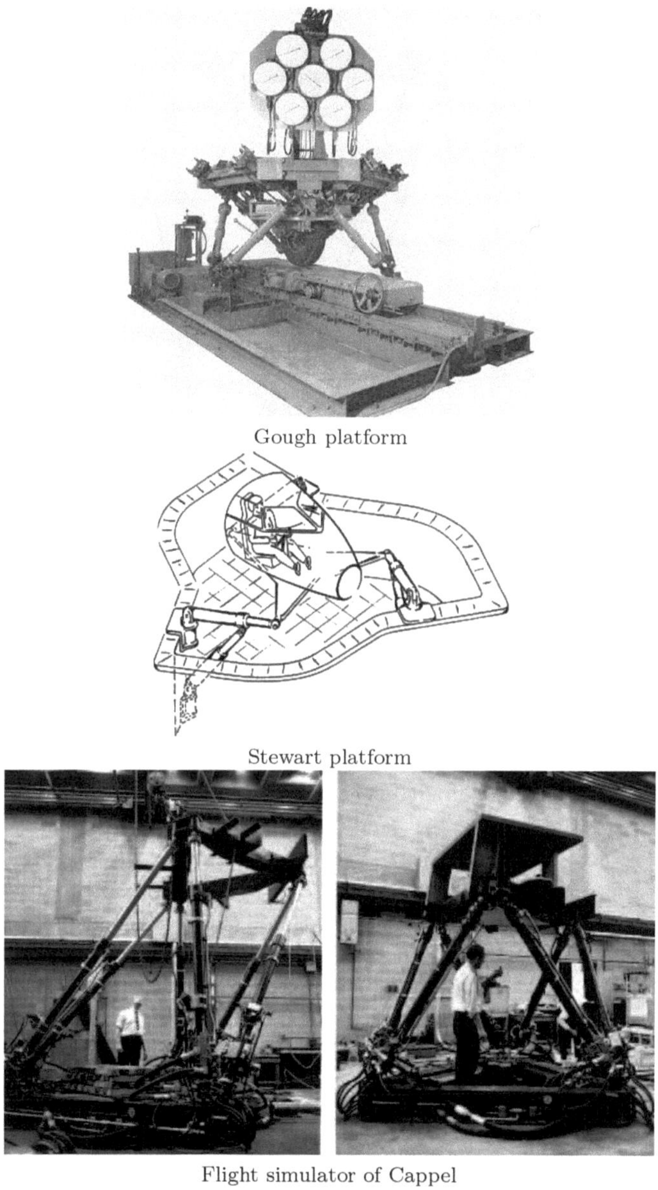

Fig. 2.3 Three parallel manipulators and a polemic discussion about their origins

of limbs. Kim and Tsai (2003) developed the Cartesian parallel manipulator. In this innovative robot, its 3 DOF are completely decoupled; in other words, the motion of each prismatic joint is controlled by only one linear actuator. Parallel manipulators meeting such features are known as linear parallel manipulators and have attracted

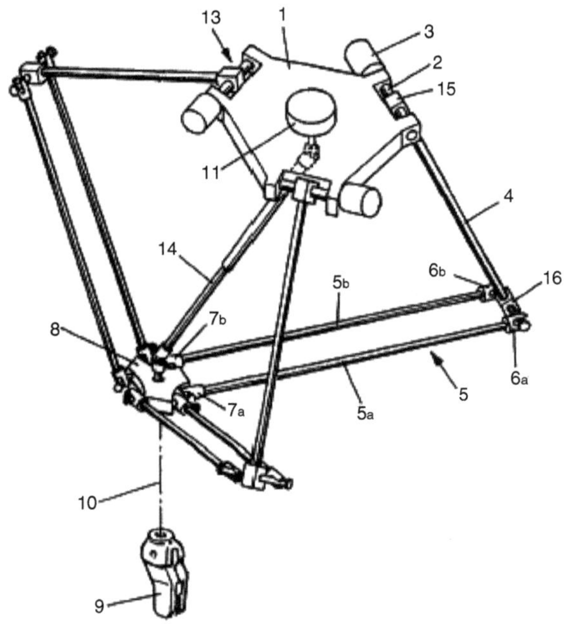

Fig. 2.4 The Delta robot (U.S. Patent No. 4,976,582)

the attention of many researchers because of their evident benefits such as simple finite kinematics and control; see, for instance, Kong and Gosselin (2002a,b,c, 2004) and Li et al. (2005a,b).

Translational parallel manipulators have remarkable histories to tell, but spherical parallel manipulators have also been successfully developed in both academia and industry. Asada and Cro Granito (1985) suggested using a spherical parallel manipulator as a wrist. Cox and Tesar (1989) designed a robot shoulder based on a parallel wrist. Canfield (1997) developed a human-like wrist named the *Carpal Wrist*. In the same year a comprehensive study of robot wrists was published by Rosheim (1989). In the early 1990s, a spherical parallel manipulator designed for the rapid orientation of a camera called the *Agile Eye* was developed by Gosselin and coworkers (Gosselin and Lavoie 1993; Gosselin and Hamel 1994; Gosselin and Cagné 1995; Gosselin et al. 1996). Because of its extraordinary performance, this ingenious parallel manipulator has gained in popularity ever since; for example, the workspace of the Agile Eye is superior to that of the human eye and can produce angular velocities as high as $1000\,^{\circ}/\mathrm{s}$ and angular accelerations of $20{,}000\,^{\circ}/\mathrm{s}^2$. However, some manufacturing problems such as the alignment of revolute axes are among the main drawbacks of the Agile Eye. In order to overcome such disadvantages, a robot called Argos was proposed, once again, at the EPFL by Vischer (1995, 1996). The first prototype of the Argos robot was formally reported a few years later by Vischer and Clavel (2000). Since the possibilities offered by parallel manipulators are immense, the variety of robot designs has certainly not ceased to grow. Researchers hope that in the near future a better performance will allow parallel manipulators to substitute for traditional serial manipulators where operations like accuracy, rigidity, and highest payload capacity are required.

Finally, this chapter contains animations of novel parallel manipulators developed at Instituto Tecnológico de Celaya, México. Details of the videos, whose designations are in bold font, can be found in the following references.

- **2(3UPS)hexapod.**

  Garcia-Murillo, M. A., 2015. Dinámica de un manipulador paralelo para tareas de maquinado considerando fuerzas de corte, doctoral dissertation, CICATA, Instituto Politécnico Nacional, México. In Spanish.
  Gallardo-Alvarado, J., Garcia-Murillo, M. A., & Castillo Castaneda, E., 2013. A 2(3-RRPS) parallel manipulator inspired by Gough-Stewart platform. *Robotica, 31*(3):381–388.
  doi: http://dx.doi.org/10.1017/S0263574712000392

- **2RPS.**

  Gallardo-Alvarado, J., & Posadas-Garcia, J. D., 2013. Mobility analysis and kinematics of the semi-general 2(3-RPS) series-parallel manipulator. *Robotics and Computer-Integrated Manufacturing, 29*(6):463–472.
  doi: http://dx.doi.org/10.1016/j.rcim.2013.05.004

- **3PPS-3RPS.**

  Gallardo-Alvarado, J., Rodriguez-Castro, R., Aguilar-Najera, C. R., & Perez-Gonzalez, L., 2012. A novel six-degrees-of-freedom series-parallel manipulator. *Journal of Mechanical Science and Technology, 26*(6):1901–1909.
  doi:10.1007/s12206-012-0408-5

- **3SPR-RRRPRP.**

  Gallardo-Alvarado, J., Garcia-Murillo, M. A., Rodriguez-Castro, R., & Alcaraz-Caracheo, L. A., 2015. A new four-degrees-of-freedom parallel manipulator. Submitted to *Advances in Robotics Research*.

- **DeLiA1, DeLiA2, DeLiA3.**

  Gallardo-Alvarado, J., 2011. DeLiA: A new redundant partially decoupled robot. *Advanced Robotics, 25*:1295–1310. doi:10.1163/016918611X574722
  Gallardo-Alvarado, J., 2010. DeLiA: A new family of redundant robot manipulators. In S. E. Shafiei (Ed.), *Advanced strategies for robot manipulators,* SCIYO, Vienna, ISBN: 978-953-307-099-5.
  doi: 10.5772/10190

- **HexapodMixed.**

  Gallardo-Alvarado, J., & Garcia-Murillo, M. A., 2015. Rigid body hyper-jerk analysis using screw theory. In M. Ceccarelli and E. E. Hernandez-Martinez (Eds.), *Multibody mechatronic systems, mechanisms and machine science,* Vol. 25, Springer, pp. 411–422, ISBN: 978-3-319-09857-9.
  doi: 10.1007/978-3-319-09858-6-39

- **LinceJJP-complex-motions, LinceJJP-drilling.**

  Gallardo-Alvarado, J., Orozco-Mendoza, H., & Rico-Martinez, J. M., 2010. A novel five-degrees-of-freedom decoupled robot. *Robotica, 28*(6):909–917. doi: 10.1017/S0263574709990749

- **SphericalPM.**

  Gallardo-Alvarado, J., Garcia-Murillo, M. A., & Perez-Gonzalez, L., 2013. Kinematics of the 3-RRRS+S parallel wrist. *Mechanics Based Design of Structures and Machines: An International Journal, 41*(4):452–467. doi: 10.1080/15397734.2013.781480

- **Stewart-type platform.**

  Gallardo-Alvarado, J., & Garcia-Murillo, M. A., 2013. A parallel manipulator inspired by the original Stewart platform. *Proceedings of the Institution of Mechanical Engineers, Part C: Journal of Mechanical Engineering Science, 27*(4):831–844. doi: 10.1177/0954406212454970

- **TPM1, TPM2, TPM3, TPM4.**

  Gallardo-Alvarado, J., Orozco-Mendoza, H., Sanchez-Rodriguez, A., & Alici, G., 2014. Kinematic analyses of novel translational parallel manipulators. *Proceedings of the Institution of Mechanical Engineers, Part C: Journal of Mechanical Engineering Science, 228*(2):330–341. doi:10.1177/0954406213484225

References

Angeles, J., Caro, S., Khan, W., & Morozov, A. (2006). The kinetostatic design of an innovative Schönflies motion generator. *Proceedings Institution of Mechanical Engineers, Part C: Journal of Mechanical Engineering Science, 220*(C7), 935–944.

Asada, H., & Cro Granito, J. A. (1985). Kinematic and static characterization of wrist joints and their optimal design. In *Proceedings IEEE International Conference on Robotics and Automation* (pp. 244–250).

Bonev, I. A., Chablat, C., & Wenger, P. (2006). Working and assembly modes of the Agile Eye. In *Proceedings of the 2006 IEEE International Conference on Robotics and Automation, Orlando, Florida, May* (pp. 2317–2322).

Bottema, O., & Roth, B. (1979). *Theoretical kinematics*. New York: Dover.

Bricard, R. (1897). Mémoire sur la théorie de lóctahèdre articulé. *Journal de Mathématiques Pures et Appliqués, 3*, 113–150.

Canfield, S. L. (1997). *Development of the Carpal Wrist; a symmetric, parallel-architecture robotic wrist*. Ph.D. thesis, Virginia Polytechnic Institute and State University Department of Mechanical Engineering.

Cappel, K. L. (1967). Motion simulator. U.S. Patent No. 3,295,224, January 3.

Clavel, R. (1990). Device for the movement and positioning of an element in space. U.S. Patent No. 4,976,582, December 11.

Clavel, R. (1991). *Conception d'un robot parallèle rapide à 4 degrés de liberté*. Ph.D. thesis, Ecole Polytechique Fédérale de Lausanne (EPFL).

Cox, D., & Tesar, D. (1989). The dynamic model of a three degree of freedom parallel robotic shoulder module. In K. Waldron (Ed.), *Advanced robotics 1989. Fourth International Conference on Advanced Robotics, Columbus* (pp. 475–487). Berlin: Springer.

Gosselin, C. M., & Gagné, M. (1995). A closed-form solution for the direct kinematics of a special class of spherical three-degree-of-freedom parallel manipulators. In J.-P. Merlet & B. Ravani (Eds.), *Computational Kinematics, Proceedings of the Second Workshop on Computational Kinematics* (pp. 231–240). Dordrecht: Kluwer.

Gosselin, C. M., & Hamel, J.-F. (1994). The agile eye: a high-performance three-degree-of-freedom camera-orienting device. In *Proceedings IEEE International Conference on Robotics and Automation, San Diego* (pp. 781–786).

Gosselin, C. M., & Lavoie, E. (1993). On the kinematic design of spherical three-degree-of-freedom parallel manipulators. *International Journal of Robotics Research, 12*(4), 394–402.

Gosselin, C. M., St.-Pierre, E., & Gagné, M. (1996). On the development of the Agile Eye. *Robotics and Automation Magazine, IEEE, 3*(4), 29–37.

Gough, V. E., & Whitehall, S. G. (1962). Universal tyre testing machine. In G. Eley (Ed.), *Proceedings of 9th International Automobile Technical Congress*, discussion pp. 250ff; Fédération Internationale des Sociétés d'Ingénieurs des Techniques de lÁutomobile (FISITA) (pp. 117–137). IMechE 1, London.

Gwinnett, J. E. (1931). Amusement devices. U.S. Patent No. 1,789,680, January 20.

Hervé, J. (1999). The Lie group of rigid body displacements, a fundamental tool for mechanism design. *Mechanism and Machine Theory, 34*(5), 719–730.

Huxley, G. L. (1960) The geometrical work of Christopher Wren. *Scripta Mathematica, 25*, 201–208.

Kim, H. S., & Tsai, L.-W. (2003). Design optimization of a Cartesian parallel manipulator. *ASME Journal of Mechanical Design, 125*(1), 43–51.

Kong, X., & Gosselin, C. M. (2002a). Type synthesis of linear translational parallel manipulators. In J. Lenarčič & F. Thomas (Eds.), *Advances in robot kinematics—theory and applications* (pp. 411–420). Dordrecht: Kluwer.

Kong, X., & Gosselin, C. M. (2002b). A class of 3-DOF translational parallel manipulators with linear input-output equations. In *Proceedings of the Workshop on Fundamental Issues and Future Research Directions for Parallel Mechanisms and Manipulators, Quebec City, Canada* (pp. 25–32).

Kong, X., & Gosselin, C. M. (2002c). Kinematics and singularity analysis of 3-CRR 3-DOF translational parallel manipulators. *International Journal of Robotics Research, 21*(9), 791–798.

Kong, X., & Gosselin, C. M. (2004). Type synthesis of 3-DOF translational parallel manipulators based on screw theory. *ASME Journal of Mechanical Design, 126*(1), 83–92.

Lee, C.-C., & Hervé, J. (2005). On the enumeration of Schönflies motion generators. In *Ninth IFToMM International Symposium on Theory of Machines and Mechanisms*, Bucharest, Romania.

Li, W., Gao, F., & Zhang, J. (2005a). R-CUBE, a decoupled parallel manipulator only with revolute joints. *Mechanism and Machine Theory, 40*(4), 467–473.

Li, W., Gao, F., & Zhang, J. (2005b). A three-DOF translational manipulator with decoupled geometry. *Robotica, 23*(6), 805–808.

Pierrot, F., & Company, O. (1999). H4: a new family of 4-DOF parallel robots. In *Proceedings 1999 IEEE/ASME International Conference on Advanced Intelligent Mechatronics, Atlanta* (pp. 508–513).

Pierrot, F., Nabat, V., Company, O., Krut, S., & Poignet, P. (2009). Optimal design of a 4-DOF parallel manipulator: from academia to industry. *IEEE Transactions on Robotics, 25*(2), 213–224.

Pollard, W. L. G. (1940). Spray painting machine. U.S. Patent No. 2,213,108, August 26.

Rosheim, M. E. (1989). *Robot wrist actuators*. New York: Wiley.

Stewart, D. (1965–1966). A platform with six degrees of freedom. *Proceedings Institution of Mechanical Engineers, 180*(1), 371–386.

Vischer, P. (1995). *Argos: a novel parallel spherical structure*. Technical Report 95-03, Ecole Polytechique Fédérale de Lausanne (EPFL).
Vischer, P. (1996). *Improving the accuracy of parallel robots*. Ph.D. thesis, document 1570, Ecole Polytechique Fédérale de Lausanne (EPFL).
Vischer, P., & Clavel, R. (2000). Argos: a novel 3-DOF parallel wrist mechanism. *International Journal of Robotics Research, 19*(1), 5–11.

Part II
Fundamentals of the Theory of Screws

Chapter 3
Mathematical Background

3.1 Introduction

This chapter presents fundamental concepts of the so-called first-order kinematic analysis of rigid bodies in light of screw theory to provide a proper foundation for the next chapters. First, a formal representation of the velocity state of a rigid body, commonly known as the twist about a screw defined by Ball (1900), is introduced based on propositions that are thoroughly proved. Next, the representation of lines is explained according to the so-called Plücker coordinates. Finally, the operations and properties of the Lie algebra $se(3)$ of the Euclidean group $SE(3)$ are revised; of special interest are the Lie product and the Klein form. The helicoidal vector field theory is also revised because of its central role in the representation of the higher-order kinematic states of rigid bodies.

3.2 Preamble

The angular and linear velocities of an arbitrary point of a rigid body, as measured from another body or reference frame, form an inseparable entity named the velocity state ($\boldsymbol{V}$). In general spatial motion, the velocity state may be considered six-dimensional, while in general planar motion the velocity state may be reduced into a three-dimensional vector. The angular and linear components of the velocity state are named, respectively, the primal part, $P(\boldsymbol{V})$, and the dual part, $D(\boldsymbol{V})$. The velocity state of a rigid body, also known as the twist about a screw (Ball 1900), may be expressed in a screw form; such a representation is the heart of screw theory, also referred to as motor algebra; see, for instance, von Mises (1924a,b) and Sugimoto (1987, 1990). In this book, six-dimensional vectors, like the velocity state, are denoted with boldface italic uppercase letters to avoid confusion with the usual three-dimensional vectors.

J. Gallardo-Alvarado, *Kinematic Analysis of Parallel Manipulators by Algebraic Screw Theory*, DOI 10.1007/978-3-319-31126-5_3

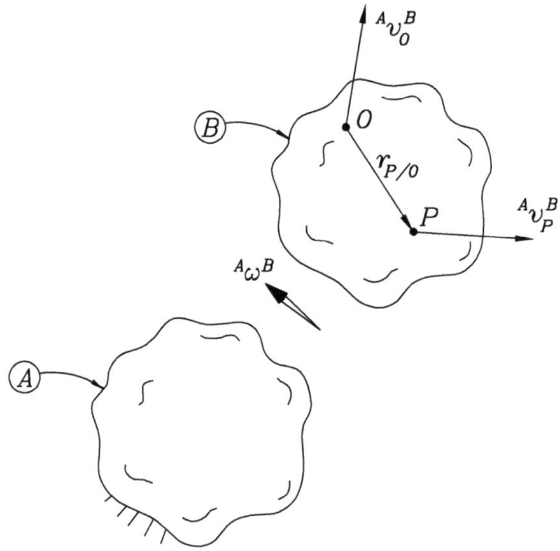

Fig. 3.1 A body *B* in motion with respect to another body *A* comprising angular and linear velocities

3.2.1 Velocity State or Twist About a Screw of a Rigid Body

The main purpose of the velocity state of a rigid body is to provide enough information to efficiently compute the linear velocity of any point of a rigid body as observed from another body or reference frame.

Figure 3.1 shows a rigid body B in motion with respect to another body or reference frame A. To compute the linear velocity of an arbitrary point P of B, it is necessary to know two fundamental vectors: (1) the angular velocity of B with respect to A, ${}^{A}\boldsymbol{\omega}^{B}$, and (2) the linear velocity of a point O of body B, ${}^{A}\boldsymbol{v}_{O}^{B}$. In fact, using elementary kinematics we can relate the velocities of points P and O as follows:

$$ {}^{A}\boldsymbol{v}_{P}^{B} = {}^{A}\boldsymbol{v}_{O}^{B} + {}^{A}\boldsymbol{\omega}^{B} \times \boldsymbol{r}_{P/O}, \tag{3.1} $$

where $\boldsymbol{r}_{P/O}$ is the position vector of point P with respect to point O expressed in the reference frame A. With these considerations in mind, it is reasonable to assume that the velocity state of a rigid body B, as observed from another body or reference frame A, may be defined as a six-dimensional vector ${}^{A}\boldsymbol{V}_{O}^{B}$ as follows:

$$ {}^{A}\boldsymbol{V}_{O}^{B} \equiv \begin{bmatrix} {}^{A}\boldsymbol{\omega}^{B} \\ {}^{A}\boldsymbol{v}_{O}^{B} \end{bmatrix}, \tag{3.2} $$

where point O is named the reference pole. Furthermore, it is worth noting that unlike the linear velocity vector, the angular velocity vector does not depend on the reference pole and therefore can be considered a rigid body property. Certainly, the angular velocity vector is the same for any point of a rigid body. Therefore, it is possible to affirm that the velocity state of a rigid body does not have a unique representation; this assumption is clarified in the next proposition. For quick reference and clarity, Eq. (3.2) as well as some repetitive expressions will be provided without labels in this book.

Proposition 3.1. *Let B be a rigid body that is observed from another body or reference frame A. Furthermore, let O and O^* be two arbitrary points of body B. Then, according to Eq. (3.2), two representations of the velocity state of B with respect to A involving points O and O^* notated respectively as*

$$^A\boldsymbol{V}_O^B = \begin{bmatrix} ^A\boldsymbol{\omega}_O^B \\ ^A\boldsymbol{v}_O^B \end{bmatrix}$$

and

$$^A\boldsymbol{V}_{O*}^B = \begin{bmatrix} ^A\boldsymbol{\omega}_{O*}^B \\ ^A\boldsymbol{v}_{O*}^B \end{bmatrix}$$

represent the same velocity state if and only if

$$^A\boldsymbol{\omega}_O^B = {}^A\boldsymbol{\omega}_{O*}^B = {}^A\boldsymbol{\omega}^B \tag{3.3}$$

and

$$^A\boldsymbol{v}_{O*}^B = {}^A\boldsymbol{v}_O^B + {}^A\boldsymbol{\omega}^B \times \boldsymbol{r}_{O*/O}, \tag{3.4}$$

where $\boldsymbol{r}_{O*/O}$ is the position vector of O^* with respect to O. Note that since the angular velocity vector is a property of the rigid body, the reference poles O and O^* are removed from Eq. (3.3).

Proof. Consider that P is an arbitrary point of body B. Two representations $^A\boldsymbol{V}_O^B$ and $^A\boldsymbol{V}_{O*}^B$ denote the same velocity state of body B as observed from body A if and only if

$$^A\boldsymbol{v}_O^B + {}^A\boldsymbol{\omega}_O^B \times \boldsymbol{r}_{P/O} = {}^A\boldsymbol{v}_P^B = {}^A\boldsymbol{v}_{O*}^B + {}^A\boldsymbol{\omega}_{O*}^B \times \boldsymbol{r}_{P/O*}. \tag{3.5}$$

Considering that $P = O$, then we get

$$^A\boldsymbol{v}_O^B + {}^A\boldsymbol{\omega}_O^B \times \boldsymbol{r}_{O/O} = {}^A\boldsymbol{v}_{O*}^B + {}^A\boldsymbol{\omega}_{O*}^B \times \boldsymbol{r}_{O/O*}. \tag{3.6}$$

Hence, because $\boldsymbol{r}_{O/O} = \boldsymbol{0}$ we have

$$^A\boldsymbol{v}_O^B = {}^A\boldsymbol{v}_{O*}^B + {}^A\boldsymbol{\omega}_{O*}^B \times \boldsymbol{r}_{O/O*}. \tag{3.7}$$

Similarly, if $P = O^*$, then one obtains

$$^A\boldsymbol{v}_O^B + {}^A\boldsymbol{\omega}_O^B \times \boldsymbol{r}_{O*/O} = {}^A\boldsymbol{v}_{O*}^B. \tag{3.8}$$

Next, subtracting Eqs. (3.7) and (3.8) yields

$$ {}^{A}\boldsymbol{\omega}_{O}^{B} \times \boldsymbol{r}_{O^{*}/O} = -{}^{A}\boldsymbol{\omega}_{O^{*}}^{B} \times \boldsymbol{r}_{O/O^{*}}, \tag{3.9} $$

or

$$ \left({}^{A}\boldsymbol{\omega}_{O}^{B} - {}^{A}\boldsymbol{\omega}_{O^{*}}^{B}\right) \times \boldsymbol{r}_{O^{*}/O} = \boldsymbol{0}. \tag{3.10} $$

Furthermore, if we take into account that in general $O \neq O^{*} \rightarrow \boldsymbol{r}_{O^{*}/O} \neq \boldsymbol{0}$, with the purpose to satisfy Eq. (3.10), it is necessary that

$$ {}^{A}\boldsymbol{\omega}_{O}^{B} - {}^{A}\boldsymbol{\omega}_{O^{*}}^{B} = \boldsymbol{0}. \tag{3.11} $$

Hence,

$$ {}^{A}\boldsymbol{\omega}_{O}^{B} = {}^{A}\boldsymbol{\omega}_{O^{*}}^{B} = {}^{A}\boldsymbol{\omega}^{B}. \tag{3.12} $$

Furthermore, according to expression (3.5) it follows that

$$ {}^{A}\boldsymbol{v}_{O}^{B} + {}^{A}\boldsymbol{\omega}^{B} \times \boldsymbol{r}_{P/O} = {}^{A}\boldsymbol{v}_{O^{*}}^{B} + {}^{A}\boldsymbol{\omega}^{B} \times \boldsymbol{r}_{P/O^{*}}, \tag{3.13} $$

which implies that

$$ {}^{A}\boldsymbol{v}_{O^{*}}^{B} = {}^{A}\boldsymbol{v}_{O}^{B} + {}^{A}\boldsymbol{\omega}^{B} \times \left(\boldsymbol{r}_{P/O} - \boldsymbol{r}_{P/O^{*}}\right). \tag{3.14} $$

Finally, gathering these results, we concluded that in fact

$$ {}^{A}\boldsymbol{v}_{O^{*}}^{B} = {}^{A}\boldsymbol{v}_{O}^{B} + {}^{A}\boldsymbol{\omega}^{B} \times \boldsymbol{r}_{O^{*}/O}, $$

and the first part of Proposition 3.1 is proved.

In order to complete the proof, according to Eqs. (3.3) and (3.4) we write

$$ {}^{A}\boldsymbol{v}_{P}^{B} = {}^{A}\boldsymbol{v}_{O}^{B} + {}^{A}\boldsymbol{\omega}^{B} \times \boldsymbol{r}_{P/O} \tag{3.15} $$

and

$$ {}^{A}\boldsymbol{v}_{P}^{B} = {}^{A}\boldsymbol{v}_{O^{*}}^{B} + {}^{A}\boldsymbol{\omega}^{B} \times \boldsymbol{r}_{P/O^{*}}. \tag{3.16} $$

Then, substituting Eq. (3.4) into Eq. (3.16) leads to

$$ \begin{aligned} {}^{A}\boldsymbol{v}_{P}^{B} &= \left({}^{A}\boldsymbol{v}_{O}^{B} + {}^{A}\boldsymbol{\omega}^{B} \times \boldsymbol{r}_{O^{*}/O}\right) + {}^{A}\boldsymbol{\omega}^{B} \times \boldsymbol{r}_{P/O^{*}} \\ &= {}^{A}\boldsymbol{v}_{O}^{B} + {}^{A}\boldsymbol{\omega}^{B} \times \left(\boldsymbol{r}_{P/O^{*}} + \boldsymbol{r}_{O^{*}/O}\right) \\ &= {}^{A}\boldsymbol{v}_{O}^{B} + {}^{A}\boldsymbol{\omega}^{B} \times \boldsymbol{r}_{P/O}. \end{aligned} \tag{3.17} $$

Therefore, the representations ${}^{A}\boldsymbol{V}_{O}^{B}$ and ${}^{A}\boldsymbol{V}_{O^{*}}^{B}$ define the same velocity state. ◇

Definition 3.1. The following definition is a formalism of Proposition 3.1. Let

$$ {}^A\boldsymbol{V}_O^B = \begin{bmatrix} {}^A\boldsymbol{\omega}_O^B \\ {}^A\boldsymbol{v}_O^B \end{bmatrix} $$

and

$$ {}^A\boldsymbol{V}_{O*}^B = \begin{bmatrix} {}^A\boldsymbol{\omega}_{O*}^B \\ {}^A\boldsymbol{v}_{O*}^B \end{bmatrix} $$

be two representations of the velocity state of a rigid body B with respect to another body or reference frame A [see Eq. (3.2)]. Then it is said that the velocity states ${}^A\boldsymbol{V}_O^B$ and ${}^A\boldsymbol{V}_{O*}^B$ are equivalent if and only if both representations have the same primal part. Indeed,

$$ {}^A\boldsymbol{\omega}_O^B = {}^A\boldsymbol{\omega}_{O*}^B, $$

and the dual part of the velocity state ${}^A\boldsymbol{V}_{O*}^B$ can be obtained based on the dual part of the velocity state ${}^A\boldsymbol{V}_O^B$ as follows:

$$ {}^A\boldsymbol{v}_{O*}^B = {}^A\boldsymbol{v}_O^B + {}^A\boldsymbol{\omega}^B \times \boldsymbol{r}_{O*/O}. $$

Definition 3.1 is without a doubt simple, compact, elegant, and logical at the level of elementary kinematics; however, it deserves to be deeply investigated by resorting to the concept of equivalence relationship.

3.2.2 *The Velocity State as an Equivalence Relationship*

Proposition 3.2. *The concept of velocity state, as given in Definition 3.1, is an equivalence relationship notated as* ($\sim$).

Proof. The relationship must be symmetric, reflexive, and transitive.

Let O, O^*, and $O^\#$ be three arbitrary points of body B. Then, three representations of the velocity state of body B, with respect to body A, according to these points may be written as

$$ {}^A\boldsymbol{V}_O^B = \begin{bmatrix} {}^A\boldsymbol{\omega}_O^B \\ {}^A\boldsymbol{v}_O^B \end{bmatrix}, \quad {}^A\boldsymbol{V}_{O*}^B = \begin{bmatrix} {}^A\boldsymbol{\omega}_{O*}^B \\ {}^A\boldsymbol{v}_{O*}^B \end{bmatrix}, \quad \text{and} \quad {}^A\boldsymbol{V}_{O\#}^B = \begin{bmatrix} {}^A\boldsymbol{\omega}_O^B \\ {}^A\boldsymbol{v}_{O\#}^B \end{bmatrix}. $$

With these considerations in mind, in what follows Proposition 3.2 may be elucidated.

1. *Reflexive.* It is necessary to prove that

$$ {}^A\boldsymbol{V}_O^B \sim {}^A\boldsymbol{V}_O^B. \tag{3.18} $$

Since equality is itself an equivalence relationship, for the primal part of the velocity state the proof is immediate. In fact,

$$ {}^{A}\boldsymbol{\omega}_{O}^{B} \sim {}^{A}\boldsymbol{\omega}_{O}^{B}. \tag{3.19}$$

Meanwhile, concerned with the dual part of the velocity state we have

$$ {}^{A}\boldsymbol{v}_{O}^{B} = {}^{A}\boldsymbol{v}_{O}^{B} + {}^{A}\boldsymbol{\omega}_{O}^{B} \times \boldsymbol{r}_{O/O} = {}^{A}\boldsymbol{v}_{O}^{B} \rightarrow {}^{A}\boldsymbol{v}_{O}^{B} \sim {}^{A}\boldsymbol{v}_{O}^{B}. \tag{3.20}$$

Therefore, it is evident that ${}^{A}\boldsymbol{V}_{O}^{B} \sim {}^{A}\boldsymbol{V}_{O}^{B}$.

2. *Symmetric.* It is sufficient to prove that

$$ {}^{A}\boldsymbol{V}_{O}^{B} \sim {}^{A}\boldsymbol{V}_{O*}^{B} \rightarrow {}^{A}\boldsymbol{V}_{O*}^{B} \sim {}^{A}\boldsymbol{V}_{O}^{B}. \tag{3.21}$$

To this end, assume that ${}^{A}\boldsymbol{V}_{O}^{B} \sim {}^{A}\boldsymbol{V}_{O*}^{B}$. Thus, evidently

$$ {}^{A}\boldsymbol{\omega}_{O}^{B} = {}^{A}\boldsymbol{\omega}_{O*}^{B} \quad \text{and} \quad {}^{A}\boldsymbol{v}_{O*}^{B} = {}^{A}\boldsymbol{v}_{O}^{B} + {}^{A}\boldsymbol{\omega}^{B} \times \boldsymbol{r}_{O*/O}. \tag{3.22}$$

On the other hand, since the equality is itself reflexive, we have

$$ {}^{A}\boldsymbol{\omega}_{O}^{B} = {}^{A}\boldsymbol{\omega}_{O*}^{B} \rightarrow {}^{A}\boldsymbol{\omega}_{O*}^{B} = {}^{A}\boldsymbol{\omega}_{O}^{B}. \tag{3.23}$$

Furthermore, the linear velocity vector of point O can be determined upon the kinematic quantities of point O^* using college kinematics as

$$ {}^{A}\boldsymbol{v}_{O}^{B} = {}^{A}\boldsymbol{v}_{O*}^{B} - {}^{A}\boldsymbol{\omega}^{B} \times \boldsymbol{r}_{O*/O} = {}^{A}\boldsymbol{v}_{O*}^{B} + {}^{A}\boldsymbol{\omega}^{B} \times \boldsymbol{r}_{O/O*} \rightarrow {}^{A}\boldsymbol{v}_{O*}^{B} \sim {}^{A}\boldsymbol{v}_{O}^{B}. \tag{3.24}$$

Hence, it follows that ${}^{A}\boldsymbol{V}_{O*}^{B} \sim {}^{A}\boldsymbol{V}_{O}^{B}$.

3. *Transitive.* It is necessary to prove the following:

$$ {}^{A}\boldsymbol{V}_{O}^{B} \sim {}^{A}\boldsymbol{V}_{O*}^{B}, \quad {}^{A}\boldsymbol{V}_{O*}^{B} \sim {}^{A}\boldsymbol{V}_{O\#}^{B} \rightarrow {}^{A}\boldsymbol{V}_{O}^{B} \sim {}^{A}\boldsymbol{V}_{O\#}^{B}. \tag{3.25}$$

This part of the proposition is left as an exercise.[1]

It is well known that an equivalence relationship induces a partition of the considered set; for details, the reader is referred to Fraleigh (1982). In this case, the contribution is limited to giving only a simple definition of the issue.

Definition 3.2 (The Velocity State of a Rigid Body is an Equivalence Relationship). Let B be a rigid body whose motion is measured according to the reference

[1]The demonstration of a proposition can be achieved using different reasonings. Chap. 16 contains demonstrations of the proposed propositions in this book. The reader is encouraged to prove the propositions by resorting to different or closer strategies.

frame A. Furthermore, let O be an arbitrary point of B. The velocity state of body B with respect to body A is given by the following equivalence class:

$$\overline{{}^{A}\boldsymbol{V}^{B}} \equiv \left\{ {}^{A}\boldsymbol{v}_{O}^{B} \middle| {}^{A}\boldsymbol{v}_{O}^{B} \sim {}^{A}\boldsymbol{V}^{B} \right\}. \tag{3.26}$$

Furthermore, any element of this equivalence class must be considered a representation of the velocity state. Hence, the six-dimensional vector ${}^{A}\boldsymbol{V}^{B}$

$${}^{A}\boldsymbol{V}^{B} = \begin{bmatrix} {}^{A}\boldsymbol{\omega}^{B} \\ {}^{A}\boldsymbol{v}_{O}^{B} \end{bmatrix}$$

is a representation of the velocity state of body B as observed from body A.

Based on Definition 3.2, it is interesting to note that any correctly defined operation involving the velocity state of a rigid body does not depend on the selected representation. After, the reference pole can be freely chosen, a natural advantage of screw theory.

3.2.3 *The Velocity State and the Associated Screw of Rigid Body Motion*

Proposition 3.3. *Let*

$${}^{A}\boldsymbol{V}^{B} = \begin{bmatrix} {}^{A}\boldsymbol{\omega}^{B} \\ {}^{A}\boldsymbol{v}_{O}^{B} \end{bmatrix}$$

be a representation of the velocity state of body B, with respect to body A. In other words, let ${}^{A}\boldsymbol{V}^{B}$ be an element of the equivalence class $\overline{{}^{A}\boldsymbol{V}^{B}}$. The proposition states that all the points of body B have the same component of linear velocity parallel to the angular velocity vector ${}^{A}\boldsymbol{\omega}^{B}$ of body B.

Proof. Let us consider P and O to be two arbitrary points of body B. The linear velocity of point P, notated as ${}^{A}\boldsymbol{v}_{P}^{B}$, can be determined upon the kinematic quantities associated to point O as follows:

$${}^{A}\boldsymbol{v}_{P}^{B} = {}^{A}\boldsymbol{v}_{O}^{B} + {}^{A}\boldsymbol{\omega}^{B} \times \boldsymbol{r}_{P/O}.$$

Moreover, assuming that

$$\hat{\boldsymbol{u}} = \frac{{}^{A}\boldsymbol{\omega}^{B}}{{}_{A}\omega_{B}} \tag{3.27}$$

is a unit vector along the vector ${}^{A}\boldsymbol{\omega}^{B}$, whose magnitude is denoted by ${}_{A}\omega_{B}$, we write

$${}^{A}\boldsymbol{v}_{P}^{B} \bullet \hat{\boldsymbol{u}} = \left({}^{A}\boldsymbol{v}_{O}^{B} + {}^{A}\boldsymbol{\omega}^{B} \times \boldsymbol{r}_{P/O}\right) \bullet \hat{\boldsymbol{u}} = {}^{A}\boldsymbol{v}_{O}^{B} \bullet \hat{\boldsymbol{u}} + \left({}^{A}\boldsymbol{\omega}^{B} \times \boldsymbol{r}_{P/O}\right) \bullet \hat{\boldsymbol{u}}. \tag{3.28}$$

Here the bullet (•) denotes the usual inner product of three-dimensional vectorial algebra. On the other hand, according to the unit vector $\hat{\boldsymbol{u}}$, we have

$$\left({}^{A}\boldsymbol{\omega}^{B} \times \boldsymbol{r}_{P/O}\right) \bullet \hat{\boldsymbol{u}} = \left({}^{A}\boldsymbol{\omega}^{B} \times \boldsymbol{r}_{P/O}\right) \bullet \frac{{}^{A}\boldsymbol{\omega}^{B}}{{}_{A}\omega_{B}} = 0. \tag{3.29}$$

Furthermore, taking into account that

$${}^{A}\boldsymbol{\omega}^{B} \times \boldsymbol{r}_{P/O} = {}^{A}\boldsymbol{v}_{P}^{B} - {}^{A}\boldsymbol{v}_{O}^{B},$$

it follows that

$$\left({}^{A}\boldsymbol{\omega}^{B} \times \boldsymbol{r}_{P/O}\right) \bullet \hat{\boldsymbol{u}} = \left({}^{A}\boldsymbol{v}_{P}^{B} - {}^{A}\boldsymbol{v}_{O}^{B}\right) \bullet \hat{\boldsymbol{u}} = 0. \tag{3.30}$$

Hence, in order to obtain a general result, we must disregard the trivial possibility $\hat{\boldsymbol{u}} = \boldsymbol{0}$ from the analysis, yielding

$${}^{A}\boldsymbol{v}_{P}^{B} \bullet \hat{\boldsymbol{u}} = {}^{A}\boldsymbol{v}_{O}^{B} \bullet \hat{\boldsymbol{u}}, \tag{3.31}$$

which allows us to affirm that the components of the linear velocities ${}^{A}\boldsymbol{v}_{P}^{B}$ and ${}^{A}\boldsymbol{v}_{O}^{B}$ along the direction of the angular velocity vector ${}^{A}\boldsymbol{\omega}^{B}$ are, in fact, the same. ◇

Proposition 3.4. *Assume that P is a point of B where the component of the linear velocity ${}^{A}\boldsymbol{v}_{P}^{B}$ perpendicular to the angular velocity ${}^{A}\boldsymbol{\omega}^{B}$ vanishes. Then there is a line passing through point P whose direction is parallel to ${}^{A}\boldsymbol{\omega}^{B}$. Furthermore, the points along this line have the same linear velocity of point P. In other words, the component of the linear velocity perpendicular to ${}^{A}\boldsymbol{\omega}^{B}$ is also null.*

Proof. The proof is left as an exercise.

Proposition 3.5 (Introduction to Chasles' Theorem).
Let

$${}^{A}\boldsymbol{V}^{B} = \begin{bmatrix} {}^{A}\boldsymbol{\omega}^{B} \\ {}^{A}\boldsymbol{v}_{O}^{B} \end{bmatrix}$$

be a representation of the velocity state of body B with respect to body A. There is a point P of body B whose linear velocity vector is parallel to the angular velocity vector ${}^{A}\boldsymbol{\omega}^{B}$. In other words, the component of the linear velocity ${}^{A}\boldsymbol{v}_{P}^{B}$ perpendicular to the angular velocity vector ${}^{A}\boldsymbol{\omega}^{B}$ vanishes.

Proof. By resorting to the unit vector $\hat{\boldsymbol{u}}$, we may express the linear velocity ${}^{A}\boldsymbol{v}_{P}^{B}$ as

$${}^{A}\boldsymbol{v}_{P}^{B} = \left({}^{A}\boldsymbol{v}_{P}^{B} \bullet \hat{\boldsymbol{u}}\right)\hat{\boldsymbol{u}}. \tag{3.32}$$

Furthermore, the fact that ${}^{A}\boldsymbol{v}_{P}^{B} = {}^{A}\boldsymbol{v}_{O}^{B} + {}^{A}\boldsymbol{\omega}^{B} \times \boldsymbol{r}_{P/O}$ gives us

$$\boldsymbol{0} = {}^{A}\boldsymbol{v}_{O}^{B} + {}^{A}\boldsymbol{\omega}^{B} \times \boldsymbol{r}_{P/O} - \left[\left({}^{A}\boldsymbol{v}_{O}^{B} + {}^{A}\boldsymbol{\omega}^{B} \times \boldsymbol{r}_{P/O}\right) \bullet \hat{\boldsymbol{u}}\right]\hat{\boldsymbol{u}}. \tag{3.33}$$

However, $\left({}^A\boldsymbol{\omega}^B \times \boldsymbol{r}_{P/O}\right) \bullet \hat{\boldsymbol{u}} = 0$. Thus,

$$\mathbf{0} = {}^A\boldsymbol{v}_O^B - \left({}^A\boldsymbol{v}_O^B \bullet \hat{\boldsymbol{u}}\right)\hat{\boldsymbol{u}} + {}^A\boldsymbol{\omega}^B \times \boldsymbol{r}_{P/O}. \tag{3.34}$$

Premultiplying Eq. (3.34) by ${}^A\boldsymbol{\omega}^B$, one obtains

$$\mathbf{0} = {}^A\boldsymbol{\omega}^B \times {}^A\boldsymbol{v}_O^B - \left({}^A\boldsymbol{v}_O^B \bullet \hat{\boldsymbol{u}}\right){}^A\boldsymbol{\omega}^B \times \hat{\boldsymbol{u}} + {}^A\boldsymbol{\omega}^B \times \left({}^A\boldsymbol{\omega}^B \times \boldsymbol{r}_{P/O}\right). \tag{3.35}$$

On the other hand, according to Wylie and Barret (1982), the triple vectorial product ${}^A\boldsymbol{\omega}^B \times \left({}^A\boldsymbol{\omega}^B \times \boldsymbol{r}_{P/O}\right)$ can be decomposed in

$$\begin{aligned} {}^A\boldsymbol{\omega}^B \times \left({}^A\boldsymbol{\omega}^B \times \boldsymbol{r}_{P/O}\right) &= \left({}^A\boldsymbol{\omega}^B \bullet \boldsymbol{r}_{P/O}\right){}^A\boldsymbol{\omega}^B - \left({}^A\boldsymbol{\omega}^B \bullet {}^A\boldsymbol{\omega}^B\right)\boldsymbol{r}_{P/O} \\ &= \left({}^A\boldsymbol{\omega}^B \bullet \boldsymbol{r}_{P/O}\right){}^A\boldsymbol{\omega}^B - {}_A\omega_B{}^2\boldsymbol{r}_{P/O}. \end{aligned} \tag{3.36}$$

Furthermore, it is evident that

$$ {}^A\boldsymbol{\omega}^B \times \hat{\boldsymbol{u}} = {}^A\boldsymbol{\omega}^B \times {}^A\boldsymbol{\omega}^B / \mid {}_A\omega_B \mid = \mathbf{0}. \tag{3.37}$$

Therefore, with the substitution of Eqs. (3.36) and (3.37) into Eq. (3.35), we get

$$\mathbf{0} = {}^A\boldsymbol{\omega}^B \times {}^A\boldsymbol{v}_O^B + \left({}^A\boldsymbol{\omega}^B \bullet \boldsymbol{r}_{P/O}\right){}^A\boldsymbol{\omega}^B - {}_A\omega_B^2\boldsymbol{r}_{P/O}, \tag{3.38}$$

where point P can be freely chosen. Without loss of generality, assuming that point P is such that ${}^A\boldsymbol{\omega}^B \bullet \boldsymbol{r}_{P/O} = 0$, according to Eq. (3.38) the position vector $\boldsymbol{r}_{P/O}$ results in

$$\boldsymbol{r}_{P/O} = \frac{{}^A\boldsymbol{\omega}^B \times {}^A\boldsymbol{v}_O^B}{{}_A\omega_B^2}. \tag{3.39}$$

Equation (3.39) admits two possibilities:

1. If ${}_A\omega_B \neq 0$, then it is possible to compute the position vector $\boldsymbol{r}_{P/O}$.
2. If ${}_A\omega_B = 0$, then ${}^A\boldsymbol{\omega}^B = \mathbf{0}$, and the point P goes to infinity in a direction perpendicular to ${}^A\boldsymbol{v}_O^B = {}^A\boldsymbol{v}_P^B$, which implies that the rigid body B possesses only infinitesimal translational motion. Hence, all the points of B have the same linear velocity.

Note that Proposition 3.5 describes the existence of a point P, fixed to body B, where its linear velocity ${}^A\boldsymbol{v}_P^B$ is given by

$${}^A\boldsymbol{v}_P^B = \lambda \frac{{}^A\boldsymbol{\omega}^B}{{}_A\omega_B} \quad \forall \lambda \in \Re. \tag{3.40}$$

In other words, there is a point whose linear velocity is parallel to the angular velocity vector ${}^A\boldsymbol{\omega}^B$. However, Proposition 3.4 establishes that given point P, all the points along the line passing though P have the same property. In fact,

$$ {}^A\boldsymbol{v}_Q^B = \lambda \frac{{}^A\boldsymbol{\omega}^B}{{}_A\omega_B}, \tag{3.41} $$

where Q is a point of the aforementioned line that is known as the instantaneous screw axis (ISA) of the motion of rigid body B with respect to body A. ◇

Proposition 3.6. *Let us consider*

$$ {}^A\boldsymbol{V}^B = \begin{bmatrix} {}^A\boldsymbol{\omega}^B \\ {}^A\boldsymbol{v}_O^B \end{bmatrix}, $$

a representation of the velocity state of body B as observed from body A. The velocity state ${}^A\boldsymbol{V}^B$ *and an infinitesimal motion of body B as measured from body A about the ISA are equivalent. Furthermore, the magnitude* ${}_A\omega_B$ *of the angular velocity* ${}^A\boldsymbol{\omega}^B$ *and the pitch h of the screw are related by*

$$ h = \frac{{}^A\boldsymbol{v}_O^B \bullet {}^A\boldsymbol{\omega}^B}{{}_A\omega_B^2}. \tag{3.42} $$

Proof. Let P be a point of the ISA. The linear velocity of point P is given (see Proposition 3.5) by

$$ {}^A\boldsymbol{v}_P^B = \left(\frac{{}^A\boldsymbol{v}_O^B \bullet {}^A\boldsymbol{\omega}^B}{{}_A\omega_B{}^2} \right) {}^A\boldsymbol{\omega}^B. \tag{3.43} $$

It must be noted that Eq. (3.43) yields the component of the linear velocity of point P along the direction of the angular velocity ${}^A\boldsymbol{\omega}^B$. However, according to Proposition 3.5, the component of ${}^A\boldsymbol{v}_P^B$ perpendicular to ${}^A\boldsymbol{\omega}^B$ vanishes. Therefore, Eq. (3.43) is, in reality, the velocity of point P. In this way, the linear velocity of an arbitrary point O of body B, undergoing an infinitesimal motion around the ISA with an associated angular velocity vector ${}^A\boldsymbol{\omega}^B$ notated as $\left({}^A\boldsymbol{v}_O^B\right)_T$, must be given by

$$ \left({}^A\boldsymbol{v}_O^B\right)_T = {}^A\boldsymbol{v}_P^B + {}^A\boldsymbol{\omega}^B \times \boldsymbol{r}_{O/P}. \tag{3.44} $$

However, according to Eq. (3.39) we write

$$ \boldsymbol{r}_{O/P} = -\boldsymbol{r}_{P/O} = -\frac{{}^A\boldsymbol{\omega}^B \times {}^A\boldsymbol{v}_O^B}{{}_A\omega_B^2}. \tag{3.45} $$

Substituting Eq. (3.45) into Eq. (3.44), one obtains

$$ \begin{aligned} \left({}^A\boldsymbol{v}_O^B\right)_T &= {}^A\boldsymbol{v}_P^B + {}^A\boldsymbol{\omega}^B \times \left(-\frac{{}^A\boldsymbol{\omega}^B \times {}^A\boldsymbol{v}_O^B}{{}_A\omega_B^2} \right) \\ &= \left(\frac{{}^A\boldsymbol{v}_O^B \bullet {}^A\boldsymbol{\omega}^B}{{}_A\omega_B{}^2} \right) {}^A\boldsymbol{\omega}^B - \frac{{}^A\boldsymbol{\omega}^B \times \left({}^A\boldsymbol{\omega}^B \times {}^A\boldsymbol{v}_O^B\right)}{{}_A\omega_B^2}. \end{aligned} \tag{3.46} $$

On the other hand, by resorting to the triple vectorial product of the usual three-dimensional vectorial algebra, we have

$$
\begin{aligned}
{}^{A}\boldsymbol{\omega}^{B} \times \left({}^{A}\boldsymbol{\omega}^{B} \times {}^{A}\boldsymbol{v}_{O}^{B}\right) &= \left({}^{A}\boldsymbol{\omega}^{B} \bullet {}^{A}\boldsymbol{v}_{O}^{B}\right) {}^{A}\boldsymbol{\omega}^{B} - \left({}^{A}\boldsymbol{\omega}^{B} \bullet {}^{A}\boldsymbol{\omega}^{B}\right) {}^{A}\boldsymbol{v}_{O}^{B} \\
&= \left({}^{A}\boldsymbol{\omega}^{B} \bullet {}^{A}\boldsymbol{v}_{O}^{B}\right) {}^{A}\boldsymbol{\omega}^{B} - {}_{A}\omega_{B}{}^{2}\, {}^{A}\boldsymbol{v}_{O}^{B}.
\end{aligned}
\tag{3.47}
$$

Therefore,

$$
\begin{aligned}
\left({}^{A}\boldsymbol{v}_{O}^{B}\right)_{T} = \left(\frac{{}^{A}\boldsymbol{v}_{O}^{B} \bullet {}^{A}\boldsymbol{\omega}^{B}}{{}_{A}\omega_{B}{}^{2}}\right) {}^{A}\boldsymbol{\omega}^{B} - \frac{\left({}^{A}\boldsymbol{\omega}^{B} \bullet {}^{A}\boldsymbol{v}_{O}^{B}\right) {}^{A}\boldsymbol{\omega}^{B} - {}_{A}\omega_{B}{}^{2}\, {}^{A}\boldsymbol{v}_{O}^{B}}{{}_{A}\omega_{B}{}^{2}} \\
= {}^{A}\boldsymbol{v}_{O}^{B}.
\end{aligned}
\tag{3.48}
$$

To compute the pitch h of the screw, it is necessary to recognize that the linear velocity of point P depends only on the translational motion of the associated screw. Indeed,

$$
{}^{A}\boldsymbol{v}_{P}^{B} = h\, {}_{A}\omega_{B}\, \hat{\boldsymbol{u}}. \tag{3.49}
$$

Finally, equating expressions (3.43) and (3.49) yields expression (3.42). ◇

3.2.4 *Plücker Coordinates of Lines*

Plücker coordinates are a way to assign six homogeneous coordinates to each line in projective 3-space $\mathbf{E}^3$. This visionary idea was introduced in geometry by Julius Plücker in the nineteenth century.

Let ${}^{A}\boldsymbol{s}^{B}$ be a unit vector along the direction of the angular velocity ${}^{A}\boldsymbol{\omega}^{B}$; then the vector ${}^{A}\boldsymbol{\omega}^{B}$ can be written as

$$
{}^{A}\boldsymbol{\omega}^{B} = {}_{A}\omega_{B}\, {}^{A}\boldsymbol{s}^{B}. \tag{3.50}
$$

Considering that ${}_{A}h_{B}$ is the pitch of a screw given by Eq. (3.42) and $\boldsymbol{r}_{O/P}$ is the position vector of two points O and P of body B, where point P is an arbitrary point located in the ISA, the moment ${}^{A}\boldsymbol{s}_{O}^{B}$ is defined as

$$
{}^{A}\boldsymbol{s}_{O}^{B} \equiv {}_{A}h_{B}\, {}^{A}\boldsymbol{s}^{B} + {}^{A}\boldsymbol{s}^{B} \times \boldsymbol{r}_{O/P}. \tag{3.51}
$$

A normalized or unit screw can be represented in Plücker coordinates by

$$
{}^{A}\$^{B} \equiv \begin{bmatrix} {}^{A}\boldsymbol{s}^{B} \\ {}^{A}\boldsymbol{s}_{O}^{B} \end{bmatrix}. \tag{3.52}
$$

However, it is important to clarify that since in general only the primal part of the screw is normalized, for instance, $|{}^{A}s^{B}| = 1$, then the term "normalized screw" must be reconsidered.

Example 3.1. Consider that XYZ is a reference frame with associated conventional unit vectors $\hat{i}\hat{j}\hat{k}$. Compute the Plücker coordinates of a line passing through points $A = (1, 5, -3)$ and $B = (-2, 6, 7)$, where the origin of the reference frame, the point $O = (0, 0, 0)$, plays the role of reference pole. Indeed, the dual part of the line is the moment produced by the primal part around the origin O of the frame XYZ. Furthermore, consider that the primal part of the line is directed from A to B.

Solution. Let $\$ = ({}^{A}\hat{s}^{B}; s_O)$ be the unknown line in Plücker coordinates. The primal part ${}^{A}\hat{s}^{B}$ is computed as

$$ {}^{A}\hat{s}^{B} = \frac{\boldsymbol{b} - \boldsymbol{a}}{|\,\boldsymbol{b} - \boldsymbol{a}\,|}, $$

where $\boldsymbol{a}$ and $\boldsymbol{b}$ are the position vectors of A and B, respectively. Taking into account that $\boldsymbol{b} - \boldsymbol{a} = -3\hat{i} + \hat{j} + 10\hat{k}$ and $|\,\boldsymbol{b} - \boldsymbol{a}\,| = \sqrt{110}$, the primal part of \$ results in

$$ {}^{A}\hat{s}^{B} = -\frac{3}{\sqrt{110}}\hat{i} + \frac{1}{\sqrt{110}}\hat{j} + \frac{10}{\sqrt{110}}\hat{k}. $$

Furthermore, the dual part $\boldsymbol{s}_O$ of \$ may be computed as

$$ \boldsymbol{s}_O = {}^{A}\hat{s}^{B} \times \boldsymbol{r}_{O/A}, $$

where $\boldsymbol{r}_{O/A}$ is a vector pointed from A to O. Hence, one obtains

$$ \boldsymbol{s}_O = (-\frac{3}{\sqrt{110}}\hat{i} + \frac{1}{\sqrt{110}}\hat{j} + \frac{10}{\sqrt{110}}\hat{k}) \times (-\hat{i} - 5\hat{j} + 3\hat{k}) = \frac{53}{\sqrt{110}}\hat{i} - \frac{1}{\sqrt{110}}\hat{j} + \frac{16}{\sqrt{110}}\hat{k}. $$

Finally, the line in Plücker coordinates meeting the geometric conditions of the example is given by

$$ \$ = \frac{1}{\sqrt{110}} \begin{bmatrix} -3 \\ 1 \\ 10 \\ 53 \\ -1 \\ 16 \end{bmatrix}. $$

Proposition 3.7 (The Velocity State of a Rigid Body is the Twist on a Screw). *A relationship between the velocity state*

$$ {}^{A}\boldsymbol{V}^{B} = \begin{bmatrix} {}^{A}\boldsymbol{\omega}^{B} \\ {}^{A}\boldsymbol{v}_{O}^{B} \end{bmatrix} $$

and the infinitesimal screw

$$^A\$^B = \begin{bmatrix} ^A\boldsymbol{s}^B \\ ^A\boldsymbol{s}^B_O \end{bmatrix}$$

may be formally established as

$$^A\boldsymbol{V}^B = {}_A\omega_B\, ^A\$^B, \tag{3.53}$$

where ${}_A\omega_B$ is the magnitude of the primal part of the six-dimensional vector $^A\boldsymbol{V}^B$, that is, ${}_A\omega_B = \mid {}^A\boldsymbol{\omega}^B \mid$.

Proof. Let us consider that

$${}_A\omega_B\, ^A\$^B = \begin{bmatrix} ^A\boldsymbol{s}^B \\ ^A\boldsymbol{s}^B_O \end{bmatrix} = \begin{bmatrix} {}_A\omega_B\, ^A\boldsymbol{s}^B \\ {}_A\omega_B\, ^A\boldsymbol{s}^B_O \end{bmatrix}, \tag{3.54}$$

where, provided that $^A\boldsymbol{s}^B_O = {}_Ah_B\, ^A\boldsymbol{s}^B + {}^A\boldsymbol{s}^B \times \boldsymbol{r}_{O/P}$, we write

$$\begin{aligned} {}_A\omega_B\, ^A\boldsymbol{s}^B_O &= {}_A\omega_B \left({}_Ah_B\, ^A\boldsymbol{s}^B + {}^A\boldsymbol{s}^B \times \boldsymbol{r}_{O/P}\right) \\ &= {}_A\omega_B\, {}_Ah_B\, ^A\boldsymbol{s}^B + {}_A\omega_B\, ^A\boldsymbol{s}^B \times \left(\frac{-{}^A\boldsymbol{\omega}^B \times {}^A\boldsymbol{v}^B_O}{{}_A\omega^2_B}\right) \\ &= \frac{^A\boldsymbol{v}^B_O \bullet {}^A\boldsymbol{\omega}^B}{{}_A\omega^2_B}\, ^A\boldsymbol{\omega}^B + {}^A\boldsymbol{\omega}^B \times \left(\frac{-{}^A\boldsymbol{\omega}^B \times {}^A\boldsymbol{v}^B_O}{{}_A\omega^2_B}\right). \end{aligned} \tag{3.55}$$

Furthermore, by resorting to the definition of triple vectorial product one obtains

$$^A\boldsymbol{\omega}^B \times \left(\frac{-{}^A\boldsymbol{\omega}^B \times {}^A\boldsymbol{v}^B_O}{{}_A\omega^2_B}\right) = \frac{-({}^A\boldsymbol{\omega}^B \bullet {}^A\boldsymbol{v}^B_O)^A\boldsymbol{\omega}^B + ({}^A\boldsymbol{\omega}^B \bullet {}^A\boldsymbol{\omega}^B)^A\boldsymbol{v}^B_O}{{}_A\omega^2_B}. \tag{3.56}$$

Hence,

$${}_A\omega_B\, ^A\boldsymbol{s}^B_O = \frac{\left({}^A\boldsymbol{\omega}^B \bullet {}^A\boldsymbol{\omega}^B\right) {}^A\boldsymbol{v}^B_O}{{}_A\omega^2_B} = {}^A\boldsymbol{v}^B_O, \tag{3.57}$$

yielding

$${}_A\omega_B{}^A\$^B = \begin{bmatrix} {}_A\omega_B{}^A\boldsymbol{s}^B \\ {}_A\omega_B{}^A\boldsymbol{s}^B_O \end{bmatrix} = \begin{bmatrix} ^A\boldsymbol{\omega}^B \\ ^A\boldsymbol{v}^B_O \end{bmatrix} = {}^A\boldsymbol{V}^B. \quad \diamond \tag{3.58}$$

For details of this subsection, the reader is referred to Bottema and Roth (1979).

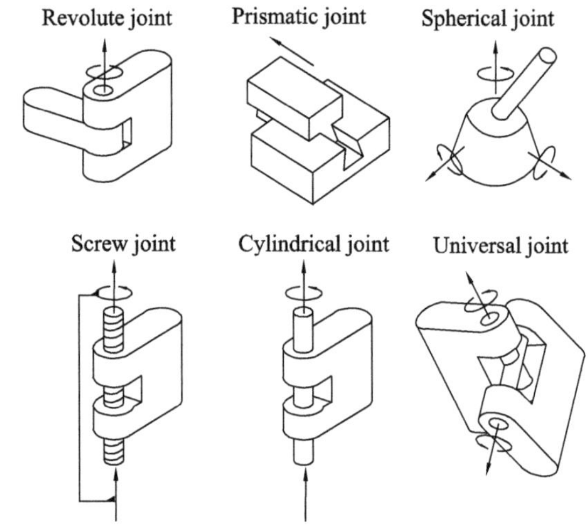

Fig. 3.2 Typical lower kinematic pairs

3.2.5 *Lower Kinematic Pairs and Their Equivalent Infinitesimal Screws*

A lower pair is a kinematic joint representing the relative motion of two links whose contact is through a surface. Most robotic mechanical systems are composed of lower kinematic pairs. The screw pair is commonly known as the helicoidal pair, whereas the basic lower kinematic pairs are the revolute and prismatic joints. Other lower kinematic joints can be generated upon particular arrangements of revolute and prismatic kinematic pairs, for instance, spherical, cylindrical, and universal joints; see Fig. 3.2.

A revolute joint is modeled as a screw where the pitch h vanishes, and therefore the associated screw is reduced to

$$^{A}\$^{B} = \begin{bmatrix} ^{A}\boldsymbol{s}^{B} \\ ^{A}\boldsymbol{s}^{B} \times \boldsymbol{r}_{O/P} \end{bmatrix}. \tag{3.59}$$

Meanwhile, in a prismatic joint the pitch h goes to infinity and the equivalent screw is given by

$$^{A}\$^{B} = \begin{bmatrix} \boldsymbol{0} \\ ^{A}\boldsymbol{s}^{B} \end{bmatrix}. \tag{3.60}$$

On the other hand, a spherical joint can be modeled by a set of three revolute joints whose axes intersect at a common point. In a standard representation that is not a rule, the axes of the revolute joints are preferred to be orthogonal. Meanwhile, a universal joint may be modeled as two revolute joints whose axes are orthonormal and intersect at a common point, whereas a cylindrical joint results from the combination of two lower kinematic pairs, one a revolute joint and the other a prismatic joint. Finally, an unusual lower kinematic joint used in robotics is the flat kinematic pair, which results from the combination of one revolute joint and two prismatic joints.

Example 3.2. Determine the infinitesimal screws associated to the kinematic pairs of the 3-RPR (where R and P stand for revolute and prismatic joints, respectively) 3-DOF planar parallel manipulator.

Solution. Consider that point O is the reference pole of the parallel manipulator at hand. The revolute joint connecting the ith limb to the fixed platform may be represented by a screw

$$\$_i = \begin{bmatrix} \hat{\boldsymbol{s}}_i \\ \boldsymbol{s}_{Oi} = \hat{\boldsymbol{s}}_i \times \boldsymbol{r}_i \end{bmatrix},$$

where $\hat{\boldsymbol{s}}_i$ is a unit vector normal to the plane of the fixed platform. Meanwhile, $\boldsymbol{s}_{Oi}$ is the moment produced by $\hat{\boldsymbol{s}}_i$ about the point O according to a vector $\boldsymbol{r}_i$ directed from a point of the axis of the screw to the point O. It is straightforward to show that a similar representation can be used to simulate the kinematic connection of the ith limb to the moving platform. Finally, the prismatic joint of the ith limb can be considered a screw denoted as

$$\$_i = \begin{bmatrix} \boldsymbol{0} \\ \hat{\boldsymbol{s}}_i \end{bmatrix},$$

where $\hat{\boldsymbol{s}}_i$ is a unit vector along the ith leg and the dual part vector of the screw representing the prismatic joint is perpendicular to the primal part vectors of the screws representing the revolute joints.

3.3 The Lie Algebra $se(3)$ of the Euclidean Group $SE(3)$

The Lie algebra $se(3)$ can be considered the algebra of the infinitesimal elements of the Euclidean group $SE(3)$; it is isomorphic to screw theory, which is also known as motor algebra.

Definition 3.3. The Lie algebra $se(3)$ of the Euclidean group $SE(3)$ is the set of equivalence classes $\overline{V}$ of the form

$$\overline{V} = \overline{(\boldsymbol{\omega}; \boldsymbol{v}_O)} = \overline{\begin{bmatrix} \boldsymbol{\omega} \\ \boldsymbol{v}_O \end{bmatrix}}. \quad (3.61)$$

Hence, the equivalence classes are the velocity states of a rigid body.

Let

$$\overline{V_1} = \overline{(\omega_1; v_{O1})},\ \overline{V_2} = \overline{(\omega_2; v_{O2})}, \overline{V_3} = \overline{(\omega_3; v_{O3})} \tag{3.62}$$

be three elements of the Lie algebra *se*(3). The Lie algebra *se*(3) has the following fundamental operations.

1. Addition:

$$\overline{V_1} + \overline{V_2} = \overline{(\omega_1; v_{O1})} + \overline{(\omega_2; v_{O2})} \equiv \overline{(\omega_1 + \omega_2; v_{O1} + v_{O2})}. \tag{3.63}$$

2. Multiplication by a scalar:

$$\lambda\overline{V_1} = \lambda\overline{(\omega_1; v_{O1})} \equiv \overline{(\lambda\omega_1; \lambda v_{O1})}, \quad \lambda \in \Re. \tag{3.64}$$

3. Lie product:

$$\left[\overline{(\omega_1; v_{O1})} \quad \overline{(\omega_2; v_{O2})}\right] = \overline{(\omega_1 \times \omega_2; \omega_1 \times v_{O2} - \omega_2 \times v_{O1})}. \tag{3.65}$$

In these equations a generic kinematic state W may be notated as

$$W = \begin{bmatrix} P \\ D \end{bmatrix} = (P; D), \tag{3.66}$$

where the triplets P and D denote the primal and dual parts, respectively, of W.

Proposition 3.8. *The operations of addition and multiplication by a scalar and Lie product given in Definition 3.3 are well defined.*

Proof. Consider that

$$(\omega_1; v_{O1}), (\omega_1; v_{O^*1}) \in \overline{V_1} \tag{3.67}$$

and

$$(\omega_2; v_{O2}), (\omega_2; v_{O^*2}) \in \overline{V_2}. \tag{3.68}$$

Then it follows that

$$v_{O^*1} = v_{O1} + \omega_1 \times r_{O^*/O}$$

and

$$v_{O^*2} = v_{O2} + \omega_2 \times r_{O^*/O},$$

where O and O^* are two points of the rigid body whose velocity state is represented by $\overline{V_1}$ and $\overline{V_2}$. Therefore, it is evident that any of the two representations can be used in order to perform the operations provided in Definition 3.3. In fact,

1. Addition:
   Taking into account that

$$(\boldsymbol{\omega}_1; \boldsymbol{v}_{O1}) + (\boldsymbol{\omega}_2; \boldsymbol{v}_{O2}) = (\boldsymbol{\omega}_1 + \boldsymbol{\omega}_2; \boldsymbol{v}_{O1} + \boldsymbol{v}_{O2}) \tag{3.69}$$

   and

$$(\boldsymbol{\omega}_1; \boldsymbol{v}_{O^*1}) + (\boldsymbol{\omega}_2; \boldsymbol{v}_{O^*2}) = (\boldsymbol{\omega}_1 + \boldsymbol{\omega}_2; \boldsymbol{v}_{O^*1} + \boldsymbol{v}_{O^*2}), \tag{3.70}$$

   it immediately emerges that

$$\boldsymbol{\omega}_1 + \boldsymbol{\omega}_2 = \boldsymbol{\omega}_1 + \boldsymbol{\omega}_2 \tag{3.71}$$

   and

$$\begin{aligned} \boldsymbol{v}_{O^*1} + \boldsymbol{v}_{O^*2} &= (\boldsymbol{v}_{O1} + \boldsymbol{\omega}_1 \times \boldsymbol{r}_{O^*/O}) + (\boldsymbol{v}_{O2} + \boldsymbol{\omega}_2 \times \boldsymbol{r}_{O^*/O}) \\ &= (\boldsymbol{v}_{O1} + \boldsymbol{v}_{O2}) + (\boldsymbol{\omega}_1 + \boldsymbol{\omega}_2) \times \boldsymbol{r}_{O^*/O}. \end{aligned} \tag{3.72}$$

2. Multiplication by a scalar:
   It is evident that

$$\lambda(\boldsymbol{\omega}_1; \boldsymbol{v}_{O1}) = (\lambda\boldsymbol{\omega}_1; \lambda\boldsymbol{v}_{O1}) \tag{3.73}$$

   and

$$\lambda(\boldsymbol{\omega}_1; \boldsymbol{v}_{O^*1}) = (\lambda\boldsymbol{\omega}_1; \lambda\boldsymbol{v}_{O^*1}). \tag{3.74}$$

   Therefore, it follows that

$$\lambda\boldsymbol{\omega}_1 = \lambda\boldsymbol{\omega}_1 \tag{3.75}$$

   and

$$\lambda\boldsymbol{v}_{O^*1} = \lambda(\boldsymbol{v}_{O1} + \boldsymbol{\omega}_1 \times \boldsymbol{r}_{O^*/O}) = (\lambda\boldsymbol{v}_{O1}) + (\lambda\boldsymbol{\omega}_1) \times \boldsymbol{r}_{O^*/O}. \tag{3.76}$$

3. Lie product:
   The application of the Lie product between $(\boldsymbol{\omega}_1; \boldsymbol{v}_{O1})$ and $(\boldsymbol{\omega}_2; \boldsymbol{v}_{O2}$ yields

$$[(\boldsymbol{\omega}_1; \boldsymbol{v}_{O1}) \quad (\boldsymbol{\omega}_2; \boldsymbol{v}_{O2})] = (\boldsymbol{\omega}_1 \times \boldsymbol{\omega}_2; \boldsymbol{\omega}_1 \times \boldsymbol{v}_{O2} - \boldsymbol{\omega}_2 \times \boldsymbol{v}_{O1}) \tag{3.77}$$

and

$$[(\boldsymbol{\omega}_1; \boldsymbol{v}_{O^*1}) \quad (\boldsymbol{\omega}_2; \boldsymbol{v}_{O^*2})] = (\boldsymbol{\omega}_1 \times \boldsymbol{\omega}_2; \boldsymbol{\omega}_1 \times \boldsymbol{v}_{O^*2} - \boldsymbol{\omega}_2 \times \boldsymbol{v}_{O^*1}), \tag{3.78}$$

where the dual part $\boldsymbol{\omega}_1 \times \boldsymbol{v}_{O^*2} - \boldsymbol{\omega}_2 \times \boldsymbol{v}_{O^*1}$ may be expanded as follows:

$$\begin{aligned} \boldsymbol{\omega}_1 \times \boldsymbol{v}_{O^*2} - \boldsymbol{\omega}_2 \times \boldsymbol{v}_{O^*1} = \\ \boldsymbol{\omega}_1 \times (\boldsymbol{v}_{O2} + \boldsymbol{\omega}_2 \times \boldsymbol{r}_{O^*/O}) - \boldsymbol{\omega}_2 \times (\boldsymbol{v}_{O1} + \boldsymbol{\omega}_1 \times \boldsymbol{r}_{O^*/O}) = \\ (\boldsymbol{\omega}_1 \times \boldsymbol{v}_{O2} - \boldsymbol{\omega}_2 \times \boldsymbol{v}_{O1}) + \boldsymbol{\omega}_1 \times (\boldsymbol{\omega}_2 \times \boldsymbol{r}_{O^*/O}) - \boldsymbol{\omega}_2 \times (\boldsymbol{\omega}_1 \times \boldsymbol{r}_{O^*/O}). \end{aligned} \tag{3.79}$$

However,

$$\begin{aligned} \boldsymbol{\omega}_1 \times (\boldsymbol{\omega}_2 \times \boldsymbol{r}_{O^*/O}) - \boldsymbol{\omega}_2 \times (\boldsymbol{\omega}_1 \times \boldsymbol{r}_{O^*/O}) = \\ \boldsymbol{\omega}_1 \times (\boldsymbol{\omega}_2 \times \boldsymbol{r}_{O^*/O}) + \boldsymbol{\omega}_2 \times (\boldsymbol{r}_{O^*/O} \times \boldsymbol{\omega}_1). \end{aligned} \tag{3.80}$$

Furthermore, according to the Jacobi identity of the triple vectorial product it follows that

$$\boldsymbol{\omega}_1 \times (\boldsymbol{\omega}_2 \times \boldsymbol{r}_{O^*/O}) + \boldsymbol{\omega}_2 \times (\boldsymbol{r}_{O^*/O} \times \boldsymbol{\omega}_1) + \boldsymbol{r}_{O^*/O} \times (\boldsymbol{\omega}_1 \times \boldsymbol{\omega}_2) = \mathbf{0}. \tag{3.81}$$

Therefore,

$$\begin{aligned} \boldsymbol{\omega}_1 \times (\boldsymbol{\omega}_2 \times \boldsymbol{r}_{O^*/O}) - \boldsymbol{\omega}_2 \times (\boldsymbol{\omega}_1 \times \boldsymbol{r}_{O^*/O}) = \\ -\boldsymbol{r}_{O^*/O} \times (\boldsymbol{\omega}_1 \times \boldsymbol{\omega}_2) = (\boldsymbol{\omega}_1 \times \boldsymbol{\omega}_2) \times \boldsymbol{r}_{O^*/O}, \end{aligned} \tag{3.82}$$

which leads to

$$\boldsymbol{\omega}_1 \times \boldsymbol{v}_{O^*2} - \boldsymbol{\omega}_2 \times \boldsymbol{v}_{O^*1} = (\boldsymbol{\omega}_1 \times \boldsymbol{v}_{O2} - \boldsymbol{\omega}_2 \times \boldsymbol{v}_{O1}) + (\boldsymbol{\omega}_1 \times \boldsymbol{\omega}_2) \times \boldsymbol{r}_{O^*/O}. \quad \diamond \tag{3.83}$$

Gathering these results based on the fact that the operations of the Lie algebra $se(3)$ are well defined, we see that there are no distinctions among the operations of the equivalence classes, the elements of $se(3)$, and the corresponding operations with any of their possible representations. Furthermore, it must be noted that a Lie algebra is a vectorial space that has the operations of addition, multiplication by a scalar, and the Lie product. Therefore, the Lie algebra $se(3)$ meets all the axioms of vectorial space. On the other hand, the Lie product plays a central role in the higher-order kinematic analyses of robot manipulators and therefore deserves to be investigated deeply. This is a justification for the next proposition.

Proposition 3.9 (Properties of the Lie Product). *Let us consider three equivalence classes given by*

$$\overline{\boldsymbol{V}_1} = (\boldsymbol{\omega}_1; \boldsymbol{v}_{O1}), \overline{\boldsymbol{V}_2} = (\boldsymbol{\omega}_1; \boldsymbol{v}_{O2}), \overline{\boldsymbol{V}_3} = (\boldsymbol{\omega}_1; \boldsymbol{v}_{O3}).$$

Furthermore, let λ_1, λ_2, λ_3 be three scalars, for instance, $\lambda_{1,2,3} \in \Re$. The Lie product, as given in Definition 3.3, of the Lie algebra $se(3)$ has three relevant properties.

1. Nilpotent:

$$\left[\overline{V_1} \quad \overline{V_1}\right] = \mathbf{0}. \tag{3.84}$$

2. Distributive:

$$\begin{aligned} \left[\overline{V_1} \quad \lambda_2\overline{V_2} + \lambda_3\overline{V_3}\right] &= \lambda_2\left[\overline{V_1} \quad \overline{V_2}\right] + \lambda_3\left[\overline{V_1} \quad \overline{V_3}\right], \\ \left[\lambda_1\overline{V_1} + \lambda_2\overline{V_2} \quad \overline{V_3}\right] &= \lambda_1\left[\overline{V_1} \quad \overline{V_3}\right] + \lambda_2\left[\overline{V_2} \quad \overline{V_3}\right]. \end{aligned} \tag{3.85}$$

3. Jacobi identity:

$$\left[\overline{V_1} \quad \left[\overline{V_2} \quad \overline{V_3}\right]\right] + \left[\overline{V_3} \quad \left[\overline{V_1} \quad \overline{V_2}\right]\right] + \left[\overline{V_2} \quad \left[\overline{V_3} \quad \overline{V_1}\right]\right] = \mathbf{0}. \tag{3.86}$$

Proof.

1. Nilpotent:

$$\begin{aligned} [\boldsymbol{V}_1 \quad \boldsymbol{V}_1] = [(\boldsymbol{\omega}_1; \boldsymbol{v}_{O1}) \quad (\boldsymbol{\omega}_1; \boldsymbol{v}_{O1})] = \\ (\boldsymbol{\omega}_1 \times \boldsymbol{\omega}_1; \boldsymbol{\omega}_1 \times \boldsymbol{v}_{O1} - \boldsymbol{\omega}_1 \times \boldsymbol{v}_{O1}) = (\mathbf{0}; \mathbf{0}) = \mathbf{0}. \end{aligned} \tag{3.87}$$

2. Distributive:

$$[\boldsymbol{V}_1 \quad \lambda_2\boldsymbol{V}_2 + \lambda_3\boldsymbol{V}_3] = [(\boldsymbol{\omega}_1; \boldsymbol{v}_{O1}) \quad \lambda_2(\boldsymbol{\omega}_2; \boldsymbol{v}_{O2}) + \lambda_3(\boldsymbol{\omega}_3; \boldsymbol{v}_{O3})] \tag{3.88}$$

$$\begin{aligned} &= [(\boldsymbol{\omega}_1; \boldsymbol{v}_{O1}) \quad (\lambda_2\boldsymbol{\omega}_2 + \lambda_3\boldsymbol{\omega}_3; \lambda_2\boldsymbol{v}_{O2} + \lambda_3\boldsymbol{v}_{O3})] \\ &= \begin{bmatrix} \boldsymbol{\omega}_1 \times (\lambda_2\boldsymbol{\omega}_2 + \lambda_3\boldsymbol{\omega}_3) \\ \boldsymbol{\omega}_1 \times (\lambda_2\boldsymbol{v}_{O2} + \lambda_3\boldsymbol{v}_{O3}) - (\lambda_2\boldsymbol{\omega}_2 + \lambda_3\boldsymbol{\omega}_3) \times \boldsymbol{v}_{O1} \end{bmatrix} \\ &= \lambda_2 \begin{bmatrix} \boldsymbol{\omega}_1 \times \boldsymbol{\omega}_2 \\ \boldsymbol{\omega}_1 \times \boldsymbol{v}_{O2} - \boldsymbol{\omega}_2 \times \boldsymbol{v}_{O1} \end{bmatrix} + \lambda_3 \begin{bmatrix} \boldsymbol{\omega}_1 \times \boldsymbol{\omega}_3 \\ \boldsymbol{\omega}_1 \times \boldsymbol{v}_{O3} - \boldsymbol{\omega}_3 \times \boldsymbol{v}_{O1} \end{bmatrix} \\ &= \lambda_2\left[\boldsymbol{V}_1 \quad \boldsymbol{V}_2\right] + \lambda_3\left[\boldsymbol{V}_1 \quad \boldsymbol{V}_3\right]. \end{aligned} \tag{3.89}$$

On the other hand,

$$\begin{aligned} [\lambda_1\boldsymbol{V}_1 + \lambda_2\boldsymbol{V}_2 \quad \boldsymbol{V}_3] &= [\lambda_1(\boldsymbol{\omega}_1; \boldsymbol{v}_{O1}) + \lambda_2(\boldsymbol{\omega}_2; \boldsymbol{v}_{O2}) \quad (\boldsymbol{\omega}_3; \boldsymbol{v}_{O3})] \\ &= [(\lambda_1\boldsymbol{\omega}_1 + \lambda_2\boldsymbol{\omega}_2; \lambda_1\boldsymbol{v}_{O1} + \lambda_2\boldsymbol{v}_{O2}) \quad (\boldsymbol{\omega}_3; \boldsymbol{v}_{O3})] \\ &= \begin{bmatrix} (\lambda_1\boldsymbol{\omega}_1 + \lambda_2\boldsymbol{\omega}_2) \times \boldsymbol{\omega}_3 \\ (\lambda_1\boldsymbol{\omega}_1 + \lambda_2\boldsymbol{\omega}_2) \times \boldsymbol{v}_{O3} - \boldsymbol{\omega}_3 \times (\lambda_1\boldsymbol{v}_{O1} + \lambda_2\boldsymbol{v}_{O2}) \end{bmatrix} \end{aligned}$$

$$= \lambda_1 \begin{bmatrix} \boldsymbol{\omega}_1 \times \boldsymbol{\omega}_3 \\ \boldsymbol{\omega}_1 \times \boldsymbol{v}_{O3} - \boldsymbol{\omega}_3 \times \boldsymbol{v}_{O1} \end{bmatrix} + \lambda_2 \begin{bmatrix} \boldsymbol{\omega}_2 \times \boldsymbol{\omega}_3 \\ \boldsymbol{\omega}_2 \times \boldsymbol{v}_{O3} - \boldsymbol{\omega}_3 \times \boldsymbol{v}_{O2} \end{bmatrix}$$

$$= \lambda_1 [\boldsymbol{V}_1 \quad \boldsymbol{V}_3] + \lambda_2 [\boldsymbol{V}_2 \quad \boldsymbol{V}_3]. \quad (3.90)$$

3. Jacobi identity. The proof is left as an exercise. ◇

In what follows other prominent properties of the Lie product are investigated.

Proposition 3.10 (The Lie Product is Noncommutative). *Let $\boldsymbol{V}_1$ and $\boldsymbol{V}_2$ be two elements of the Lie algebra $se(3)$. The Lie product between $\boldsymbol{V}_1$ and $\boldsymbol{V}_2$ is noncommutative. Indeed,*

$$[\boldsymbol{V}_1 \quad \boldsymbol{V}_2] = -[\boldsymbol{V}_2 \quad \boldsymbol{V}_1]. \quad (3.91)$$

Proof. Let us consider

$$[\boldsymbol{V}_1 + \boldsymbol{V}_2 \quad \boldsymbol{V}_2 + \boldsymbol{V}_1] = [\boldsymbol{V}_1 \quad \boldsymbol{V}_1] + [\boldsymbol{V}_1 \quad \boldsymbol{V}_2] + [\boldsymbol{V}_2 \quad \boldsymbol{V}_1] + [\boldsymbol{V}_2 \quad \boldsymbol{V}_2] = \boldsymbol{0}. \quad (3.92)$$

Thus, according to Proposition 3.9 it is evident that

$$[\boldsymbol{V}_1 \quad \boldsymbol{V}_1] = [\boldsymbol{V}_2 \quad \boldsymbol{V}_2] = \boldsymbol{0}, \quad (3.93)$$

which leads to

$$[\boldsymbol{V}_1 \quad \boldsymbol{V}_2] + [\boldsymbol{V}_2 \quad \boldsymbol{V}_1] = \boldsymbol{0}. \quad (3.94)$$

Finally, one obtains

$$[\boldsymbol{V}_2 \quad \boldsymbol{V}_1] = -[\boldsymbol{V}_1 \quad \boldsymbol{V}_2]. \quad \diamond \quad (3.95)$$

Following this trend, it is straightforward to demonstrate that the Lie product is nonassociative. In fact,

$$[\boldsymbol{V}_1 \quad [\boldsymbol{V}_2 \quad \boldsymbol{V}_3]] \neq [[\boldsymbol{V}_1 \quad \boldsymbol{V}_2] \quad \boldsymbol{V}_3]. \quad (3.96)$$

Overall, the Lie algebra $se(3)$ is distributive and noncommutative and meets the Jacobi identity. In addition, the structure of the Lie algebra $se(3)$ associated to the operations of addition, multiplication by a scalar, and Lie product allows us to assign structures of orthogonal space, which plays a central role in a wide variety of applications dealing with spatial kinematics.

The last operations of the Lie algebra $se(3)$ to be investigated are the Killing and Klein forms. In particular, the Klein form allows us to considerably simplify the velocity analysis of parallel manipulators.

Definition 3.4 (Killing and Klein Forms). Consider the Lie algebra $se(3)$ of the Euclidean group $SE(3)$. The Killing and Klein forms of the Lie algebra $se(3)$ are two bilinear symmetric forms, also known as scalar products.

Let $\overline{\boldsymbol{V}_1}$ and $\overline{\boldsymbol{V}_2}$ be two elements of $se(3)$. The Killing form, notated as $\langle *; * \rangle$, is defined as

$$\langle *; * \rangle : e(3) \times e(3) \to \Re \quad \langle \overline{\boldsymbol{V}_1}, \overline{\boldsymbol{V}_2} \rangle = \langle (\boldsymbol{\omega}_1; \boldsymbol{v}_{O1}), (\boldsymbol{\omega}_2; \boldsymbol{v}_{O2}) \rangle = \boldsymbol{\omega}_1 \bullet \boldsymbol{\omega}_2, \tag{3.97}$$

where the bullet ($\bullet$) denotes the usual inner product of three-dimensional vectorial algebra. Furthermore, $(\boldsymbol{\omega}_1; \boldsymbol{v}_{O1})$ and $(\boldsymbol{\omega}_2; \boldsymbol{v}_{O2})$ are two representations of the velocity states $\overline{\boldsymbol{V}_1}$ and $\overline{\boldsymbol{V}_2}$, respectively. Meanwhile, the Klein form, notated as $\{;\}$, is defined as

$$\begin{aligned} \{*; *\} : e(3) \times e(3) \to \Re \quad \{\overline{\boldsymbol{V}_1}, \overline{\boldsymbol{V}_2}\} &= \{(\boldsymbol{\omega}_1; \boldsymbol{v}_{O1}), (\boldsymbol{\omega}_2; \boldsymbol{v}_{O2})\} \\ &= \boldsymbol{\omega}_1 \bullet \boldsymbol{v}_{O2} + \boldsymbol{\omega}_2 \bullet \boldsymbol{v}_{O1}. \end{aligned} \tag{3.98}$$

It is said that two screws $\$_1$ and $\$_2$ are reciprocal if the application of the Klein form between them yields $\{\$_1; \$_2\} = 0$.

Proposition 3.11. *The Killing and Klein forms given in Definition 3.4 are well defined.*

Proof. Let

$$(\boldsymbol{\omega}_1; \boldsymbol{v}_{O1}), (\boldsymbol{\omega}_1; \boldsymbol{v}_{O^*1}) \in \overline{\boldsymbol{V}_1} \tag{3.99}$$

and

$$(\boldsymbol{\omega}_2; \boldsymbol{v}_{O2}), (\boldsymbol{\omega}_2; \boldsymbol{v}_{O^*2}) \in \overline{\boldsymbol{V}_2} \tag{3.100}$$

be two elements of the Lie algebra $se(3)$. The first part of the proof is devoted to the Killing form. To this end we write

$$\boldsymbol{v}_{O^*1} = \boldsymbol{v}_{O1} + \boldsymbol{\omega}_1 \times \boldsymbol{r}_{O^*/O}$$

and

$$\boldsymbol{v}_{O^*2} = \boldsymbol{v}_{O2} + \boldsymbol{\omega}_2 \times \boldsymbol{r}_{O^*/O},$$

where O and O^* are two points of the rigid body. Two representations of the velocity state of the rigid body are given by $\overline{\boldsymbol{V}_1}$ and $\overline{\boldsymbol{V}_2}$. The definition of Killing form leads to

$$\langle (\boldsymbol{\omega}_1; \boldsymbol{v}_{O^*1}), (\boldsymbol{\omega}_2; \boldsymbol{v}_{O^*2}) \rangle = \boldsymbol{\omega}_1 \bullet \boldsymbol{\omega}_2 = \langle (\boldsymbol{\omega}_1; \boldsymbol{v}_{O1}), (\boldsymbol{\omega}_2; \boldsymbol{v}_{O2}) \rangle. \tag{3.101}$$

Finally, the definition of Klein form gives us

$$\begin{aligned}\{(\boldsymbol{\omega}_1; \boldsymbol{v}_{O^*1}), (\boldsymbol{\omega}_2; \boldsymbol{v}_{O^*2})\} &= \boldsymbol{\omega}_1 \bullet \boldsymbol{v}_{O^*2} + \boldsymbol{\omega}_2 \bullet \boldsymbol{v}_{O^*1} \\ &= \boldsymbol{\omega}_1 \bullet (\boldsymbol{v}_{O2} + \boldsymbol{\omega}_2 \times \boldsymbol{r}_{O^*/O}) + \boldsymbol{\omega}_2 \bullet (\boldsymbol{v}_{O1} + \boldsymbol{\omega}_1 \times \boldsymbol{r}_{O^*/O}) \\ &= (\boldsymbol{\omega}_1 \bullet \boldsymbol{v}_{O2} + \boldsymbol{\omega}_2 \bullet \boldsymbol{v}_{O1}) + \boldsymbol{\omega}_1 \bullet (\boldsymbol{\omega}_2 \times \boldsymbol{r}_{O^*/O}) + \boldsymbol{\omega}_2 \bullet (\boldsymbol{\omega}_1 \times \boldsymbol{r}_{O^*/O}) \\ &= (\boldsymbol{\omega}_1 \bullet \boldsymbol{v}_{O2} + \boldsymbol{\omega}_2 \bullet \boldsymbol{v}_{O1}) + \boldsymbol{\omega}_1 \bullet (\boldsymbol{\omega}_2 \times \boldsymbol{r}_{O^*/O}) - \boldsymbol{\omega}_1 \bullet (\boldsymbol{\omega}_2 \times \boldsymbol{r}_{O^*/O}) \\ &= \boldsymbol{\omega}_1 \bullet \boldsymbol{v}_{O2} + \boldsymbol{\omega}_2 \bullet \boldsymbol{v}_{O1} = \{(\boldsymbol{\omega}_1; \boldsymbol{v}_{O1}), \boldsymbol{\omega}_2; \boldsymbol{v}_{O2})\}. \quad \diamond \end{aligned} \tag{3.102}$$

Once it has been proved that the Killing and Klein forms are well defined, the next step consists of demonstrating that these forms are symmetric and bilinear. The Killing form is a degenerate and semidefinite form that in the literature is related with the inner product of the velocity state and does not depend on the reference pole. Similarly, the Klein form is a nondegenerate and indefinite form commonly known as the reciprocal product. More information about the Killing and Klein forms can be found in Rico-Martínez and Duffy (1992). Finally, considering that a screw is a six-dimensional vector, then to obtain a compact representation of the input–output equations concerned with the infinitesimal kinematics of robot manipulators, it is strongly advised to introduce the so-called operator of polarity $\boldsymbol{\Delta}$ (Lipkin and Duffy 1985). This entity is a matrix possessing the following property:

$$\boldsymbol{\Delta}\boldsymbol{\Delta} = \mathbf{I}_6, \tag{3.103}$$

where $\mathbf{I}$ is the 6×6 identity matrix. Thus, a possible representation of matrix $\boldsymbol{\Delta}$ is given by

$$\boldsymbol{\Delta} = \begin{bmatrix} \mathbf{0} & \mathbf{I}_3 \\ \mathbf{I}_3 & \mathbf{0}. \end{bmatrix} \tag{3.104}$$

Example 3.3. It is known that the properties of reciprocal screws allow us to considerably simplify the velocity analysis of parallel manipulators; for example, by resorting to reciprocal screw theory, it is possible to avoid the computation of passive joint velocity rates of this class of robots. This example provides some cases of reciprocal screws.

Solution. Let $\$_1 = (\hat{s}_1, \boldsymbol{s}_{O1})$ and $\$_2 = (\hat{s}_2, \boldsymbol{s}_{O2})$ be two elements of the Lie algebra $se(3)$ of the Euclidean group $SE(3)$. It is said that the screws $\$_1$ and $\$_2$ are reciprocal if the Klein form between them vanishes. Some cases follow:

1. The screws $\$_1$ and $\$_2$ represent revolute joints whose primal parts intersect at a common point O. If O is chosen as the reference pole, then the dual parts of both screws vanish, yielding $\{\$_1; \$_2\} = \hat{s}_1 \bullet \mathbf{0} + \hat{s}_2 \bullet \mathbf{0} = 0$.
2. The screws $\$_1$ and $\$_2$ represent prismatic joints. It is evident that $\{\$_1; \$_2\} = \{(\mathbf{0}; \hat{s}_1); (\mathbf{0}; \hat{s}_2)\} = \mathbf{0} \bullet \hat{s}_2 + \mathbf{0} \bullet \hat{s}_1 = 0$.

3. The screw $\$_1$ represents a revolute joint whereas the screw $\$_2$ represents a prismatic joint, where the primal part of $\$_1$ is orthogonal to the dual part of $\$_2$, that is, $\hat{s}_1 \perp \hat{s}_2$. Under these conditions it follows that $\{\$_1; \$_2\} = \{(\hat{s}_1; s_{O1}); (\mathbf{0}; \hat{s}_2)\} = \hat{s}_1 \bullet \hat{s}_2 + \mathbf{0} \bullet \hat{s}_{O1} = 0$.

3.3.1 The Lie Product and the Euclidean Motion

Clifford (1873) defined a motor as a magnitude associated to a screw. On the other hand, Dimentberg (1965) defined the motor as a rotor—a line in the space—associated to a free vector. More recently, Chevallier (1991) considered that a screw is a vector field that satisfies the conditions of helical field. This last consideration was taken into account by Gallardo-Alvarado et al. (2008) in an attempt to approach the jerk analysis of a rigid body.

The Lie product, also known as the motor product or dual vector product, is a fundamental operation of the Lie algebra $se(3)$ and deserves special attention. Let A and B be two lines located, respectively, at planes S_A and S_B. von Mises (1924a) defined a motor under the following assumptions:

1. An ordered pair of lines determines a motor with an axis N.
2. Any rotation along the axis N of the lines yields the same motor.

If the lines A and B belong to the same plane, then the axis N is normal to this plane and passes through the intersection of both lines. Moreover, to specify a motor it is necessary to know the angle and separation between the lines. Furthermore, because four parameters are necessary to specify the axis N, a total of six parameters is required to characterize a motor.

Let $\boldsymbol{G} = (\boldsymbol{g}; \boldsymbol{g}_O)$ and $\boldsymbol{M} = (\boldsymbol{m}; \boldsymbol{m}_O)$ be two infinitesimal screws; von Mises defined the motor product as a special vectorial arrangement given by

$$[\boldsymbol{G} \quad \boldsymbol{M}] = [(\boldsymbol{g}; \boldsymbol{g}_O) \quad (\boldsymbol{m}; \boldsymbol{m}_O)] = (\boldsymbol{g} \times \boldsymbol{m}; \boldsymbol{g} \times \boldsymbol{m}_O + \boldsymbol{g}_O \times \boldsymbol{m}), \tag{3.105}$$

where $\times$ denotes the usual cross product of three-dimensional vectorial algebra. Note that Eq. (3.105) does not require the unit dual ϵ commonly used in Clifford's algebras, where $\epsilon^2 = 0$. On the other hand, let $\$_i$ and $\$_j$ be two elements of the Lie algebra $se(3)$. Hausner and Schwartz (1968) considered that the Lie product can be defined in terms of Euclidean displacements as follows:

$$[\$_i \quad \$_j] \equiv \frac{d}{dt}\{\mathbf{M}_i(t)\$_j[\mathbf{M}_i(t)^{-1}]\}|_{t=0}, \tag{3.106}$$

where $\$_i$ is a tangent vector of the Euclidean motion $\mathbf{M}_i(t)$ at time $t = 0$. In reality, both concepts are equivalent, which the next proposition clarifies.

Proposition 3.12. *Let* $\$_j = (\boldsymbol{\omega}_j; \boldsymbol{v}_{Oj})$ *and* $\$_k = (\boldsymbol{\omega}_k; \boldsymbol{v}_{Ok}) \in se(3)$ *be two infinitesimal screws, where* $\$_j$ *is chosen as the tangent vector to the Euclidean motion* $\mathbf{M}_j(t)$ *at time* $t = 0$. *The Lie product is equivalent to*

$$[\$_j \quad \$_k] \equiv \frac{d}{dt}\{\mathbf{M}_j(t)\$_j[\mathbf{M}_j(t)^{-1}]\}|_{t=0} = (\boldsymbol{\omega}_j \times \boldsymbol{\omega}_k; \boldsymbol{\omega}_j \times \boldsymbol{v}_{Ok} - \boldsymbol{\omega}_k \times \boldsymbol{v}_{Oj}). \qquad (3.107)$$

Proof. Let $OXYZ$ be an arbitrary reference frame. The Euclidean motion $\mathbf{M}_j(t)$ can be represented as

$$\mathbf{M}_j(t) = \begin{bmatrix} \mathbf{R}_{\theta,\hat{\boldsymbol{u}}} & \boldsymbol{t} \\ 0_{3\times 1} & 1 \end{bmatrix}, \qquad (3.108)$$

where $\mathbf{R}_{\theta,\hat{\boldsymbol{u}}}$ is the rotation matrix according to a unit vector $\hat{u}$. Meanwhile, $\boldsymbol{t} \in \Re$ is the vector displacement of origin O.

Assume that

$$\$_k = \begin{bmatrix} \mathbf{M}_k & \boldsymbol{v}_{Ok} \\ 0 & 0 \end{bmatrix}, \qquad (3.109)$$

where $\mathbf{M}_k$ is a skew-symmetric matrix given by

$$\mathbf{M}_k = \begin{bmatrix} 0 & -\omega_{kz} & \omega_{ky} \\ \omega_{kz} & 0 & -\omega_{kz} \\ -\omega_{ky} & \omega_{kz} & 0 \end{bmatrix} \qquad (3.110)$$

with $\boldsymbol{\omega}_k = (\omega_{kx}, \omega_{ky}, \omega_{kz})$ and $\boldsymbol{v}_{Ok} \in \Re^3$. After a few computations it follows that

$$\begin{aligned} [\$_j \quad \$_k] &\equiv \frac{d}{dt}\{\mathbf{M}_j(t)\$_k[\mathbf{M}_j(t)]^{-1}\}|_{t=0} \\ &= \begin{bmatrix} \mathbf{M}_j\mathbf{M}_k - \mathbf{M}_k\mathbf{M}_j & \mathbf{M}_j\boldsymbol{v}_{Ok} - \mathbf{M}_k\boldsymbol{v}_{Oj} \\ 0 & 0 \end{bmatrix}. \end{aligned} \qquad (3.111)$$

On the other hand, the velocity vector $\boldsymbol{v}_{Oj}$ may clearly be obtained as the time derivative of vector $\boldsymbol{t}$. Indeed,

$$\boldsymbol{v}_{Oj} = \frac{d}{dt}\boldsymbol{t}|_{t=0}, \qquad (3.112)$$

while $\mathbf{M}_j$ is a skew-symmetric matrix given by

$$\mathbf{M}_j = \begin{bmatrix} 0 & -\omega_{jz} & \omega_{jy} \\ \omega_{jz} & 0 & -\omega_{jx} \\ -\omega_{jy} & \omega_{jx} & 0 \end{bmatrix}, \qquad (3.113)$$

where ω_{jx}, ω_{jy}, and ω_{jz} are the components of the angular velocity vector $\boldsymbol{\omega}_j$. Indeed,

$$\boldsymbol{\omega}_j = \omega_{jx}\hat{\boldsymbol{i}} + \omega_{jy}\hat{\boldsymbol{j}} + \omega_{jz}\hat{\boldsymbol{k}}. \tag{3.114}$$

Finally, the isomorphism between the Lie algebra $se(3)$ and the screw theory (Karger 1996; Murray et al. 1994) allows us to conclude that

$$\frac{d}{dt}\{\mathbf{M}_j(t)\$_k[\mathbf{M}_j(t)]^{-1}\}|_{t=0} = \begin{bmatrix} \boldsymbol{\omega}_j \times \boldsymbol{\omega}_k \\ \boldsymbol{\omega}_j \times \boldsymbol{v}_{Ok} - \boldsymbol{\omega}_k \times \boldsymbol{v}_{Oj} \end{bmatrix} = [\$_j \quad \$_k]. \quad \diamond \tag{3.115}$$

3.3.2 Geometric Interpretation of the Lie Product

Rico and Duffy (1996) developed a geometric interpretation of the Lie product that is close to the one reported by Brand (1947).

Let $\$_1$ and $\$_2$ be two reciprocal screws whose primal parts are normalized (see Fig. 3.3), where O is the reference pole. Then according to Eq. (3.51), it follows that

$$\$_i = \begin{bmatrix} \boldsymbol{s}_i \\ \boldsymbol{r}_i \times \boldsymbol{s}_i + h_i\boldsymbol{s}_i \end{bmatrix}, \quad i = 1, 2. \tag{3.116}$$

Applying the Lie product and the Jacobi identity of the triple vectorial product leads to

$$[\$_1 \quad \$_2] = \begin{bmatrix} \boldsymbol{s}_1 \times \boldsymbol{s}_2 \\ \boldsymbol{s}_1 \times (\boldsymbol{r}_2 \times \boldsymbol{s}_2 + h_2\boldsymbol{s}_2) - \boldsymbol{s}_2 \times (\boldsymbol{r}_1 \times \boldsymbol{s}_1 + h_1\boldsymbol{s}_1) \end{bmatrix} =$$

$$\begin{bmatrix} \boldsymbol{s}_1 \times \boldsymbol{s}_2 \\ \boldsymbol{s}_1 \times (\boldsymbol{r}_2 \times \boldsymbol{s}_2) - \boldsymbol{s}_2 \times (\boldsymbol{r}_1 \times \boldsymbol{s}_1) + (h_1 + h_2)(\boldsymbol{s}_1 \times \boldsymbol{s}_2) \end{bmatrix} =$$

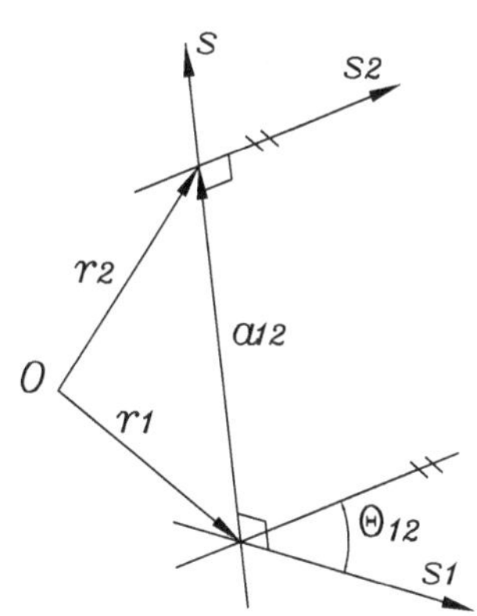

Fig. 3.3 Geometric interpretation of the Lie product

$$\begin{bmatrix} s_1 \times s_2 \\ s_1 \times (r_2 \times s_2) + r_1 \times (s_1 \times s_2) + s_1 \times (s_2 \times r_1) + (h_1 + h_2)(s_1 \times s_2) \end{bmatrix} =$$

$$\begin{bmatrix} s_1 \times s_2 \\ s_1 \times (s_2 \times (r_1 - r_2)) + r_1 \times (s_1 \times s_2) + (h_1 + h_2)(s_1 \times s_2) \end{bmatrix}. \tag{3.117}$$

Consider the vector $\boldsymbol{a}_{12} = r_2 - r_1 = a_{12}s$. Substituting $\boldsymbol{a}_{12}$ into Eq. (3.117) and resorting to the triple vectorial product, we obtain

$$[\$_1 \quad \$_2] =$$

$$\begin{bmatrix} s_1 \times s_2 \\ s_1 \times (s_2 \times (-a_{12}s)) + r_1 \times (s_1 \times s_2) + (h_1 + h_2)(s_1 \times s_2) \end{bmatrix} =$$

$$\begin{bmatrix} s_1 \times s_2 \\ a_{12}s(s_1 \bullet s_2) - a_{12}s_2(s \bullet s_1) + r_1 \times (s_1 \times s_2) + (h_1 + h_2)(s_1 \times s_2) \end{bmatrix} =$$

$$\begin{bmatrix} s_1 \times s_2 \\ a_{12}s(s_1 \bullet s_2) + r_1 \times (s_1 \times s_2) + (h_1 + h_2)(s_1 \times s_2) \end{bmatrix}. \tag{3.118}$$

On the other hand, consider that

$$s_1 \times s_2 = \sin(\alpha_{12})s \tag{3.119}$$

and

$$s_1 \bullet s_2 = \cos(\alpha_{12})s. \tag{3.120}$$

Clearly, s is a unit vector normal to the plane shaped by the unit vectors s_1 and s_2. Substituting Eqs. (3.119) and (3.120) into Eq. (3.118) and reducing terms, we obtain

$$[\$_1 \quad \$_2] = \sin(\alpha_{12}) \begin{bmatrix} s \\ r_1 \times s + [a_{12}\cot(\alpha_{12}) + (h_1 + h_2)]s \end{bmatrix}. \tag{3.121}$$

Expression (3.121) indicates that the Lie product between the screws $\$_1$ and $\$_2$ yields another screw whose primal part s is normal to the plane formed by the primal parts of $\$_1$ and $\$_2$ that passes through the common perpendicular of both screws. Furthermore, the pitch of the resulting screw is computed as $a_{12}\cot(\alpha_{12}) + (h_1 + h_2)$.

3.4 Helicoidal Vector Fields

A linear line complex is a set of lines whose Plücker coordinates satisfy a linear relation, Huang et al. (1991). It has been reported (see Hunt 1978; Phillips 1984, 1990) that in dealing with infinitesimal kinematics, there exists a linear line complex associated to the instantaneous motion of a rigid body. In Chevallier (1991) a twist is considered as the geometrical representation of a field with helicoidal properties. The theory of helicoidal vector fields plays a central role in determining the kinematic states of a rigid body. In this subsection we investigate its fundamentals, which provides a proper foundation for the next chapters.

Definition 3.5 (Helicoidal Vector Field).
Let E be the physical space, and let $\Re^3$ be the three-dimensional real vector space equipped with the usual cross product. A helicoidal vector field is a field given by

$$X : E \to \Re^3 \quad X(P) = X(P) \quad \forall P \in E, \tag{3.122}$$

where $X(P) \in \Re^3$. In addition, there is a vector $\boldsymbol{a} \in \Re^3$ named the vector of the helicoidal field such that for any $Q \in E$ we have

$$X(Q) = X(P) + \boldsymbol{a} \times \boldsymbol{r}_{Q/P}. \tag{3.123}$$

Proposition 3.13. *Let X be a helicoidal field. Then X can be determined in accordance with the vector $\boldsymbol{a}$ of the helicoidal field and the helicoidal field $X(P)$ of point P.*

Proof. If Q is an arbitrary point of the space E, then applying Definition 3.5 leads to

$$X(Q) = X(P) + \boldsymbol{a} \times \boldsymbol{r}_{Q/P}. \tag{3.124}$$

Therefore, expression (3.124) allows us to determine the vector field of point Q based on the vector field of point P and the vector $\boldsymbol{a}$. ◇

Notice that since one of the two vectors that determines the helicoidal field depends on the physical space E, a helicoidal field can support more than one representation. Hence, it is necessary to establish under which conditions a pair of helicoidal fields are equivalent.

Proposition 3.14. *Let X_1 and X_2 be two helicoidal fields defined, respectively, by $(\boldsymbol{a}, X_1(P))$ and $(\boldsymbol{c}, X_2(Q))$, where $\boldsymbol{a}, X_1(P), \boldsymbol{c}, X_2(Q) \in \Re^3$ and $P, Q \in E$ with $P \neq Q$. It is said that $X_1 = X_2$ if and only if*

$$\boldsymbol{a} = \boldsymbol{c} \text{ and } X_2(Q) = X_1(P) + \boldsymbol{a} \times \boldsymbol{r}_{Q/P}, \tag{3.125}$$

where $\boldsymbol{a} = \boldsymbol{c}$ is the vector of the helicoidal field.

Proof. In what follows the proposition is proved by analyzing all its possibilities. First, assume that the conditions of the proposition are satisfied. Let M be an arbitrary point of E. Then two helicoidal fields may be expressed as

$$X_1(M) = X_1(P) + \boldsymbol{a} \times \boldsymbol{r}_{M/P} \quad \text{and} \quad X_2(M) = X_2(Q) + \boldsymbol{c} \times \boldsymbol{r}_{M/Q}. \tag{3.126}$$

However, in agreement with Definition 3.5 we have

$$X_2(M) = X_2(Q) + \boldsymbol{a} \times \boldsymbol{r}_{M/Q} \tag{3.127}$$

and

$$X_2(Q) = X_1(P) + \boldsymbol{a} \times \boldsymbol{r}_{Q/P}. \tag{3.128}$$

Hence,

$$X_2(M) = X_1(P) + \boldsymbol{a} \times \boldsymbol{r}_{Q/P} + \boldsymbol{a} \times \boldsymbol{r}_{M/Q}, \tag{3.129}$$

or

$$X_2(M) = X_1(P) + \boldsymbol{a} \times (\boldsymbol{r}_{M/Q} + \boldsymbol{r}_{Q/P}). \tag{3.130}$$

Furthermore, since $\boldsymbol{r}_{M/P} = \boldsymbol{r}_{M/Q} + \boldsymbol{r}_{Q/P}$, then it follows that

$$X_2(M) = X_1(P) + \boldsymbol{a} \times \boldsymbol{r}_{M/P}, \tag{3.131}$$

which leads to

$$X_2(M) = X_1(M). \tag{3.132}$$

Removing point M from Eq. (3.132), it follows that $X_2 = X_1$.

On the other hand, if $X_2 = X_1$, then

$$X_2(M) = X_1(M) \forall M \in E. \tag{3.133}$$

Assuming that $M = P$, we obtain

$$X_1(P) = X_2(P) = X_2(Q) + \boldsymbol{c} \times \boldsymbol{r}_{P/Q}. \tag{3.134}$$

Similarly, if $M = Q$, then

$$X_1(P) + \boldsymbol{a} \times \boldsymbol{r}_{Q/P} = X_1(Q) = X_2(Q). \tag{3.135}$$

Gathering these results, we get

$$X_1(P) + \boldsymbol{a} \times \boldsymbol{r}_{Q/P} + \boldsymbol{c} \times \boldsymbol{r}_{P/Q} = X_1(P), \tag{3.136}$$

or

$$\boldsymbol{a} \times \boldsymbol{r}_{Q/P} - \boldsymbol{c} \times \boldsymbol{r}_{Q/P} = \boldsymbol{0}. \tag{3.137}$$

Hence, taking into account that P, Q, and E are arbitrary points of E, then necessarily $\boldsymbol{a} = \boldsymbol{c}$, which implies that the first condition is that the vector of the helicoidal fields must be the same. Furthermore, assuming that M is an arbitrary point of E, then it is necessary that

$$X_1(P) + \boldsymbol{a} \times \boldsymbol{r}_{M/P} = X_1(M) = X_2(M) = X_2(Q) + \boldsymbol{a} \times \boldsymbol{r}_{M/Q}. \tag{3.138}$$

Therefore,

$$X_2(Q) = X_1(P) + \boldsymbol{a} \times (\boldsymbol{r}_{M/P} - \boldsymbol{r}_{M/Q}) = X_1(P) + \boldsymbol{a} \times (\boldsymbol{r}_{Q/M} + \boldsymbol{r}_{M/P}), \tag{3.139}$$

or

$$X_2(Q) = X_1(P) + \boldsymbol{a} \times \boldsymbol{r}_{Q/P}. \quad \diamond \tag{3.140}$$

Proposition 3.15. *Let* $X_1 = (\boldsymbol{a}, X_1(P))$ *and* $X_2 = (\boldsymbol{c}, X_2(Q))$ *be a pair of helicoidal fields where* $P, Q \in E$. *The following relationship*

$$X_1 \sim X_2 \longleftrightarrow \text{ and } X_2(Q) = X_1(P) + \boldsymbol{a} \times \boldsymbol{r}_{Q/P} \tag{3.141}$$

is an equivalence relationship.

Proof. The relationship must be reflexive, symmetric, and transitive.

- *Reflexive.* $X_1 \sim X_2$.
  It is evident that $\boldsymbol{a} = \boldsymbol{a}$. In addition, according to Definition 3.5 we write

$$X_1(Q) = X_1(P) + \boldsymbol{a} \times \boldsymbol{r}_{Q/P}.$$

- *Symmetric.* $X_1 \sim X_2 \rightarrow X_2 \sim X_1$.
  It is evident that if $\boldsymbol{a} = \boldsymbol{c}$, then $\boldsymbol{c} = \boldsymbol{a}$. Furthermore, if $X_1 \sim X_2$, then

$$X_2(Q) = X_1(P) + \boldsymbol{a} \times \boldsymbol{r}_{Q/P}.$$

  On the other hand, since $X_1(P) = X_2(M) \forall M \in E$, then $X_1(Q) = X_2(Q)$ and $X_2(P) = X_1(P)$, which leads to

$$X_1(Q) = X_2(P) + \boldsymbol{a} \times \boldsymbol{r}_{Q/P}.$$

  However, since $X_1 \sim X_2$, then necessarily $\boldsymbol{c} = \boldsymbol{a}$, which yields

$$X_1(Q) = X_2(P) + \boldsymbol{c} \times \boldsymbol{r}_{Q/P}.$$

- *Transitive.* $X_1 \sim X_2, X_2 \sim X_3 \rightarrow X_1 \sim X_3$.
  Let $X_3 = (\boldsymbol{d}, X_3(P))$ be a helicoidal field, such that $X_1 \sim X_2$ y $X_2 \sim X_3$. Thus,

$$\boldsymbol{a} = \boldsymbol{c}, \boldsymbol{c} = \boldsymbol{d} \rightarrow \boldsymbol{a} = \boldsymbol{d}.$$

Furthermore,

$$X_2(Q) = X_1(P) + \boldsymbol{a} \times \boldsymbol{r}_{Q/P} \text{ and } X_3(M) = X_2(Q) + \boldsymbol{c} \times \boldsymbol{r}_{M/Q}.$$

Then

$$X_3(M) = X_2(Q) + \boldsymbol{c} \times \boldsymbol{r}_{M/Q} = X_1(P) + \boldsymbol{a} \times \boldsymbol{r}_{Q/P} + \boldsymbol{c} \times \boldsymbol{r}_{M/Q},$$

or

$$X_3(M) = X_1(P) + \boldsymbol{a} \times (\boldsymbol{r}_{Q/P} + \boldsymbol{r}_{M/Q}).$$

Hence,

$$X_3(M) = X_1(P) + \boldsymbol{a} \times \boldsymbol{r}_{M/P},$$

which proves that $X_1 \sim X_3$. ◇

Finally, it is well known (see Fraleigh 1982) that an equivalence relationship induces a partition of the corresponding set where every subset is an equivalence class of an element. With these considerations in mind, we define the equivalence class of X_1, $\overline{X}_1$ by

$$\overline{X}_1 = \{X \mid X \sim X_1\}. \tag{3.142}$$

Example 3.4. Determine the angular velocity tensor of a rigid body based on the concept of helicoidal vector field.

Solution. Let

$$V_1 = \begin{bmatrix} \boldsymbol{\omega} \\ \boldsymbol{v}_1 \end{bmatrix}$$

and

$$V_2 = \begin{bmatrix} \boldsymbol{\omega} \\ \boldsymbol{v}_2 \end{bmatrix}$$

be two representations of the velocity state of a rigid body, where

$$\boldsymbol{\omega} = \omega_X \hat{\boldsymbol{i}} + \omega_Y \hat{\boldsymbol{j}} + \omega_Z \hat{\boldsymbol{k}}$$

is the angular velocity of the body as measured from another body or reference frame XYZ with associated unit vectors $\hat{\boldsymbol{i}}\hat{\boldsymbol{j}}\hat{\boldsymbol{k}}$; that is, $\boldsymbol{\omega}$ is the primal part of the velocity state. Meanwhile,

$$\boldsymbol{v}_1 = v_{1X}\hat{\boldsymbol{i}} + v_{1Y}\hat{\boldsymbol{j}} + v_{1Z}\hat{\boldsymbol{k}}$$

and

$$\boldsymbol{v}_2 = v_{2X}\hat{\boldsymbol{i}} + v_{2Y}\hat{\boldsymbol{j}} + v_{2Z}\hat{\boldsymbol{k}}$$

are the velocity vectors of two different points P_1 and P_2 of the rigid body under study; that is, $\boldsymbol{v}_1$ and $\boldsymbol{v}_2$ are the dual parts of the corresponding velocity states. Then, since $\boldsymbol{\omega}$ is the vector of the helicoidal field, the dual part of the velocity state V_2 can be written upon the dual part of the velocity state V_1 as follows:

$$\boldsymbol{v}_2 = \boldsymbol{v}_1 + \boldsymbol{\omega} \times \boldsymbol{r}_{2/1},$$

where $\boldsymbol{r}_{2/1} = r_X\hat{\boldsymbol{i}} + r_Y\hat{\boldsymbol{j}} + r_Z\hat{\boldsymbol{k}}$ is the position vector of point P_2 with respect to point P_1. After a few computations, the following relationship is easily obtained:

$$\Omega\, \boldsymbol{r}_{2/1} = \boldsymbol{v}_2 - \boldsymbol{v}_1, \tag{3.143}$$

where

$$\Omega = \begin{bmatrix} 0 & -\omega_Z & \omega_Y \\ \omega_Z & 0 & -\omega_X \\ -\omega_Y & \omega_X & 0 \end{bmatrix} \tag{3.144}$$

is called the tensor of angular velocity of the rigid body.

3.5 Exercises

1. Complete the proof of Proposition 3.2.
2. Prove Proposition 3.4.
3. Prove that the screw representing a prismatic joint, in Eq. (3.60), is well established.
4. Which condition must satisfy revolute joints in order to model spherical and universal joints?
5. The plane joint is another type of lower kinematic pair. How is it modeled with infinitesimal screws?
6. Complete the proof of Proposition 3.9.

7. Give two examples of reciprocal screws.
8. Prove that the Lie product is nonassociative.
9. Prove that any point of the ISA of a screw can be used to compute the dual part of the screw.
10. Identify the infinitesimal screws associated to the kinematic pairs of Gough's tyre testing machine.

References

Ball, R. S. (1900). *A treatise on the theory of screws*. Cambridge: Cambridge University Press (reprinted 1998).

Bottema, O., & Roth, B. (1979). *Theoretical kinematics*. New York: Dover.

Brand, L. (1947). *Vector and tensor analysis*. New York: Wiley.

Chevallier, D. P. (1991). Lie algebras, modules, dual quaternions and algebraic methods in kinematics. *Mechanism and Machine Theory, 26*(6), 613–627.

Clifford, W. K. (1873). Preliminary sketch of bi-quaternions. *Proceedings London Mathematical Society, 4*(64/65), 381–395.

Dimentberg, F. M. (1965). *Screw calculus and its applications to mechanics*. Moscow (in Russian). English translation: AD680993 (Clearinghouse for Federal Scientific & Technical Information, Springfield, Moscow, Russia).

Fraleigh, J. B. (1982). *A first course in abstract algebra* (3rd ed.). Reading, MA: Addison-Wesley.

Gallardo-Alvarado, J., Orozco-Mendoza, H., & Rodríguez-Castro, R. (2008). Finding the jerk properties of multibody systems using helicoidal vector fields. *Proceedings Institution of Mechanical Engineers, Part C: Journal of Mechanical Engineering Science, 222*(11), 2217–2229.

Hausner, M., & Schwartz, J. T. (1968). *Lie groups, Lie algebras*. New York: Gordon and Breach.

Huang, C., Kuo, W., & Ravani, B. (2008). On the linear line complex and helicoidal vector field associated with homologous lines of a finite displacement. *Mechanism and Machine Theory, 43*(2), 138–146.

Hunt, K.H. (1978). *Kinematic Geometry of Mechanisms*. Oxford: Oxford University Press.

Karger, A. (1996). Singularity analysis of serial robot-manipulators. *ASME Journal of Mechanical Design, 118*(4), 520–525.

Lipkin, H., & Duffy, J. (1985). The elliptic polarity of screws. *ASME Journal of Mechanisms, Transmissions and Automation in Design, 107*(3), 377–388.

Murray, R. M., Li, Z., & Sastry, S. S. (1994). *A mathematical introduction to robotic manipulation*. Boca Raton: CRC Press.

Phillips, J. (1984). *Freedom in machinery. Introducing screw theory* (Vol. 1). Cambridge: Cambridge University Press.

Phillips, J. (1990). *Freedom in machinery. Screw theory exemplified* (Vol. 2). Cambridge: Cambridge University Press.

Rico, J. M., & Duffy, J. (1996). An efficient inverse acceleration analysis of in-parallel manipulators. In *Proceedings 1996 ASME Design Engineering Technical Conference and Computers in Engineering Conference*.

Rico-Martínez, J. M., & Duffy, J. (1992). Orthogonal spaces and screw systems. *Mechanism and Machine Theory, 27*(4), 451–458.

Sugimoto, K. (1987). Kinematic and dynamic analysis of parallel manipulators by means of motor algebra. *ASME Journal of Mechanisms, Transmissions and Automation in Design, 109*(1), 3–7.

Sugimoto, K. (1990). Existence criteria for overconstrained mechanisms. An extension of motor algebra. *ASME Journal of Mechanical Design, 112*(3), 295–298.

von Mises, R. (1924a). Motorrechnung, ein neues hilfsmittel der mechanik. *Zeitschrift für Angewandte Mathematik und Mechanik, 4*(2), 155–181.

von Mises, R. (1924b). Anwendungen der motorrenchung. *Zeitschrift für Angewandte Mathematik und Mechanik, 4*(3), 193–213.

Wylie, C. R., & Barret, L. C. (1982). *Advanced engineering mathematics* (5th ed.). New York: McGraw-Hill.

Chapter 4
Velocity Analysis

4.1 Introduction

A vector is defined as a physical entity with both magnitude and direction. Vectors are often used to represent forces, moments, and, as we're discussing in this book, angular and linear kinematic quantities. The time derivatives of vectors with respect to different reference frames play a fundamental role in the development of this chapter. On the other hand, the notation used here is an adaptation of that introduced by Kane and Levinson (1985).

The velocity analysis of a mechanism consists of formulating equations involving the individual kinematic characteristics of its components and whose solution allows us to measure the angular and linear velocities of any of its elements. A relevant application of velocity analysis includes computing the angular and linear velocities of a rigid body like the end-effector of a serial manipulator, the moving platform of a parallel manipulator, or the end-effector platform of a hyper-redundant manipulator. Formulating the velocity analysis of kinematic chains by means using screw theory and linear algebra dates back to the 1970s and has been well appreciated by the kinematician community.

4.2 Fundamental Equations of Velocity

This section starts by validating the definition of the velocity state of a rigid body. The angular velocity of a rigid body as measured from another body or reference frame provides incomplete information about the characterization of rigid body motion. Hence, to completely understand the motion of a rigid body, it is necessary to add a second entity, called the linear velocity of an arbitrary point of the body under study. The angular velocity of a rigid body and the linear velocity of an arbitrary point of the rigid body form an inseparable entity called the velocity state

J. Gallardo-Alvarado, *Kinematic Analysis of Parallel Manipulators by Algebraic Screw Theory*, DOI 10.1007/978-3-319-31126-5_4

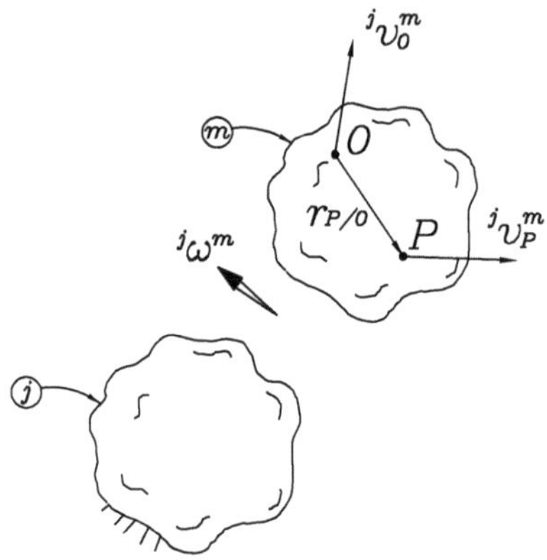

Fig. 4.1 Bodies *m* and *j* in relative motion; angular and linear velocities

of a rigid body. The next proposition is a proof that the velocity state is, in fact, a helicoidal vector field—a transcendental concept whose application allowed screw theory to be extended to the higher-order analyses of kinematic chains.

Proposition 4.1. *Let m and j be two rigid bodies in relative motion (see Fig. 4.1), and let P and O be two points of body m. A representation of the velocity state of body m as observed from body j, the six-dimensional vector* $^j\boldsymbol{V}^m$*, is a helicoidal field with the property to assign to each point* $P \in m$ *its corresponding linear velocity measured from body j, the vector* $^j\boldsymbol{v}_P^m$*. Indeed,*

$$^j\boldsymbol{V}^m : E \to \Re^3 \quad ^j\boldsymbol{V}^m(P) = {}^j\boldsymbol{v}_P^m \; \forall P \in E. \tag{4.1}$$

Proof. Let $^j\boldsymbol{v}_P^m$ and $^j\boldsymbol{v}_O^m$ be the linear velocities of points *P* and *O*, respectively. By resorting to college kinematics, we may express the linear velocity vector $^j\boldsymbol{v}_P^m$ based on the linear velocity vector $^j\boldsymbol{v}_O^m$ as

$$^j\boldsymbol{v}_P^m = {}^j\boldsymbol{v}_O^m + {}^j\boldsymbol{\omega}^m \times \boldsymbol{r}_{P/O}, \tag{4.2}$$

where $^j\boldsymbol{\omega}^m$ is the angular velocity vector of body *m* as measured from body *j* and $\boldsymbol{r}_{P/O}$ is the position vector of point *P* with respect to point *O*. Hence, in accordance with Definition 3.5, $^j\boldsymbol{V}^m$ is a helicoidal field where $^j\boldsymbol{\omega}^m$ is the vector of the field. ◇

It is interesting to note that while subscripts are used to complete the information concerning linear velocities, the angular velocity is free of such additional notation. The reason for this simplification is that the angular velocity vector is a property of a rigid body; that is, this vector is the same for any point of the considered rigid body. On the other hand, expression (4.2) provides the means to express the velocity state of rigid body *m* as observed from body *j*, for instance,

$$^j\boldsymbol{V}_P^m = \begin{bmatrix} ^j\boldsymbol{\omega}^m \\ ^j\boldsymbol{v}_P^m \end{bmatrix}.$$

Similarly,

$$^j\boldsymbol{V}_O^m = \begin{bmatrix} ^j\boldsymbol{\omega}^m \\ ^j\boldsymbol{v}_O^m \end{bmatrix}.$$

Clearly, the velocity state supports more than one representation, where each representation can be obtained from another representation via the application of the helicoidal vector field theory. It is an accepted practice to establish that the velocity state of a rigid body is a six-dimensional vector composed of two concatenated independent three-dimensional vectors. The first vector is termed the primal vector of the velocity state, whereas the second vector is named the dual part. Naturally, although the primal and dual parts are independent vectors, both entities are required to characterize the motion of a rigid body.

Example 4.1. Let m be a body in motion with respect to another body j, and let XYZ be a conventional reference frame attached to body j with associated unit vectors $\hat{\boldsymbol{i}}\hat{\boldsymbol{j}}\hat{\boldsymbol{k}}$. Furthermore, consider that A and B are two points of body m whose coordinates expressed in the XYZ reference frame are given by $A = (5, 2, 3)$ and $B = (-3, 1, -4)$. If the velocity state of body m as observed from body j, taking point A as the reference pole, is given by

$$^{j}\boldsymbol{V}_A^m = \begin{bmatrix} 2 \\ 4 \\ -1 \\ -3 \\ 7 \\ 8 \end{bmatrix},$$

compute the velocity state of body m with respect to body j taking point B as the reference pole by resorting to the theory of helicoidal vector fields. In other words, compute the six-dimensional vector

$$^{j}\boldsymbol{V}_B^m = \begin{bmatrix} ^{j}\boldsymbol{\omega}^m \\ ^{j}\boldsymbol{v}_B^m \end{bmatrix}.$$

Solution. Assume that

$$^{j}\boldsymbol{V}_A^m = \begin{bmatrix} ^{j}\boldsymbol{\omega}^m \\ ^{j}\boldsymbol{v}_A^m \end{bmatrix}.$$

Then it follows that

$$^{j}\boldsymbol{\omega}^m = 2\hat{\boldsymbol{i}} + 4\hat{\boldsymbol{j}} - \hat{\boldsymbol{k}}$$

and

$$^{j}\boldsymbol{v}_A^m = -3\hat{\boldsymbol{i}} + 7\hat{\boldsymbol{j}} + 8\hat{\boldsymbol{k}}.$$

Furthermore, since the angular velocity $^{j}\boldsymbol{\omega}^m$ is the vector of the helicoidal field, then the dual part of the velocity state $^{j}\boldsymbol{V}_B^m$, the three-dimensional vector $D(^{j}\boldsymbol{V}_B^m) = {}^{j}\boldsymbol{v}_B^m$, may be computed based on the vectors of the velocity state ${}^{j}\boldsymbol{V}_{A}^{m}$ as

$$D({}^{j}\boldsymbol{V}_{B}^{m}) = {}^{j}\boldsymbol{v}_{A}^{m} + {}^{j}\boldsymbol{\omega}^{m} \times \boldsymbol{r}_{B/A},$$

where

$$\boldsymbol{r}_{B/A} = -8\hat{\boldsymbol{i}} - \hat{\boldsymbol{j}} - 7\hat{\boldsymbol{k}}$$

is a vector pointed from A to B. Hence, the dual part of ${}^{j}\boldsymbol{V}_{B}^{m}$ is directly determined as

$$D({}^{j}\boldsymbol{V}_{B}^{m}) = -32\hat{\boldsymbol{i}} + 29\hat{\boldsymbol{j}} + 38\hat{\boldsymbol{k}}.$$

Finally, the six-dimensional vector ${}^{j}\boldsymbol{V}_{B}^{m}$ is obtained as

$${}^{j}\boldsymbol{V}_{B}^{m} = \begin{bmatrix} 2 \\ 4 \\ -1 \\ -32 \\ 29 \\ 38 \end{bmatrix}.$$

Example 4.2. In this example we revisit Chasles' theorem (see Proposition 3.5). Without a doubt, applications of the instantaneous screw axis (ISA) of a rigid body are interesting issues. In this example, given the angular and linear velocities of an arbitrary point Q embedded in a rigid body m, one is required to determine the relative position vector of a point P on the ISA as observed from point Q; for example, it is necessary to compute the vector $\boldsymbol{r}_{P/Q}$.

Solution. To obtain the relative vector $\boldsymbol{r}_{P/Q}$, let m be a rigid body in motion with respect to another body or reference frame j, and let P and Q be two points of body m. From college kinematics it is evident that the velocities of points P and Q, the vectors ${}^{j}\boldsymbol{v}_{P}^{m}$ and ${}^{j}\boldsymbol{v}_{Q}^{m}$, respectively, are related by

$${}^{j}\boldsymbol{\omega}^{m} \times \boldsymbol{r}_{P/Q} = {}^{j}\boldsymbol{v}_{P}^{m} - {}^{j}\boldsymbol{v}_{Q}^{m}, \tag{4.3}$$

where $\boldsymbol{r}_{P/Q}$ is a vector pointed from Q to P. Assuming that the instantaneous velocity of point P vanishes, that is, ${}^{j}\boldsymbol{v}_{P}^{m} = \boldsymbol{0}$, Eq. (4.3) reduces to

$${}^{j}\boldsymbol{\omega}^{m} \times \boldsymbol{r}_{P/Q} = -{}^{j}\boldsymbol{v}_{Q}^{m}. \tag{4.4}$$

Note that an infinite number of points P satisfy Eq. (4.4). To find one point P, premultiply Eq. (4.4) by ${}^{j}\boldsymbol{\omega}^{m}$ to obtain

$${}^{j}\boldsymbol{\omega}^{m} \times ({}^{j}\boldsymbol{\omega}^{m} \times \boldsymbol{r}_{P/Q}) = -{}^{j}\boldsymbol{\omega}^{m} \times {}^{j}\boldsymbol{v}_{Q}^{m}. \tag{4.5}$$

However, it is known that the triple cross product ${}^j\boldsymbol{\omega}^m \times ({}^j\boldsymbol{\omega}^m \times \boldsymbol{r}_{P/Q})$ may be decomposed in

$$ {}^j\boldsymbol{\omega}^m \times ({}^j\boldsymbol{\omega}^m \times \boldsymbol{r}_{P/Q}) = ({}^j\boldsymbol{\omega}^m \bullet \boldsymbol{r}_{P/Q})\boldsymbol{r}_{P/Q} - {}_j\omega_m^2 \boldsymbol{r}_{P/Q}, \tag{4.6} $$

where ${}_j\omega_m$ denotes the magnitude of the angular velocity vector, for example, ${}_j\omega_m = | {}^j\boldsymbol{\omega}^m |$. On the other hand, taking into account Chasles' theorem establishes that

Given a rigid body moving with a general displacement, a set of its points, located on a line L, undergo identical displacements of minimum magnitude. Moreover, line L and the minimum magnitude displacement are parallel to the axis of the rotation involved.

Then it is possible to ensure that there is a point P fixed to body m whose linear velocity ${}^j\boldsymbol{v}_P^m$ is parallel to the angular velocity vector ${}^j\boldsymbol{\omega}^m$. In other words, the perpendicular component of the linear velocity of P vanishes. It is straightforward to demonstrate that such a point is on the ISA. Furthermore, point P can be chosen, without loss of generality, in such a way that the vector ${}^j\boldsymbol{\omega}^m$ is normal to the vector $\boldsymbol{r}_{P/Q}$. Therefore,

$$ {}^j\boldsymbol{\omega}^m \bullet \boldsymbol{r}_{P/Q} = 0. \tag{4.7} $$

On the other hand, taking into account that

$$ {}^j\boldsymbol{\omega}^m \times \boldsymbol{r}_{P/Q} = -{}^j\boldsymbol{\omega}^m \times \boldsymbol{r}_{Q/P} = {}^j\boldsymbol{v}_Q^m, $$

we find after a few computations that the relative vector $\boldsymbol{r}_{P/Q}$ results in

$$ \boldsymbol{r}_{P/Q} = \frac{{}^j\boldsymbol{\omega}^m \times {}^j\boldsymbol{v}_Q^m}{{}_j\omega_m^2}. \tag{4.8} $$

Proposition 4.1 does not require that the involved bodies be physically connected to each other. In what follows the angular velocity vector is investigated in the case where two rigid bodies are connected via a helical pair.

With reference to Fig. 4.2, let k and l be two rigid bodies. The vector ${}^k\boldsymbol{\omega}^l$ indicates the angular velocity of body l as observed from body k whose magnitude, or joint-velocity rate, is notated as ${}_k\omega_l$. Similarly, the linear velocity, or velocity, for brevity, of a point O^l embedded in l measured from body k is denoted by ${}^k\boldsymbol{v}_O^l$. Furthermore, if the bodies k and l are connected to each other through a helical pair with pitch ${}_kh_l$ and whose direction is indicated by the unit vector ${}^k\boldsymbol{s}^l$ that passes through point B^l of body l, then the angular velocity ${}^k\boldsymbol{\omega}^l$ may be expressed as

$$ {}^k\boldsymbol{\omega}^l = {}_k\omega_l {}^k\boldsymbol{s}^l. \tag{4.9} $$

Assuming that the angular velocity vector ${}^k\boldsymbol{\omega}^l$ is a normalized or unit vector, that is, ${}_k\omega_l = 1$, then the moment produced by vector ${}^k\boldsymbol{\omega}^l$ about point O^l embedded in body l, the vector ${}^k\boldsymbol{s}_O^l$, is the linear velocity vector ${}^k\boldsymbol{v}_O^l$ of point O^l as observed from body k, which can be computed [see Eq. (3.51)] as

$$ {}^k\boldsymbol{s}_O^l = {}_kh_l {}^k\boldsymbol{s}^l + {}^k\boldsymbol{s}^l \times \boldsymbol{q}, \tag{4.10} $$

where $\boldsymbol{q}$ is a position vector directed from point B^l to point O^l.

Fig. 4.2 Bodies k and l connected to each other through a helical pair

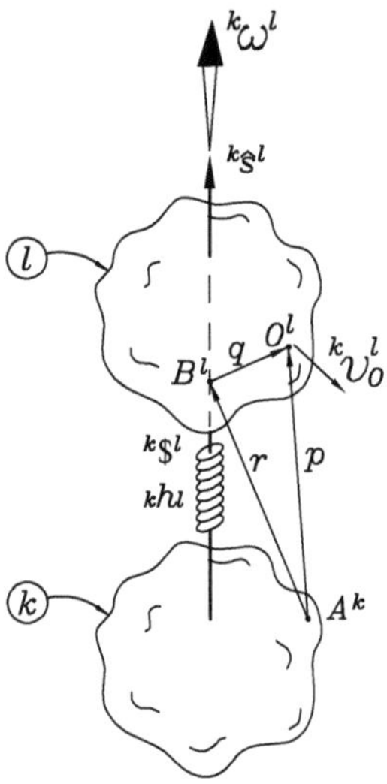

Fig. 4.3 A right-hand orthonormal basis $\{\boldsymbol{b}_1, \boldsymbol{b}_2, \boldsymbol{b}_3\}$ attached to a rigid body

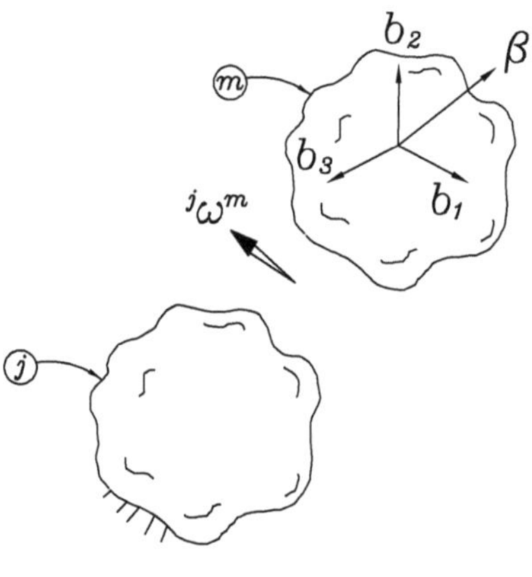

With these considerations in mind, the following definition of angular velocity vector represents the beginning of understanding how the time derivative of a vector depends on the selected reference frame. The procedures are systematic and require the explicit presentation of all vectors capable of being derived with respect to time; otherwise, some vectors must be overlooked, leading to wrong equations.

Definition 4.1. Let $\{\boldsymbol{b}_1, \boldsymbol{b}_2, \boldsymbol{b}_3\}$ be a right-hand orthonormal basis of $\Re^3$ fixed to body m (see Fig. 4.3). Then the angular velocity of body m with respect to body j is defined as

$$ {}^j\boldsymbol{\omega}^m \equiv \boldsymbol{b}_1 \frac{{}^j d\boldsymbol{b}_2}{dt} \bullet \boldsymbol{b}_3 + \boldsymbol{b}_2 \frac{{}^j d\boldsymbol{b}_3}{dt} \bullet \boldsymbol{b}_1 + \boldsymbol{b}_3 \frac{{}^j d\boldsymbol{b}_1}{dt} \bullet \boldsymbol{b}_2. \tag{4.11} $$

Definition 4.1 is the basis to formulating the following proposition.

Proposition 4.2. *With reference to Fig. 4.3, let m and j be two rigid bodies. If* ${}^j\boldsymbol{\omega}^m$ *is the angular velocity of body m with respect to body j, then the time derivative of a vector* $\boldsymbol{\beta}$ *bounded in body m as observed from another body or reference frame j may be expressed as*

$$\frac{{}^jd}{dt}\boldsymbol{\beta} = {}^j\boldsymbol{\omega}^m \times \boldsymbol{\beta}. \tag{4.12}$$

Proof. The cross product between vectors ${}^j\boldsymbol{\omega}^m$ and $\boldsymbol{b}_1$ is given by

$${}^j\boldsymbol{\omega}^m \times \boldsymbol{b}_1 = \left(\boldsymbol{b}_1\frac{{}^jd\boldsymbol{b}_2}{dt}\bullet\boldsymbol{b}_3 + \boldsymbol{b}_2\frac{{}^jd\boldsymbol{b}_3}{dt}\bullet\boldsymbol{b}_1 + \boldsymbol{b}_3\frac{{}^jd\boldsymbol{b}_1}{dt}\bullet\boldsymbol{b}_2\right)\times\boldsymbol{b}_1. \tag{4.13}$$

By reducing terms, we have

$${}^j\boldsymbol{\omega}^m \times \boldsymbol{b}_1 = -\boldsymbol{b}_3\frac{{}^jd\boldsymbol{b}_3}{dt}\bullet\boldsymbol{b}_1 + \boldsymbol{b}_2\frac{{}^jd\boldsymbol{b}_1}{dt}\bullet\boldsymbol{b}_2. \tag{4.14}$$

However, considering that $\boldsymbol{b}_1 \bullet \boldsymbol{b}_1 = 1$, then

$$\frac{{}^jd}{dt}(\boldsymbol{b}_1\bullet\boldsymbol{b}_1) = \frac{{}^jd\boldsymbol{b}_1}{dt}\bullet\boldsymbol{b}_1 + \boldsymbol{b}_1\bullet\frac{{}^jd\boldsymbol{b}_1}{dt} = 0 \rightarrow \frac{{}^jd\boldsymbol{b}_1}{dt}\bullet\boldsymbol{b}_1 = 0, \tag{4.15}$$

which gives us

$$\boldsymbol{b}_1\frac{{}^jd\boldsymbol{b}_1}{dt}\bullet\boldsymbol{b}_1 = \boldsymbol{0}. \tag{4.16}$$

On the other hand, since $\boldsymbol{b}_3 \bullet \boldsymbol{b}_1 = 0$, then

$$\frac{{}^jd}{dt}(\boldsymbol{b}_3\bullet\boldsymbol{b}_1) = \frac{{}^jd\boldsymbol{b}_3}{dt}\bullet\boldsymbol{b}_1 + \boldsymbol{b}_3\bullet\frac{{}^jd\boldsymbol{b}_1}{dt} = 0. \tag{4.17}$$

Therefore,

$$\frac{{}^jd\boldsymbol{b}_3}{dt}\bullet\boldsymbol{b}_1 = -\frac{{}^jd\boldsymbol{b}_1}{dt}\bullet\boldsymbol{b}_3. \tag{4.18}$$

Substituting Eqs. (4.16) and (4.18) into Eq. (4.14) leads to

$${}^j\boldsymbol{\omega}^m \times \boldsymbol{b}_1 = \frac{{}^jd\boldsymbol{b}_1}{dt}\bullet\boldsymbol{b}_1\boldsymbol{b}_1 + \frac{{}^jd\boldsymbol{b}_1}{dt}\bullet\boldsymbol{b}_2\boldsymbol{b}_2 + \frac{{}^jd\boldsymbol{b}_1}{dt}\bullet\boldsymbol{b}_3\boldsymbol{b}_3. \tag{4.19}$$

Note that the right side of Eq. (4.19) is precisely the decomposition of vector $\frac{{}^jd\boldsymbol{b}_1}{dt}$ in accordance with the orthonormal basis $\{\boldsymbol{b}_1, \boldsymbol{b}_2, \boldsymbol{b}_3\}$. Indeed,

$$\frac{{}^jd\boldsymbol{b}_1}{dt}\bullet\boldsymbol{b}_1\boldsymbol{b}_1 + \frac{{}^jd\boldsymbol{b}_1}{dt}\bullet\boldsymbol{b}_2\boldsymbol{b}_2 + \frac{{}^jd\boldsymbol{b}_1}{dt}\bullet\boldsymbol{b}_3\boldsymbol{b}_3 = \frac{{}^jd\boldsymbol{b}_1}{dt}. \tag{4.20}$$

Hence,

$$ {}^j\boldsymbol{\omega}^m \times \boldsymbol{b}_1 = \frac{{}^j d\boldsymbol{b}_1}{dt}. \tag{4.21} $$

Following this trend, it is straightforward to demonstrate that

$$ {}^j\boldsymbol{\omega}^m \times \boldsymbol{b}_2 = \frac{{}^j d}{dt}\boldsymbol{b}_2 \tag{4.22} $$

and

$$ {}^j\boldsymbol{\omega}^m \times \boldsymbol{b}_3 = \frac{{}^j d}{dt}\boldsymbol{b}_3. \tag{4.23} $$

Furthermore, taking into account that the vector $\boldsymbol{\beta}$ can be expressed according to the orthonormal basis $\{\boldsymbol{b}_1, \boldsymbol{b}_2, \boldsymbol{b}_3\}$ as

$$ \boldsymbol{\beta} = \beta_1\boldsymbol{b}_1 + \beta_2\boldsymbol{b}_2 + \beta_3\boldsymbol{b}_3 \quad \forall \quad \beta_i(i = 1, 2, 3) \in \Re, \tag{4.24} $$

then the cross product between vectors ${}^j\boldsymbol{\omega}^m$ and $\boldsymbol{\beta}$ leads to

$$ {}^j\boldsymbol{\omega}^m \times \boldsymbol{\beta} = \beta_1{}^j\boldsymbol{\omega}^m \times \boldsymbol{b}_1 + \beta_2{}^j\boldsymbol{\omega}^m \times \boldsymbol{b}_2 + \beta_3{}^j\boldsymbol{\omega}^m \times \boldsymbol{b}_3. \tag{4.25} $$

Hence,

$$ {}^j\boldsymbol{\omega}^m \times \boldsymbol{\beta} = \beta_1\frac{{}^j d\boldsymbol{b}_1}{dt} + \beta_2\frac{{}^j d\boldsymbol{b}_2}{dt} + \beta_3\frac{{}^j d\boldsymbol{b}_3}{dt}. \tag{4.26} $$

However, it is evident that

$$ \frac{{}^j d}{dt}\beta_1 = \frac{{}^j d}{dt}\beta_2 = \frac{{}^j d}{dt}\beta_3 = 0. \tag{4.27} $$

Thus, the right side of Eq. (4.26) may be expressed as

$$ \beta_1\frac{{}^j d\boldsymbol{b}_1}{dt} + \beta_2\frac{{}^j d\boldsymbol{b}_2}{dt} + \beta_3\frac{{}^j d\boldsymbol{b}_3}{dt} = \frac{{}^j d}{dt}\left(\beta_1\boldsymbol{b}_1 + \beta_2\boldsymbol{b}_2 + \beta_3\boldsymbol{b}_3\right) \tag{4.28} $$

or

$$ \beta_1\frac{{}^j d\boldsymbol{b}_1}{dt} + \beta_2\frac{{}^j d\boldsymbol{b}_2}{dt} + \beta_3\frac{{}^j d\boldsymbol{b}_3}{dt} = \frac{{}^j d\boldsymbol{\beta}}{dt}. \tag{4.29} $$

Finally, substituting Eq. (4.29) into Eq. (4.26) yields ${}^j\boldsymbol{\omega}^m \times \boldsymbol{\beta} = \frac{{}^j d\boldsymbol{\beta}}{dt}$. ◇

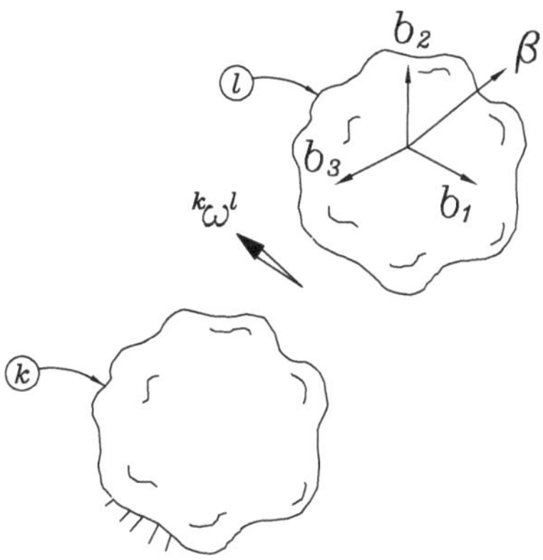

Fig. 4.4 A free vector $\boldsymbol{\beta}$ in a body l

The time derivative of a vector using two reference frames k and l (see Fig. 4.4) may be related by the angular velocity $^k\boldsymbol{\omega}^l$. To this end, it is convenient to note that the simplest case occurs when the vector is embedded in the body where the reference frame is located; for example, $\frac{^ld}{dt}\boldsymbol{\beta} = \mathbf{0}$ if $\boldsymbol{\beta}$ is attached to body l. The next proposition investigates the case where $\boldsymbol{\beta}$ is a free vector.

Proposition 4.3. *Let k and l be two arbitrary bodies and let $\boldsymbol{\beta}$ be a free vector in body l (see Fig. 4.4). The time derivative of vector $\boldsymbol{\beta}$ with respect to body k may be expressed as*

$$\frac{^kd}{dt}\boldsymbol{\beta} = \frac{^ld}{dt}\boldsymbol{\beta} + {^k\boldsymbol{\omega}^l} \times \boldsymbol{\beta}. \tag{4.30}$$

Proof. Consider that the vector $\boldsymbol{\beta}$ may be expressed according to the orthonormal basis $\{\boldsymbol{b}_1, \boldsymbol{b}_2, \boldsymbol{b}_3\}$ as $\boldsymbol{\beta} = \beta_1\boldsymbol{b}_1 + \beta_2\boldsymbol{b}_2 + \beta_3\boldsymbol{b}_3$. Hence, the time derivative of vector $\boldsymbol{\beta}$ with respect to body k is given by

$$\frac{^kd\boldsymbol{\beta}}{dt} = \frac{^kd}{dt}\left(\beta_1\boldsymbol{b}_1 + \beta_2\boldsymbol{b}_2 + \beta_3\boldsymbol{b}_3\right), \tag{4.31}$$

which leads to

$$\frac{^kd\boldsymbol{\beta}}{dt} = \frac{^kd\beta_1}{dt}\boldsymbol{b}_1 + \frac{^kd\beta_2}{dt}\boldsymbol{b}_2 + \frac{^kd\beta_3}{dt}\boldsymbol{b}_3 + \beta_1\frac{^kd\boldsymbol{b}_1}{dt} + \beta_2\frac{^kd\boldsymbol{b}_2}{dt} + \beta_3\frac{^kd\boldsymbol{b}_3}{dt}. \tag{4.32}$$

However, according to Proposition 4.2 it is evident that

$$\frac{^kd}{dt}\boldsymbol{b}_i = {^k\boldsymbol{\omega}^l} \times \boldsymbol{b}_i, \quad i = 1, 2, 3. \tag{4.33}$$

Furthermore, since the time derivatives of $\beta_i (i = 1, 2, 3)$ do not depend on the reference frame, then

$$\frac{^kd\beta_i}{dt} = \frac{^ld\beta_i}{dt}. \tag{4.34}$$

Hence, expression (4.32) can be rewritten as

$$\frac{^kd}{dt}\boldsymbol{\beta} = \frac{^ld\beta_1}{dt}\boldsymbol{b}_1 + \frac{^ld\beta_2}{dt}\boldsymbol{b}_2 + \frac{^ld\beta_3}{dt}\boldsymbol{b}_3 + \beta_1{}^k\boldsymbol{\omega}^l \times \boldsymbol{b}_1 + \beta_2{}^k\boldsymbol{\omega}^l \times \boldsymbol{b}_2 + \beta_3{}^k\boldsymbol{\omega}^l \times \boldsymbol{b}_3. \tag{4.35}$$

On the other hand, note that

$$\frac{^ld}{dt}\boldsymbol{b}_i = \boldsymbol{0}. \tag{4.36}$$

Thus,

$$\frac{^kd}{dt}\boldsymbol{\beta} = \frac{^ld}{dt}\left(\beta_1\boldsymbol{b}_1 + \beta_2\boldsymbol{b}_2 + \beta_3\boldsymbol{b}_3\right) + {}^k\boldsymbol{\omega}^l \times \left(\beta_1\boldsymbol{b}_1 + \beta_2\boldsymbol{b}_2 + \beta_3\boldsymbol{b}_3\right), \tag{4.37}$$

which implies that in fact

$$\frac{^kd}{dt}\boldsymbol{\beta} = \frac{^ld}{dt}\boldsymbol{\beta} + {}^k\boldsymbol{\omega}^l \times \boldsymbol{\beta}. \quad \diamond$$

In what follows Propositions 4.2 and 4.3 are used to determine the angular and linear velocities of several bodies in relative motion.

With reference to Fig. 4.5, consider a multibody system formed by three rigid bodies labeled j, k, and m. Furthermore, let $\boldsymbol{\beta}$ be an arbitrary vector. We are required to find the angular and linear velocities of point O in body m as observed from body j considering the existence of body k. To accomplish this, the time derivatives of $\boldsymbol{\beta}$ as observed from bodies j and k are expressed, respectively, as

$$\frac{^jd}{dt}\boldsymbol{\beta} = {}^j\boldsymbol{\omega}^m \times \boldsymbol{\beta} \tag{4.38}$$

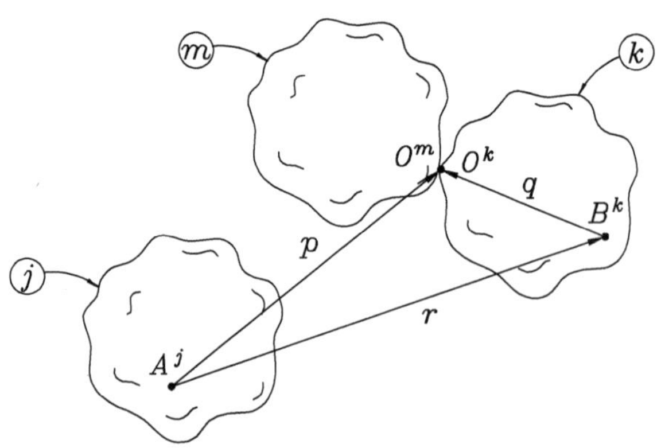

Fig. 4.5 Bodies j, k, and m in relative motion

and

$$\frac{{}^k d}{dt}\boldsymbol{\beta} = {}^k\boldsymbol{\omega}^m \times \boldsymbol{\beta}. \tag{4.39}$$

However, clearly

$$\frac{{}^j d}{dt}\boldsymbol{\beta} = \frac{{}^k d}{dt}\boldsymbol{\beta} + {}^j\boldsymbol{\omega}^k \times \boldsymbol{\beta}. \tag{4.40}$$

Therefore, substituting Eqs. (4.38) and (4.39) into Eq. (4.40) leads to

$${}^j\boldsymbol{\omega}^m \times \boldsymbol{\beta} = {}^k\boldsymbol{\omega}^m \times \boldsymbol{\beta} + {}^j\boldsymbol{\omega}^k \times \boldsymbol{\beta} \tag{4.41}$$

or

$${}^j\boldsymbol{\omega}^m \times \boldsymbol{\beta} = \left({}^j\boldsymbol{\omega}^k + {}^k\boldsymbol{\omega}^m\right) \times \boldsymbol{\beta}. \tag{4.42}$$

If we take into account that $\boldsymbol{\beta}$ is an arbitrary free vector, then Eq. (4.41) is valid if and only if

$${}^j\boldsymbol{\omega}^m = {}^j\boldsymbol{\omega}^k + {}^k\boldsymbol{\omega}^m. \tag{4.43}$$

To obtain the linear velocity of a point O^m fixed in body m with respect to body j, the vector ${}^j\boldsymbol{v}_O^m$, consider that O^k is a point of body k that is instantaneously coincident with point O^m, whereas the points A^j and B^k are embedded in bodies j and k. Furthermore, a closure equation for vector $\boldsymbol{p}$ is given by

$$\boldsymbol{p} = \boldsymbol{r} + \boldsymbol{q}, \tag{4.44}$$

where $\boldsymbol{p}$ is the position vector of O^m with respect to A^j, $\boldsymbol{r}$ is the position vector of B^k with respect to A^j, and $\boldsymbol{q}$ is a vector with head O^m and tail in B^k. After, the velocity vector ${}^j\boldsymbol{v}_O^m$ can be obtained as the time derivative of Eq. (4.44) with respect to body j, that is, ${}^j\boldsymbol{v}_O^m \equiv \frac{{}^j d}{dt}(\boldsymbol{p})$. Therefore,

$${}^j\boldsymbol{v}_O^m = \frac{{}^j d}{dt}(\boldsymbol{r}) + \frac{{}^j d}{dt}(\boldsymbol{q}). \tag{4.45}$$

However,

$$\frac{{}^j d}{dt}(\boldsymbol{r}) = {}^j\boldsymbol{v}_B^k. \tag{4.46}$$

On the other hand, from Proposition 4.3 we have

$$\frac{{}^j d}{dt}(\boldsymbol{q}) = {}^j\boldsymbol{\omega}^k \times \boldsymbol{q} + \frac{{}^k d}{dt}(\boldsymbol{q}), \tag{4.47}$$

where $\frac{^k d}{dt}(\boldsymbol{q}) = {}^k\boldsymbol{v}_O^m$. Hence, substituting Eqs. (4.46) and (4.47) into Eq. (4.45) leads to

$$ {}^j\boldsymbol{v}_O^m = {}^j\boldsymbol{v}_B^k + {}^j\boldsymbol{\omega}^k \times \boldsymbol{q} + {}^k\boldsymbol{v}_O^m. \tag{4.48}$$

Furthermore, from college kinematics it is evident that

$$ {}^j\boldsymbol{v}_O^k = {}^j\boldsymbol{v}_B^k + {}^j\boldsymbol{\omega}^k \times \boldsymbol{q}. \tag{4.49}$$

Therefore, substituting Eq. (4.49) into Eq. (4.48), we finally get

$$ {}^j\boldsymbol{v}_O^m = {}^j\boldsymbol{v}_O^k + {}^k\boldsymbol{v}_O^m. \tag{4.50}$$

It is interesting to note that while the linear velocity ${}^j\boldsymbol{v}_O^m$ depends on the chosen point O^m, the angular velocity ${}^j\boldsymbol{\omega}^m$ does not depend on the reference pole. Certainly, the angular velocity vector is a property of a rigid body. On the other hand, Eqs. (4.43) and (4.50) can be used to determine the velocity state of body m with respect to body j, the six-dimensional vector ${}^j\boldsymbol{V}_O^m$, taking into account the motion of body k. To this end consider that

$$ {}^j\boldsymbol{V}_O^k = \begin{bmatrix} {}^j\boldsymbol{\omega}^k \\ {}^j\boldsymbol{v}_O^k \end{bmatrix} \tag{4.51}$$

and

$$ {}^k\boldsymbol{V}_O^m = \begin{bmatrix} {}^k\boldsymbol{\omega}^m \\ {}^k\boldsymbol{v}_O^m \end{bmatrix}. \tag{4.52}$$

Hence,

$$ {}^j\boldsymbol{V}_O^k + {}^k\boldsymbol{V}_O^m = \begin{bmatrix} {}^j\boldsymbol{\omega}^k + {}^k\boldsymbol{\omega}^m \\ {}^j\boldsymbol{v}_O^k + {}^k\boldsymbol{v}_O^m \end{bmatrix}. \tag{4.53}$$

Later, substituting Eqs. (4.43) and (4.50) into Eq. (4.53) yields

$$ {}^j\boldsymbol{V}_O^k + {}^k\boldsymbol{V}_O^m = \begin{bmatrix} {}^j\boldsymbol{\omega}^m \\ {}^j\boldsymbol{v}_O^m \end{bmatrix}. \tag{4.54}$$

However, since

$$ {}^j\boldsymbol{V}_O^m = \begin{bmatrix} {}^j\boldsymbol{\omega}^m \\ {}^j\boldsymbol{v}_O^m \end{bmatrix}, $$

then

$$ {}^j\boldsymbol{V}_O^m = {}^j\boldsymbol{V}_O^k + {}^k\boldsymbol{V}_O^m. \tag{4.55}$$

The next proposition is a generalization of Eq. (4.55).

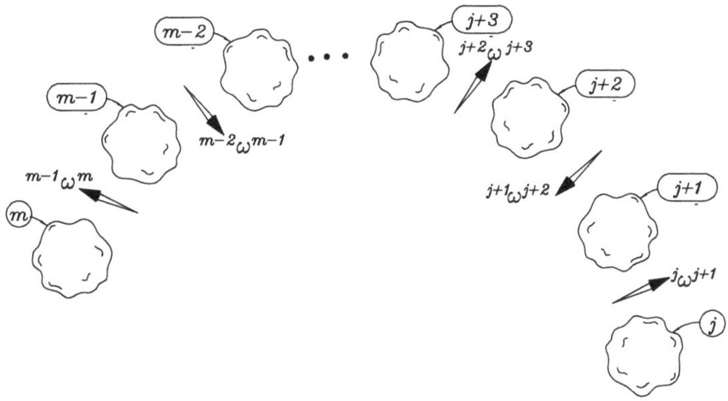

Fig. 4.6 Multibody mechanical system

Proposition 4.4. *With reference to Fig. 4.6, consider a multibody mechanical system shaped with rigid bodies labeled j, j + 1 ,..., m − 1, and m. The velocity state of body m as observed from body j may be expressed as*

$$ {}^{j}\mathbf{V}^{m}_{O} = {}^{j}\mathbf{V}^{j+1}_{O} + {}^{j+1}\mathbf{V}^{j+2}_{O} + \ldots + {}^{m-2}\mathbf{V}^{m-1}_{O} + {}^{m-1}\mathbf{V}^{m}_{O}. \tag{4.56} $$

Proof. By substituting $j = m - 2$ and $k = m - 1$ into Eq. (4.55), we have

$$ {}^{m-2}\mathbf{V}^{m}_{O} = {}^{m-2}\mathbf{V}^{m-1}_{O} + {}^{m-1}\mathbf{V}^{m}_{O}. \tag{4.57} $$

Furthermore, by the induction hypothesis we write

$$ {}^{j+1}\mathbf{V}^{m}_{O} = {}^{j+1}\mathbf{V}^{j+2}_{O} + \ldots + {}^{m-2}\mathbf{V}^{m-1}_{O} + {}^{m-1}\mathbf{V}^{m}_{O}. \tag{4.58} $$

On the other hand, by substituting $k = j + 1$ into Eq. (4.55) we obtain

$$ {}^{j}\mathbf{V}^{m}_{O} = {}^{j}\mathbf{V}^{j+1}_{O} + {}^{j+1}\mathbf{V}^{m}_{O}. \tag{4.59} $$

Finally, the substitution of Eq (4.58) into Eq. (4.59) leads to

$$ {}^{j}\mathbf{V}^{m}_{O} = {}^{j}\mathbf{V}^{j+1}_{O} + {}^{j+1}\mathbf{V}^{m}_{O} = {}^{j}\mathbf{V}^{j+1}_{O} + {}^{j+1}\mathbf{V}^{j+2}_{O} + \ldots + {}^{m-2}\mathbf{V}^{m-1}_{O} + {}^{m-1}\mathbf{V}^{m}_{O}. \quad \diamond $$

Note that Eq. (4.56) is available for multibody systems whose rigid bodies do not need to be physically interconnected. The velocity analysis of kinematic chains serially connected by helical pairs is revealed in the next section. With that in mind, it is worth emphasizing that the union of the concept of Plücker coordinates with the concept of velocity state of a rigid body allows one to elegantly formulate the velocity analysis of serial kinematic chains.

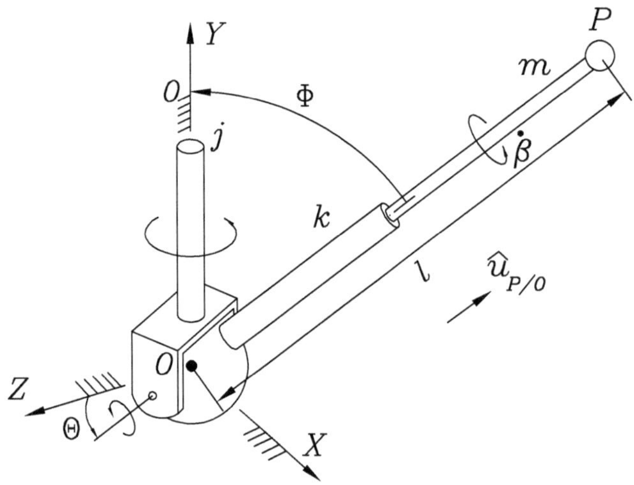

Fig. 4.7 Example 4.3. Actuating mechanism for a telescoping antenna on a spacecraft

Example 4.3. Figure 4.7 shows an actuating mechanism used as a telescoping antenna on a spacecraft. The supporting shaft j rotates about the fixed Y-axis while the orientation of the telescoping antenna is determined according to the angles θ and ϕ. At the same time, the antenna extends according to a variable length l while body m rotates freely with respect to body k along the unit axis $\hat{\boldsymbol{u}}_{P/O}$. Determine symbolically the velocity state of body m as measured from the body as well as the angular and linear velocities of point P.

Solution. Let XYZ be a reference frame attached to the fixed body 0 whose origin is located at point O, with associated unit vectors $\hat{\boldsymbol{i}}\hat{\boldsymbol{j}}\hat{\boldsymbol{k}}$. Furthermore, let

$$\hat{\boldsymbol{u}}_{P/O} = \sin(\phi)\cos(\theta)\hat{\boldsymbol{i}} + \cos(\phi)\hat{\boldsymbol{j}} - \sin(\phi)\sin(\theta)\hat{\boldsymbol{k}}$$

be a unit vector pointed from O to P. To approach the velocity analysis, consider that

- ${}^0\boldsymbol{\omega}^j = \dot{\theta}\hat{\boldsymbol{j}}$;
- ${}^j\boldsymbol{\omega}^k = \dot{\phi}[\sin(\theta)\hat{\boldsymbol{i}} + \cos(\theta)\hat{\boldsymbol{k}}]$;
- ${}^k\boldsymbol{\omega}^m = \dot{\beta}\hat{\boldsymbol{u}}_{P/O}$;
- ${}^0\boldsymbol{v}_P^j = {}^0\boldsymbol{\omega}^j \times \boldsymbol{r}_{P/O} = l\,{}^0\boldsymbol{\omega}^j \times \hat{\boldsymbol{u}}_{P/O}$;
- ${}^j\boldsymbol{v}_P^k = {}^j\boldsymbol{\omega}^k \times \boldsymbol{r}_{P/O} = l\,{}^j\boldsymbol{\omega}^k \times \hat{\boldsymbol{u}}_{P/O}$;
- ${}^k\boldsymbol{v}_P^m = \dot{l}\hat{\boldsymbol{u}}_{P/O}$.

Next, the velocity state of point P fixed to body m as measured from the Earth (0), the vector

$${}^0\boldsymbol{V}_P^m = \begin{bmatrix} {}^0\boldsymbol{\omega}^m \\ {}^0\boldsymbol{v}_P^m \end{bmatrix},$$

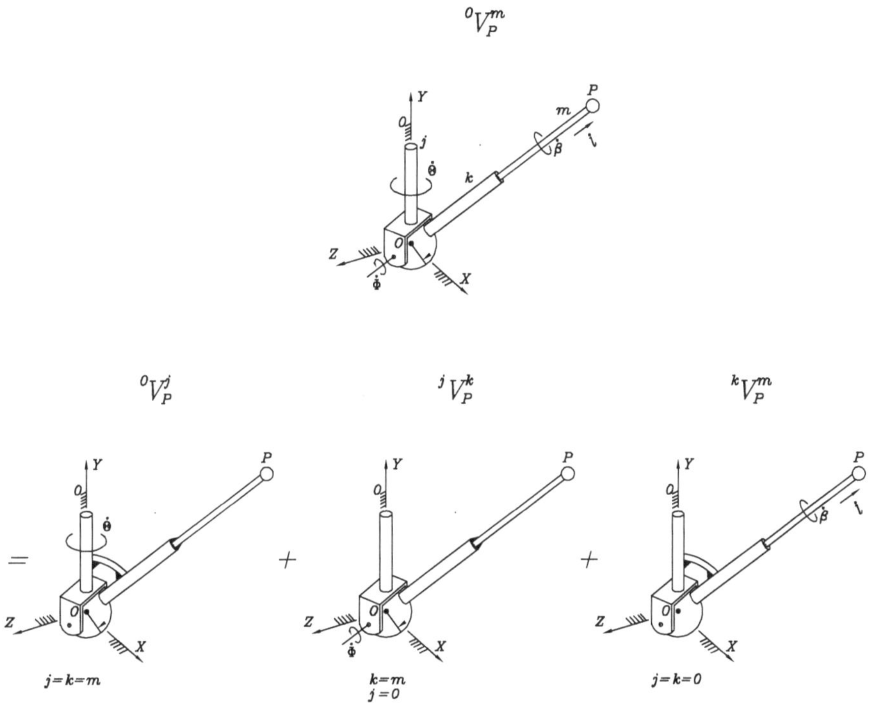

Fig. 4.8 Example 4.3. Decomposition of the velocity analysis

may be expressed as

$$\begin{aligned} {}^0\boldsymbol{V}_P^m &= {}^0\boldsymbol{V}_P^j + {}^j\boldsymbol{V}_P^k + {}^k\boldsymbol{V}_P^m \\ &= \begin{bmatrix} {}^0\boldsymbol{\omega}^j \\ {}^0\boldsymbol{v}_P^j \end{bmatrix} + \begin{bmatrix} {}^j\boldsymbol{\omega}^k \\ {}^j\boldsymbol{v}_P^k \end{bmatrix} + \begin{bmatrix} {}^k\boldsymbol{\omega}^m \\ {}^k\boldsymbol{v}_P^m \end{bmatrix} \\ &= \begin{bmatrix} \dot{\phi}\sin(\theta) + \dot{\beta}\sin(\phi)\cos(\theta) \\ \dot{\theta} + \dot{\beta}\cos(\phi) \\ \dot{\phi}\cos(\theta) - \dot{\beta}\sin(\phi)\sin(\theta) \\ -l\dot{\theta}\sin(\phi)\sin(\theta) - l\dot{\phi}\cos(\theta)\cos(\phi) + \dot{l}\sin(\phi)\cos(\theta) \\ l\dot{\phi}\sin(\phi) + \dot{l}\cos(\phi) \\ -l\dot{\theta}\sin(\phi)\cos(\theta) + l\dot{\phi}\sin(\theta)\cos(\phi) - \dot{l}\sin(\phi)\sin(\theta) \end{bmatrix}. \end{aligned} \tag{4.60}$$

For clarity, Fig. 4.8 shows the decomposition of the velocity analysis of the open kinematic chain at hand. After, the angular velocity of body m as measured from body 0 is obtained as the primal part of the six-dimensional vector ${}^0\boldsymbol{V}_P^m$ as follows:

$${}^0\boldsymbol{\omega}^m = \left[\dot{\phi}\sin(\theta) + \dot{\beta}\sin(\phi)\cos(\theta)\right]\hat{\boldsymbol{i}} +$$

$$[\dot{\theta} + \dot{\beta}\cos(\phi)]\hat{\boldsymbol{j}} +$$
$$[\dot{\phi}\cos(\theta) - \dot{\beta}\sin(\phi)\sin(\theta)]\,\hat{\boldsymbol{k}}. \tag{4.61}$$

Meanwhile, the velocity of point P is precisely the dual part of ${}^{0}\boldsymbol{V}_{P}^{m}$. Indeed,

$$\begin{aligned} {}^{0}\boldsymbol{v}_{P}^{m} = \big[-l\dot{\theta}\sin(\phi)\sin(\theta) - l\dot{\phi}\cos(\theta)\cos(\phi) + \dot{l}\sin(\phi)\cos(\theta)\big]\hat{\boldsymbol{i}} + \\ \big[l\dot{\phi}\sin(\phi) + \dot{l}\cos(\phi)\big]\hat{\boldsymbol{j}} + \\ \big[-l\dot{\theta}\sin(\phi)\cos(\theta) + l\dot{\phi}\sin(\theta)\cos(\phi) - \dot{l}\sin(\phi)\sin(\theta)\big]\hat{\boldsymbol{k}}. \end{aligned} \tag{4.62}$$

4.3 Equations of Velocity in Screw Form

In this section the connection between the velocity state of a rigid body and the concept of twist about of screw formulated by Ball (1900) is formally presented. Furthermore, the fundamentals to express the velocity analysis of serial chains in screw form are revealed, assuming that the involved links are connected by infinitesimal screws (helical pairs). The section is complemented by illustrative examples.

Proposition 4.5. *With reference to Fig. 4.2, consider two rigid bodies k and l interconnected by a helical pair ${}^{k}\$^{l}$ denoted in Plücker coordinates as*

$$ {}^{k}\$^{l} = \begin{bmatrix} {}^{k}\boldsymbol{s}^{l} \\ {}^{k}\boldsymbol{s}_{O}^{l} \end{bmatrix}, $$

where ${}^{k}\boldsymbol{s}^{l}$ is a unit vector along the direction of the axis of the helical pair whereas ${}^{k}\boldsymbol{s}_{O}^{l}$ is the moment measured from body k produced by ${}^{k}\boldsymbol{s}^{l}$ about point O^{l}. The velocity state ${}^{k}\boldsymbol{V}_{O}^{l}$ of body l, with respect to body k, is given in screw form by

$$ {}^{k}\boldsymbol{V}_{O}^{l} = {}_{k}\omega_{l}\,{}^{k}\$^{l}. \tag{4.63}$$

In other words, the velocity state of a rigid body is a twist on a screw.

Proof. It is evident that the angular velocity vector ${}^{k}\boldsymbol{\omega}^{l}$ may be expressed as

$$ {}^{k}\boldsymbol{\omega}^{l} = {}_{k}\omega_{l}\,{}^{k}\boldsymbol{s}^{l}. \tag{4.64}$$

On the other hand, a closure equation for the position vector $\boldsymbol{p}$ is given by

$$ \boldsymbol{p} = \boldsymbol{r} + \boldsymbol{q}, \tag{4.65}$$

where

- $\boldsymbol{p}$ is the position vector of point O^l with respect to point A^k;
- $\boldsymbol{r}$ is the vector with tail in point A^k and head in point B^l;
- $\boldsymbol{q}$ is the position vector of point O^l with respect to point B^l.

Hence, taking into account that ${}^k\boldsymbol{v}_O^l \equiv \frac{{}^k d}{dt}(\boldsymbol{p})$, we have

$$ {}^k\boldsymbol{v}_O^l = \frac{{}^k d}{dt}(\boldsymbol{r}) + \frac{{}^k d}{dt}(\boldsymbol{q}), \tag{4.66} $$

where

$$ \frac{{}^k d}{dt}\boldsymbol{r} = {}^k\boldsymbol{v}_B^l. \tag{4.67} $$

Furthermore, the velocity ${}^k\boldsymbol{v}_B^l$ is a vector along the axis of the helical pair and therefore may be expressed in accordance with the corresponding pitch ${}_kh_l$ as follows:

$$ {}^k\boldsymbol{v}_B^l = {}_kh_{lk}\omega_l{}^k\boldsymbol{s}^l. \tag{4.68} $$

On the other hand, from college kinematics we have

$$ \frac{{}^k d}{dt}\boldsymbol{q} = {}^k\boldsymbol{\omega}^l \times \boldsymbol{q}, \tag{4.69} $$

or

$$ \frac{{}^k d}{dt}\boldsymbol{q} = {}_k\omega_l{}^k\boldsymbol{s}^l \times \boldsymbol{q}. \tag{4.70} $$

Gathering these results, it follows that

$$ {}^k\boldsymbol{v}_O^l = {}_kh_{lk}\omega_l{}^k\boldsymbol{s}^l + {}_k\omega_l{}^k\boldsymbol{s}^l \times \boldsymbol{q}. \tag{4.71} $$

Rearranging terms, we get

$$ {}^k\boldsymbol{v}_O^l = {}_k\omega_l\left({}^k\boldsymbol{s}^l \times \boldsymbol{q} + {}_kh_{lk}\omega_l{}^k\boldsymbol{s}^l.\right) \tag{4.72} $$

However, it was well proved that

$$ {}^k\boldsymbol{s}_O^l = {}^k\boldsymbol{s}^l \times \boldsymbol{q} + {}_kh_{lk}\omega_l{}^k\boldsymbol{s}^l. $$

Thus, the velocity vector ${}^k\boldsymbol{v}_O^l$ is given by

$$ {}^k\boldsymbol{v}_O^l = {}_k\omega_l{}^k\boldsymbol{s}_O^l. \tag{4.73} $$

Then, taking into account that

$$ {}^{k}\mathbf{V}^{l}_{O} = \begin{bmatrix} {}^{k}\boldsymbol{\omega}^{l} \\ {}^{k}\boldsymbol{v}^{l}_{O} \end{bmatrix}, \tag{4.74} $$

substituting Eqs. (4.64) and (4.73) into Eq. (4.74) leads to

$$ {}^{k}\mathbf{V}^{l}_{O} = \begin{bmatrix} {}_{k}\omega_{l}{}^{k}\mathbf{s}^{l} \\ {}_{k}\omega_{l}{}^{k}\mathbf{s}^{l}_{O} \end{bmatrix}, \tag{4.75} $$

or

$$ {}^{k}\mathbf{V}^{l}_{O} = {}_{k}\omega_{l} \begin{bmatrix} {}^{k}\mathbf{s}^{l} \\ {}^{k}\mathbf{s}^{l}_{O} \end{bmatrix}. \tag{4.76} $$

However, it was shown that in Plücker coordinates we have

$$ \begin{bmatrix} {}^{k}\mathbf{s}^{l} \\ {}^{k}\mathbf{s}^{l}_{O} \end{bmatrix} = {}^{k}\$^{l}. $$

Hence, evidently

$$ {}^{k}\mathbf{V}^{l}_{O} = {}_{k}\omega_{l} \begin{bmatrix} {}^{k}\mathbf{s}^{l} \\ {}^{k}\mathbf{s}^{l}_{O} \end{bmatrix} = {}_{k}\omega_{l}{}^{k}\$^{l}. \tag{4.77} $$

Certainly, the velocity state of a rigid body is, in reality, a twist on a screw. ◇

It should be noted that to avoid an involved nomenclature, a subscript below the symbol \$ is not considered to describe an infinitesimal screw; for example, in this book it is unnecessary to use $\$_O$ to indicate that O is the reference pole although certainly a more precise designation could be ${}^{k}\$^{l}_{O}$ instead of ${}^{k}\$^{l}$. On the other hand, expression (4.63) gives us the opportunity to formulate the velocity analysis equations of serial robot manipulators in screw form. This formulation can be extended without significant effort to parallel manipulators.

Proposition 4.6. *With reference to Fig. (4.9), consider an open kinematic chain formed by rigid bodies or links labeled $j, j+1, \ldots, m-2, m-1, m$. Furthermore, consider that adjacent bodies k and $k+1$ are interconnected by a helical pair ${}^{k}\$^{k+1}$. Then the velocity state of body m as observed from body j, the six-dimensional vector ${}^{j}\mathbf{V}^{m}_{O}$, considering point O^m of body m as the reference pole, may be expressed in screw form as*

$$ {}^{j}\mathbf{V}^{m}_{O} = {}_{j}\omega_{j+1}{}^{j}\$^{j+1} + {}_{j+1}\omega_{j+2}{}^{j+1}\$^{j+2} + \ldots + {}_{m-2}\omega_{m-1}{}^{m-2}\$^{m-1} + {}_{m-1}\omega_{m}{}^{m-1}\$^{m}. \tag{4.78} $$

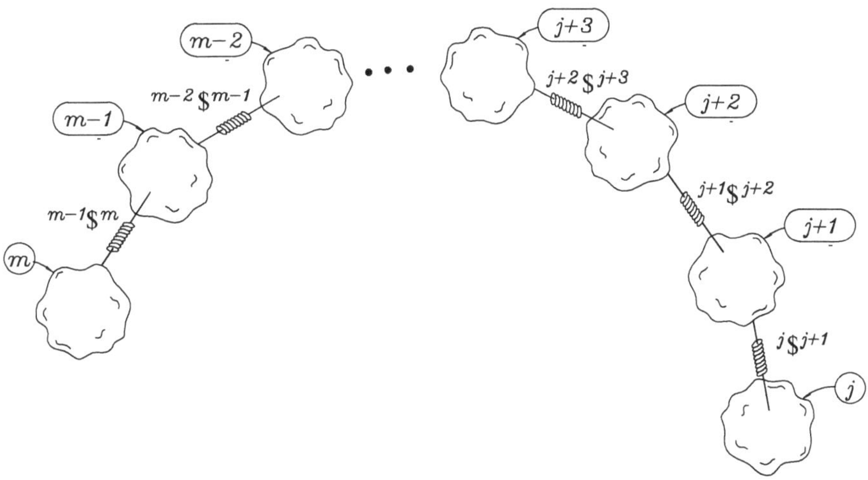

Fig. 4.9 Serial kinematic chain

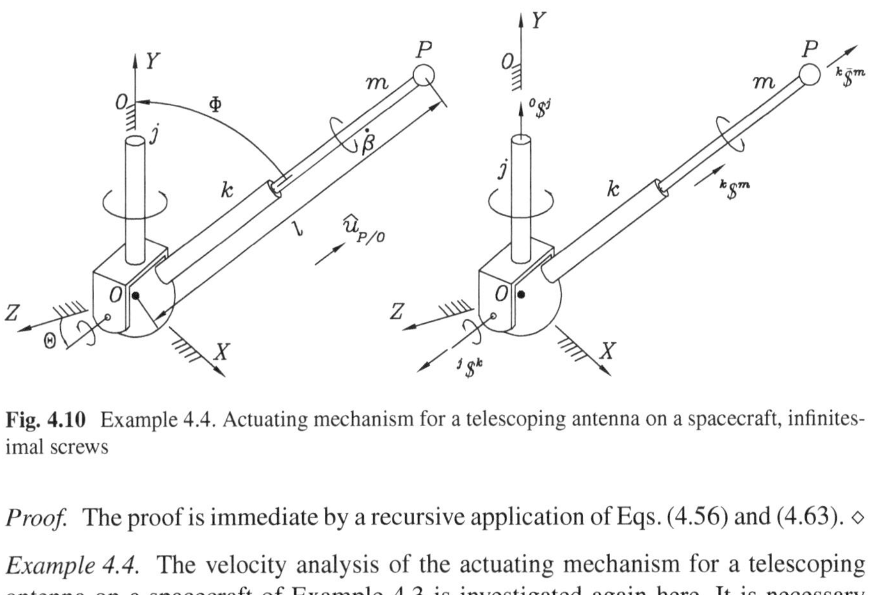

Fig. 4.10 Example 4.4. Actuating mechanism for a telescoping antenna on a spacecraft, infinitesimal screws

Proof. The proof is immediate by a recursive application of Eqs. (4.56) and (4.63). ⋄

Example 4.4. The velocity analysis of the actuating mechanism for a telescoping antenna on a spacecraft of Example 4.3 is investigated again here. It is necessary to determine the equation of velocity in screw form of body m as measured from body 0. After, we must apply that equation to compute the angular velocity of m with respect to body 0 as well as the velocity of point P.

With reference to Fig. 4.10 to simplify the computation of the Plücker coordinates of the involved infinitesimal screws, the origin O of the fixed reference frame XYZ is chosen as the reference pole. The velocity state of body m with respect to body 0, the vector ${}^0\boldsymbol{V}_O^m$, is given by

$$\dot{\theta}\,{}^0\$^j + \dot{\phi}\,{}^j\$^k + \dot{\beta}\,{}^k\$^m + \dot{l}\,{}^k\overline{\$}^m = {}^0\boldsymbol{V}_O^m, \tag{4.79}$$

where ${}^k\$^m$ is the screw representing the revolute joint associated to the revolute joint connecting the bodies m and k while ${}^k\overline{\$}^m$ represents the screw associated to the translational motion of body m as observed from body k. The infinitesimal screws in Plücker coordinates of the infinitesimal screws are given by

$$
{}^0\$^j = \begin{bmatrix} 0 \\ 1 \\ 0 \\ 0 \\ 0 \\ 0 \end{bmatrix}, \; {}^j\$^k = \begin{bmatrix} \sin(\theta) \\ 0 \\ \cos(\theta) \\ 0 \\ 0 \\ 0 \end{bmatrix}, \; {}^k\$^m = \begin{bmatrix} \sin(\phi)\cos(\theta) \\ \cos(\phi) \\ -\sin(\phi)\sin(\theta) \\ 0 \\ 0 \\ 0 \end{bmatrix}, \; {}^k\overline{\$}^m = \begin{bmatrix} 0 \\ 0 \\ 0 \\ \sin(\phi)\cos(\theta) \\ \cos(\phi) \\ -\sin(\phi)\sin(\theta) \end{bmatrix}.
$$

Hence,

$$
\begin{aligned}
{}^0\boldsymbol{V}_O^m &= \dot{\theta} \begin{bmatrix} 0 \\ 1 \\ 0 \\ 0 \\ 0 \\ 0 \end{bmatrix} + \dot{\phi} \begin{bmatrix} \sin(\theta) \\ 0 \\ \cos(\theta) \\ 0 \\ 0 \\ 0 \end{bmatrix} + \dot{\beta} \begin{bmatrix} \sin(\phi)\cos(\theta) \\ \cos(\phi) \\ -\sin(\phi)\sin(\theta) \\ 0 \\ 0 \\ 0 \end{bmatrix} + \dot{l} \begin{bmatrix} 0 \\ 0 \\ 0 \\ \sin(\phi)\cos(\theta) \\ \cos(\phi) \\ -\sin(\phi)\sin(\theta) \end{bmatrix} \\
&= \begin{bmatrix} \dot{\phi}\sin(\theta) + \dot{\beta}\sin(\phi)\cos(\theta) \\ \dot{\theta} + \dot{\beta}\cos(\phi) \\ \dot{\phi}\cos(\theta) - \dot{\beta}\sin(\phi)\sin(\theta) \\ \dot{l}\sin(\phi)\cos(\theta) \\ \dot{l}\cos(\phi) \\ -\dot{l}\sin(\phi)\sin(\theta) \end{bmatrix}.
\end{aligned} \tag{4.80}
$$

Thus, the angular velocity vector of m as observed from body 0, namely, the primal part of ${}^0\boldsymbol{V}_O^m$, results in

$$
{}^0\boldsymbol{\omega}^m = \begin{bmatrix} \dot{\phi}\sin(\theta) + \dot{\beta}\sin(\phi)\cos(\theta) \\ \dot{\theta} + \dot{\beta}\cos(\phi)] \\ \dot{\phi}\cos(\theta) - \dot{\beta}\sin(\phi)\sin(\theta) \end{bmatrix}, \tag{4.81}
$$

while the velocity of point O, fixed in body m as measured from the fixed body 0, namely the dual part of ${}^0\boldsymbol{V}_O^m$, is given by

$$
{}^0\boldsymbol{v}_O^m = \begin{bmatrix} \dot{l}\sin(\phi)\cos(\theta) \\ \dot{l}\cos(\phi) \\ -\dot{l}\sin(\phi)\sin(\theta)\hat{\boldsymbol{k}} \end{bmatrix}. \tag{4.82}
$$

On the other hand, by resorting to elementary kinematics, we may determine the velocity of point P as

$$^{0}\boldsymbol{v}_P^m = {}^{0}\boldsymbol{v}_O^m + {}^{0}\boldsymbol{\omega}^m \times \boldsymbol{r}_{P/O}.$$

After a few computations we get

$$^{0}\boldsymbol{v}_P^m = \left[-l\dot{\theta}\sin(\phi)\sin(\theta) - l\dot{\phi}\cos(\theta)\cos(\phi) + \dot{l}\sin(\phi)\cos(\theta)\right]\hat{\boldsymbol{i}} +$$

$$\left[l\dot{\phi}\sin(\phi) + \dot{l}\cos(\phi)\right]\hat{\boldsymbol{j}} + \left[-l\dot{\theta}\sin(\phi)\cos(\theta) + l\dot{\phi}\sin(\theta)\cos(\phi) - \dot{l}\sin(\phi)\sin(\theta)\right]\hat{\boldsymbol{k}},$$

which confirms the solution obtained in Example 4.3.

It is straightforward to demonstrate that the same vector $^{0}\boldsymbol{v}_P^m$ may be obtained by applying the theory of helicoidal vector fields in which $^{0}\boldsymbol{\omega}^m$ acts as the vector of the field. In that way, it is worth noting how easy it is to compute the Plücker coordinates of infinitesimal screws when the reference pole is properly selected.

Example 4.5. One of the most popular examples of a constrained linkage is the four-bar linkage mechanism. In spite of its simple topology, this planar closed kinematic chain is commonly applied in aerospace mechanical linkages as well in robot manipulators. For instance, the celebrated Delta robot (Clavel 1991) is considered a spatial generalization of a four-bar linkage mechanism. In this example the position and velocity analyses of the famous four-bar linkage mechanism are presented.

Solution. As shown in Fig. 4.11, the input angle or generalized coordinate q denotes the orientation of the input crank a_1 measured counterclockwise from the X-axis, whereas the orientation of the coupler bar a_2 is indicated by the angle β measured counterclockwise from the X-axis. Furthermore, the angle γ denotes the orientation of the output crank a_3 measured clockwise from the X-axis.

To approach the displacement analysis of the four-bar linkage, we consider it a parallel manipulator where the coupler bar is assumed to be the moving platform, body 2, while the crank a_0 is assumed to be the fixed platform, body 0. Then one closure equation may be expressed based on the parameter a_2 as follows:

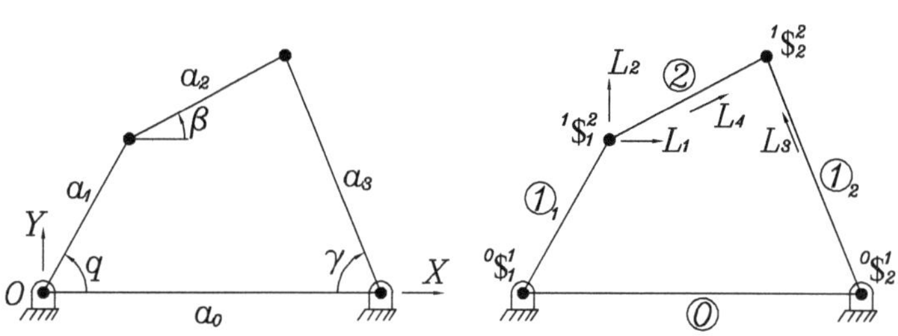

Fig. 4.11 Four-bar linkage mechanism

$$(a_0 - a_1 \cos q - a_3 \cos \gamma)^2 + (a_3 \sin \gamma - a_1 \sin q)^2 = a_2^2. \tag{4.83}$$

After a few computations, a quadratic equation in the unknown $\sin \gamma$ results in

$$(\tilde{A}^2 + \tilde{B}^2) \sin^2 \gamma - 2\tilde{B}\tilde{C} \sin \gamma + \tilde{C}^2 - \tilde{A}^2 = 0, \tag{4.84}$$

where the coefficients are determined upon the input angle q and the parameters a_0, a_1, a_2, and a_3 as

$$\begin{aligned} \tilde{A} &= -2a_0 a_3 + 2a_1 a_3 \cos q, \\ \tilde{B} &= -2a_1 a_3 \sin q, \\ \tilde{C} &= -a_0^2 + 2a_0 a_1 \cos q - a_1^2 + a_2^2 + a_3^2. \end{aligned} \tag{4.85}$$

Equation (4.84) indicates that given the input angle q there are at most two different poses that the moving platform can reach with respect to the fixed platform. Furthermore, the angle β may be computed taking into account that

$$a_1 \sin q + a_2 \sin \beta = a_3 \sin \gamma. \tag{4.86}$$

In what follows the velocity analysis of the mechanism is addressed by resorting to reciprocal screw theory. Let $\boldsymbol{\omega} = \omega_X \hat{\boldsymbol{i}} + \omega_Y \hat{\boldsymbol{j}} + \omega_Z \hat{\boldsymbol{k}}$ be the angular velocity vector of the moving platform with respect to the fixed platform, where $\hat{\boldsymbol{i}}, \hat{\boldsymbol{j}}$, and $\hat{\boldsymbol{k}}$ are unit vectors associated to the X-, Y-, and Z-axes, respectively. Furthermore, let $\boldsymbol{v}_O = v_{OX}\hat{\boldsymbol{i}} + v_{OY}\hat{\boldsymbol{j}} + v_{OZ}\hat{\boldsymbol{k}}$ be the velocity vector of the point embedded in the moving platform that is instantaneously coincident with the origin O of the reference frame XYZ attached to the fixed platform. Then the velocity state of the moving platform with respect to the fixed platform, the vector

$${}^0\boldsymbol{V}_O^2 = \begin{bmatrix} \boldsymbol{\omega} \\ \boldsymbol{v}_O \end{bmatrix},$$

may be expressed through any of the two connector chains as

$${}^0\boldsymbol{V}_O^2 = {}_0\omega_1^i\,{}^0\$_i^1 + {}_1\omega_2^i\,{}^1\$_i^2, \quad i = 1, 2, \tag{4.87}$$

where ${}_0\omega_1^1 = \dot{q}$. Meanwhile, the screws in Eq. (4.87) are given in Plücker coordinates by

$$
{}^{0}\$_{1}^{1} = \begin{bmatrix} 0 \\ 0 \\ 1 \\ 0 \\ 0 \\ 0 \end{bmatrix}, {}^{1}\$_{1}^{2} = \begin{bmatrix} 0 \\ 0 \\ 1 \\ a_1 \sin q \\ -a_1 \cos q \\ 0 \end{bmatrix}, {}^{0}\$_{2}^{1} = \begin{bmatrix} 0 \\ 0 \\ 1 \\ 0 \\ -a_0 \\ 0 \end{bmatrix}, {}^{1}\$_{2}^{2} = \begin{bmatrix} 0 \\ 0 \\ 1 \\ a_3 \sin \gamma \\ -a_1 \cos q - a_2 \cos \beta \\ 0 \end{bmatrix}. \tag{4.88}
$$

On the other hand, referring to Fig. 4.11, consider three lines $\boldsymbol{L}_1$, $\boldsymbol{L}_2$, and $\boldsymbol{L}_3$ whose Plücker coordinates are given by

$$
\boldsymbol{L}_1 = \begin{bmatrix} 1 \\ 0 \\ 0 \\ 0 \\ 0 \\ -a_1 \sin q \end{bmatrix}, \boldsymbol{L}_2 = \begin{bmatrix} 0 \\ 1 \\ 0 \\ 0 \\ 0 \\ a_1 \cos q \end{bmatrix}, \boldsymbol{L}_3 = \begin{bmatrix} -\cos \gamma \\ \sin \gamma \\ 0 \\ 0 \\ 0 \\ a_0 \sin \gamma \end{bmatrix}. \tag{4.89}
$$

Later on, by systematically applying the Klein form of the lines $\boldsymbol{L}_i (i = 1, 2, 3)$ to both sides of Eq. (4.87) and reducing terms, we obtain

$$
\{\boldsymbol{L}_1; {}^{0}\boldsymbol{V}_{O}^{2}\} = -a_1 \dot{q} \sin q, \{\boldsymbol{L}_2; {}^{0}\boldsymbol{V}_{O}^{2}\} = a_1 \dot{q} \cos q, \{\boldsymbol{L}_3; {}^{0}\boldsymbol{V}_{O}^{2}\} = 0. \tag{4.90}
$$

Finally, casting Eq. (4.90) into matrix-vector form, we find that the input–output equation of velocity of the four-bar linkage results in

$$
\mathbf{J}^T \boldsymbol{\Delta} {}^{0}\boldsymbol{V}_{O}^{2} = \mathbf{q}, \tag{4.91}
$$

where

$$
\mathbf{J} = [\boldsymbol{L}_1 \; \boldsymbol{L}_2 \; \boldsymbol{L}_3] \tag{4.92}
$$

is the 3×3 reduced screw-coordinate Jacobian matrix of the manipulator where the third, fourth, and fifth rows were removed from the six-dimensional vectors $\boldsymbol{L}_i (i = 1, 2, 3)$ because of the planar arrangement of the links of the mechanism at hand. Similarly, the velocity state ${}^{0}\boldsymbol{V}_{O}^{2}$ is reduced into a three-dimensional vector considering that $\omega_X = \omega_Y = v_Z = 0$, whereas

$$
\mathbf{q} = \begin{bmatrix} -a_1 \dot{q} \sin q \\ a_1 \dot{q} \cos q \\ 0 \end{bmatrix} \tag{4.93}
$$

is the driver matrix of the robot and

$$
\boldsymbol{\Delta} = \begin{bmatrix} 0 & 1 & 0 \\ 0 & 0 & 1 \\ 1 & 0 & 0 \end{bmatrix} \tag{4.94}
$$

is an operator of polarity.

Finally, it should be noted that if the intent of the analysis is to find a relationship between the angular velocities $\dot{q}$ and $\dot{\gamma}$, then it is possible to resort to the reciprocal line $\boldsymbol{L}_4 = \left[\cos\beta \ \sin\beta \ 0; 0 \ \ 0 \ a_1 \sin(\beta - q)\right]$. In fact, keeping in mind that according to Eq. (4.87) we have

$$ {}_0\omega_1^{1\,0}\$_1^1 + {}_1\omega_2^{1\,1}\$_1^2 = {}_0\omega_1^{2\,0}\$_2^1 + {}_1\omega_2^{2\,1}\$_2^2, \tag{4.95} $$

then by applying the Klein form of the line $\boldsymbol{L}_4$ to both sides of Eq. (4.95) and reducing terms, we obtain

$$ \dot{q}\left\{{}^0\$_1^1; \boldsymbol{L}_4\right\} = \dot{\gamma}\left\{{}^0\$_2^1; \boldsymbol{L}_4\right\}. \tag{4.96} $$

Expression (4.96) can also be called the input–output equation of velocity of the four-bar planar mechanism. Furthermore, Eq. (4.96) provides the means to explain the singular configurations of the mechanism, also referred as the *illness regions* of the parallel manipulator. Expression (4.96) clearly indicates under which conditions it is not possible to realize the velocity analysis of the mechanism. In general, these undesirable configurations must be taken into proper account at the design stage of robots. Otherwise, damage may be caused to the manipulator at hand.

Example 4.6. This example consists of computing the velocity and displacement analyses of the crank-slider linkage mechanism, a planar mechanism widely used in reciprocating engines where cranks (with connecting rods) are used to transform the back-and-forth motion of the pistons into rotary motion, and vice versa. The cranks usually are incorporated into a crankshaft. Prominent applications of this mechanism include internal combustion engines, steam trains, pumps, compressors, and of course robotics applications.

Solution. A crank-slider linkage is a four-bar mechanism where one of the pin joints is substituted by a sliding joint (see Fig. 4.12). In a standard arrangement, the line connecting two adjacent pin joints is parallel to the axis of the sliding joint. The purpose of the planar mechanism at hand is to convert the rotational motion of the crankshaft (1) into oscillatory motion of the slider (3), or vice versa.

With reference to Fig. 4.12, d denotes the nominal position of the slider with respect to the origin O of the reference frame XY, which is quickly determined as

$$ d^2 = a_1^2 + a_2^2 - 2a_1a_2\cos\beta, \tag{4.97} $$

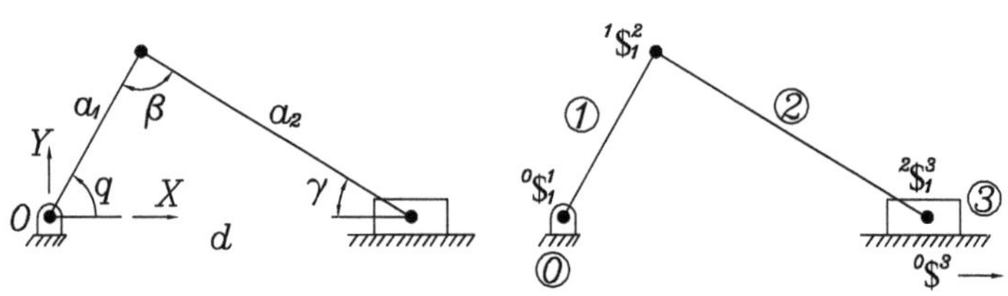

Fig. 4.12 Crank-slider linkage mechanism

where $\beta = \pi - q - \gamma$ is the orientation of the coupler bar with respect to the crankshaft in which $\gamma = \arcsin(a_1 \sin q / a_2)$ is the orientation of the coupler bar measured clockwise from the X-axis. Furthermore, the velocity state of the slider (3) as observed from the base link (0), the vector

$$ {}^0\mathbf{V}_O^3 = \left[0\ 0\ 0\ \dot{d}\ 0\ 0\right]^T, \tag{4.98}$$

may be expressed as

$$\dot{q}\,{}^0\$_1^1 + {}_1\omega_2^1\,{}^1\$_1^2 + {}_2\omega_3^1\,{}^2\$_1^3 = \dot{d}\,{}^0\$^3 = {}^0\mathbf{V}_O^3, \tag{4.99}$$

where the screws of Eq. (4.99) are given by

$$ {}^0\$_1^1 = \begin{bmatrix} 0\\0\\1\\0\\0\\0 \end{bmatrix}, {}^1\$_1^2 = \begin{bmatrix} 0\\0\\1\\a_1 \sin q\\-a_1 \cos q\\0 \end{bmatrix}, {}^2\$_1^3 = \begin{bmatrix} 0\\0\\1\\0\\-d\\0 \end{bmatrix}, {}^0\$^3 = \begin{bmatrix} 0\\0\\0\\1\\0\\0 \end{bmatrix}.$$

After reducing and rearranging terms in matrix form Eq. (4.99), we find that the input–output equation of velocity of the crank-slider linkage mechanism results in

$$\mathbf{A\Omega} = \mathbf{I\dot{q}}, \tag{4.100}$$

where

- The first-order coefficient matrix $\mathbf{A}$ of the mechanism is given by

$$\mathbf{A} = \begin{bmatrix} -1 & -1 & 0\\ a_1 \sin q & 0 & -1\\ a_1 \cos q & d & 0 \end{bmatrix}; \tag{4.101}$$

- $\mathbf{I}$ is the identity matrix of order 3;
- $\mathbf{\Omega}$ is a matrix containing the passive joint-rate velocities of the mechanism as follows:

$$\mathbf{\Omega} = \left[{}_1\omega_2^1\ {}_2\omega_3^1\ \dot{d}\right]^T. \tag{4.102}$$

- Finally, the first-order driver matrix of the manipulator $\mathbf{\dot{q}}$ is defined as

$$\mathbf{\dot{q}} = \left[\dot{q}\ 0\ 0\right]^T. \tag{4.103}$$

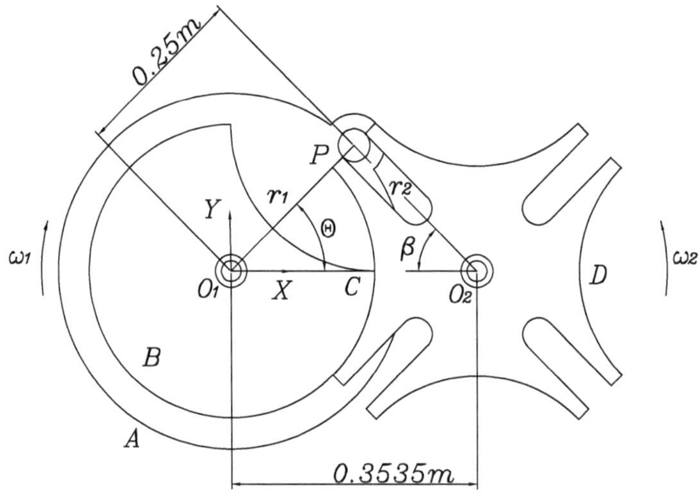

Fig. 4.13 Geneva wheel and its geometry scheme

Expression (4.100) allows us to determine singular configurations of the crank-slider linkage mechanism. The issue is left as an exercise.

Example 4.7. The Geneva wheel or Maltese cross (see Fig. 4.13) is a mechanism where a driven body undergoes intermittent rotary motion produced by the continuous rotation of a driver body. The name derives from the devices used mainly in mechanical watches, and Geneva is an important center of watchmaking. Prominent applications of this one-degree-of-freedom planar mechanism include movie projectors, the pen change mechanism in plotters, indexing tables in assembly lines, automated sampling devices, tool changers for CNC machines, and banknote counting. The rotating drive wheel A has a pin P and a plate B. The motion of the mechanism is such that wheel D turns one fourth of a revolution for each revolution of the pin P. At the engagement position shown in Fig. 4.13, we have $\theta = 30°$. For a constant clockwise angular velocity $\omega_1 = 2$ rad/s of wheel A, it is required to compute the corresponding angular velocity ω_2 of the driven wheel D and the relative velocity of point P as measured from wheel D. The intention of this example is twofold: (1) The velocity analysis is carried out using the method exposed in standard books; and (2) the velocity analysis is carried out using screw theory.

Solution. Let XYZ be a reference frame whose origin is point O_1. One classical solution of the velocity analysis of the Geneva wheel is as follows. Using elementary kinematics, the velocity of pin P is given by

$$\boldsymbol{v}_P = \boldsymbol{\omega}_1 \times \boldsymbol{r}_{P/O_1}, \tag{4.104}$$

where $\boldsymbol{r}_{P/O_1}$ is a vector pointed from O_1 to point P. Furthermore, the same vector $\boldsymbol{v}_P$ may be obtained upon the driven wheel as

$$\boldsymbol{v}_P = \boldsymbol{v}'_P + \boldsymbol{v}''_P, \tag{4.105}$$

where $\boldsymbol{v}'_P$ is the velocity of the point of body D which is instantaneously coincident with point P while $\boldsymbol{v}''_P$ is the velocity of point P relative to the wheel D. Indeed,

$$\boldsymbol{v}'_P = \boldsymbol{\omega}_2 \times \boldsymbol{r}_{P/O_2}, \quad \boldsymbol{v}''_P = v''_P \hat{\boldsymbol{u}}_{P/O_2}, \tag{4.106}$$

where $\boldsymbol{r}_{P/O_2}$ is the vector pointed from O_2 to point P, $v''_P = \mid \boldsymbol{v}''_P \mid$, and $\hat{\boldsymbol{u}}_{P/O_2}$ is a unit vector pointed from P to O_2. Hence, equating (4.104) and (4.105), it follows that

$$\boldsymbol{\omega}_1 \times \boldsymbol{r}_{P/O_1} = \boldsymbol{\omega}_2 \times \boldsymbol{r}_{P/O_2} + v''_P \hat{\boldsymbol{u}}_{P/O_2}. \tag{4.107}$$

It is straightforward to show that expression (4.107) leads to a linear system of two equations in the unknowns w_2 and v''_P. Solving such a system, one obtains the general expressions

$$\omega_2 = \omega_1 r_1 \cos(\theta + \beta)/r_2, \quad \boldsymbol{v}''_P = \omega_1 r_1 \sin(\theta + \beta), \tag{4.108}$$

where r_2 and β are computed according to the geometry of the triangle $\Delta O_1 CP$; that is, $r_2 = 0.1854\,\text{m}$ and $\beta = 42.3789°$. Hence with the substitution of numerical data, one finally obtains $\omega_2 = 0.8161\,\text{rad/s}$ and $v''_P = 0.4765\,\text{m/s}$.

In what follows the velocity analysis of the Geneva wheel is approached using screw theory. To this end, the modeling of the screws is depicted in Fig. 4.14.

While $\$_1$, $\$_2$, and $\$_4$ are infinitesimal screws associated to revolute joints, the screw $\$_3$ is associated to the prismatic joint simulating the adjustable distance

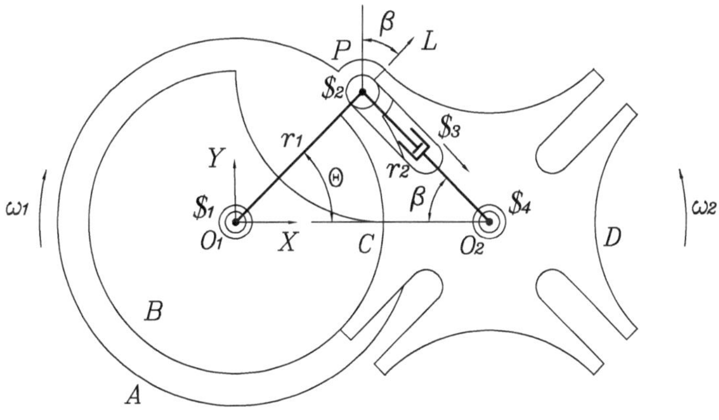

Fig. 4.14 Screws associated to the kinematic pairs of the Geneva wheel

between points P and O_2. According to the closed kinematic chain O_1PO_2, it is evident that the equation of velocity in screw form of the Geneva wheel may be written as follows:

$$v_1\$_1 + v_2\$_2 + v_3\$_3 + v_4\$_4 = \mathbf{0}, \tag{4.109}$$

where the involved infinitesimal screws, taking point O_1 as the reference pole, are computed as

$$\$_1 = \begin{bmatrix} 0 \\ 0 \\ 1 \\ 0 \\ 0 \\ 0 \end{bmatrix}, \$_2 = \begin{bmatrix} 0 \\ 0 \\ 1 \\ r_1 \sin(\theta) \\ -r_1 \cos(\theta) \\ 0 \end{bmatrix}, \$_3 = \begin{bmatrix} 0 \\ 0 \\ 0 \\ \cos(\beta) \\ -\sin(\beta) \\ 0 \end{bmatrix}, \$_4 = \begin{bmatrix} 0 \\ 0 \\ 1 \\ 0 \\ -C \\ 0 \end{bmatrix}. \tag{4.110}$$

Thus, after reducing terms, we obtain a linear system of three equations in the unknown joint-velocity rates v_2, v_3, and v_4:

$$\left. \begin{array}{l} v_1 + v_2 + v_4 = 0 \\ r_1 \sin(\theta) v_2 + \cos(\beta) v_3 = 0 \\ r_1 \cos(\theta) v_2 + \sin(\beta) v_3 - C v_4 = 0 \end{array} \right\}. \tag{4.111}$$

Next, the angular velocity $\omega_2 = -v_4$ of the driven wheel results in

$$\omega_2 = -\frac{v_1 r_1 (\cos(\theta)\cos(\beta) - \sin(\theta)\sin(\beta))}{-r_1\cos(\theta)\cos(\beta) + C\cos(\beta) + r_1 \sin(\theta)\sin(\beta)}. \tag{4.112}$$

However, since $v_1 = -\omega_1$ and $C = r_1\cos(\theta) + r_2\cos(\beta)$ while $\beta = \arcsin(r_1\sin(\theta)/r_2)$, then

$$-r_1\cos(\theta)\cos(\beta) + C\cos(\beta) + r_1\sin(\theta)\sin(\beta) = r_2, \tag{4.113}$$

yielding

$$\omega_2 = \omega_1 r_1 \cos(\theta + \beta)/r_2,$$

which confirms the solution obtained using the classical method seen in standard books.

On the other hand, the computation of the angular velocity of the driven wheel can be shortened by resorting to reciprocal screw theory. To do this, consider a line $\boldsymbol{L}$ in Plücker coordinates whose primal part is normal to the line PO_2. Applying the Klein form of $\boldsymbol{L}$ to both sides of Eq. (4.109) and reducing terms, for example, $\{\$_2; \boldsymbol{L}\} = \{\$_3; \boldsymbol{L}\} = 0$, leads to

$$v_1 \{\$_1; \boldsymbol{L}\} + v_4 \{\$_1; \boldsymbol{L}\} = 0. \tag{4.114}$$

Hence,

$$v_4 = -\frac{\{\$_1; \boldsymbol{L}\}}{\{\$_1; \boldsymbol{L}\}} v_1. \tag{4.115}$$

Furthermore, since $v_4 = -\omega_2$ and $v_1 = -\omega_1$, we get

$$\omega_2 = -\frac{\{\$_1; \boldsymbol{L}\}}{\{\$_1; \boldsymbol{L}\}} \omega_1 = \omega_1 r_1 \cos(\theta + \beta)/r_2. \tag{4.116}$$

To determine the singular configurations of the mechanism, we obtain the input–output equation of the Geneva wheel based on Eq. (4.111) as follows:

$$\mathbf{A}\boldsymbol{\Omega} = \dot{\mathbf{q}}, \tag{4.117}$$

where

- $\mathbf{A}$ is the first-order coefficient matrix of the manipulator. Indeed,

$$\mathbf{A} = \begin{bmatrix} 1 & 0 & 1 \\ r_1 \sin(\theta) & \cos(\beta) & 0 \\ r_1 \cos(\theta) & \sin(\beta) & -C \end{bmatrix}; \tag{4.118}$$

- $\boldsymbol{\Omega} = \begin{bmatrix} v_2 \; v_3 \; v_4 \end{bmatrix}^T$ is a matrix containing the joint-velocity rates of the mechanism;
- $\dot{\mathbf{q}} = \begin{bmatrix} -v_1 \; 0 \; 0 \end{bmatrix}^T$ is the first-order driver matrix of the Geneva wheel.

According to Eq. (4.117), the mechanism is at a singular configuration when $\det(\mathbf{A}) = 0$. The singular surfaces in loci form of the Geneva mechanism considering $r_1 = 0.25$, $C = 0.3535$, $-2\pi \le \theta \le 2\pi$, and $-2\pi \le \beta \le 2\pi$ are shown in Fig. 4.15.

4.4 Exercises

1. Explore the possibility, if any, to obtain the angular velocity as the time derivative of a vector.
2. In your own words, give an explanation of Eq. (4.11).
3. Let m and j be two rigid bodies, and let ${}^j\boldsymbol{\omega}^m$ be the angular velocity vector of m with respect to j. If $\boldsymbol{\beta}$ is a bound vector in body m, then prove, without using Proposition 4.2, that

$$\frac{{}^j d}{dt}\boldsymbol{\beta} = {}^j\boldsymbol{\omega}^m \times \boldsymbol{\beta}.$$

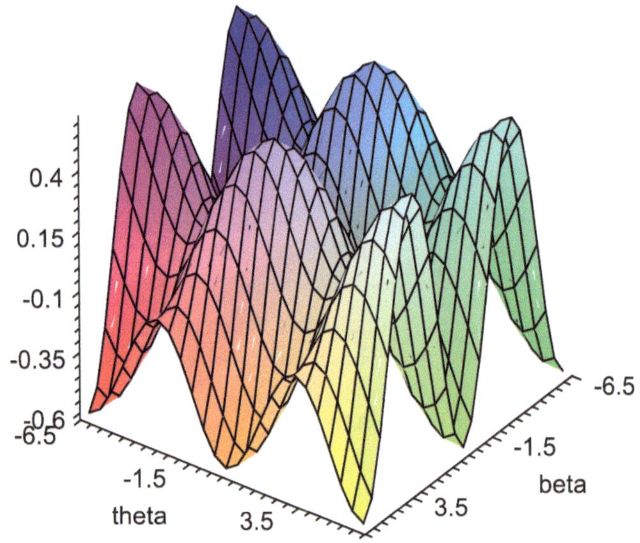

Fig. 4.15 Singularities in loci form of the Geneva wheel

4. Although this book is devoted mainly to parallel manipulators, the material provided in this chapter is available for serial manipulators. With that in mind, determine the velocity state of the end-effector as measured from the base link of the SCARA robot.
5. Carry out the singularity analysis of the four-bar planar mechanism.
6. Based on screw theory, determine the velocity of the middle of the coupler platform of the four-bar planar mechanism.
7. Determine in loci form the singular regions of the crank-slider linkage mechanism. In other words, find an analytic expression to compute the singularities of the mechanism.
8. Using classical methods, Eq. (4.107) leads to a linear system of two equations in the unknowns ω_2 and v_P''; determine such a system.
9. Apply another classical method to solve the velocity analysis of the Geneva wheel, Example 4.7. *Hint*: Obtain two closure equations from the triangle O_1O_2P indicated in Fig. 4.13. Finding the corresponding time derivatives of such equations allows us to solve the velocity analysis.
10. In Example 4.7 we considered that $v_4 = -\omega_2$. Give a justification, if any, for this assumption.
11. The Geneva wheel of Example 4.7 is a little case that illustrates how easy it is to formulate the velocity analysis of parallel manipulators when reciprocal screw theory is used. Determine the input–output equation of velocity of the planar parallel manipulators shown in Fig. 4.16 by resorting to reciprocal screw theory.

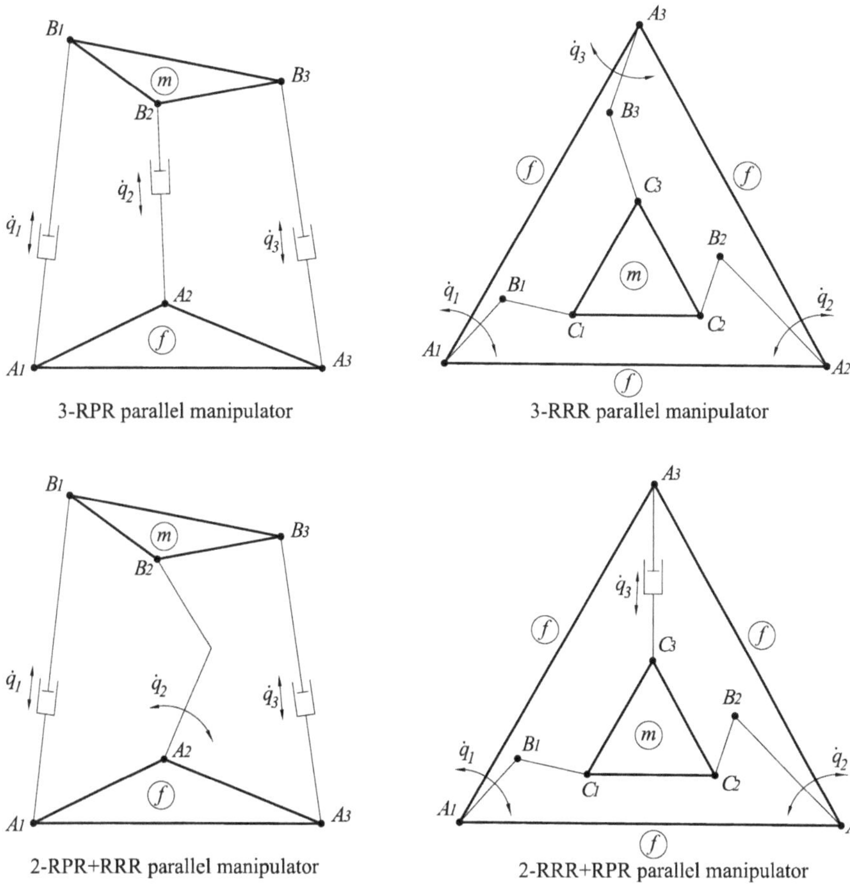

Fig. 4.16 Nonredundant planar parallel manipulators

Fig. 4.17 Six-bar dwell linkage

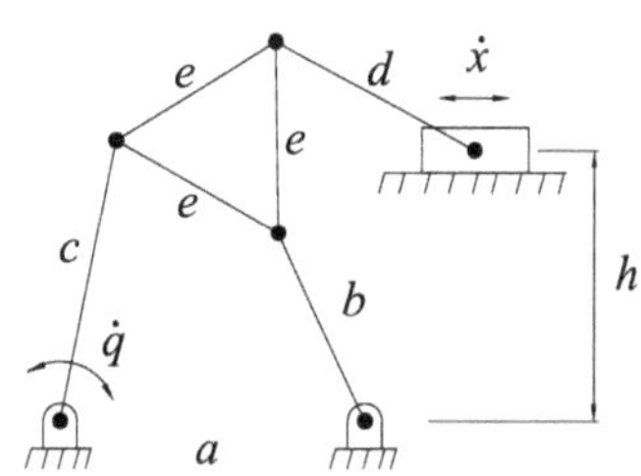

12. The link c of the six-bar dwell mechanism in Fig. 4.17 rotates with a generalized velocity $\dot{q}$. The topology of the mechanism is such that the angular velocity $\dot{q}$ produces an oscillatory linear velocity $\dot{x}$. Determine the input–output equation of velocity of the planar mechanism.
13. The mechanism shown in Fig. 4.18 is used to produce high torque in the shaft at B and is a good example to which to apply screw theory. The gear unit is pivoted

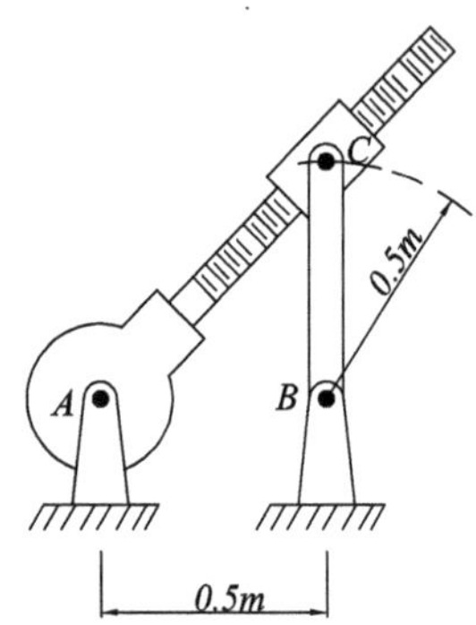

Fig. 4.18 High-torque mechanism

at A and produces a constant angular velocity of 5 rad/s counterclockwise as observed from A to C over the right-hand screw. If the pitch of the screw is six single threads per inch of length, determine the joint-velocity rate of link BC.

References

Ball, R. S. (1900). *A treatise on the theory of screws*. Cambridge: Cambridge University Press (reprinted 1998).

Clavel, R. (1990). Device for the movement and positioning of an element in space. U.S. Patent No. 4,976,582, December 11.

Clavel, R. (1991). *Conception d'un robot parallèle rapide à 4 degrés de liberté*. Ph.D. thesis, Ecole Polytechique Fédérale de Lausanne (EPFL).

Kane, T. R., & Levinson, D. A. (1985). *Dynamics: theory and applications*. New York: McGraw-Hill.

Part III
Higher-Order Kinematic Analyses of Rigid Body

Chapter 5
Acceleration Analysis

5.1 Introduction

Acceleration is the time rate of the velocity and has a direct relationship with the forces and moments generated in a multibody mechanical system. Acceleration analysis consists of formulating equations involving the angular and linear accelerations of each element of a multibody mechanical system. The solution of acceleration analysis allows us to determine the angular and linear accelerations of any body of the system under study. Pioneering contributions modeling the acceleration of rigid bodies based on six-dimensional vectors date back to 1947 when Brand introduced in kinematics the concept of acceleration motor or, for brevity, accelerator. However, despite the monumental work of Ball (1900), which was well received in the kinematician community, the possibilities to extend it to acceleration analysis were overlooked by scientists and engineers. For example, while it was easy to accept the definition of twist about a screw, the definition of accelerator appears to be more complicated to understand because both kinematic entities must represent helicoidal vector fields. Certainly, while the primal part of the accelerator results in the time rate of the primal part of the velocity state, the result is not so obvious when the computation of the dual part of the accelerator must be achieved. Hence, since the contribution of Brand (1947), except for some isolated works, the acceleration analysis of kinematic chains using only screw theory remained with the status of "open kinematic problem" for more than a half-century. For instance, Kieffer (1990, 1994) and Legnani et al. (1996) had assumed that one of the main drawbacks of the screw theory was the complexity to extend it to acceleration analysis. With that in mind, prominent contributors include Sugimoto (1987, 1989, 1990), Ridley et al. (1992), and Narasimhan and Kumar (1994). However, as far as we are aware, the first contribution approaching the acceleration analysis of kinematic chains using only screw theory is credited to Rico and Duffy (1996).

J. Gallardo-Alvarado, *Kinematic Analysis of Parallel Manipulators by Algebraic Screw Theory*, DOI 10.1007/978-3-319-31126-5_5

This chapter presents the acceleration analysis of kinematic chains using screw theory. In this book this chapter is the beginning of the extension of screw theory from velocity analysis to the higher-order analyses of kinematic chains.

5.2 Fundamental Equations of Acceleration

This section starts with a formal definition of acceleration.

Definition 5.1. Let k and l be two rigid bodies (see Fig. 4.2). The angular acceleration vector of body l as observed from body k, the vector ${}^k\boldsymbol{\alpha}^l$, is defined as follows:

$${}^k\boldsymbol{\alpha}^l \equiv \frac{{}^k d}{dt}({}^k\boldsymbol{\omega}^l), \tag{5.1}$$

where ${}^k\boldsymbol{\omega}^l$ is the angular velocity vector of body l as measured from body k. Furthermore, let O^l be an arbitrary point embedded in body l. Then the linear acceleration vector of point O^l as measured from body k is defined as

$${}^k\boldsymbol{a}^l_O \equiv \frac{{}^k d}{dt}({}^k\boldsymbol{v}^l_O), \tag{5.2}$$

where ${}^k\boldsymbol{v}^l_O$ is the linear velocity vector of point O^l as observed from body k.

It is noticeable that unlike the angular velocity vector, the angular acceleration vector can be obtained as shown in Eq. (5.1) upon the time derivative of a vector. On the other hand, since ${}^k\boldsymbol{\omega}^l = {}_k\omega_l{}^k\boldsymbol{s}^l$, it follows that the vector ${}^k\boldsymbol{\alpha}^l$ may be expressed as

$${}^k\boldsymbol{\alpha}^l = {}_k\alpha_l{}^k\boldsymbol{s}^l, \tag{5.3}$$

where the scalar ${}_k\alpha_l \equiv \frac{{}^k d}{dt}({}_k\omega_l)$ is the magnitude of the vector ${}^k\boldsymbol{\alpha}^l$ or joint-acceleration rate between bodies l and k.

Kane and Levinson (1985) reported that in general the angular acceleration of a rigid body m as observed from another body or reference frame j, denoted as the vector ${}^j\boldsymbol{\alpha}^m$, is different from the vectorial sum of the joint-acceleration rates ${}^k\boldsymbol{\alpha}^{k+1}(k = j, j+1, \ldots, m-2, m-1)$ of intermediate bodies between m and j (see Fig. 4.6). In fact, in general,

$${}^j\boldsymbol{\alpha}^m \neq {}^j\boldsymbol{\alpha}^{j+1} + {}^{j+1}\boldsymbol{\alpha}^{j+2} + \ldots + {}^{m-2}\boldsymbol{\alpha}^{m-1} + {}^{m-1}\boldsymbol{\alpha}^m. \tag{5.4}$$

In order to clarify this point, the next proposition investigates the relationship between the joint-acceleration rates of a multibody mechanical system formed with three rigid bodies. Then the obtained results are extended to multibody mechanical systems where the number of bodies is not a limitation.

Proposition 5.1. *Let j, k, and m be three rigid bodies or reference frames (see Fig. 4.5). If O^m is a point embedded in body m that is instantaneously coincident with* *the point O^k of body k, then the angular and linear accelerations of body m, with point O chosen as the reference pole, with respect to body j are given respectively by*

$$^j\boldsymbol{\alpha}^m = {}^j\boldsymbol{\alpha}^k + {}^k\boldsymbol{\alpha}^m + {}^j\boldsymbol{\omega}^k \times {}^k\boldsymbol{\omega}^m \tag{5.5}$$

and

$$^j\boldsymbol{a}_O^m = {}^j\boldsymbol{a}_O^k + {}^k\boldsymbol{a}_O^m + 2{}^j\boldsymbol{\omega}^k \times {}^k\boldsymbol{v}_O^m. \tag{5.6}$$

Proof. Taking into account that ${}^j\boldsymbol{\omega}^m = {}^j\boldsymbol{\omega}^k + {}^k\boldsymbol{\omega}^m$ [see Eq. (4.43)], according to Definition 5.1 we have

$$^j\boldsymbol{\alpha}^m \equiv \frac{^jd}{dt}({}^j\boldsymbol{\omega}^m) = \frac{^jd}{dt}({}^j\boldsymbol{\omega}^k + {}^k\boldsymbol{\omega}^m) = \frac{^jd}{dt}\left({}^j\boldsymbol{\omega}^k\right) + \frac{^jd}{dt}\left({}^k\boldsymbol{\omega}^m\right). \tag{5.7}$$

Furthermore, clearly $\frac{^jd}{dt}\left({}^j\boldsymbol{\omega}^k\right) = {}^j\boldsymbol{\alpha}^k$, while the application of Proposition 4.3 to $\frac{^jd}{dt}\left({}^k\boldsymbol{\omega}^m\right)$ leads to

$$\frac{^jd}{dt}\left({}^k\boldsymbol{\omega}^m\right) = {}^j\boldsymbol{\omega}^k \times {}^k\boldsymbol{\omega}^m + \frac{^kd}{dt}\left({}^k\boldsymbol{\omega}^m\right), \tag{5.8}$$

where $\frac{^kd}{dt}\left({}^k\boldsymbol{\omega}^m\right) = {}^k\boldsymbol{\alpha}^m$. Hence,

$$^j\boldsymbol{\alpha}^m = {}^j\boldsymbol{\alpha}^k + {}^j\boldsymbol{\omega}^k \times {}^k\boldsymbol{\omega}^m + {}^k\boldsymbol{\alpha}^m = {}^j\boldsymbol{\alpha}^k + {}^k\boldsymbol{\alpha}^m + {}^j\boldsymbol{\omega}^k \times {}^k\boldsymbol{\omega}^m$$

and the first part of the proof is finished.

To prove Eq. (5.6), consider that ${}^j\boldsymbol{a}_O^m \equiv \frac{^jd}{dt}\left({}^j\boldsymbol{v}_O^m\right)$, where [see Eq. (4.48)] ${}^j\boldsymbol{v}_O^m = {}^j\boldsymbol{v}_{\tilde{B}}^k + {}^j\boldsymbol{\omega}^k \times \boldsymbol{q} + {}^k\boldsymbol{v}_O^m$. Hence,

$$\begin{aligned} {}^j\boldsymbol{a}_O^m &= \frac{^jd}{dt}\left({}^j\boldsymbol{v}_{\tilde{B}}^k + {}^j\boldsymbol{\omega}^k \times \boldsymbol{q} + {}^k\boldsymbol{v}_O^m\right) \\ &= \frac{^jd}{dt}\left({}^j\boldsymbol{v}_{\tilde{B}}^k\right) + \frac{^jd}{dt}\left({}^j\boldsymbol{\omega}^k\right) \times \boldsymbol{q} + {}^j\boldsymbol{\omega}^k \times \frac{^jd}{dt}(\boldsymbol{q}) + \frac{^jd}{dt}\left({}^k\boldsymbol{v}_O^m\right), \end{aligned} \tag{5.9}$$

where

$$\left.\begin{array}{l} \frac{^jd}{dt}\left({}^j\boldsymbol{v}_{\tilde{B}}^k\right) = {}^j\boldsymbol{a}_{\tilde{B}}^k \\ \frac{^jd}{dt}\left({}^j\boldsymbol{\omega}^k\right) = {}^j\boldsymbol{\alpha}^k \\ \frac{^jd}{dt}(\boldsymbol{q}) = {}^j\boldsymbol{\omega}^k \times \boldsymbol{q} + \frac{^kd}{dt}(\boldsymbol{q}) = {}^j\boldsymbol{\omega}^k \times \boldsymbol{q} + {}^k\boldsymbol{v}_O^m \\ \frac{^jd}{dt}\left({}^k\boldsymbol{v}_O^m\right) = {}^j\boldsymbol{\omega}^k \times {}^k\boldsymbol{v}_O^m + \frac{^kd}{dt}\left({}^k\boldsymbol{v}_O^m\right) = {}^j\boldsymbol{\omega}^k \times {}^k\boldsymbol{v}_O^m + {}^k\boldsymbol{a}_O^m \end{array}\right\}. \tag{5.10}$$

Substituting expressions (5.10) into Eq. (5.9) and reducing terms, we get

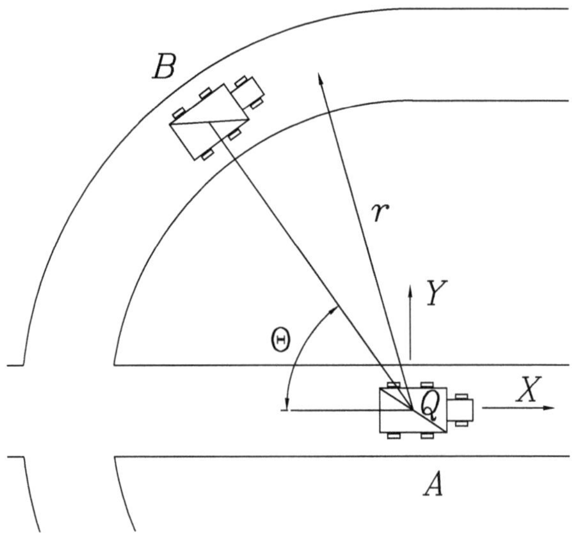

Fig. 5.1 Example 5.1. Two trucks in relative motion

$$ {}^j\boldsymbol{a}_O^m = {}^j\boldsymbol{a}_{\tilde{B}}^k + {}^j\boldsymbol{\alpha}^k \times \boldsymbol{q} + {}^j\boldsymbol{\omega}^k \times \left({}^j\boldsymbol{\omega}^k \times \boldsymbol{q}\right) + 2{}^j\boldsymbol{\omega}^k \times {}^k\boldsymbol{v}_O^m + {}^k\boldsymbol{a}_O^m. \tag{5.11} $$

On the other hand, from college kinematics it is evident that

$$ {}^j\boldsymbol{a}_{\tilde{B}}^k + {}^j\boldsymbol{\alpha}^k \times \boldsymbol{q} + {}^j\boldsymbol{\omega}^k \times \left({}^j\boldsymbol{\omega}^k \times \boldsymbol{q}\right) = {}^j\boldsymbol{a}_O^k. \tag{5.12} $$

Finally, substituting Eq. (5.12) into Eq. (5.11) yields expression (5.6). ◇

Postulate 1. *The Coriolis acceleration, notated as a_{cor}, is given by the following terms contained in Eq. (5.11):*

$$ a_{cor} = 2{}^j\boldsymbol{\omega}^k \times {}^k\boldsymbol{v}_O^m + {}^k\boldsymbol{a}_O^m. \tag{5.13} $$

Example 5.1. Truck B is traveling at the constant speed of 50 km/h over a circular curve of radius $r = 325$ m (see Fig. 5.1). At the same time, truck A is accelerating in the direction of its motion at the rate 3 m/s^2 and passes the center of the circle. Determine the acceleration of truck B as observed from truck A considering $\theta = 30°$. *Hint*: As an initial step identify bodies j, k, and m (see Fig. 4.5).

Solution. Let XY be a reference frame attached to the center of the curve with associated unit vectors $\hat{\boldsymbol{i}}\hat{\boldsymbol{j}}$. In this example one can observe three bodies in relative motion. To apply Eq. (5.6), consider that body $j = 0$ is the Earth while $k = A$ and $m = B$ denote the labels of the trucks. Hence,

$$ {}^0\boldsymbol{a}^B = {}^0\boldsymbol{a}^A + {}^A\boldsymbol{a}^B + 2{}^0\boldsymbol{\omega}^A \times {}^A\boldsymbol{v}^B, \tag{5.14} $$

where, because of the translational motion of body A as observed from body 0, the vector ${}^0\boldsymbol{\omega}^A$ vanishes. Furthermore, ${}^0\boldsymbol{a}^A = 3\hat{\boldsymbol{i}}$ while

$$ {}^0\boldsymbol{a}^B = \frac{v_B^2}{r}[\cos(\theta)\hat{\boldsymbol{i}} - \sin(\theta)\hat{\boldsymbol{j}}]. \tag{5.15} $$

Thus, the relative acceleration between bodies A and B, the vector ${}^A\boldsymbol{a}^B$, is given by

$$ {}^A\boldsymbol{a}^B = {}^0\boldsymbol{a}^B - {}^0\boldsymbol{a}^A = \frac{v_B^2}{r}[\cos(\theta)\hat{\boldsymbol{i}} - \sin(\theta)\hat{\boldsymbol{j}}] - {}^0\boldsymbol{a}^A, \tag{5.16} $$

where $v_B = 50\,\text{km/h} = 13.88\,\text{m/s}$ and $r = 325\,\text{m}$. Therefore, data substitution yields

$$ {}^A\boldsymbol{a}^B = 0.593[\cos(30°)\hat{\boldsymbol{i}} - \sin(30°)\hat{\boldsymbol{j}}] - 3\hat{\boldsymbol{i}} = -2.486\hat{\boldsymbol{i}} - 0.296\hat{\boldsymbol{j}}. $$

5.3 Equations of Acceleration in Screw Form

The interpretation of the velocity state of a rigid body or twist on a screw is simple; therefore, since its introduction by Ball in 1900 it has been well received in the kinematician community. On the other hand, the interpretation of the reduced acceleration state of a rigid body is more elusive and deserves to be deeply investigated. At a glance, it seems reasonable to define the reduced acceleration state of rigid body m as observed from another rigid body j in accordance with the angular acceleration vector ${}^j\boldsymbol{\alpha}^m$ of m with respect to j as follows:

$$ {}^j\boldsymbol{A}_O^m \equiv \begin{bmatrix} {}^j\boldsymbol{\alpha}^m \\ {}^j\boldsymbol{a}_O^m \end{bmatrix}, \tag{5.17} $$

where O^m is a point of body m termed the reference pole, whereas ${}^j\boldsymbol{a}_O^m$ denotes the acceleration of O^m measured from body j. However, it is obvious that ${}^j\boldsymbol{A}_O^m$, as defined in Eq. (5.17), does not satisfy the conditions of helicoidal vector fields; for instance, if P^m is an arbitrary point of m with $P^m \neq O^m$, then in general

$$ {}^j\boldsymbol{a}_P^m \neq {}^j\boldsymbol{a}_O^m + {}^j\boldsymbol{\alpha}^m \times \boldsymbol{r}_{P/O}, \tag{5.18} $$

where ${}^j\boldsymbol{a}_P^m$ is the acceleration of point P^m measured from body j.

Many researchers have tried unsuccessfully to elucidate the acceleration analysis of robot manipulators based on the weak argument that the reduced acceleration state, or accelerator, of a rigid body can be obtained as a simple time derivative of the velocity state, such as

$$ \boldsymbol{A} = \frac{d}{dt}\boldsymbol{V} = \frac{d}{dt}\begin{bmatrix} \boldsymbol{\omega} \\ \boldsymbol{v} \end{bmatrix} = \begin{bmatrix} \frac{d}{dt}\boldsymbol{\omega} \\ \frac{d}{dt}\boldsymbol{v} \end{bmatrix} = \begin{bmatrix} \boldsymbol{\alpha} \\ \boldsymbol{a} \end{bmatrix}. \tag{5.19} $$

It is sad to write, but perhaps for such a reason many efforts to investigate the kinematics of robot manipulators using screw theory are limited to the so-called first-order kinematic analysis. In fact, it is a common practice that once velocity analysis has been addressed using screw theory, then acceleration analysis is reported as simple time derivatives of kinematic entities such as the angular and linear velocities of a rigid body, which is a reasonable strategy, but this is not, in reality, screw theory. In this area it is not strange to find phrases in the specialized literature such as

Unfortunately, screw theory is usually explained following descriptive definitions rather than short axiomatic lines of reasoning.

Screw theory allows simple geometrical interpretation, but it is restricted to speed and infinitesimal displacement analysis.

In order to find a proper, complete, and overall true representation of the reduced acceleration state, we must take into account two fundamental conditions: (1) The reduced acceleration state must be a helicoidal field where the angular acceleration ${}^j\boldsymbol{\alpha}^m$ is the vector of the helicoidal field; (2) the vector assigned by the reduced acceleration state must include the linear acceleration vector of a body embedded in the rigid body m. With these clues in mind, it is possible to introduce the following proposition.

Proposition 5.2. *Let m be a rigid body whose motions are measured from another rigid body j; in other words, j plays the role of reference coordinate system or simply reference frame. Furthermore, let O^m be a point of the rigid body m. Then a representation of the reduced acceleration state of body m with respect to body j, the six-dimensional vector ${}^j\boldsymbol{A}^m_O$, is given by*

$$ {}^j\boldsymbol{A}^m_O = \begin{bmatrix} {}^j\boldsymbol{\alpha}^m \\ {}^j\boldsymbol{a}^m_O - {}^j\boldsymbol{\omega}^m \times {}^j\boldsymbol{v}^m_O . \end{bmatrix} \tag{5.20} $$

Furthermore, the reduced acceleration state is a helicoidal vector field.

Proof. Let P and Q be two points of the rigid body m. In accordance with the clues given at the beginning of this section, it is strongly advisable to write two representations of the reduced acceleration state as follows:

$$ {}^j\boldsymbol{A}^m_P = \begin{bmatrix} {}^j\boldsymbol{\alpha}^m \\ {}^j\boldsymbol{a}^m_P + \boldsymbol{F}(P) \end{bmatrix}, \quad {}^j\boldsymbol{A}^m_Q = \begin{bmatrix} {}^j\boldsymbol{\alpha}^m \\ {}^j\boldsymbol{a}^m_Q + \boldsymbol{F}(Q) \end{bmatrix}, \tag{5.21} $$

where $\boldsymbol{F}$ is, at this moment, an unknown helicoidal field. Assuming that expressions (5.21) satisfy the conditions of helicoidal vector fields, we have

$$ {}^j\boldsymbol{a}^m_P + \boldsymbol{F}(P) = {}^j\boldsymbol{a}^m_Q + \boldsymbol{F}(Q) + {}^j\boldsymbol{\alpha}^m \times \boldsymbol{r}_{P/Q}, \tag{5.22} $$

or

$$ {}^j\boldsymbol{a}^m_P - {}^j\boldsymbol{a}^m_Q - {}^j\boldsymbol{\alpha}^m \times \boldsymbol{r}_{P/Q} = \boldsymbol{F}(Q) - \boldsymbol{F}(P). \tag{5.23} $$

However, from college kinematics it is evident that

$$ {}^{j}\boldsymbol{a}_{P}^{m} = {}^{j}\boldsymbol{a}_{Q}^{m} + {}^{j}\boldsymbol{\alpha}^{m} \times \boldsymbol{r}_{P/Q} + {}^{j}\boldsymbol{\omega}^{m} \times ({}^{j}\boldsymbol{\omega}^{m} \times \boldsymbol{r}_{P/Q}), \tag{5.24} $$

where

$$ {}^{j}\boldsymbol{\omega}^{m} \times \boldsymbol{r}_{P/Q} = {}^{j}\boldsymbol{v}_{P}^{m} - {}^{j}\boldsymbol{v}_{Q}^{m}. \tag{5.25} $$

Hence,

$$ {}^{j}\boldsymbol{a}_{P}^{m} - {}^{j}\boldsymbol{a}_{Q}^{m} - {}^{j}\boldsymbol{\alpha}^{m} \times \boldsymbol{r}_{P/Q} = {}^{j}\boldsymbol{\omega}^{m} \times ({}^{j}\boldsymbol{v}_{P}^{m} - {}^{j}\boldsymbol{v}_{Q}^{m}). \tag{5.26} $$

The substitution of Eq. (5.26) into Eq. (5.23) leads to

$$ \boldsymbol{F}(Q) - \boldsymbol{F}(P) = {}^{j}\boldsymbol{\omega}^{m} \times ({}^{j}\boldsymbol{v}_{P}^{m} - {}^{j}\boldsymbol{v}_{Q}^{m}). \tag{5.27} $$

Therefore, it is possible to write two vectorial equations as follows:

$$ \boldsymbol{F}(P) = -{}^{j}\boldsymbol{\omega}^{m} \times {}^{j}\boldsymbol{v}_{P}^{m} + \boldsymbol{C} \tag{5.28} $$

and

$$ \boldsymbol{F}(Q) = -{}^{j}\boldsymbol{\omega}^{m} \times {}^{j}\boldsymbol{v}_{Q}^{m} + \boldsymbol{C}. \tag{5.29} $$

According to expressions (5.28) and (5.29), the simplest result arises when $\boldsymbol{C} = \boldsymbol{0}$. Hence, expression (5.20) is obtained by removing the reference poles P and Q. $\diamond$

Once the correctness of Proposition 5.2 has been proved, the reduced acceleration state of a rigid body is formally introduced in the next definition; perhaps these are repetitive lines, but they are necessary in order to follow the natural sequence of this chapter.

Definition 5.2 (The Reduced Acceleration State of a Rigid Body). Let j and m be two rigid bodies or reference frames and let O^m be a point of body m. The reduced acceleration state, also known as a motor acceleration (Brand 1947) or accelerator, for brevity, is defined as

$$ {}^{j}\boldsymbol{A}_{O}^{m} \equiv \begin{bmatrix} {}^{j}\boldsymbol{\alpha}^{m} \\ {}^{j}\boldsymbol{a}_{O}^{m} - {}^{j}\boldsymbol{\omega}^{m} \times {}^{j}\boldsymbol{v}_{O}^{m} \end{bmatrix}. \tag{5.30} $$

Definition 5.2 and Proposition 5.1 allow us to obtain the reduced acceleration state of a multibody mechanical system formed from three rigid bodies. In fact, substituting Eqs. (4.41), (4.50), (5.5), and (5.6) into Eq. (5.30) leads to

$$ {}^{j}\boldsymbol{A}_{O}^{m} = \begin{bmatrix} {}^{j}\boldsymbol{\alpha}^{k} + {}^{k}\boldsymbol{\alpha}^{m} + {}^{j}\boldsymbol{\omega}^{k} \times {}^{k}\boldsymbol{\omega}^{m} \\ \left({}^{j}\boldsymbol{a}_{O}^{k} + {}^{k}\boldsymbol{a}_{O}^{m} + 2{}^{j}\boldsymbol{\omega}^{k} \times {}^{k}\boldsymbol{v}_{O}^{m}\right) - \left({}^{j}\boldsymbol{\omega}^{k} + {}^{k}\boldsymbol{\omega}^{m}\right) \times \left({}^{j}\boldsymbol{v}_{O}^{k} + {}^{k}\boldsymbol{v}_{O}^{m}\right) \end{bmatrix}, \tag{5.31} $$

or

$$
{}^j\boldsymbol{A}_O^m = \begin{bmatrix} {}^j\boldsymbol{\alpha}^k \\ {}^j\boldsymbol{a}_O^k - {}^j\boldsymbol{\omega}^k \times {}^j\boldsymbol{v}_O^k \end{bmatrix} + \begin{bmatrix} {}^k\boldsymbol{\alpha}^m \\ {}^k\boldsymbol{a}_O^m - {}^k\boldsymbol{\omega}^m \times {}^k\boldsymbol{v}_O^m \end{bmatrix} + \begin{bmatrix} {}^j\boldsymbol{\omega}^k \times {}^k\boldsymbol{\omega}^m \\ {}^j\boldsymbol{\omega}^k \times {}^k\boldsymbol{v}_O^m - {}^k\boldsymbol{\omega}^m \times {}^j\boldsymbol{v}_O^k \end{bmatrix}. \tag{5.32}
$$

Therefore, taking into account that

$$
\begin{bmatrix} {}^j\boldsymbol{\alpha}^k \\ {}^j\boldsymbol{a}_O^k - {}^j\boldsymbol{\omega}^k \times {}^j\boldsymbol{v}_O^k \end{bmatrix} = {}^j\boldsymbol{A}_O^k, \tag{5.33}
$$

we meanwhile have

$$
\begin{bmatrix} {}^k\boldsymbol{\alpha}^m \\ {}^k\boldsymbol{a}_O^m - {}^k\boldsymbol{\omega}^m \times {}^k\boldsymbol{v}_O^m \end{bmatrix} = {}^k\boldsymbol{A}_O^m \tag{5.34}
$$

and

$$
\begin{bmatrix} {}^j\boldsymbol{\omega}^k \times {}^k\boldsymbol{\omega}^m \\ {}^j\boldsymbol{\omega}^k \times {}^k\boldsymbol{v}_O^m - {}^k\boldsymbol{\omega}^m \times {}^j\boldsymbol{v}_O^k \end{bmatrix} = \begin{bmatrix} {}^j\boldsymbol{V}_O^k & {}^k\boldsymbol{V}_O^m \end{bmatrix}. \tag{5.35}
$$

The substitution of Eqs. (5.33)–(5.35) into Eq. (5.32) leads to

$$
{}^j\boldsymbol{A}_O^m = {}^j\boldsymbol{A}_O^k + {}^k\boldsymbol{A}_O^m + \begin{bmatrix} {}^j\boldsymbol{V}_O^k & {}^k\boldsymbol{V}_O^m \end{bmatrix}. \tag{5.36}
$$

Equation (5.36) can be extended recursively to multibody mechanical systems. In other words, the number of bodies is not a limitation, which is formally proved in the next proposition.

Proposition 5.3. *Consider a multibody mechanical system composed of serially connected bodies labeled $j, j+1, \ldots, m-1, m$. The reduced acceleration state of body m as observed from body j is given by*

$$
\begin{aligned}
{}^j\boldsymbol{A}_O^m = {}^j\boldsymbol{A}_O^{j+1} + {}^{j+1}\boldsymbol{A}_O^{j+2} + \ldots + {}^{m-2}\boldsymbol{A}_O^{m-1} + {}^{m-1}\boldsymbol{A}_O^m + \\
\begin{bmatrix} {}^j\boldsymbol{V}_O^{j+1} & {}^{j+1}\boldsymbol{V}_O^m \end{bmatrix} + \begin{bmatrix} {}^{j+1}\boldsymbol{V}_O^{j+2} & {}^{j+2}\boldsymbol{V}_O^m \end{bmatrix} + \ldots + \begin{bmatrix} {}^{m-2}\boldsymbol{V}_O^{m-1} & {}^{m-1}\boldsymbol{V}_O^m \end{bmatrix}.
\end{aligned} \tag{5.37}
$$

Proof. If j and k are substituted, respectively, by $m-2$ and $m-1$ in Eq. (5.36), then one obtains

$$
{}^{m-2}\boldsymbol{A}_O^m = {}^{m-2}\boldsymbol{A}_O^{m-1} + {}^{m-1}\boldsymbol{A}_O^m + \begin{bmatrix} {}^{m-2}\boldsymbol{V}_O^{m-1} & {}^{m-1}\boldsymbol{V}_O^m \end{bmatrix}. \tag{5.38}
$$

On the other hand, by the induction hypothesis, the reduced acceleration state ${}^{j+1}\boldsymbol{A}_O^m$ may be written as follows:

$$
\begin{aligned}
{}^{j+1}\boldsymbol{A}_O^m = {}& {}^{j+1}\boldsymbol{A}_O^{j+2} + {}^{j+2}\boldsymbol{A}_O^{j+3} + \ldots + {}^{m-2}\boldsymbol{A}_O^{m-1} + {}^{m-1}\boldsymbol{A}_O^m \\
& + \left[{}^{j+1}\boldsymbol{V}_O^{j+2} \quad {}^{j+2}\boldsymbol{V}_O^m\right] + \ldots + \left[{}^{m-2}\boldsymbol{V}_O^{m-1} \quad {}^{m-1}\boldsymbol{V}_O^m\right].
\end{aligned} \tag{5.39}
$$

Furthermore, consider that $k = j + 1$ and substitute it into Eq. (5.36) to get

$$
{}^{j}\boldsymbol{A}_O^m = {}^{j}\boldsymbol{A}_O^{j+1} + {}^{j+1}\boldsymbol{A}_O^m + \left[{}^{j}\boldsymbol{V}_O^{j+1} \quad {}^{j+1}\boldsymbol{V}_O^m\right]. \tag{5.40}
$$

Finally, the substitution of Eq. (5.39) into Eq. (5.40) yields Eq. (5.37). ◇

The velocity state is a twist on a screw [see Eq. (4.63)]. At this point a natural question arises: Can it be the same for the reduced acceleration state? To answer this query, consider two rigid bodies k and l connected to each other by a helical pair ${}^k\$^l$ (see Fig. 4.2). Furthermore, consider that the reduced acceleration state ${}^k\boldsymbol{A}_O^l$ is given by

$$
{}^k\boldsymbol{A}_O^l = \begin{bmatrix} {}^k\boldsymbol{\alpha}^l \\ {}^k\boldsymbol{a}_O^l - {}^k\boldsymbol{\omega}^l \times {}^k\boldsymbol{v}_O^l \end{bmatrix}, \tag{5.41}
$$

where ${}^k\boldsymbol{\alpha}^l = {}_k\alpha_l {}^k\boldsymbol{s}^l$ [see Eq. (5.1)]. On the other hand, from the velocity analysis we know that

$$
{}^k\boldsymbol{v}_O^l = {}_kh_l {}^k\boldsymbol{\omega}^l + {}^k\boldsymbol{\omega}^l \times \boldsymbol{q}. \tag{5.42}
$$

Therefore, taking into account that ${}^k\boldsymbol{a}_O^l \equiv \frac{{}^kd}{dt}({}^k\boldsymbol{v}_O^l)$, then the linear acceleration ${}^k\boldsymbol{a}_O^l$ results in

$$
{}^k\boldsymbol{a}_O^l = {}_kh_l {}^k\boldsymbol{\alpha}^l + {}^k\boldsymbol{\alpha}^l \times \boldsymbol{q} + {}^k\boldsymbol{\omega}^l \times \left({}^k\boldsymbol{\omega}^l \times \boldsymbol{q}\right). \tag{5.43}
$$

However, from Eq. (5.42) we have

$$
{}^k\boldsymbol{\omega}^l \times \boldsymbol{q} = {}^k\boldsymbol{v}_O^l - {}_kh_l {}^k\boldsymbol{\omega}^l. \tag{5.44}
$$

Hence, substituting Eq. (5.44) into Eq. (5.43) leads to

$$
{}^k\boldsymbol{a}_O^l = {}_kh_l {}^k\boldsymbol{\alpha}^l + {}^k\boldsymbol{\alpha}^l \times \boldsymbol{q} + {}^k\boldsymbol{\omega}^l \times \left({}^k\boldsymbol{v}_O^l - {}_kh_l {}^k\boldsymbol{\omega}^l\right) \tag{5.45}
$$

or

$$
{}^k\boldsymbol{a}_O^l = {}_kh_l {}^k\boldsymbol{\alpha}^l + {}^k\boldsymbol{\alpha}^l \times \boldsymbol{q} + {}^k\boldsymbol{\omega}^l \times {}^k\boldsymbol{v}_O^l. \tag{5.46}
$$

Next, the dual part of the reduced acceleration state ${}^k\boldsymbol{A}_O^l$ results in

$$
{}^k\boldsymbol{a}_O^l - {}^k\boldsymbol{\omega}^l \times {}^k\boldsymbol{v}_O^l = {}_kh_l {}^k\boldsymbol{\alpha}^l + {}^k\boldsymbol{\alpha}^l \times \boldsymbol{q}. \tag{5.47}
$$

Hence, substituting Eq. (5.3) into Eq. (5.47) yields

$$ {}^{k}\boldsymbol{a}^{l}_{O} - {}^{k}\boldsymbol{\omega}^{l} \times {}^{k}\boldsymbol{v}^{l}_{O} = {}_{k}\alpha_{l}({}_{k}h_{l}{}^{k}\boldsymbol{s}^{l} + {}^{k}\boldsymbol{s}^{l} \times \boldsymbol{q}), \tag{5.48} $$

where

$$ {}_{k}h_{l}{}^{k}\boldsymbol{s}^{l} + {}^{k}\boldsymbol{s}^{l} \times \boldsymbol{q} = {}^{k}\boldsymbol{s}^{l}_{O}. \tag{5.49} $$

Thus, the dual part of the accelerator is given by

$$ {}^{k}\boldsymbol{a}^{l}_{O} - {}^{k}\boldsymbol{\omega}^{l} \times {}^{k}\boldsymbol{v}^{l}_{O} = {}_{k}\alpha_{l}{}^{k}\boldsymbol{s}^{l}_{O}. \tag{5.50} $$

Substituting expressions (5.3) and (5.50) into expression (5.41) yields

$$ {}^{k}\boldsymbol{A}^{l}_{O} = \begin{bmatrix} {}_{k}\alpha_{l}{}^{k}\boldsymbol{s}^{l} \\ {}_{k}\alpha_{l}{}^{k}\boldsymbol{s}^{l}_{O} \end{bmatrix} = {}_{k}\alpha_{l} \begin{bmatrix} {}^{k}\boldsymbol{s}^{l} \\ {}^{k}\boldsymbol{s}^{l}_{O} \end{bmatrix} = {}_{k}\alpha_{l}{}^{k}\$^{l}. \tag{5.51} $$

With these results in mind, the reduced acceleration state can be considered a twist on a screw involving the angular acceleration vector. The fact that the reduced acceleration state may be understood as a twist on a screw opened the possibility to extend screw theory to the so-called higher-order analyses of kinematic chains. This problem seemed to be unsolvable until the mid-1990s, when Rico and Duffy (1996) published a monumental work entitled *An Application of Screw Algebra to the Acceleration Analysis of Serial Chains.* Certainly, most contributions concerned with the subject have ensured that screw theory will be confined to velocity analysis because of the difficulty in expressing acceleration analysis in screw form.

The following proposition, a contribution of Rico and Duffy (1996), is one of the highlights of modern screw theory.

Proposition 5.4. *Let us consider a serial chain composed of $j, j+1, j+2, \ldots, m-2, m-1, m$ rigid bodies or links, where adjacent bodies are interconnected by helical pairs. The reduced acceleration state of body m with respect to body j may be expressed in screw form as follows:*

$$ \begin{aligned} {}^{j}\boldsymbol{A}^{m}_{O} = {}&{}_{j}\alpha_{j+1}{}^{j}\$^{j+1} + {}_{j+1}\alpha_{j+2}{}^{j+1}\$^{j+2} + \ldots + {}_{m-1}\alpha_{m}{}^{m-1}\$^{m} \\ &+ \left[{}_{j}\omega_{j+1}{}^{j}\$^{j+1} \quad {}_{j+1}\omega_{j+2}{}^{j+1}\$^{j+2} + \ldots + {}_{m-1}\omega_{m}{}^{m-1}\$^{m}\right] \\ &+ \left[{}_{j+1}\omega_{j+2}{}^{j+1}\$^{j+2} \quad {}_{j+2}\omega_{j+3}{}^{j+2}\$^{j+3} + \ldots + {}_{m-1}\omega_{m}{}^{m-1}\$^{m}\right] \\ &+ \ldots + \left[{}_{m-2}\omega_{m-1}{}^{m-2}\$^{m-1} \quad {}_{m-1}\omega_{m}{}^{m-1}\$^{m}\right], \end{aligned} \tag{5.52} $$

where ${}_{k}\alpha_{k+1} = \frac{d}{dt}{}_{k}\omega_{k+1}$ is the joint-acceleration rate between consecutive bodies k and $k+1$ interconnected by an infinitesimal screw or helical pair ${}^{k}\$^{l}$.

Proof. The proof is immediate by a systematic substitution of Eqs. (4.63), (4.78), and (5.51) into Eq. (5.37). ◇

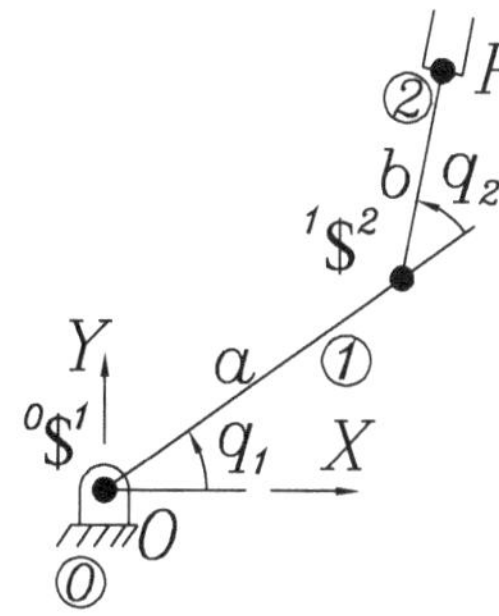

Fig. 5.2 Example 5.2. Planar two-degree-of-freedom serial manipulator

For simplicity, Eq. (5.52) may be rewritten as follows:

$$ {}^j\mathbf{A}_O^m = {}_j\alpha_{j+1}{}^j\$^{j+1} + {}_{j+1}\alpha_{j+2}{}^{j+1}\$^{j+2} + \dots + {}_{m-1}\alpha_m{}^{m-1}\$^m + {}^j\mathcal{A}_O^m, \tag{5.53} $$

where

$$ \begin{aligned} {}^j\mathcal{A}_O^m = {} & \left[{}_j\omega_{j+1}{}^j\$^{j+1} \quad {}_{j+1}\omega_{j+2}{}^{j+1}\$^{j+2} + \dots + {}_{m-1}\omega_m{}^{m-1}\$^m\right] \\ & + \left[{}_{j+1}\omega_{j+2}{}^{j+1}\$^{j+2} \quad {}_{j+2}\omega_{j+3}{}^{j+2}\$^{j+3} + \dots + {}_{m-1}\omega_m{}^{m-1}\$^m\right] \\ & + \dots + \left[{}_{m-2}\omega_{m-1}{}^{m-2}\$^{m-1} \quad {}_{m-1}\omega_m{}^{m-1}\$^m\right] \end{aligned} \tag{5.54} $$

is termed the *Lie screw of acceleration* of body m as measured from body j taking point O as the reference pole.

The generality and elegance of Eq. (5.53) are worth noting. Expression (5.53) represents the first formulation of the acceleration analysis of kinematic chains and therefore is a remarkable advance in the extension of screw theory from the so-called first-order to the higher-order analyses of kinematic chains. In particular, it was successfully applied in the acceleration analysis of a wide variety of robot manipulators with parallel kinematic topologies. However, it is interesting to mention that still today there is a certain lack of credibility among some kinematicians about the correctness of Eq. (5.53) even though it has been repeatedly verified with the aid of commercially available software.

Example 5.2. The example consists of computing the reduced acceleration state of the end-effector of the planar two-degree-of-freedom serial manipulator shown in Fig. 5.2. As an intermediate step, it is necessary to compute the velocity analysis of the robot. Although this book is devoted mainly to the analysis of parallel manipulators, the inclusion in some passages of the kinematic analysis of serial manipulators is necessary to clarify some key concepts concerning the analysis of parallel manipulators.

With reference to Fig. 5.2, the velocity state of the end-effector, body labeled 2, as measured from the base link, body labeled 0, considering P as the reference pole, the six-dimensional vector

$$\mathbf{V}_P = \begin{bmatrix} {}^0\boldsymbol{\omega}^2 \\ \boldsymbol{v}_P \end{bmatrix}$$

may be written in screw form as follows:

$$\dot{q}_1 {}^0\$^1 + \dot{q}_2 {}^1\$^2 = {}^0\boldsymbol{V}_P^2, \tag{5.55}$$

where $\dot{q}_i (i = 1, 2)$ are the generalized speeds of the robot. Furthermore, the Plücker coordinates of the screws ${}^0\$^1$ and ${}^1\$^2$ are given by

$${}^0\$^1 = \begin{bmatrix} 0 \\ 0 \\ 1 \\ c_1 \\ c_2 \\ 0 \end{bmatrix}, \quad {}^1\$^2 = \begin{bmatrix} 0 \\ 0 \\ 1 \\ c_3 \\ c_4 \\ 0 \end{bmatrix}, \tag{5.56}$$

where

- $c_1 = -a \sin q_1 - b \sin(q_1 + q_2)$;
- $c_2 = a \cos q_1 + b \cos(q_1 + q_2)$;
- $c_3 = -b \sin(q_1 + q_2)$;
- $c_4 = b \cos(q_1 + q_2)$.

Hence, the velocity state ${}^0\boldsymbol{V}_P^2$ results in

$${}^0\boldsymbol{V}_P^2 = \begin{bmatrix} 0 \\ 0 \\ \dot{q}_1 + \dot{q}_2 \\ c_1\dot{q}_1 + c_3\dot{q}_2 \\ c_2\dot{q}_1 + c_4\dot{q}_2 \\ 0 \end{bmatrix}. \tag{5.57}$$

Clearly, from the primal part of ${}^0\boldsymbol{V}_P^2$ the angular velocity vector results in ${}^0\boldsymbol{\omega}^2$:

$${}^0\boldsymbol{\omega}^2 = (\dot{q}_1 + \dot{q}_2)\hat{\boldsymbol{k}}. \tag{5.58}$$

Meanwhile, from the dual part of ${}^0\boldsymbol{V}_P^2$ it follows that the velocity of point P is given by

$$\boldsymbol{v}_P = (c_1\dot{q}_1 + c_3\dot{q}_2)\hat{\boldsymbol{i}} + (c_2\dot{q}_1 + c_4\dot{q}_2)\hat{\boldsymbol{j}}. \tag{5.59}$$

In what follows, the acceleration analysis of the mechanism is approached using screw theory. The reduced acceleration state of the end-effector 2 as observed from the base link 0 considering P as the reference pole, the vector

$$^{0}\boldsymbol{A}_{P}^{2} = \begin{bmatrix} ^{0}\boldsymbol{\alpha}^{2} \\ \boldsymbol{a}_{P} - {}^{0}\boldsymbol{\omega}^{2} \times \boldsymbol{v}_{P} \end{bmatrix}$$

may be expressed in screw form as follows:

$$\ddot{q}_1{}^0\$^1 + \ddot{q}_2{}^1\$^2 + \left[\dot{q}_1{}^0\$^1 \quad \dot{q}_2{}^1\$^2\right] = {}^0\boldsymbol{A}_P^2. \tag{5.60}$$

Here $\ddot{q}_i(i = 1, 2)$ are the generalized accelerations of the robot. After a few computations, Eq. (5.60) leads to

$$^{0}\boldsymbol{A}_{P}^{2} = \begin{bmatrix} 0 \\ 0 \\ \ddot{q}_1 + \ddot{q}_2 \\ c_1\ddot{q}_1 + c_3\ddot{q}_2 + (c_2 - c_4)\dot{q}_1\dot{q}_2 \\ c_2\ddot{q}_1 + c_4\ddot{q}_2 + (c_3 - c_1)\dot{q}_1\dot{q}_2 \\ 0 \end{bmatrix}. \tag{5.61}$$

Hence, the primal part of the accelerator $^0\boldsymbol{A}_P^2$ gives us the angular acceleration vector as

$$^{0}\boldsymbol{\alpha}^{2} = (\ddot{q}_1 + \ddot{q}_2)\hat{\boldsymbol{k}}. \tag{5.62}$$

Meanwhile, from the corresponding dual part of the reduced acceleration state $^0\boldsymbol{A}_P^2$, it follows that the acceleration vector of point P is given by

$$\begin{aligned} \boldsymbol{a}_P = {} & (c_1\ddot{q}_1 + c_3\ddot{q}_2 - 2c_4\dot{q}_1\dot{q}_2 - c_2\dot{q}_1^2 - c_4\dot{q}_2^2)\hat{\boldsymbol{i}} + \\ & (c_2\ddot{q}_1 + c_4\ddot{q}_2 + 2c_3\dot{q}_1\dot{q}_2 + c_1\dot{q}_1^2 + c_3\dot{q}_2^2)\hat{\boldsymbol{j}}. \end{aligned} \tag{5.63}$$

It must be noted that the acceleration $\boldsymbol{a}_P$ was obtained by adding the cross product $^0\boldsymbol{\omega}^2 \times \boldsymbol{v}_P$ to the dual part of the reduced acceleration state $^0\boldsymbol{A}_P^2$; for details, see Eq. (5.20).

Example 5.2 gives us an important clue to simplify the kinematic analysis of mechanisms when screw theory is used: the proper selection of the reference pole.

Example 5.3. The Geneva wheel of Example 4.7, Chap. 4, is reconsidered here. We need to compute the angular acceleration of the driven wheel D while taking into account that the driver wheel A is rotating at the constant angular velocity $\omega_1 = 2$ rad/s clockwise. To do this, consider the data and numerical results obtained from the corresponding velocity analysis.

Solution. For clarity, the figure of the Geneva wheel as well as the results of the velocity analysis are reproduced in Fig. 5.3.

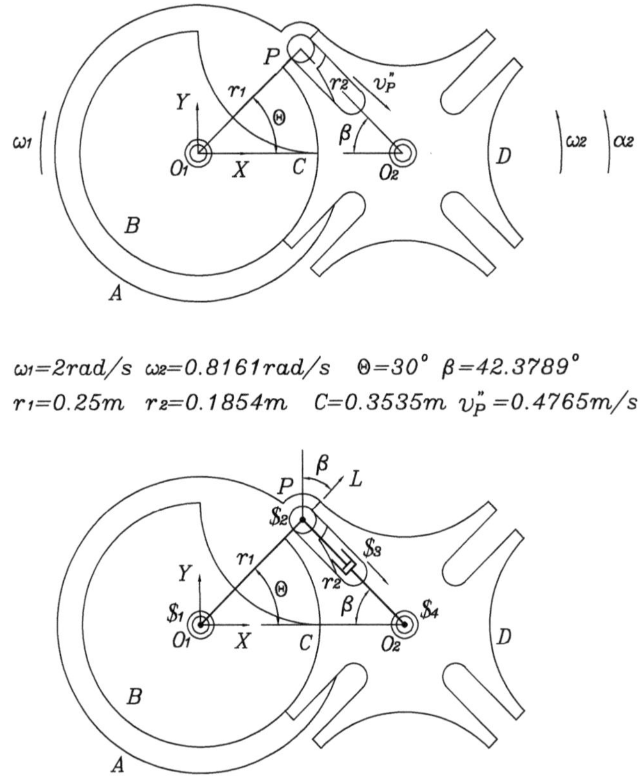

Fig. 5.3 Example 5.3. Acceleration analysis of the Geneva wheel

One classical solution of the acceleration analysis of the Geneva wheel is as follows. Using elementary kinematics, since the angular velocity ω_1 of the driver wheel is constant, then the acceleration of pin P is given by

$$\boldsymbol{a}_P = \boldsymbol{\omega}_1 \times (\boldsymbol{\omega}_1 \times \boldsymbol{r}_{P/O_1}). \tag{5.64}$$

Furthermore, the acceleration vector $\boldsymbol{a}_P$ may be obtained upon the driven wheel as

$$\boldsymbol{a}_P = \boldsymbol{a}'_P + \boldsymbol{a}''_P + 2\boldsymbol{\omega}_2 \times \boldsymbol{v}''_P, \tag{5.65}$$

where $\boldsymbol{a}'_P$ is the acceleration of the point of body D which is instantaneously coincident with point P. Meanwhile, $\boldsymbol{a}''_P$ is the acceleration of point P relative to the wheel D and $2\boldsymbol{\omega}_2 \times \boldsymbol{v}''_P$ is the Coriolis acceleration. Indeed,

$$\boldsymbol{a}'_P = \boldsymbol{\alpha}_2 \times \boldsymbol{r}_{P/O_2} + \boldsymbol{\omega}_2 \times (\boldsymbol{\omega}_2 \times \boldsymbol{r}_{P/O_2}), \quad \boldsymbol{a}''_P = a''_P \hat{\boldsymbol{u}}_{P/O_2}. \tag{5.66}$$

Hence, equating (5.64) and (5.65), we find that

$$\boldsymbol{\omega}_1 \times (\boldsymbol{\omega}_1 \times \boldsymbol{r}_{P/O_1}) = \boldsymbol{\alpha}_2 \times \boldsymbol{r}_{P/O_2} + \boldsymbol{\omega}_2 \times (\boldsymbol{\omega}_2 \times \boldsymbol{r}_{P/O_2}) + a''_P \hat{\boldsymbol{u}}_{P/O_2} + 2\boldsymbol{\omega}_2 \times \boldsymbol{v}''_P. \quad (5.67)$$

After a few computations, substituting data in Eq. (5.67) leads to a linear system of two equations in the unknowns α_2 and a''_P as follows:

$$0.125\alpha_2 - 0.7387a''_P = 1.481, \qquad 0.137\alpha_2 + 0.674a''_P = 0.991. \quad (5.68)$$

Solving Eq. (5.68), we get $\alpha_2 = 9.333\,\text{rad/s}^2$ and $a''_P = -0.426\,\text{m/s}^2$.

In what follows the acceleration analysis of the Geneva wheel is approached by screw theory.

According to the closed kinematic chain O_1PO_2, it is evident that the equation of acceleration in screw form of the Geneva wheel may be written as follows:

$$\dot{v}_1\$_1 + \dot{v}_2\$_2 + \dot{v}_3\$_3 + \dot{v}_4\$_4 + \mathcal{A} = \mathbf{0}, \quad (5.69)$$

where the Lie screw of acceleration $\mathcal{A}$ is given by

$$\mathcal{A} = \left[v_1\$_1 \quad v_2\$_2 + v_3\$_3 + v_4\$_4\right] + \left[v_2\$_2 \quad v_3\$_3 + v_4\$_4\right] + \left[v_3\$_3 \quad v_4\$_4\right]. \quad (5.70)$$

However, from the velocity analysis it is evident that

$$v_2\$_2 + v_3\$_3 + v_4\$_4 = -v_1\$_1. \quad (5.71)$$

Thus, we write

$$\left[v_1\$_1 \quad v_2\$_2 + v_3\$_3 + v_4\$_4\right] = \left[v_1\$_1 \quad -v_1\$_1\right] = \mathbf{0}. \quad (5.72)$$

Hence, the Lie screw of acceleration $\mathcal{A}$ is reduced to

$$\mathcal{A} = \left[v_2\$_2 \quad v_3\$_3 + v_4\$_4\right] + \left[v_3\$_3 \quad v_4\$_4\right]. \quad (5.73)$$

To compute the angular acceleration of the driven wheel, we apply the Klein form of $\boldsymbol{L}$ to both sides of Eq. (5.69) and take into account that $\{\$_2; \boldsymbol{L}\} = \{\$_3; \boldsymbol{L}\} = 0$; this leads to

$$\dot{v}_1 \{\$_1; \boldsymbol{L}\} + \dot{v}_4 \{\$_4; \boldsymbol{L}\} + \{\mathcal{A}; \boldsymbol{L}\} = 0. \quad (5.74)$$

On the other hand, let us consider that the angular acceleration of the driver wheel vanishes because it is rotating with constant angular velocity, that is, $\dot{v}_1 = -\alpha_1 = 0$. Furthermore, it is evident that $\alpha_2 = -\dot{v}_4$. Thus, the angular acceleration α_2 results in

$$\alpha_2 = \frac{\{\mathcal{A}; \boldsymbol{L}\}}{\{\$_4; \boldsymbol{L}\}}. \quad (5.75)$$

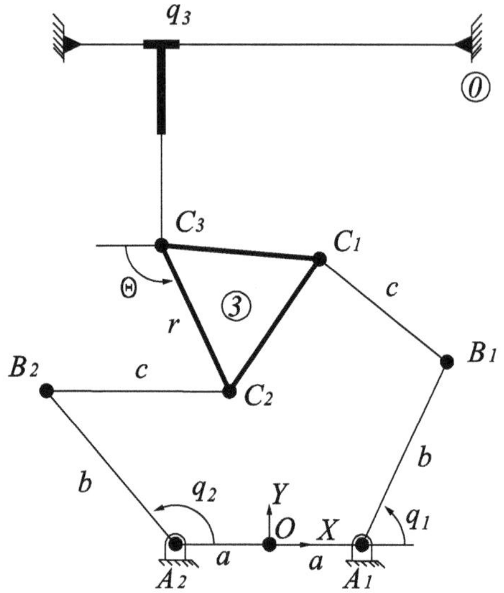

Fig. 5.4 Example 5.4. 2RRR+PPR planar parallel manipulator and its geometric scheme

To compute α_2, consider that from the velocity analysis we have $v_1 = -2\,\text{rad/s}$, $v_2 = -1.4080v_1$, $v_3 = 0.2382v_1$, and $v_4 = 0.4080v_1$. Thus, the substitution of data in Eq. (5.75) yields $\alpha_2 = 9.333\,\text{rad/s}^2$.

Example 5.4. With reference to Fig. 5.4, the example consists of solving the acceleration analysis of a 2RRR+PPR (where R and P stand for revolute and prismatic joint, respectively) three-degree-of-freedom planar parallel manipulator. For the sake of completeness, the displacement and velocity analyses of the robot are also included. Originally, the author conceived of this robot manipulator with two purposes in mind: (1) simplifying the forward displacement analysis of the 3RRR planar parallel manipulator; (2) proposing a kinematically partially decoupled new parallel manipulator.

Before starting the kinematic analysis of the chosen mechanism, it is interesting to exemplify how the kinematic characteristics of a robot manipulator can be improved by reconsidering its topology. To this end, the characteristics of the two-degree-of-freedom planar parallel manipulator introduced by Nam and Park (2006) are investigated.

The left mechanism of Fig. 5.5 was introduced by Nam and Park (2006) and is composed of a moving platform (3) connected to the fixed platform (0) by two R̲RR-type active legs, where actuated revolute joints are underlined, and one RP-type central passive kinematic chain. The mobility of the 2R̲RR+RP parallel manipulator is explained as one translation, vector $\boldsymbol{v}$, along the passive leg and one rotation, angle θ, of the moving platform. Hence, arbitrary cylindrical curves of the moving platform are available for this robot manipulator within the workspace. It is difficult to visualize practical applications of such a mechanism because of the variable orientation of the central kinematic chain. An improved mechanism may be obtained

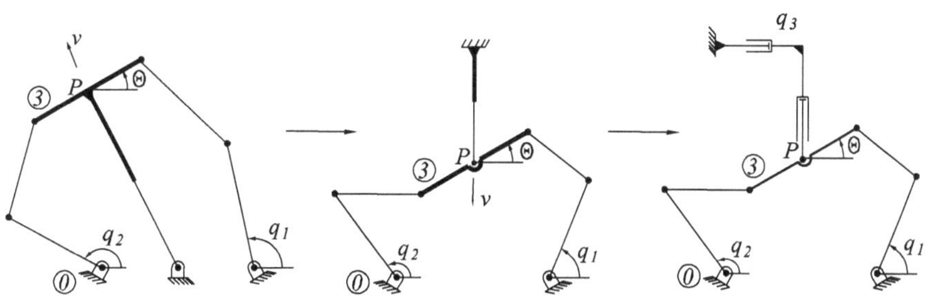

2RRR+RP parallel manipulator 2RRR+PR parallel manipulator 2RRR+PPR parallel manipulator

Fig. 5.5 The gradual improvement of a planar parallel manipulator

by an inversion of the central leg, yielding a 2RRR+PR parallel manipulator, central mechanism of Fig. 5.5, where the vector $\boldsymbol{v}$ keeps a constant orientation with respect to the fixed platform, and therefore its possible applications involving the inverse kinematics are clearer than that selected for the original mechanism. Furthermore, to generate a three-degree-of-freedom parallel manipulator, the passive PR-type chain could be substituted by an active PPR-type leg where the actuated prismatic joint is underlined, right mechanism of Fig. 5.5. This manipulator is close to the PreXYT robot (Yu et al. 2008). This brief note shows how the performance of a limited robot manipulator can be improved by introducing simple changes.

After this simple reflection, in what follows the kinematics of the parallel manipulator shown in Fig. 5.4 is presented. The robot of the example is equipped with two actuated revolute joints, q_1 and q_2, and one actuated prismatic joint, q_3. The moving platform is an equilateral triangle of side r connected to the limbs by revolute joints whose nominal positions are denoted by points $C_i (i = 1, 2, 3)$ located, respectively, by vectors $\boldsymbol{c}_i (i = 1, 2, 3)$. Point P, located by vector $\boldsymbol{p}$, denotes the geometric center of the moving platform and is named the control point of the mechanism, which is calculated as

$$\boldsymbol{p} = (\boldsymbol{c}_1 + \boldsymbol{c}_2 + \boldsymbol{c}_3)/3. \tag{5.76}$$

The pose of the moving platform is completed considering the inclination angle θ. Finally, the link lengths of the RRR-type connector chain are denoted by parameters a, b, and c.

In order to approach the displacement analysis of the robot, consider that XY is a reference frame attached to the fixed platform whose origin O is located at the middle of the line connecting points A_1 and A_2. Furthermore, to formulate the closure equations of the robot, consider that $C_3 = (q_3, y)$. Thus, provided that the moving platform is an equilateral triangle of side r, we have

$$C_2 = (q_3 - r\cos\theta, y - r\sin\theta) \tag{5.77}$$

and

$$C_1 = (q_3 - r\cos(\theta + \pi/3), y - r\sin(\theta + \pi/3)). \tag{5.78}$$

The forward displacement analysis of the manipulator consists of determining the pose, coordinates of point P and angle θ, of the moving platform as measured from the fixed platform given the generalized coordinates $q_i (i = 1, \ldots, 3)$. Thus, to determine the unknowns y and θ we may write two closure equations as follows:

$$(\boldsymbol{c}_i - \boldsymbol{b}_i) \bullet (\boldsymbol{c}_i - \boldsymbol{b}_i) = c^2, \quad i = 1, 2. \tag{5.79}$$

After a few computations, the unknown y is easily obtained in terms of $\sin\theta$ and $\cos\theta$. Then the substitution of y into any of the expressions (5.79) yields a nonlinear equation in the unknown θ as follows:

$$K_1\cos^2(\theta)\sin(\theta) + K_2\cos(\theta)\sin(\theta) + K_3\cos^3(\theta) + K_4\cos^2(\theta) + K_5\cos(\theta) + K_6\sin(\theta) + K_7 = 0, \tag{5.80}$$

where the coefficients $K_i (i = 1, 2, \ldots, 7)$ are computed according to the parameters and generalized coordinates of the mechanism. To solve Eq. (5.80) from the tangent half-angle

$$t = \tan\left(\frac{\theta}{2}\right), \tag{5.81}$$

we have

$$\sin(\theta) = \frac{2t}{1+t^2}, \quad \cos(\theta) = \frac{1-t^2}{1+t^2}. \tag{5.82}$$

With the substitution of these identities into Eq. (5.80), a sixth-order polynomial equation in the unknown t is obtained as

$$\sum_{i=0}^{6} k_i t^i = 0, \tag{5.83}$$

where the coefficients k_i are given by

- $k_0 = K_3 + K_4 + K_5 + K_7$;
- $k_1 = 2(K_1 + K_2 + K_6)$;
- $k_2 = -3K_3 - K_4 + K_5 + 3K_7$;
- $k_3 = 4(-K_1 + K_6)$;
- $k_4 = 3K_3 - K_4 - K_5 + 3K_7$;
- $k_5 = 2(K_1 - K_2 + K_6)$;
- $k_6 = -K_3 + K_4 - K_5 + K_7$.

Equation (5.83) is termed *the characteristic equation of the manipulator*, which indicates that the moving platform can reach at most six different poses with respect to the fixed platform given a set of generalized coordinates. It is worth noting how easy it is to solve the forward displacement analysis of the robot at hand when compared with the forward displacement analysis of classical nonredundant parallel manipulators; see, for instance, Gosselin et al. (1992). Furthermore, still some authors prefer today to implement a numerical technique to complete the forward displacement analysis of planar parallel manipulators (Zubizarreta et al. 2012). On the other hand, the inverse displacement analysis consists of finding the generalized coordinates $q_i (i = 1, 2, 3, 4)$ given the pose of the moving platform with respect to the fixed platform. Note that the computation of the coordinates of points $C_i (i = 1, 2, 3)$ is straightforward given P and θ. Furthermore, once point $C_3 = (x, y)$ is computed, one obtains $q_3 = x$. In order to determine the angles q_1, q_2, compute the angle q_1 by solving the triangle $A_1B_1C_1$; the procedure is straightforward. Clearly, two values are available for q_1. Similarly, the angle q_2 is obtained by solving the triangle $A_2B_2C_2$.

Once the displacement analysis of the robot was solved, the input–output equations of velocity and acceleration of the robot are systematically obtained by resorting to reciprocal screw theory. To satisfy an algebraic requirement, the 2(R̲RRR)+P̲PR parallel manipulator is considered a 2(R*R̲RS)+C̲PS robot, where R* as well as the spherical and cylindrical joints are virtual kinematic pairs; that is, selected joint rates of these imaginary kinematic pairs are null. Next, the model of the screws is depicted in Fig. 5.6. The screws associated to imaginary kinematic pairs between consecutive bodies k and $k+1$ in the ith limb are notated as ${}^{k}S_i^{k+1}$. Furthermore, $\boldsymbol{L}$ denotes a line in Plücker coordinates introduced to simplify the infinitesimal kinematics analyses of the robot.

Let $\boldsymbol{\omega} = \dot{\theta}\hat{\boldsymbol{k}}$ be the angular velocity vector of the moving platform as measured from the fixed platform and let $\boldsymbol{v}_P = v_X\hat{\boldsymbol{i}} + v_Y\hat{\boldsymbol{j}}$ be the velocity of point P, namely the reference pole. The velocity state of the moving platform as measured from the fixed platform, the vector

$$\boldsymbol{V}_P = \begin{bmatrix} \boldsymbol{\omega} \\ \boldsymbol{v}_P \end{bmatrix},$$

may be written in screw form through the R*R̲RS-type kinematic chains as follows:

$${}_0\omega_1^i\,{}^0\$_i^1 + {}_1\omega_2^i\,{}^1\$_i^2 + {}_2\omega_3^i\,{}^2\$_i^3 + {}_3\omega_4^i\,{}^3\$_i^4 + {}_4\omega_5^i\,{}^4S_i^5 + {}_5\omega_6^i\,{}^5S_i^6 = \boldsymbol{V}_P, \quad i = 1, 2. \tag{5.84}$$

In this equation, the virtual joint rates are null, for example, ${}_0\omega_1^i = {}_4\omega_5^i = {}_5\omega_6^i = 0$. Meanwhile, ${}_1\omega_2^i = \dot{q}_i$ is the ith generalized speed. The systematic application of the Klein form of the ith line $\boldsymbol{L}_i$ to both sides of Eq. (5.84), after reducing terms, leads to

$$\dot{q}_i \left\{ {}^1\$_i^2; \boldsymbol{L}_i \right\} = \left\{ \boldsymbol{V}_P; \boldsymbol{L}_i \right\}. \tag{5.85}$$

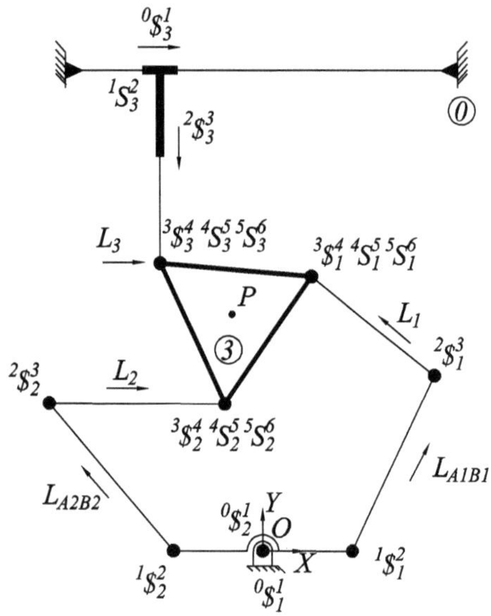

Fig. 5.6 2RRR+PPR planar parallel manipulator. Infinitesimal screws

Similarly, the velocity state $\boldsymbol{V}_P$ may be written in screw form through the C̲PS-type kinematic chain as follows:

$$ {}_0\omega_1^3\,{}^0\$_3^1 + {}_1\omega_2^3\,{}^1S_3^2 + {}_2\omega_3^3\,{}^2\$_3^3 + {}_3\omega_4^3\,{}^3\$_3^4 + {}_4\omega_5^3\,{}^4S_3^5 + {}_5\omega_6^3\,{}^5S_3^6 = \boldsymbol{V}_P. \tag{5.86}$$

Here ${}_0\omega_1^3 = \dot{q}_3$ while ${}_1\omega_2^3 = {}_4\omega_5^3 = {}_5\omega_6^3 = 0$. Applying the Klein form of the line $\boldsymbol{L}_3$ to both sides of Eq. (5.86), after reducing terms, leads to

$$\dot{q}_3 \left\{{}^0\$_3^1; \boldsymbol{L}_3\right\} = \left\{\boldsymbol{V}_P; \boldsymbol{L}_3\right\}. \tag{5.87}$$

Let us consider that the dimension of the lines $\boldsymbol{L}_i$ and the velocity state $\boldsymbol{V}_P$ must be consistent; for example, they must be three-dimensional vectors. Gathering these results, we find that the input–output equation of velocity of the manipulator results in

$$\tilde{\mathbf{A}}\boldsymbol{V}_P = \tilde{\mathbf{B}}\dot{\mathbf{q}}. \tag{5.88}$$

Here

$$\tilde{\mathbf{A}} = \mathbf{J}^T\boldsymbol{\Delta} \tag{5.89}$$

is named the active matrix of the robot in which

$$\mathbf{J} = \left[\boldsymbol{L}_1\ \boldsymbol{L}_2\ \boldsymbol{L}_3\right] \tag{5.90}$$

is the 3×3 Jacobian matrix. Meanwhile,

$$\mathbf{\Delta} = \begin{bmatrix} 0 & 1 & 0 \\ 0 & 0 & 1 \\ 1 & 0 & 0 \end{bmatrix}$$

is an operator of polarity. Furthermore,

$$\dot{\mathbf{q}} = \begin{bmatrix} \dot{q}_1 & \dot{q}_2 & \dot{q}_3 \end{bmatrix}^T \tag{5.91}$$

is the first-order driver matrix of the manipulator, while matrix $\tilde{\mathbf{B}}$ is the first-order coefficient matrix of the robot, which is given by

$$\tilde{\mathbf{B}} = diag\left[\left\{{}^1\$_1^2; \boldsymbol{L}_1\right\} \left\{{}^1\$_2^2; \boldsymbol{L}_2\right\} \left\{{}^0\$_3^1; \boldsymbol{L}_3\right\}\right]. \tag{5.92}$$

The forward velocity analysis consists of computing the velocity state $\boldsymbol{V}_P$ for a prescribed set of generalized speeds $\dot{q}_i (i = 1, \ldots, 3)$. To this end, matrix $\tilde{\mathbf{A}}$ must be invertible. On the other hand, the inverse velocity analysis consists of computing the joint-velocity rates $\dot{q}_i (i = 1, \ldots, 3)$ given the velocity state $\boldsymbol{V}_P$.

In what follows the acceleration analysis of the example is presented. Let $\boldsymbol{\alpha} = \ddot{\theta}\hat{\boldsymbol{k}}$ be the angular acceleration vector of the moving platform as measured from the fixed platform, and let $\boldsymbol{a}_P = a_X\hat{\boldsymbol{i}} + a_Y\hat{\boldsymbol{j}}$ be the acceleration of point P. The reduced acceleration state of the moving platform as measured from the fixed platform, the vector

$$\boldsymbol{A}_P = \begin{bmatrix} \boldsymbol{\alpha} \\ \boldsymbol{a}_P - \boldsymbol{\omega} \times \boldsymbol{v}_P \end{bmatrix},$$

may be written in screw form through the R*<u>R</u>RS-type kinematic chains as follows:

$${}_0\dot{\omega}_1^i\,{}^0\$_i^1 + {}_1\dot{\omega}_2^i\,{}^1\$_i^2 + {}_2\dot{\omega}_3^i\,{}^2\$_i^3 + {}_3\dot{\omega}_4^i\,{}^3\$_i^4 + {}_4\dot{\omega}_5^i\,{}^4S_i^5 + {}_5\dot{\omega}_6^i\,{}^5S_i^6 + \tilde{\boldsymbol{L}}_i = \boldsymbol{A}_P, \quad i = 1, 2. \tag{5.93}$$

Here ${}_0\dot{\omega}_1^i = {}_4\dot{\omega}_5^i = {}_5\dot{\omega}_6^i = 0$, while ${}_1\dot{\omega}_2^i = \ddot{q}_i$. Furthermore, the ith Lie screw of acceleration $\tilde{\boldsymbol{L}}_i$ is given by

$$\begin{aligned} \tilde{\boldsymbol{L}}_i = & \left[{}_0\omega_1^i\,{}^0\$_i^1 \quad {}_1\omega_2^i\,{}^1\$_i^2 + {}_2\omega_3^i\,{}^2\$_i^3 + \ldots + {}_5\omega_6^i\,{}^5S_i^6\right] \\ & + \left[{}_1\omega_2^i\,{}^1\$_i^2 \quad {}_2\omega_3^i\,{}^2\$_i^3 + {}_3\omega_4^i\,{}^3\$_i^4 + \ldots + {}_5\omega_6^i\,{}^5S_i^6\right] \\ & + \ldots + \left[{}_4\omega_5^i\,{}^4S_i^5 \quad {}_5\omega_6^i\,{}^5S_i^6\right]. \end{aligned} \tag{5.94}$$

Taking into account that ${}_0\omega_1^i = {}_4\omega_5^i = {}_5\omega_6^i = 0$ and ${}_0\omega_1^i = \dot{q}_i$, the six-dimensional vector $\tilde{\boldsymbol{L}}_i$ is reduced to

$$\begin{aligned} \tilde{\boldsymbol{L}}_i = & \left[\dot{q}_i\,{}^1\$_i^2 \quad {}_2\omega_3^i\,{}^2\$_i^3 + {}_3\omega_4^i\,{}^3\$_i^4\right] \\ & + \left[{}_2\omega_3^i\,{}^2\$_i^3 \quad {}_3\omega_4^i\,{}^3\$_i^4\right]. \end{aligned} \tag{5.95}$$

Furthermore, the passive joint-velocity rate ${}_3\omega_4^i$ is computed as

$$ {}_3\omega_4^i = \frac{\{\boldsymbol{V}_P;\boldsymbol{L}_{A_iB_i}\}}{\{{}^3\$_i^4;\boldsymbol{L}_{A_iB_i}\}}, \tag{5.96}$$

where $\boldsymbol{L}_{A_iB_i}$ is the line in Plücker coordinates pointing from point A_i to point B_i. Meanwhile, the passive joint-velocity rate ${}_2\omega_3^i$ is computed according to

$$ \dot{q}_i + {}_2\omega_3^i + {}_3\omega_4^i = \boldsymbol{\omega}. \tag{5.97}$$

Similarly, the reduced acceleration state $\boldsymbol{A}_P$ may be written in screw form through the C̲PS-type kinematic chain as follows:

$$ {}_0\dot{\omega}_1^3\,{}^0\$_3^1 + {}_1\dot{\omega}_2^3\,{}^1S_3^2 + {}_2\dot{\omega}_3^3\,{}^2\$_3^3 + {}_3\dot{\omega}_4^3\,{}^3\$_3^4 + {}_4\dot{\omega}_5^3\,{}^4S_3^5 + {}_5\dot{\omega}_6^3\,{}^5S_3^6 + \tilde{\boldsymbol{L}}_3 = \boldsymbol{A}_P. \tag{5.98}$$

Here ${}_0\dot{\omega}_1^3 = \ddot{q}_3$, while ${}_1\dot{\omega}_2^3 = {}_4\dot{\omega}_5^3 = {}_5\dot{\omega}_6^3 = 0$. Furthermore, the Lie screw of acceleration $\tilde{\boldsymbol{L}}_3$ is given by

$$\begin{aligned} \tilde{\boldsymbol{L}}_3 = & \left[{}_0\omega_1^3\,{}^0\$_3^1 \quad {}_1\omega_2^3\,{}^1S_3^2 + {}_2\omega_3^3\,{}^2\$_3^3 + \ldots + {}_5\omega_6^3\,{}^5S_3^6\right] \\ & + \left[{}_1\omega_2^3\,{}^1S_3^2 \quad {}_2\omega_3^3\,{}^2\$_3^3 + \ldots + {}_5\omega_6^3\,{}^5S_3^6\right] \\ & + \ldots + \left[{}_4\omega_5^3\,{}^4S_3^5 \quad {}_5\omega_6^3\,{}^5S_3^6\right]. \end{aligned} \tag{5.99}$$

Taking into account that ${}_0\omega_1^3 = \dot{q}_3$ and ${}_4\omega_5^3 = {}_5\omega_6^3 = 0$, the six-dimensional vector $\tilde{\boldsymbol{L}}_3$ is reduced to

$$ \tilde{\boldsymbol{L}}_3 = \left[\dot{q}_3\,{}^0\$_3^1 \quad {}_2\omega_3^3\,{}^2\$_3^3 + {}_3\omega_4^3\,{}^3\$_3^4\right] + \left[{}_2\omega_3^3\,{}^2\$_3^3 \quad {}_3\omega_4^3\,{}^3\$_3^4\right], \tag{5.100}$$

where the passive joint-velocity rates are computed based on the following expression:

$$ \dot{q}_3\,{}^0\$_3^1 + {}_2\omega_3^3\,{}^2\$_3^3 + {}_3\omega_4^3\,{}^3\$_3^4 = \boldsymbol{V}_P. \tag{5.101}$$

Finally, following the trend of the velocity analysis, the input–output equation of acceleration results in

$$ \tilde{\mathbf{A}}\boldsymbol{A}_P = \tilde{\mathbf{B}}\ddot{\mathbf{q}} + \tilde{\mathbf{C}}, \tag{5.102}$$

where

$$ \ddot{\mathbf{q}} = \left[\ddot{q}_1\ \ddot{q}_2\ \ddot{q}_3\right]^T \tag{5.103}$$

is the second-order driver matrix of the manipulator. Meanwhile,

$$ \tilde{\mathbf{C}} = \left[\left\{\tilde{\boldsymbol{L}}_1;\boldsymbol{L}_1\right\}\ \left\{\tilde{\boldsymbol{L}}_2;\boldsymbol{L}_2\right\}\ \left\{\tilde{\boldsymbol{L}}_3;\boldsymbol{L}_3\right\}\right]^T \tag{5.104}$$

is the complementary matrix of acceleration. Since the terms of the Coriolis acceleration are self-contained in matrix $\tilde{\mathbf{C}}$, it could be named something like the Coriolis/centripetal matrix.

Example 5.5. This example shows that although rigid bodies are physically disconnected, it is possible, in some cases, to simulate the kinematic analysis of multibody mechanical systems by introducing kinematically equivalent mechanisms. With reference to Fig. 5.7, truck A is rounding a curve of radius $r_1 = 120\,\text{m}$ at a constant speed of 70 km/h while truck B is rounding a curve of radius $r_2 = 130\,\text{m}$ at a constant speed of 75 km/h. We must determine the velocity and acceleration that car B appears to have for the driver of car A when $\theta_A = 30°$ and $\theta_B = 45°$.

Solution. The standard solution is presented first. To this end, let XY be a reference frame attached to the center of curve 1, the point O_1, with associated unit vectors $\hat{i}\hat{j}$. If we use college kinematics, the velocity of truck B as observed from body A, the vector $\boldsymbol{v}_{B/A}$, is given by

$$\begin{aligned} \boldsymbol{v}_{B/A} &= \boldsymbol{v}_B - \boldsymbol{v}_A \\ &= v_B\left[\sin(\theta_B)\hat{\boldsymbol{i}} + \cos(\theta_B)\hat{\boldsymbol{j}}\right] - v_A\left[\sin(\theta_A)\hat{\boldsymbol{i}} + \cos(\theta_A)\hat{\boldsymbol{j}}\right] \\ &= [v_B\sin(\theta_B) - v_A\sin(\theta_A)]\,\hat{\boldsymbol{i}} + [v_B\cos(\theta_B) - v_A\cos(\theta_A)]\hat{\boldsymbol{j}}, \end{aligned} \tag{5.105}$$

where $v_A = 70\,\text{km/h} = 19.44\,\text{m/s}$, $v_B = 75\,\text{km/h} = 20.83\,\text{m/s}$, $\theta_A = 30°$, $\theta_B = 45°$. Thus, it follows that $\boldsymbol{v}_{B/A} = 5\hat{\boldsymbol{i}} - 2.101\hat{\boldsymbol{j}}$, or $v_{B/A} = 5.423\,\text{m/s}$. On the other hand, assuming that both trucks move at a constant velocity, the acceleration of truck B as observed from truck A, the vector $\boldsymbol{a}_{B/A}$, is given by

$$\boldsymbol{a}_{B/A} = \boldsymbol{a}_B - \boldsymbol{a}_A.$$

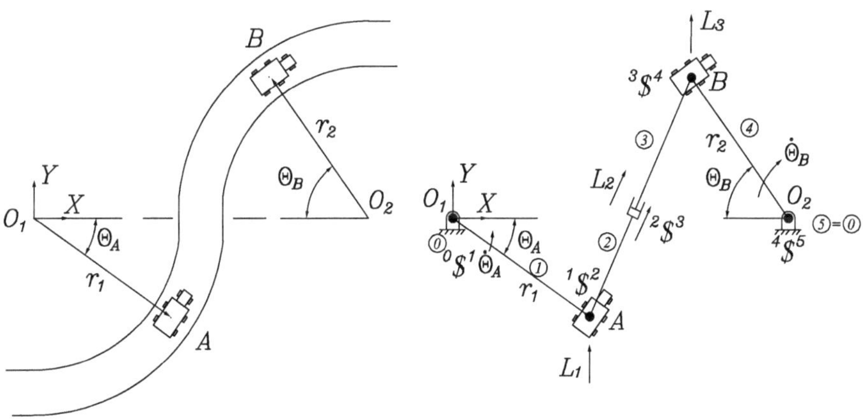

Fig. 5.7 Example 5.5. Trucks in relative motion and its kinematically equivalent mechanism

Hence,

$$\begin{aligned} \boldsymbol{a}_{B/A} &= \frac{v_B^2}{r_2}\left[\cos(\theta_B)\hat{\boldsymbol{i}} - \sin(\theta_B)\hat{\boldsymbol{j}}\right] - \frac{v_A^2}{r_1}\left[-\cos(\theta_A)\hat{\boldsymbol{i}} + \sin(\theta_A)\hat{\boldsymbol{j}}\right] \\ &= \left[\frac{v_A^2}{r_1}\cos(\theta_A) + \frac{v_B^2}{r_2}\cos(\theta_B)\right]\hat{\boldsymbol{i}} - \left[\frac{v_A^2}{r_1}\sin(\theta_A) + \frac{v_B^2}{r_2}\sin(\theta_B)\right]\hat{\boldsymbol{j}}, \end{aligned} \tag{5.106}$$

where $r_1 = 120\,\text{m}$ and $r_2 = 130\,\text{m}$. Then one obtains $\boldsymbol{a}_{B/A} = 5.089\hat{\boldsymbol{i}} - 3.936\hat{\boldsymbol{j}}$ or $a_{B/A} = 6.433\,\text{m/s}^2$.

In what follows the velocity and acceleration analyses of Example 5.5 are solved using screw theory. To help, the multibody system is modeled as an RRPRR closed kinematic chain. Furthermore, unless otherwise specified, point O_1 is chosen as the reference pole.

The equation of the closed kinematic chain may be written in screw form as follows:

$${}_0\omega_1{}^0\$^1 + {}_1\omega_2{}^1\$^2 + {}_2\omega_3{}^2\$^3 + {}_3\omega_4{}^3\$^4 + {}_4\omega_5{}^4\$^5 = \boldsymbol{0}, \tag{5.107}$$

where ${}_0\omega_1 = \dot{\theta}_A = v_A/r_1 = 0.162\,\text{rad/s}$ and ${}_4\omega_5 = \dot{\theta}_B = v_B/r_2 = 0.154\,\text{rad/s}$. Meanwhile, the Plücker coordinates of the screws are given by

$$\begin{aligned} {}^0\$^1 &= \begin{bmatrix} 0 \\ 0 \\ 1 \\ 0 \\ 0 \\ 0 \end{bmatrix}, \quad {}^1\$^2 = \begin{bmatrix} 0 \\ 0 \\ 1 \\ -r_1\sin(\theta_A) \\ -r_1\cos(\theta_A) \\ 0 \end{bmatrix}, \quad {}^2\$^3 = \begin{bmatrix} 0 \\ 0 \\ 0 \\ \hat{\boldsymbol{u}}_{B/A} \end{bmatrix}, \\ {}^3\$^4 &= \begin{bmatrix} 0 \\ 0 \\ 1 \\ r_2\sin(\theta_B) \\ r_2\cos(\theta_B) - r_1 - r_2 \\ 0 \end{bmatrix}, \quad {}^4\$^5 = \begin{bmatrix} 0 \\ 0 \\ 1 \\ 0 \\ -r_1 - r_2 \\ 0 \end{bmatrix}, \end{aligned} \tag{5.108}$$

where $\hat{\boldsymbol{u}}_{B/A}$ is a unit vector pointed from A to B, for example, $\hat{\boldsymbol{u}}_{B/A} = 0.335\hat{\boldsymbol{i}} + 0.941\hat{\boldsymbol{j}}$. On the other hand, to compute the passive joint-velocity rates of the closed kinematic chain, let us consider the following lines in Plücker coordinates:

$$L_1 = \begin{bmatrix} 0 \\ 1 \\ 0 \\ 0 \\ 0 \\ r_1 \cos(\theta_A) \end{bmatrix}, \quad L_2 = \begin{bmatrix} \hat{\boldsymbol{u}}_{B/A} \\ -\hat{\boldsymbol{u}}_{B/A} \times \boldsymbol{r}_1 \end{bmatrix}, \quad L_3 = \begin{bmatrix} 0 \\ 1 \\ 0 \\ 0 \\ 0 \\ r_1 + r_2 - r_2 \cos(\theta_A) \end{bmatrix}. \tag{5.109}$$

Applying the Klein form of the line $\boldsymbol{L}_2$ to both sides of Eq. (5.107) and reducing terms, that is, $\{\boldsymbol{L}_2; {}^1\$^2\} = \{\boldsymbol{L}_2; {}^3\$^4\} = 0$, yields

$$_0\omega_1 \{\boldsymbol{L}_2; {}^0\$^1\} + {}_2\omega_3 \{\boldsymbol{L}_2; {}^2\$^3\} + {}_4\omega_5 \{\boldsymbol{L}_2; {}^4\$^5\} = 0. \tag{5.110}$$

Thus, the passive joint-velocity rate $_2\omega_3$ is obtained as

$$_2\omega_3 = -\frac{\{\boldsymbol{L}_2; {}_0\omega_1 {}^0\$^1 + {}_4\omega_5 {}^4\$^5\}}{\{\boldsymbol{L}_2; {}^2\$^3\}} = -0.303\, m/s. \tag{5.111}$$

Similarly, the remaining passive joint-velocity rates of the closed kinematic chain result in

$$_1\omega_2 = -\frac{\{\boldsymbol{L}_3; {}_0\omega_1 {}^0\$^1 + {}_2\omega_3 {}^2\$^3 + {}_4\omega_5 {}^4\$^5\}}{\{\boldsymbol{L}_3; {}^1\$^2\}} = -0.195\,\text{rad/s}$$

and

$$_3\omega_4 = -\frac{\{\boldsymbol{L}_1; {}_0\omega_1 {}^0\$^1 + {}_2\omega_3 {}^2\$^3 + {}_4\omega_5 {}^4\$^5\}}{\{\boldsymbol{L}_1; {}^3\$^4\}} = -0.126\,\text{rad/s}.$$

In order to compute the velocity of truck A, consider that the velocity state of body 1 as measured from body 0, the vector ${}^0\boldsymbol{V}^1_{O_1}$, is given by

$${}^0\boldsymbol{V}^1_{O_1} = {}_0\omega_1 {}^0\$^1 = \begin{bmatrix} 0 \\ 0 \\ {}_0\omega_1 \\ 0 \\ 0 \\ 0 \end{bmatrix}.$$

Note that the dual part of ${}^0\boldsymbol{V}^1_{O_1}$ vanishes; therefore, by applying the theory of helicoidal vector fields we get

$$\boldsymbol{v}_A = \boldsymbol{0} + {}^0\boldsymbol{\omega}^1 \times \boldsymbol{r}_{A/O_1} = 9.722\hat{\boldsymbol{i}} + 16.839\hat{\boldsymbol{j}}.$$

To compute the velocity of truck B, consider that the velocity state of body 3 as measured from body 0, the vector ${}^0\boldsymbol{V}^3_{O_1}$, may be written in screw form as follows:

$$
{}^{0}\boldsymbol{V}^{3}_{O_1} = {}_{0}\omega_{1}{}^{0}\$^{1} + {}_{1}\omega_{2}{}^{1}\$^{2} + {}_{1}\omega_{2}{}^{1}\$^{2} = \begin{bmatrix} 0.0 \\ 0 \\ -0.033 \\ 11.638 \\ 20.049 \\ 0 \end{bmatrix}.
$$

Hence, by applying the theory of helicoidal vector fields, we see that the velocity of truck B results in

$$
\boldsymbol{v}_B = 11.638\hat{\boldsymbol{i}} + 20.049\hat{\boldsymbol{j}} - 0.033\hat{\boldsymbol{k}} \times \boldsymbol{r}_{B/O_1} = 14.731\hat{\boldsymbol{i}} + 14.731\hat{\boldsymbol{j}}.
$$

Finally, the velocity of truck B as observed from truck A results in

$$
\boldsymbol{v}_{B/A} = \boldsymbol{v}_B - \boldsymbol{v}_A = 5\hat{\boldsymbol{i}} - 2.107\hat{\boldsymbol{j}}.
$$

On the other hand, concerning with the acceleration analysis of the example, it is evident that the reduced acceleration state of body 1 as measured from body 0, the six-dimensional vector ${}^{0}\boldsymbol{A}^{1}_{O_1}$, may be written in screw form as

$$
{}^{0}\boldsymbol{A}^{1}_{O_1} = \begin{bmatrix} {}^{0}\boldsymbol{\alpha}^{1} \\ \boldsymbol{a}_{O_1} - {}^{0}\boldsymbol{\omega}^{1} \times \boldsymbol{v}_{O_1} \end{bmatrix} = {}_{0}\alpha_{1}{}^{0}\$^{1}.
$$

However, since ${}_{0}\alpha_1 = \ddot{\theta}_A = 0$, then the dual part of the reduced acceleration state, in this problem, vanishes for any point of body 1. Indeed,

$$
\boldsymbol{a}_A - {}^{0}\boldsymbol{\omega}^{1} \times \boldsymbol{v}_A = \boldsymbol{0}.
$$

Thus,

$$
\boldsymbol{a}_A = {}^{0}\boldsymbol{\omega}^{1} \times \boldsymbol{v}_A = {}^{0}\boldsymbol{\omega}^{1} \times ({}^{0}\boldsymbol{\omega}^{1} \times \boldsymbol{r}_{A/O_1}) = -2.728\hat{\boldsymbol{i}} + 1.575\hat{\boldsymbol{j}}.
$$

To compute the acceleration of truck B, let us consider that the reduced acceleration state of link 3 as measured from body 0, the six-dimensional vector ${}^{0}\boldsymbol{A}^{3}_{O_1}$, may be written in screw form as follows:

$$
{}^{0}\boldsymbol{A}^{3}_{O_1} = {}_{0}\alpha_{1}{}^{0}\$^{1} + {}_{1}\alpha_{2}{}^{1}\$^{2} + {}_{2}\alpha_{3}{}^{2}\$^{3} + {}^{0}\mathcal{A}^{3}_{O_1} = \begin{bmatrix} {}^{0}\boldsymbol{\alpha}^{3} \\ {}^{0}\boldsymbol{a}^{3}_{O_1} - {}^{0}\boldsymbol{\omega}^{3} \times {}^{0}\boldsymbol{v}^{3}_{O_1} \end{bmatrix}, \tag{5.112}
$$

where the Lie screw of acceleration ${}^{0}\mathcal{A}^{3}_{O_1}$ is given by

$$
{}^{0}\mathcal{A}^{3}_{O_1} = \left[{}_{0}\omega_{1}{}^{0}\$^{1} \quad {}_{1}\omega_{2}{}^{1}\$^{2} + {}_{2}\omega_{3}{}^{2}\$^{3}\right] + \left[{}_{1}\omega_{2}{}^{1}\$^{2} \quad {}_{2}\omega_{3}{}^{2}\$^{3}\right]
$$

$$= \begin{bmatrix} 0 \\ 0 \\ 0 \\ -3.304 \\ 1.905 \\ 0 \end{bmatrix}. \tag{5.113}$$

Following the trend of the velocity analysis, the passive joint-acceleration rates of the closed kinematic chain result in

$$
\begin{aligned}
{}_2\alpha_3 &= -\frac{\{\boldsymbol{L}_2; {}_0\alpha_1{}^0\$^1 + {}_4\alpha_5{}^4\$^5 + {}^0\mathcal{A}^5_{O_1}\}}{\{\boldsymbol{L}_2; {}^2\$^3\}} = -1.816\,\mathrm{m/s^2},\\
{}_1\alpha_2 &= -\frac{\{\boldsymbol{L}_3; {}_0\alpha_1{}^0\$^1 + {}_2\alpha_3{}^2\$^3 + {}_4\alpha_5{}^4\$^5 + {}^0\mathcal{A}^5_{O_1}\}}{\{\boldsymbol{L}_3; {}^1\$^2\}} = -0.038\,\mathrm{rad/s^2},\\
{}_3\alpha_4 &= -\frac{\{\boldsymbol{L}_1; {}_0\alpha_1{}^0\$^1 + {}_2\alpha_3{}^2\$^3 + {}_4\alpha_5{}^4\$^5 + {}^0\mathcal{A}^5_{O_1}\}}{\{\boldsymbol{L}_1; {}^3\$^4\}} = 0.038\,\mathrm{rad/s^2},
\end{aligned} \tag{5.114}
$$

where

$$
\begin{aligned}
{}^0\mathcal{A}^5_{O_1} = &\left[{}_0\omega_1{}^0\$^1 \quad {}_1\omega_2{}^1\$^2 + {}_2\omega_3{}^2\$^3 + {}_3\omega_4{}^3\$^4 + {}_4\omega_5{}^4\$^5\right] + \\
&\left[{}_1\omega_2{}^1\$^2 \quad {}_2\omega_3{}^2\$^3 + {}_3\omega_4{}^3\$^4 + {}_4\omega_5{}^4\$^5\right] + \\
&\left[{}_2\omega_3{}^2\$^3 \quad {}_3\omega_4{}^3\$^4 + {}_4\omega_5{}^4\$^5\right] + \left[{}_3\omega_4{}^3\$^4 \quad {}_4\omega_5{}^4\$^5\right] \\
= &\begin{bmatrix} 0 \\ 0 \\ 0 \\ -5.169 \\ 3.771 \\ 0 \end{bmatrix}.
\end{aligned} \tag{5.115}
$$

Thus, the reduced acceleration state ${}^0\boldsymbol{A}^3_{O_1}$ [see Eq. (5.112)] results in ${}^0\boldsymbol{A}^3_{O_1} = \left[0\ 0\ -0.038\ -1.631\ 4.148\ 0\right]$. Furthermore, based on the theory of helicoidal vector fields, the reduced acceleration state of body 3 as measured from body 0 taking point B as the reference pole, the vector ${}^0\boldsymbol{A}^3_B$, is obtained as

$$
{}^{0}\boldsymbol{A}_{B}^{3} = \begin{bmatrix} 0 \\ 0 \\ -0.038 \\ -1.631\hat{\boldsymbol{i}} + 4.148\hat{\boldsymbol{j}} - 0.038\hat{\boldsymbol{k}} \times \boldsymbol{r}_{B/O_1} \end{bmatrix} = \begin{bmatrix} 0 \\ 0 \\ -0.038 \\ 1.865 \\ -1.865 \\ 0 \end{bmatrix}.
$$

Thus, the acceleration of truck b results in

$$
\boldsymbol{a}_B = {}^{0}\boldsymbol{a}_B^{3} = 1.865\hat{\boldsymbol{i}} - 1.865\hat{\boldsymbol{j}} + {}^{0}\boldsymbol{\omega}^{3} \times \boldsymbol{v}_B = 2.360\hat{\boldsymbol{i}} - 2.36\hat{\boldsymbol{j}}.
$$

Finally, the relative acceleration between trucks A and B is obtained as

$$
\boldsymbol{a}_{B/A} = \boldsymbol{a}_B - \boldsymbol{a}_A = -2.728\hat{\boldsymbol{i}} + 1.575\hat{\boldsymbol{j}} - 2.360\hat{\boldsymbol{i}} + 2.36\hat{\boldsymbol{j}} = 5.089\hat{\boldsymbol{i}} - 3.936\hat{\boldsymbol{j}}.
$$

Example 5.5 seems to be a very involved method to compute the simple difference of acceleration of two bodies in relative motion when using screw theory. In fact, it is rather difficult to find any advantage of screw theory when it is compared with the solution that the reader can find in any standard book of kinematics for this example. However, the benefits of including this example may be understood in two ways: (1) Physically disconnected bodies may be modeled as kinematically equivalent mechanisms, and it is available for spatial kinematics too; (2) the systematic application of reciprocal screw theory in the acceleration analysis of mechanisms allows us to avoid the computation of the passive joint-acceleration rates, a tedious task when traditional methods are used.

The last example to be treated in this chapter deals with the acceleration analysis of a popular parallel manipulator named Hexaglide, a concept credited to the Institute of Machine Tools of the Swiss Federal Institute of Technology in Zurich, whose purpose was the development of a high-speed milling machine (Wiegand et al. 1996). One can find several interesting examples of robot manipulators whose architectures are based on the Hexaglide-type architecture: Hexaglide robot (Honegger et al. 1991), HexaM milling machine (ParalleMic 1991), LINAPOD (Boye and Pritschow 2005), neurosurgical robot (Narayana et al. 2000), parallel engraving machine (Kong and Zhang 2005), orthopedic surgery robot (Schwartz et al. 2009), parallel machine for masticatory studies (Narayanan 2011), wind tunnel platform to check floating wind turbines (Bayati et al. 2014), and so on. Practically all issues concerned with the control and dynamics of the Hexaglide robot have been extensively investigated. Honegger et al. (1991) implemented nonlinear adaptive control to identify parameters associated to the dynamic equation of the Hexaglide. Abtahi et al. (2008) introduced a self-calibration method for the Hexaglide robot based only on the data of the generalized coordinates. The motion/force transmissibility of the 6-PUS (where U stands for universal joint) was investigated by Wang et al. (2010), who concluded that the transmission wrench in each leg is a pure force, the main advantage of parallel manipulators equipped with linear actuators. Jong et al. (2012) considered some relevant characteristics of the Hexaglide such as motor acceleration and torque as well as workspace accuracy for application in safe human interaction. The workspace and stiffness analyses of a Hexaglide-type robot were approached by Wang et al. (2013) by resorting to screw theory.

Example 5.6. With reference to Fig. 5.8, we need to determine the input–output equations of velocity and acceleration of the Hexaglide-type robot illustrated there.

Let $\boldsymbol{\omega}$ be the angular velocity vector of the moving platform (6) as measured from the fixed platform (0). Furthermore, let $\boldsymbol{v}_C$ be the velocity vector of point C embedded in the moving platform where point C is chosen as the reference pole of the manipulator. Then the velocity state of the moving platform as observed from the fixed platform, the six-dimensional vector $\boldsymbol{V}_C = (\boldsymbol{\omega}, \boldsymbol{v}_C)$, may be expressed in screw form through the ith limb of the manipulator as follows:

$$ {}_0\omega_1^i\,{}^0\$_i^1 + {}_1\omega_2^i\,{}^1\$_i^2 + {}_2\omega_3^i\,{}^2\$_i^3 + {}_3\omega_4^i\,{}^3\$_i^4 + {}_4\omega_5^i\,{}^4\$_i^5 + {}_5\omega_6^i\,{}^5\$_i^6 = \boldsymbol{V}_C, \qquad (5.116) $$

where the screw ${}^0\$_i^1$ is associated to the active prismatic joint, that is, ${}_0\omega_1^i = \dot{q}_i$. Meanwhile, the triplet of concurrent screws ${}^1\$_i^2$, ${}^2\$_i^3$, and ${}^3\$_i^4$ simulates a spherical joint. Furthermore, the pair of concurrent and orthogonal screws ${}^4\$_i^5$ and ${}^5\$_i^6$ are associated to the ith universal joint. On the other hand, consider that $\$_i$ is a line in Plücker coordinates directed from point A_i to point B_i. Then applying the Klein form, $\{*; *\}$, of the line $\$_i$ to both sides of Eq. (5.116) in the same limb and reducing terms leads to

$$ \left\{\$_i; {}^0\$_i^1\right\} = \{\$_i; \boldsymbol{V}_C\}\,. \qquad (5.117) $$

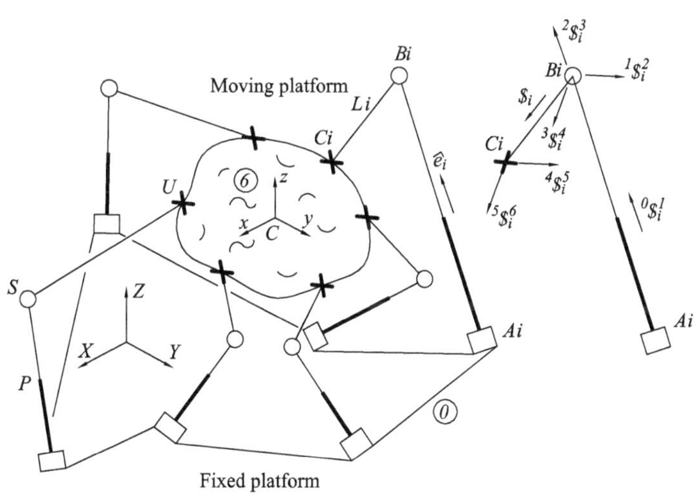

Fig. 5.8 6-PSU parallel manipulator. A Hexaglide-type robot

In other words, the line $\$_i$ is reciprocal to all screws in the same limb, excluding the screw representing the active prismatic joint. Next, casting Eq. (5.117) into matrix-vector form, we see that the input–output equation of velocity of the manipulator results in

$$\mathbf{M}\,\boldsymbol{V}_C = \dot{\mathbf{Q}}, \tag{5.118}$$

where

$$\dot{\mathbf{Q}} = \left[\dot{q}_1 \left\{\$_1; {}^0\$_1^1\right\}\ \dot{q}_2 \left\{\$_2; {}^0\$_2^1\right\} \cdots \dot{q}_6 \left\{\$_6; {}^0\$_6^1\right\}\right]^T \tag{5.119}$$

is the first-order driver matrix of the manipulator while $\mathbf{M}$ is the first-order matrix of the manipulator, which is given by

$$\mathbf{M} = \mathbf{J}^T\,\boldsymbol{\Delta}, \tag{5.120}$$

where

$$\mathbf{J} = \left[\$_1\ \$_2 \cdots \$_6\right]^T \tag{5.121}$$

is the Jacobian of the robot. Meanwhile,

$$\boldsymbol{\Delta} = \begin{bmatrix} \mathbf{0} & \mathbf{I}_3 \\ \mathbf{I}_3 & \mathbf{0} \end{bmatrix}$$

is an operator of polarity.

According to Eq. (5.118), the moving platform cannot perform arbitrary velocity states $\boldsymbol{V}$ when matrix $\mathbf{M}$ is singular. In other words, the manipulator is at a singular configuration when $\dim(\mathbf{J}) < 6$, which implies that $\mathbf{M}$ is a rank-deficient matrix (Mu and Wu 2010).

In what follows the acceleration analysis of the robot is approached following the trend of the velocity analysis. Let $\boldsymbol{\alpha}$ be the angular acceleration vector of the moving platform (6) as measured from the fixed platform (0). Furthermore, let $\boldsymbol{a}_C$ be the acceleration vector of point C. Then the reduced acceleration state of the moving platform as observed from the fixed platform, the six-dimensional vector $\boldsymbol{A}_C = (\boldsymbol{\alpha}, \boldsymbol{a}_C - \boldsymbol{\omega}\times\boldsymbol{v}_C)$, may be expressed in screw form through the ith limb of the manipulator as follows:

$${}_0\alpha_1^i\,{}^0\$_i^1 + {}_1\alpha_2^i\,{}^1\$_i^2 + {}_2\alpha_3^i\,{}^2\$_i^3 + {}_3\alpha_4^i\,{}^3\$_i^4 + {}_4\alpha_5^i\,{}^4\$_i^5 + {}_5\alpha_6^i\,{}^5\$_i^6 + \mathcal{L}_i = \boldsymbol{A}_C, \tag{5.122}$$

where ${}_*\alpha_*^i$ is the ith joint acceleration rate, that is, ${}_*\alpha_*^i = \frac{d}{dt}({}_*\omega_*^i)$. Meanwhile, $\mathcal{L}_i$ is the ith Lie screw of acceleration, which is computed as

$$\mathcal{L}_i = \left[{}_0\omega_1^i\,{}^0\$_i^1 \quad {}_1\omega_2^i\,{}^1\$_i^2 + {}_2\omega_3^i\,{}^2\$_i^3 + {}_3\omega_4^i\,{}^3\$_i^4 + {}_4\omega_5^i\,{}^4\$_i^5 + {}_5\omega_6^i\,{}^5\$_i^6\right] +$$

$$
\begin{aligned}
&\left[{}_1\omega_2^i \, {}^1\$_i^2 \quad {}_2\omega_3^i \, {}^2\$_i^3 + {}_3\omega_4^i \, {}^3\$_i^4 + {}_4\omega_5^i \, {}^4\$_i^5 + {}_5\omega_6^i \, {}^5\$_i^6\right] + \\
&\left[{}_2\omega_3^i \, {}^2\$_i^3 \quad {}_3\omega_4^i \, {}^3\$_i^4 + {}_4\omega_5^i \, {}^4\$_i^5 + {}_5\omega_6^i \, {}^5\$_i^6\right] + \\
&\left[{}_3\omega_4^i \, {}^3\$_i^4 \quad {}_4\omega_5^i \, {}^4\$_i^5 + {}_5\omega_6^i \, {}^5\$_i^6\right] + \left[{}_4\omega_5^i \, {}^4\$_i^5 \quad {}_5\omega_6^i \, {}^5\$_i^6\right].
\end{aligned} \tag{5.123}
$$

Next, by resorting to reciprocal screw theory, the input–output equation of acceleration of the parallel manipulator results in

$$\mathbf{M}\boldsymbol{A}_C = \ddot{\mathbf{Q}} + \mathcal{L}, \tag{5.124}$$

where

$$\ddot{\mathbf{Q}} = \left[\ddot{q}_1 \left\{\$_1; {}^0\$_1^1\right\} \ \ddot{q}_2 \left\{\$_2; {}^0\$_2^1\right\} \cdots \ddot{q}_6 \left\{\$_6; {}^0\$_6^1\right\}\right]^T \tag{5.125}$$

is the second-order driver matrix of the manipulator while

$$\mathcal{L} = \left[\{\$_1; \mathcal{L}_1\} \ \{\$_2; \mathcal{L}_2\} \cdots \{\$_6; \mathcal{L}_6\}\right]^T \tag{5.126}$$

is the complementary matrix of acceleration. It is worth emphasizing that the terms of the Coriolis acceleration are self-contained in matrix $\mathcal{L}$.

5.4 Exercises

1. It was demonstrated that ${}^j\boldsymbol{v}_O^m = {}^j\boldsymbol{v}_O^k + {}^k\boldsymbol{v}_O^m$. Prove that even though such an expression is correct, the acceleration vector ${}^j\boldsymbol{a}_O^m$, where ${}^j\boldsymbol{a}_O^m \equiv \frac{{}^jd}{dt}({}^j\boldsymbol{v}_O^m)$, cannot be determined correctly based on such an expression.
2. Determine the input–output equation of acceleration of the four-bar planar mechanism by resorting to screw theory. Afterward, determine the angular acceleration of the driven link given the angular acceleration of the driver link. *Hint*: Write the equation of acceleration in screw form and apply the appropriate Klein form to both sides of such an equation.
3. Using a standard method, determine the symbolic equation of the angular acceleration of the driven wheel of the Geneva mechanism considering that the driver wheel rotates at a constant angular velocity ω_1 counterclockwise.
4. Carry out the singularity analysis of the 2RRR+PPR parallel manipulator based on the input–output equation of velocity.
5. Figure 5.9 shows a planar timing mechanism. The motion of pin B is controlled by the linear slot A, which is being elevated at a constant velocity upward of $1.5\,\mathrm{m/s}$ as measured from the fixed reference frame. If $r = 0.3\,\mathrm{m}$ and $\theta = 30°$, compute (a) the acceleration of pin B as measured from the Earth, (b) the relative acceleration of pin B as observed from the circular slot. *Hint*: Subdivide

Fig. 5.9 Planar timing mechanism

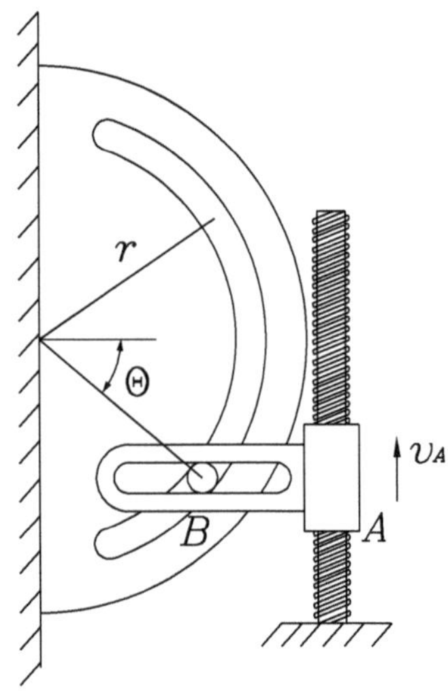

the timing mechanism into two different planar mechanisms and identify the corresponding j, k, and m bodies for each mechanism.

6. Determine the general input–output equations of velocity and acceleration of the timing mechanism considering that the active kinematic pair is the lead screw whose pitch is h. *Hint*: The timing mechanism may be modeled as a PPR closed kinematic chain where the first prismatic joint is related to the active kinematic joint through the pitch h of the lead screw.

References

Abtahi, M., Pendar, M., Alasty, A., & Vossoughi, G. (2008). Calibration of Hexaglide parallel manipulators using only input joint variables. In *Proceedings ASME International Mechanical Engineering Congress & Exposition, Boston*. 2008 paper IMECE2008-66162, seven pages.

Ball, R. S. (1900). *A treatise on the theory of screws*. Cambridge: Cambridge University Press (reprinted 1998).

Bayati, I., Belloli, M., Ferrari, D., Fossati, F., & Giberti, H. (2014). Design of a 6-DoF robotic platform for wind tunnel tests of floating wind turbines. *Energy Procedia, 53*, 313–323.

Bonev, I. A. (1991). http://www.parallemic.org/WhosWhoCompaniesProfile002.html

Boye, T., & Pritschow, G. (2005). New transformation and analysis of an N-DOF LINAPOD with six struts for higher accuracy. *Robotica, 23*, 555–560.

Brand, L. (1947). *Vector and tensor analysis*. New York: Wiley.

Gosselin, C. M., Sefrioui, J., & Richard, M. J. (1992). Solutions polynomiales au probléme de la cinématique directe des manipulateurs parallèles plans a trois degrés de liberté [polynomial solutions to the direct kinematic problem of planar three degree-of-freedom parallel manipulators]. *Mechanism and Machine Theory, 27*(2), 107–119.

Honegger, M., Codourey, A., & Burdet, E. (1991). Adaptive control of the Hexaglide, a 6 DOF parallel manipulator. Technical Report No. Institution of Robotics, ETH Zurich, Switzerland.

Jong, J. J., Stienen, A. H. A., van der Wijk, V., Wessels, M., & van der Kooij, H. (2012). A method for evaluation and comparison of parallel robots for safe human interaction, applied to robotic TMS. In *Proceedings Fourth IEEE RAS/EMBS International Conference on Biomedical Robotics and Biomechatronics, Rome* (pp. 986–991).

Kane, T. R., & Levinson, D. A. (1985). *Dynamics: theory and applications*. New York: McGraw-Hill.

Kieffer, J. (1990). *Selected topics in mechanisms and robotics: singularities, inverse kinematics and collision detection*. Ph.D. thesis, University of Illinois at Chicago.

Kieffer, J. (1994). Differential analysis of bifurcations and isolated singularities for robots and mechanisms. *IEEE Transactions on Robotics and Automation, 10*(1), 1–10.

Kong, L.-F., & Zhang, S.-H. (2005). A novel parallel engraving machine based on 6-PUS mechanism and related technologies. In V. Kordic, A. Lazinica, M. Merdan (Eds.), *Cutting edge robotics* (pp. 629–648). Berlin: Pro Literatur.

Legnani, G., Casolo, F., Righettini, P., & Zappa, B. (1996). A homogeneus matrix approach to 3D kinematics and dynamics. Part 1: theory. *Mechanism and Machine Theory, 31*(5), 573–587.

Mu, X., & Wu, C. Q. (2010). On post-impact angular velocities and resultant impulses with rank-deficient Jacobian matrices using Newton's impact law. *International Journal of Applied Mechanics, 2*(3), 569–584.

Nam, Y.-J., & Park, M.-K. (2006). Kinematics and optimization of 2-DOF parallel manipulator with revolute actuators and a passive leg. *Journal of Mechanical Science and Technology, 20*(6), 828–839.

Narasimhan, S., & Kumar, V. (1994). A second analysis of manipulator kinematics in singular configurations. In G. R. Pennock (Ed.), *Proceedings Robotics: Kinematics, Dynamics and Controls* (DE-Vol. 72, pp. 477–484). New York: ASME.

Narayana, S., Fox, P. T., Tandon, N., Lancaster, J. L., Iyer, M. B., & Constantine, W. (2000). Use of neurosurgical robot for aiming and holding in cortical TMS experiments. *NeuroImage, 11*, S471.

Narayanan, M. S., Kannan, S., Zhou, X., Mendel, F., & Krovi, V. (2011). Parallel architecture manipulators for use in masticatory studies. *International Journal of Intelligent Mechatronics and Robotics, 1*(4), 100–122.

Rico, J. M., & Duffy, J. (1996). An application of screw algebra to the acceleration analysis of serial chains. *Mechanism and Machine Theory, 31*(4), 445–457.

Ridley, P. R., Bokelberg, E. H., & Hunt, K. H. (1992). Spatial motion II: acceleration and the differential geometry of screws. *Mechanism and Machine Theory, 27*(1), 17–35.

Schwarz, M. L., Wagner, A., El-Shenawy, A., Gundling, R., Köpfle, A., Handel, H., Badreddin, E., Männer, R., Scharf, H.-P., Götz, M., Schill, M., & Pott, P. P. (2009). A handheld robot for orthopedic surgery—ITD. In O. Dössel, W. C. Schlegel (Eds.), *World Congress on Medical Physics and Biomedical Engineering* (Vol. 25/6, pp. 99–102). Berlin: Springer.

Sugimoto, K. (1987). Kinematic and dynamic analysis of parallel manipulators by means of motor algebra. *ASME Journal of Mechanisms, Transmissions and Automation in Design, 109*(1), 3–7.

Sugimoto, K. (1989). Computational scheme for dynamic analysis of parallel manipulators. *ASME Journal of Mechanisms, Transmissions and Automation in Design, 111*(1), 29–33.

Sugimoto, K. (1990). Existence criteria for overconstrained mechanisms. An extension of motor algebra. *ASME Journal of Mechanical Design, 112*(3), 295–298.

Wang, D., Fan, R., & Chen, W. (2013). Stiffness analysis of a hexaglide parallel loading mechanism. *Mechanism and Machine Theory, 70*, 454–473.

Wang, J., Wu, C., & Liu, X.-J. (2010). Performance evaluation of parallel manipulators: motion/force transmissibility and its index. *Mechanism and Machine Theory, 45*(10), 1462–1476.

Wiegand, A., Hebsacker, M., & Honegger, M. (1996). Parallele Kinematik und Linearmotoren: Hexaglide - ein neues, hochdynamisches Werkzeugmaschinenkonzept. Technische Rundschau Transfer Nr. 25.

Yu, A., Bonev, I. A., & Zsombor-Murray, P. (2008). Geometric approach to the accuracy analysis of a class of 3-DOF planar parallel robots. *Mechanism and Machine Theory, 43*(3), 364–375.

Zubizarreta, A., Marcos, M., Cabanes, I., Pinto, C., & Portillo, E. (2012). Redundant sensor based control of the 3RRR parallel robot. *Mechanism and Machine Theory, 54*, 1–17.

Chapter 6
Jerk Analysis

6.1 Introduction

Jerk, a research field of great interest in quite diverse academic communities, is the time rate of change of acceleration and is related to the rate of change of force, namely an impulse that is considered a *hammer blow force*. Suh (1971) and Schot (1978) investigated the geometric parameters of the radii of torsion of spatial curves. It is known that there is a direct relationship between jerk and the movements of the human body (Morasso 1981; Flash and Hogan 1985; Uno et al. 2007). Consider, for instance, that Crossman and Goodeve (1983) have shown that when high spatial precision is required, many movements related with the hands present irregularities and multiple velocity peaks. Viviani and coworkers (Viviani and Schneider 1991; Viviani and Flash 1995) proved that there is a correlation among perception, motion planning, and jerk in humans. On the other hand, Sparis and Moroutsos (1984) applied jerk analysis to improve and control the position analysis of planar mechanisms when the mechanism is near a singular configuration. Žefran (1995) presented a coordinate-free formulation to generate shortest paths, and minimum acceleration and jerk properties of rigid bodies undergoing spherical motions, which correspond to the subgroup $SO(3)$ of rotational motions of the Euclidean group $SE(3)$, and general motions of $SE(3)$. Urbinati and Pennestri (1998) developed a computer program for the jerk analysis of a rigid body with respect to an inertial reference frame. Their method is based on the representation of the Euclidean displacements of a rigid body in terms of the Euler parameters and the time derivatives of the corresponding rotation matrices associated with the motion of the rigid body. Gielen et al. (1997) noted that the characteristic pattern of cerebellar ataxia, related with jerk and submovements, is contained in the trajectory of the hand during repeated arm movements. In order to understand how the central nervous system controls the kinematics of rapid finger and hand movement, Novàk et al. (2000) proposed an objective algorithm to identify overlapping submovements, detecting appreciable inflections in the acceleration traces that are evidently related with jerk.

J. Gallardo-Alvarado, *Kinematic Analysis of Parallel Manipulators by Algebraic Screw Theory*, DOI 10.1007/978-3-319-31126-5_6

To this end, Novàk et al. (2000) studied the motions of subjects turning a knob and concluded that in many trials subjects turned the knob with a single, smooth, and regular motion, as indicated by the angular position and velocity trajectories, but in other cases, subjects produced irregularities in the kinematics, which were considered discrete corrective submovements, detecting appreciable inflections in the acceleration traces. Thus, it is reasonable to assume that jerk is responsible for such irregularities. Goldvasser et al. (2001) investigated high curvature analysis and integrated absolute jerk to differentiate between healthy and cerebellopathy patients performing pointing tasks. Cheng (2002) introduced a procedure to obtain jerk curves that are expressed in explicit parametric forms. Without a doubt, jerk analysis with applications in human motions has been the motive behind a remarkable research field (Panjabi and White 1980; Levin 1995; Willems et al. 1996; Ziddiqui et al. 2006; Ishii et al. 2006; Konz et al. 2006; Howard et al. 2007; Lessard et al. 2007; Uno et al. 2007; Li and Xu 2007; Jones et al. 2008; Zhu et al. 2008; Tsoi and Xie 2010). Furthermore, interesting mechanical applications of jerk include, among many others, the following: (1) the classical design of cam-follower systems; (2) the correct characterization of singularities in closed kinematic chains; (3) the synthesis of spherical and in general spatial linkages; (4) the computation of impact forces caused by rapid changes in acceleration; (5) the improvement of path-planning trajectories in the neighborhood of singular regions. Following that trend, Zahraee and Neff (1996) reported the jerk analysis of some commercial cams, whereas Nguyen and Kim (2007) introduced a method to smooth cam profiles using modified spline curves. Lee and Lin (1999) have shown that in milling machining operations, undesirable chattering effects can be reduced by controlling the continuity of jerks associated with the trajectory of the cutter. Finally, as shown by Erkorkmaz and Altintas (2001) and Macfarlane and Croft (1986b), it is possible to control vibrations within kinematic structures by minimizing jerk. Mandal and Nazcar (2009) designed cam displacement functions based on jerk analysis. To the best knowledge of the authors, the first contribution approaching the jerk analysis of rigid bodies using screw theory is that of Rico et al. (1999).

This chapter presents the jerk analysis of a rigid body using screw theory and its application to parallel manipulators. To avoid disrupting the sequence of the proofs of some propositions given in this chapter, some equations previously reported are reproduced with different labels than those originally assigned.

6.2 Fundamental Jerk Equations

Definition 6.1. Let k and l be two rigid bodies or reference frames. The angular jerk of body l with respect to body k, the vector ${}^k\boldsymbol{\rho}^l$, is defined as follows:

$$ {}^k\boldsymbol{\rho}^l \equiv \frac{{}^kd}{dt}\left({}^k\boldsymbol{\alpha}^l\right), \tag{6.1}$$

where ${}^k\boldsymbol{\alpha}^l$ is the angular acceleration vector of body l as measured from body k. Furthermore, the linear jerk of any point O^l of body l as observed from body j, the vector ${}^k\boldsymbol{\rho}^l_O$, is defined as follows:

$$ {}^k\boldsymbol{\rho}^l_O \equiv \frac{{}^kd}{dt}\left({}^k\boldsymbol{a}^l_O\right). \tag{6.2}$$

On the other hand, it is straightforward to demonstrate that the angular jerk vector may be expressed as

$$ {}^k\boldsymbol{\rho}^l = {}_k\rho_l{}^k\hat{\boldsymbol{s}}^l, \tag{6.3}$$

where the scalar ${}_k\rho_l \equiv \frac{{}^kd}{dt}({}_k\alpha_l)$ is the magnitude of vector ${}^k\boldsymbol{\rho}^l$, also called the joint-jerk rate between bodies k and l.

Definition 6.1 allows us to find a relationship involving the linear jerks of two points, P and Q, embedded in the same body. To this end, from college kinematics it is evident that

$$ {}^j\boldsymbol{a}^m_P = {}^j\boldsymbol{a}^m_Q + {}^j\boldsymbol{\alpha}^m \times \boldsymbol{r}_{P/Q} + {}^j\boldsymbol{\omega}^m \times \left({}^j\boldsymbol{\omega}^m \times \boldsymbol{r}_{P/Q}\right), \tag{6.4}$$

where $\boldsymbol{r}_{P/Q}$ is a vector bounded in body m denoting the position of point P with respect to point Q. Hence, in accordance with Definition 6.1 and Eq. (6.4), we write

$$\begin{aligned} {}^j\boldsymbol{\rho}^m_P &= \frac{{}^jd}{dt}\left({}^j\boldsymbol{a}^m_P\right) \\ &= \frac{{}^jd}{dt}\left({}^j\boldsymbol{a}^m_Q\right) + \frac{{}^jd}{dt}\left({}^j\boldsymbol{\alpha}^m \times \boldsymbol{r}_{P/Q}\right) + \frac{{}^jd}{dt}\left[{}^j\boldsymbol{\omega}^m \times \left({}^j\boldsymbol{\omega}^m \times \boldsymbol{r}_{P/Q}\right)\right]. \end{aligned} \tag{6.5}$$

The time derivatives given in Eq. (6.5) are displayed below:

$$\left.\begin{aligned} &\tfrac{{}^jd}{dt}\left({}^j\boldsymbol{a}^m_Q\right) = {}^j\boldsymbol{\rho}^m_Q, \\ &\tfrac{{}^jd}{dt}\left({}^j\boldsymbol{\alpha}^m \times \boldsymbol{r}_{P/Q}\right) = {}^j\boldsymbol{\alpha}^m \times ({}^j\boldsymbol{\omega}^m \times \boldsymbol{r}_{P/Q}) + {}^j\boldsymbol{\rho}^m \times \boldsymbol{r}_{P/Q}, \\ &\tfrac{{}^jd}{dt}\left[{}^j\boldsymbol{\omega}^m \times \left({}^j\boldsymbol{\omega}^m \times \boldsymbol{r}_{P/Q}\right)\right] = {}^j\boldsymbol{\alpha}^m \times ({}^j\boldsymbol{\omega}^m \times \boldsymbol{r}_{P/Q}) + \\ &{}^j\boldsymbol{\omega}^m \times \left[{}^j\boldsymbol{\alpha}^m \times \boldsymbol{r}_{P/Q} + {}^j\boldsymbol{\omega}^m \times ({}^j\boldsymbol{\omega}^m \times \boldsymbol{r}_{P/Q})\right]. \end{aligned}\right\} \tag{6.6}$$

Therefore, substituting Eqs. (6.6) into Eq. (6.5) and reducing terms yields

$$\begin{aligned} {}^j\boldsymbol{\rho}^m_P &= {}^j\boldsymbol{\rho}^m_Q + {}^j\boldsymbol{\rho}^m \times \boldsymbol{r}_{P/Q} + 2{}^j\boldsymbol{\alpha}^m \times \left({}^j\boldsymbol{\omega}^m \times \boldsymbol{r}_{P/Q}\right) + \\ &{}^j\boldsymbol{\omega}^m \times \left({}^j\boldsymbol{\alpha}^m \times \boldsymbol{r}_{P/Q}\right) + {}^j\boldsymbol{\omega}^m \times \left[{}^j\boldsymbol{\omega}^m \times \left({}^j\boldsymbol{\omega}^m \times \boldsymbol{r}_{P/Q}\right)\right]. \end{aligned} \tag{6.7}$$

Example 6.1. The rocket is fired vertically and tracked by radar as shown in Fig. 6.1. Determine general expressions to compute the velocity, acceleration, and jerk of rocket m as measured from the radar j.

Solution. According to the fixed reference frame XY, which has associated unit vectors $\hat{\boldsymbol{i}}$ and $\hat{\boldsymbol{j}}$, the position vector of point P of the rocket may be established as follows:

$$\boldsymbol{r} = r\cos(\theta)\hat{\boldsymbol{i}} + r\sin(\theta)\hat{\boldsymbol{j}}. \tag{6.8}$$

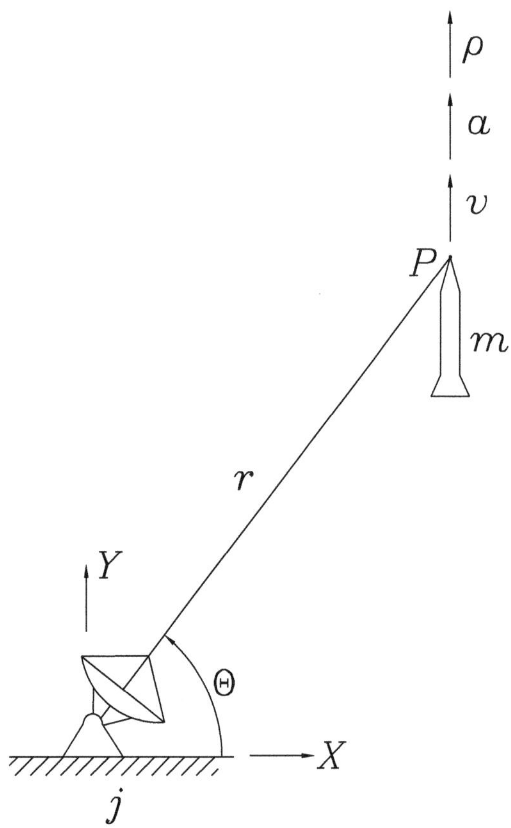

Fig. 6.1 Example 6.1. Rocket fired vertically

Hence, the velocity vector $\boldsymbol{v}$ of the rocket is obtained as

$$\boldsymbol{v} = \frac{d}{dt}\boldsymbol{r} = \Big[\dot{r}\cos(\theta) - r\dot{\theta}\sin(\theta)\Big]\hat{\boldsymbol{i}} + \Big[\dot{r}\sin(\theta) + r\dot{\theta}\cos(\theta)\Big]\hat{\boldsymbol{j}}. \tag{6.9}$$

However, since the rocket moves vertically, then evidently the component along the X-axis of $\boldsymbol{v}$ vanishes, yielding

$$\dot{r}\cos(\theta) - r\dot{\theta}\sin(\theta) = 0. \tag{6.10}$$

Therefore,

$$\boldsymbol{v} = \Big[\dot{r}\sin(\theta) + r\dot{\theta}\cos(\theta)\Big]\hat{\boldsymbol{j}}, \tag{6.11}$$

where

$$\dot{r} = r\dot{\theta}\tan(\theta). \tag{6.12}$$

Furthermore, the acceleration vector $\boldsymbol{a}$ of the rocket is obtained as

$$\begin{aligned}
\boldsymbol{a} &= \frac{d}{dt}\boldsymbol{v} \\
&= \frac{d}{dt}\Big[\dot{r}\sin(\theta) + r\dot{\theta}\cos(\theta)\Big]\hat{\boldsymbol{j}} \\
&= \Big[(\ddot{r} - r\dot{\theta}^2)\sin(\theta) + (r\ddot{\theta} + 2\dot{r}\dot{\theta})\cos(\theta)\Big]\hat{\boldsymbol{j}},
\end{aligned} \tag{6.13}$$

where

$$\ddot{r} = r\dot{\theta}^2\tan^2(\theta) + \Big[(\dot{r}\dot{\theta} + r\ddot{\theta})\Big]\tan(\theta) + r\dot{\theta}^2. \tag{6.14}$$

Finally, the jerk vector $\boldsymbol{\rho}$ of the rocket results in

$$\begin{aligned}
\boldsymbol{\rho} &= \frac{d}{dt}\boldsymbol{a} \\
&= \frac{d}{dt}\Big[(\ddot{r} - r\dot{\theta}^2)\sin(\theta) + (r\ddot{\theta} + 2\dot{r}\dot{\theta})\cos(\theta)\Big]\hat{\boldsymbol{j}} \\
&= \Big[(\dddot{r} - 3r\dot{\theta}\ddot{\theta} - 3\dot{r}\dot{\theta}^2)\sin(\theta) + (3\dot{r}\ddot{\theta} - r\dot{\theta}^3 + 3\ddot{r}\dot{\theta} + r\dddot{\theta})\cos(\theta)\Big]\hat{\boldsymbol{j}},
\end{aligned} \tag{6.15}$$

where

$$\dddot{r} = 2r\dot{\theta}^3\tan^3(\theta) + (3r\dot{\theta}\ddot{\theta} + 2\dot{r}\dot{\theta}^2)(1 + \tan^2(\theta)) + (2r\dot{\theta}^3 + \ddot{r}\dot{\theta} + 2\dot{r}\ddot{\theta} + r\dddot{\theta})\tan(\theta). \tag{6.16}$$

Proposition 6.1. *With reference to Fig. 4.5, let j, k, and m be three rigid bodies or reference frames. Furthermore, let O^m and O^k be two points, the first one fixed to body m and the second one attached to body k, where point O^m is instantaneously coincident with point O^k. The angular and linear jerks notated as ${}^j\boldsymbol{\rho}^m$ and ${}^j\boldsymbol{\rho}_O^m$, respectively, of body m with respect to body j are given by*

$$ {}^j\boldsymbol{\rho}^m = {}^j\boldsymbol{\rho}^k + {}^k\boldsymbol{\rho}^m + 2{}^j\boldsymbol{\omega}^k \times {}^k\boldsymbol{\alpha}^m + {}^j\boldsymbol{\alpha}^k \times {}^k\boldsymbol{\omega}^m + {}^j\boldsymbol{\omega}^k \times \left({}^j\boldsymbol{\omega}^k \times {}^k\boldsymbol{\omega}^m\right) \tag{6.17} $$

and

$$ {}^j\boldsymbol{\rho}_O^m = {}^j\boldsymbol{\rho}_O^k + {}^k\boldsymbol{\rho}_O^m + 3{}^j\boldsymbol{\alpha}^k \times {}^k\boldsymbol{v}_O^m + 3{}^j\boldsymbol{\omega}^k \times {}^k\boldsymbol{a}_O^m + 3{}^j\boldsymbol{\omega}^k \times \left({}^j\boldsymbol{\omega}^k \times {}^k\boldsymbol{v}_O^m\right). \tag{6.18} $$

Proof. From the acceleration analysis it is known that

$$ {}^j\boldsymbol{\alpha}^m = {}^j\boldsymbol{\alpha}^k + {}^k\boldsymbol{\alpha}^m + {}^j\boldsymbol{\omega}^k \times {}^k\boldsymbol{\omega}^m. \tag{6.19} $$

Hence, based on Definition 6.1 and Eq. (6.19), we write

$$ \begin{aligned} {}^j\boldsymbol{\rho}^m &= \frac{{}^jd}{dt}\left({}^j\boldsymbol{\alpha}^m\right) \\ &= \frac{{}^jd}{dt}\left({}^j\boldsymbol{\alpha}^k\right) + \frac{{}^jd}{dt}\left({}^k\boldsymbol{\alpha}^m\right) + \frac{{}^jd}{dt}\left({}^j\boldsymbol{\omega}^k \times {}^k\boldsymbol{\omega}^m\right), \end{aligned} \tag{6.20} $$

where

$$ \left.\begin{aligned} \frac{{}^jd}{dt}\left({}^j\boldsymbol{\alpha}^k\right) &= {}^j\boldsymbol{\rho}^k \\ \frac{{}^jd}{dt}\left({}^k\boldsymbol{\alpha}^m\right) &= {}^j\boldsymbol{\omega}^k \times {}^k\boldsymbol{\alpha}^m + {}^k\boldsymbol{\rho}^m \\ \frac{{}^jd}{dt}\left({}^j\boldsymbol{\omega}^k \times {}^k\boldsymbol{\omega}^m\right) &= {}^j\boldsymbol{\alpha}^k \times {}^k\boldsymbol{\omega}^m + {}^j\boldsymbol{\omega}^k \times \left({}^j\boldsymbol{\omega}^k \times {}^k\boldsymbol{\omega}^m + {}^k\boldsymbol{\alpha}^m\right) \end{aligned}\right\}. \tag{6.21} $$

Substituting Eqs. (6.21) into Eq. (6.20) and reducing terms yields Eq. (6.17).

To prove the second part of the proposition, we reproduce Eq. (5.11) here for the sake of clarity:

$$ {}^j\boldsymbol{a}_O^m = {}^j\boldsymbol{a}_{\tilde{B}}^k + {}^j\boldsymbol{\alpha}^k \times \boldsymbol{q} + {}^j\boldsymbol{\omega}^k \times \left({}^j\boldsymbol{\omega}^k \times \boldsymbol{q}\right) + 2{}^j\boldsymbol{\omega}^k \times {}^k\boldsymbol{v}_O^m + {}^k\boldsymbol{a}_O^m. \tag{6.22} $$

Therefore, the linear jerk ${}^j\boldsymbol{\rho}_O^m$ is given by

$$ \begin{aligned} {}^j\boldsymbol{\rho}_O^m &= \frac{{}^jd}{dt}\left({}^j\boldsymbol{a}_O^m\right) \\ &= \frac{{}^jd}{dt}\left({}^j\boldsymbol{a}_{\tilde{B}}^k\right) + \frac{{}^jd}{dt}\left({}^j\boldsymbol{\alpha}^k \times \boldsymbol{q}\right) + \frac{{}^jd}{dt}\left[{}^j\boldsymbol{\omega}^k \times \left({}^j\boldsymbol{\omega}^k \times \boldsymbol{q}\right)\right] + 2\frac{{}^jd}{dt}\left({}^j\boldsymbol{\omega}^k \times {}^k\boldsymbol{v}_O^m\right) + \frac{{}^jd}{dt}\left({}^k\boldsymbol{a}_O^m\right). \end{aligned} \tag{6.23} $$

The time derivatives indicated in Eq. (6.23) are displayed below:

$$\left.\begin{array}{l}
\frac{^jd}{dt}\left({}^j\boldsymbol{a}^k_{\tilde{B}}\right) = {}^j\boldsymbol{\rho}^k_{\tilde{B}},\\
\frac{^jd}{dt}\left({}^j\boldsymbol{\alpha}^k \times \boldsymbol{q}\right) = {}^j\boldsymbol{\rho}^k \times \boldsymbol{q} + {}^j\boldsymbol{\alpha}^k \times \left({}^j\boldsymbol{\omega}^k \times \boldsymbol{q}\right),\\
\frac{^jd}{dt}\left[{}^j\boldsymbol{\omega}^k \times \left({}^j\boldsymbol{\omega}^k \times \boldsymbol{q}\right)\right] = {}^j\boldsymbol{\alpha}^k \times \left({}^j\boldsymbol{\omega}^k \times \boldsymbol{q}\right) + {}^j\boldsymbol{\omega}^k \times \left[{}^j\boldsymbol{\alpha}^k \times \boldsymbol{q} + {}^j\boldsymbol{\omega}^k \times \left({}^j\boldsymbol{\omega}^k \times \boldsymbol{q}\right)\right],\\
2\frac{^jd}{dt}\left({}^j\boldsymbol{\omega}^k \times {}^k\boldsymbol{v}^m_O\right) = 2\left[{}^j\boldsymbol{\alpha}^k \times {}^k\boldsymbol{v}^m_O + {}^j\boldsymbol{\omega}^k \times \left({}^j\boldsymbol{\omega}^k \times {}^k\boldsymbol{v}^m_O + {}^k\boldsymbol{a}^m_O\right)\right],\\
\frac{^jd}{dt}\left({}^k\boldsymbol{a}^m_O\right) = {}^j\boldsymbol{\omega}^k \times {}^k\boldsymbol{a}^m_O + {}^k\boldsymbol{\rho}^m_O
\end{array}\right\}. \tag{6.24}$$

However, the following relationship arises by resorting to Eq. (6.7):

$${}^j\boldsymbol{\rho}^k_O = {}^j\boldsymbol{\rho}^k_{\tilde{B}} + {}^j\boldsymbol{\rho}^k \times \boldsymbol{q} + 2{}^j\boldsymbol{\alpha}^k \times \left({}^j\boldsymbol{\omega}^k \times \boldsymbol{q}\right) + {}^j\boldsymbol{\omega}^k \times \left({}^j\boldsymbol{\alpha}^k \times \boldsymbol{q}\right) + {}^j\boldsymbol{\omega}^k \times \left[{}^j\boldsymbol{\omega}^k \times \left({}^j\boldsymbol{\omega}^k \times \boldsymbol{q}\right)\right]. \tag{6.25}$$

Thus,

$${}^j\boldsymbol{\rho}^k_{\tilde{B}} = {}^j\boldsymbol{\rho}^k_O - {}^j\boldsymbol{\rho}^k \times \boldsymbol{q} - 2{}^j\boldsymbol{\alpha}^k \times \left({}^j\boldsymbol{\omega}^k \times \boldsymbol{q}\right) - {}^j\boldsymbol{\omega}^k \times \left({}^j\boldsymbol{\alpha}^k \times \boldsymbol{q}\right) - {}^j\boldsymbol{\omega}^k \times \left[{}^j\boldsymbol{\omega}^k \times \left({}^j\boldsymbol{\omega}^k \times \boldsymbol{q}\right)\right]. \tag{6.26}$$

Afterward, substituting Eqs. (6.26) and (6.24) into Eq. (6.23) and reducing terms, we obtain Eq. (6.18). ◇

Postulate 2. *Some vectors of Eq. (6.18) may be grouped as follows:*

$$\boldsymbol{j}_{cor} = 3{}^j\boldsymbol{\alpha}^k \times {}^k\boldsymbol{v}^m_O + 3{}^j\boldsymbol{\omega}^k \times {}^k\boldsymbol{a}^m_O + 3{}^j\boldsymbol{\omega}^k \times \left({}^j\boldsymbol{\omega}^k \times {}^k\boldsymbol{v}^m_O\right), \tag{6.27}$$

where $\boldsymbol{j}_{cor}$ *is termed the Coriolis jerk.*

Example 6.2. Truck B is traveling at the speed of 50 km/h and accelerates at the constant rate 2 m/s^2 over a circular curve of radius $r = 325$ m (see Fig. 6.2). At the same time, truck A is traveling at the speed 80 km/h and accelerates in the direction of its motion at the constant rate of 3 m/s^2. Determine the jerk of truck B as observed from truck A when $\theta = 30°$.

Solution. Let XY be a reference frame attached to the center of the curve with associated unit vectors $\hat{i}\hat{j}$. In this example it is possible to observe three bodies in relative motion. To apply Eq. (6.17), let us consider body j as the Earth, while body k is truck A and body m is truck B; for example, $j = 0$, $k = A$, and $m = B$. Hence,

$${}^0\boldsymbol{\rho}^B_O = {}^0\boldsymbol{\rho}^A_O + {}^A\boldsymbol{\rho}^B_O + 3{}^0\boldsymbol{\alpha}^A \times {}^A\boldsymbol{v}^B_O + 3{}^0\boldsymbol{\omega}^A \times {}^A\boldsymbol{a}^B_O + 3{}^0\boldsymbol{\omega}^A \times \left({}^0\boldsymbol{\omega}^A \times {}^A\boldsymbol{v}^B_O\right). \tag{6.28}$$

However, since body A is in translational motion, as observed from the Earth, it follows that ${}^0\boldsymbol{\omega}^A = {}^0\boldsymbol{\alpha} = \boldsymbol{0}$. Thus, Eq. 6.28 is reduced to

$${}^0\boldsymbol{\rho}^B_O = {}^0\boldsymbol{\rho}^A_O + {}^A\boldsymbol{\rho}^B_O. \tag{6.29}$$

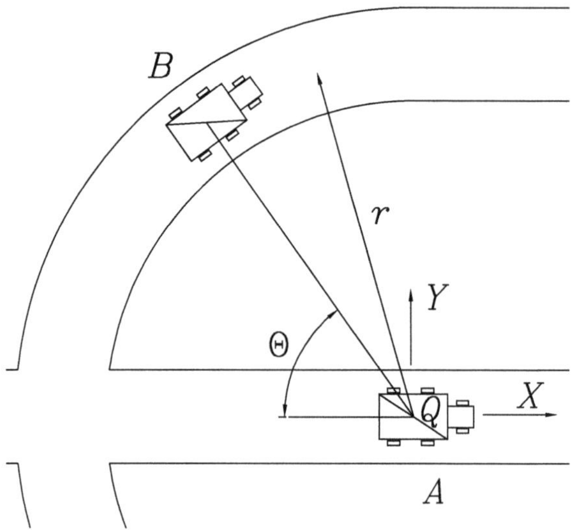

Fig. 6.2 Example 6.2. Two trucks in relative motion

Later, the jerk of truck B as measured from truck A, the vector ${}^{A}\boldsymbol{\rho}_{O}^{B}$, is given by

$$ {}^{A}\boldsymbol{\rho}_{O}^{B} = {}^{0}\boldsymbol{\rho}_{O}^{B} - {}^{0}\boldsymbol{\rho}_{O}^{A}. \tag{6.30}$$

Here the jerk of truck A, the vector ${}^{0}\boldsymbol{\rho}_{O}^{A} \equiv \frac{{}^{0}d}{dt}({}^{0}\boldsymbol{a}_{O}^{A})$, vanishes because body A accelerates at a constant rate, that is, ${}^{0}\boldsymbol{\rho}_{O}^{A} = \mathbf{0}$. Furthermore, according to Eq. (6.7), let us consider the center of the curve as point Q. Then the jerk of truck B, the vector ${}^{0}\boldsymbol{\rho}_{O}^{B}$, is given by

$$\begin{aligned} {}^{0}\boldsymbol{\rho}_{O}^{B} = \; & {}^{0}\boldsymbol{\rho}_{Q}^{B} + {}^{0}\boldsymbol{\rho}^{B} \times \boldsymbol{r}_{B/Q} + 2{}^{0}\boldsymbol{\alpha}^{B} \times \left({}^{0}\boldsymbol{\omega}^{B} \times \boldsymbol{r}_{B/Q}\right) + \\ & {}^{0}\boldsymbol{\omega}^{B} \times \left({}^{0}\boldsymbol{\alpha}^{B} \times \boldsymbol{r}_{B/Q}\right) + {}^{0}\boldsymbol{\omega}^{B} \times \left[{}^{0}\boldsymbol{\omega}^{B} \times \left({}^{0}\boldsymbol{\omega}^{B} \times \boldsymbol{r}_{B/Q}\right)\right]. \end{aligned} \tag{6.31}$$

To determine ${}^{0}\boldsymbol{\rho}_{O}^{B}$, the following facts must be considered:

- Body B moves on a circular trajectory of radius $\mid \boldsymbol{r}_{B/Q} \mid = r$. Then ${}^{0}\boldsymbol{\rho}_{Q}^{B} = \mathbf{0}$.
- ${}^{0}\boldsymbol{\omega}^{B} = \omega\hat{\boldsymbol{k}} = -v_B/r\hat{\boldsymbol{k}} = -0.042\hat{\boldsymbol{k}}$.
- ${}^{0}\boldsymbol{\alpha}^{B} = \alpha\hat{\boldsymbol{k}} = -2/325\,\hat{\boldsymbol{k}} = -0.006\,\hat{\boldsymbol{k}}$.
- Body B accelerates at a constant rate on the curve. Then ${}^{0}\boldsymbol{\rho}^{B} = \mathbf{0}$.
- $\boldsymbol{r}_{B/Q} = r_X\hat{\boldsymbol{i}} + \hat{\boldsymbol{j}} = -281.45\hat{\boldsymbol{i}} + 162.5\hat{\boldsymbol{j}}$.

Hence, according to Eq. (6.30), one obtains

$$\begin{aligned} {}^{A}\boldsymbol{\rho}_{O}^{B} &= {}^{0}\boldsymbol{\rho}_{O}^{B} \\ &= (\omega^3 r_Y - 3\omega\alpha r_X)\hat{\boldsymbol{i}} - (\omega^3 r_X - 3\omega\alpha r_Y)\hat{\boldsymbol{j}} \\ &= 0.2\hat{\boldsymbol{i}} - 0.143\hat{\boldsymbol{j}} \end{aligned} \tag{6.32}$$

or $\mid {}^{A}\boldsymbol{\rho}_{O}^{B} \mid = 0.245\ \text{m/s}^3$.

6.3 Jerk Equations in Screw Form

The representation of the reduced jerk state, or jerkor for brevity, of a rigid body meeting the conditions of helicoidal vector fields is a mandatory prerequisite to approach the jerk analysis of robot manipulators when screw theory is used.

Given two bodies m and j in relative motion, for us to determine a proper and true representation of the reduced jerk state of body m as measured from body j, the six-dimensional vector ${}^j\boldsymbol{J}^m_O$, we need to account for the conditions of helicoidal vector fields. In that case the vector of the helicoidal field must be the angular jerk ${}^j\boldsymbol{\rho}^m$. Meanwhile, the vector assigned by the reduced jerk state must include the linear jerk ${}^j\boldsymbol{\rho}^m_O$ of a point O^m of body m. Hence, given two points P and Q embedded in body m, we may write two respective representations of the reduced jerk state as follows:

$$ {}^j\boldsymbol{J}^m_P = \begin{bmatrix} {}^j\boldsymbol{\rho}^m \\ {}^j\boldsymbol{\rho}^m_P + \boldsymbol{F}(P) \end{bmatrix} \tag{6.33} $$

and

$$ {}^j\boldsymbol{J}^m_Q = \begin{bmatrix} {}^j\boldsymbol{\rho}^m \\ {}^j\boldsymbol{\rho}^m_Q + \boldsymbol{F}(Q) \end{bmatrix}. \tag{6.34} $$

Here $\boldsymbol{F}$ is, in this part of the section, an unknown entity.

Assuming that Eqs. (6.33 and (6.34) satisfy the properties of helicoidal vector fields, we write the following relationship:

$$ {}^j\boldsymbol{\rho}^m_P + \boldsymbol{F}(P) = {}^j\boldsymbol{\rho}^m_Q + \boldsymbol{F}(Q) + {}^j\boldsymbol{\rho}^m \times \boldsymbol{r}_{P/Q}. \tag{6.35} $$

Furthermore, the substitution of Eq. (6.7) into Eq. (6.35) yields

$$ \begin{aligned} &{}^j\boldsymbol{\rho}^m_Q + {}^j\boldsymbol{\rho}^m \times \boldsymbol{r}_{P/Q} + 2{}^j\boldsymbol{\alpha}^m \times \left({}^j\boldsymbol{\omega}^m \times \boldsymbol{r}_{P/Q}\right) + {}^j\boldsymbol{\omega}^m \times \left({}^j\boldsymbol{\alpha}^m \times \boldsymbol{r}_{P/Q}\right) + \\ &{}^j\boldsymbol{\omega}^m \times \left[{}^j\boldsymbol{\omega}^m \times \left({}^j\boldsymbol{\omega}^m \times \boldsymbol{r}_{P/Q}\right)\right] + \boldsymbol{F}(P) = {}^j\boldsymbol{\rho}^m_Q + \boldsymbol{F}(Q) + {}^j\boldsymbol{\rho}^m \times \boldsymbol{r}_{P/Q}. \end{aligned} \tag{6.36} $$

Thus,

$$ \begin{aligned} &2{}^j\boldsymbol{\alpha}^m \times ({}^j\boldsymbol{\omega}^m \times \boldsymbol{r}_{P/Q}) + {}^j\boldsymbol{\omega}^m \times ({}^j\boldsymbol{\alpha}^m \times \boldsymbol{r}_{P/Q}) + \\ &{}^j\boldsymbol{\omega}^m \times [{}^j\boldsymbol{\omega}^m \times ({}^j\boldsymbol{\omega}^m \times \boldsymbol{r}_{P/Q})] + \boldsymbol{F}(P) = \boldsymbol{F}(Q), \end{aligned} \tag{6.37} $$

where

$$ {}^j\boldsymbol{\omega}^m \times \boldsymbol{r}_{P/Q} = {}^j\boldsymbol{v}^m_P - {}^j\boldsymbol{v}^m_Q. \tag{6.38} $$

Meanwhile,

$$^j\boldsymbol{\alpha}^m \times \boldsymbol{r}_{P/Q} = {}^j\boldsymbol{a}_P^m - {}^j\boldsymbol{a}_Q^m - {}^j\boldsymbol{\omega}^m \times ({}^j\boldsymbol{\omega}^m \times \boldsymbol{r}_{P/Q}). \tag{6.39}$$

Substituting Eqs. (6.38) and (6.39) into Eq. (6.37) yields

$$\boldsymbol{F}(P) - \boldsymbol{F}(Q) = -(2^j\boldsymbol{\alpha}^m \times {}^j\boldsymbol{v}_P^m + {}^j\boldsymbol{\omega}^m \times {}^j\boldsymbol{a}_P^m) + (2^j\boldsymbol{\alpha}^m \times {}^j\boldsymbol{v}_Q^m + {}^j\boldsymbol{\omega}^m \times {}^j\boldsymbol{a}_Q^m). \tag{6.40}$$

It is a natural conclusion that two vector equations can be written upon Eq. (6.40) as follows:

$$\boldsymbol{F}(P) = -(2^j\boldsymbol{\alpha}^m \times {}^j\boldsymbol{v}_P^m + {}^j\boldsymbol{\omega}^m \times {}^j\boldsymbol{a}_P^m) + \boldsymbol{C} \tag{6.41}$$

and

$$\boldsymbol{F}(Q) = -(2^j\boldsymbol{\alpha}^m \times {}^j\boldsymbol{v}_Q^m + {}^j\boldsymbol{\omega}^m \times {}^j\boldsymbol{a}_Q^m) + \boldsymbol{C}. \tag{6.42}$$

Finally, if the goal of the analysis is to find the entity $\boldsymbol{F}$, the simplest solution arises when $\boldsymbol{C} = \boldsymbol{0}$ in Eqs. (6.41) and (6.42). Therefore, the following definition is well conceived.

Definition 6.2. Let m and j be two rigid bodies in relative motion and let O be a point embedded in body m. The reduced jerk state, or jerkor for brevity, of body m as observed from body j, taking point O as the reference pole, the six-dimensional vector ${}^j\boldsymbol{J}_O^m$, is defined as follows:

$$^j\boldsymbol{J}_O^m \equiv \begin{bmatrix} {}^j\boldsymbol{\rho}^m \\ {}^j\boldsymbol{\rho}_O^m - 2^j\boldsymbol{\alpha}^m \times {}^j\boldsymbol{v}_O^m - {}^j\boldsymbol{\omega}^m \times {}^j\boldsymbol{a}_O^m \end{bmatrix}. \tag{6.43}$$

Proposition 6.2. *With reference to Fig. 4.1, let j, k, and m be three rigid bodies in relative motion. The reduced jerk state of body m as observed from body j taking point O as the reference pole, the six-dimensional vector* ${}^j\boldsymbol{J}_O^m$, *may be expressed as*

$$^j\boldsymbol{J}_O^m = {}^j\boldsymbol{J}_O^k + {}^k\boldsymbol{J}_O^m + 2\left[{}^j\mathbf{V}_O^k\ {}^k\mathbf{A}_O^m\right] + \left[{}^j\mathbf{A}_O^k\ {}^k\mathbf{V}_O^m\right] + \left[{}^j\mathbf{V}_O^k\left[{}^j\mathbf{V}_O^k\ {}^k\mathbf{V}_O^m\right]\right], \tag{6.44}$$

where ${}^j\boldsymbol{J}_O^k$ *is the jerkor of body k with respect to body j. Meanwhile,* ${}^k\boldsymbol{J}_O^m$ *is the jerkor of body m measured from body k.*

Proof. Substituting Eqs. (6.17) and (6.18) into Eq. (6.43) leads to

$$^j\boldsymbol{J}_O^m = \begin{bmatrix} {}^j\boldsymbol{\rho}^k + {}^k\boldsymbol{\rho}^m + 2^j\boldsymbol{\omega}^k \times {}^k\boldsymbol{\alpha}^m + {}^j\boldsymbol{\alpha}^k \times {}^k\boldsymbol{\omega}^m + {}^j\boldsymbol{\omega}^k \times \left({}^j\boldsymbol{\omega}^k \times {}^k\boldsymbol{\omega}^m\right) \\ \\ {}^j\boldsymbol{\rho}_O^k + {}^k\boldsymbol{\rho}_O^m + 3^j\boldsymbol{\alpha}^k \times {}^k\boldsymbol{v}_O^m + 3^j\boldsymbol{\omega}^k \times {}^k\boldsymbol{a}_O^m + 3^j\boldsymbol{\omega}^k \times \left({}^j\boldsymbol{\omega}^m \times {}^k\boldsymbol{v}_O^m\right) \\ -2\left({}^j\boldsymbol{\alpha}^k + {}^k\boldsymbol{\alpha}^m + {}^j\boldsymbol{\omega}^k \times {}^k\boldsymbol{\omega}^m\right) \times \left({}^j\boldsymbol{v}_O^k + {}^k\boldsymbol{v}_O^m\right) \\ -\left({}^j\boldsymbol{\omega}^k + {}^k\boldsymbol{\omega}^m\right) \times \left({}^j\boldsymbol{a}_O^k + {}^k\boldsymbol{a}_O^m + 2^j\boldsymbol{\omega}^k \times {}^k\boldsymbol{v}_O^m\right) \end{bmatrix}. \tag{6.45}$$

Expanding and grouping terms in Eq. (6.45), we have

$$
\begin{aligned}
{}^{j}\boldsymbol{J}_{O}^{m} = {} & \begin{bmatrix} {}^{j}\boldsymbol{\rho}^{k} \\ {}^{j}\boldsymbol{\rho}_{O}^{k} - 2{}^{j}\boldsymbol{\alpha}^{k} \times {}^{j}\boldsymbol{v}_{O}^{k} - {}^{j}\boldsymbol{\omega}^{k} \times {}^{j}\boldsymbol{a}_{O}^{k} \end{bmatrix} + \\
& \begin{bmatrix} {}^{k}\boldsymbol{\rho}^{m} \\ {}^{k}\boldsymbol{\rho}_{O}^{m} - 2{}^{k}\boldsymbol{\alpha}^{m} \times {}^{k}\boldsymbol{v}_{O}^{m} - {}^{k}\boldsymbol{\omega}^{m} \times {}^{k}\boldsymbol{a}_{O}^{m} \end{bmatrix} + \\
& 2\begin{bmatrix} {}^{j}\boldsymbol{\omega}^{k} \times {}^{k}\boldsymbol{\alpha}^{m} \\ {}^{j}\boldsymbol{\omega}^{k} \times \left({}^{k}\boldsymbol{a}_{O}^{m} - {}^{k}\boldsymbol{\omega}^{m} \times {}^{k}\boldsymbol{v}_{O}^{m}\right) - {}^{k}\boldsymbol{\alpha}^{m} \times {}^{j}\boldsymbol{v}_{O}^{k} \end{bmatrix} + \\
& \begin{bmatrix} {}^{j}\boldsymbol{\alpha}^{k} \times {}^{k}\omega^{m} \\ {}^{j}\boldsymbol{\alpha}^{k} \times {}^{k}\boldsymbol{v}_{O}^{m} - {}^{k}\omega^{m} \times \left({}^{j}\boldsymbol{a}_{O}^{k} - {}^{j}\boldsymbol{\omega}^{k} \times {}^{j}\boldsymbol{v}_{O}^{k}\right) \end{bmatrix} + \\
& \begin{bmatrix} {}^{j}\boldsymbol{\omega}^{k} \times \left({}^{j}\boldsymbol{\omega}^{k} \times {}^{k}\boldsymbol{\omega}^{m}\right) \\ {}^{j}\boldsymbol{\omega}^{k} \times \left({}^{j}\boldsymbol{\omega}^{k} \times {}^{k}\boldsymbol{v}_{O}^{m} - {}^{k}\boldsymbol{\omega}^{m} \times {}^{j}\boldsymbol{v}_{O}^{k}\right) - \left({}^{j}\boldsymbol{\omega}^{k} \times {}^{k}\boldsymbol{\omega}^{m}\right) \times {}^{j}\boldsymbol{v}_{O}^{k} \end{bmatrix}.
\end{aligned}
\tag{6.46}
$$

On the other hand, it is straightforward to demonstrate that the terms of the right side of Eq. (6.46) are equivalent to

$$
\left.
\begin{aligned}
& \begin{bmatrix} {}^{j}\boldsymbol{\rho}^{k} \\ {}^{j}\boldsymbol{\rho}_{O}^{k} - 2{}^{j}\boldsymbol{\alpha}^{k} \times {}^{j}\boldsymbol{v}_{O}^{k} - {}^{j}\boldsymbol{\omega}^{k} \times {}^{j}\boldsymbol{a}_{O}^{k} \end{bmatrix} = {}^{j}\boldsymbol{J}_{O}^{k}, \\
& \begin{bmatrix} {}^{k}\boldsymbol{\rho}^{m} \\ {}^{k}\boldsymbol{\rho}_{O}^{m} - 2{}^{k}\boldsymbol{\alpha}^{m} \times {}^{k}\boldsymbol{v}_{O}^{m} - {}^{k}\boldsymbol{\omega}^{m} \times {}^{k}\boldsymbol{a}_{O}^{m} \end{bmatrix} = {}^{k}\boldsymbol{J}_{O}^{m}, \\
& 2\begin{bmatrix} {}^{j}\boldsymbol{\omega}^{k} \times {}^{k}\boldsymbol{\alpha}^{m} \\ {}^{j}\boldsymbol{\omega}^{k} \times \left({}^{k}\boldsymbol{a}_{O}^{m} - {}^{k}\boldsymbol{\omega}^{m} \times {}^{k}\boldsymbol{v}_{O}^{m}\right) - {}^{k}\boldsymbol{\alpha}^{m} \times {}^{j}\boldsymbol{v}_{O}^{k} \end{bmatrix} = 2\left[{}^{j}\boldsymbol{V}_{O}^{k} \;\; {}^{k}\boldsymbol{A}_{O}^{m}\right], \\
& \begin{bmatrix} {}^{j}\boldsymbol{\alpha}^{k} \times {}^{k}\omega^{m} \\ {}^{j}\boldsymbol{\alpha}^{k} \times {}^{k}\boldsymbol{v}_{O}^{m} - {}^{k}\omega^{m} \times \left({}^{j}\boldsymbol{a}_{O}^{k} - {}^{j}\boldsymbol{\omega}^{k} \times {}^{j}\boldsymbol{v}_{O}^{k}\right) \end{bmatrix} = \left[{}^{j}\boldsymbol{A}_{O}^{k} \;\; {}^{k}\boldsymbol{V}_{O}^{m}\right], \\
& \begin{bmatrix} {}^{j}\boldsymbol{\omega}^{k} \times \left({}^{j}\boldsymbol{\omega}^{k} \times {}^{k}\boldsymbol{\omega}^{m}\right) \\ {}^{j}\boldsymbol{\omega}^{k} \times \left({}^{j}\boldsymbol{\omega}^{k} \times {}^{k}\boldsymbol{v}_{O}^{m} - {}^{k}\boldsymbol{\omega}^{m} \times {}^{j}\boldsymbol{v}_{O}^{k}\right) - \left({}^{j}\boldsymbol{\omega}^{k} \times {}^{k}\boldsymbol{\omega}^{m}\right) \times {}^{j}\boldsymbol{v}_{O}^{k} \end{bmatrix} = \\
& \left[{}^{j}\boldsymbol{V}_{O}^{k} \left[{}^{j}\boldsymbol{V}_{O}^{k} \; {}^{k}\boldsymbol{V}_{O}^{m}\right]\right]
\end{aligned}
\right\}.
\tag{6.47}
$$

Finally, substituting Eqs. (6.47) into Eq. (6.46) yields the result of the proposition. ◇

Example 6.3. The robot arm shown in Fig. 6.3 is elevating and extending simultaneously while it is rotating about a fixed vertical axis with a constant angular velocity $\dot{\beta} = 6\,\text{rad/s}$ along the Y-axis. At a given instant, $\theta = 30°$, $\dot{\theta} = 4\,\text{rad/s}$, $\ddot{\theta} = 2\,\text{rad/s}^2$, $\dddot{\theta} = 3\,\text{rad/s}^3$, $l = 1.5\,\text{m}$, $\dot{l} = 0.25\,\text{m/s}$, $\ddot{l} = 0.1\,\text{m/s}^2$, and $\dddot{l} = 0.15\text{m/s}^3$. Determine the angular and linear jerks of the gripped body P as measured from the fixed reference frame.

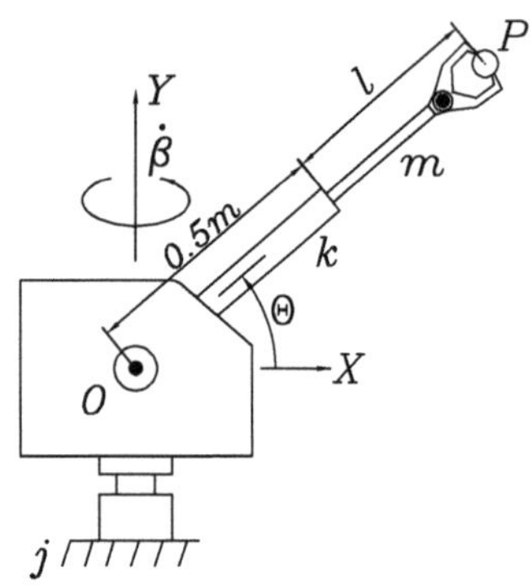

Fig. 6.3 Example 6.3. Three-degree-of-freedom robot arm

Solution. As an initial step, body j is chosen as the fixed reference frame whose origin is located at point O. Meanwhile, k denotes the arm of the robot, whereas m denotes the corresponding forearm. Taking into account that body k is in rotational motion with respect to body j, one can assume that $k = m$. Next, the angular velocity vector of body k as observed from body j is given by

$$ {}^j\boldsymbol{\omega}^k = \dot{\beta}\hat{\boldsymbol{j}} + \dot{\theta}\hat{\boldsymbol{k}}. \tag{6.48}$$

Similarly, the angular acceleration vector ${}^j\boldsymbol{\alpha}^k$ is denoted as

$$ {}^j\boldsymbol{\alpha}^k = \ddot{\theta}\hat{\boldsymbol{k}}, \tag{6.49}$$

while the angular jerk ${}^j\boldsymbol{\rho}^k$ is notated as

$$ {}^j\boldsymbol{\rho}^k = \dddot{\theta}\hat{\boldsymbol{k}}. \tag{6.50}$$

Furthermore, it is evident that body m is in translational motion as observed from body k, namely, $j = k$, which implies that the angular velocity, acceleration, and jerk of body m as measured from body k vanish. In other words, ${}^k\boldsymbol{\omega}^m = {}^k\boldsymbol{\alpha}^m = {}^k\boldsymbol{\rho}^m = \mathbf{0}$. Therefore, clearly the angular jerk of body m as measured from body j, the vector ${}^j\boldsymbol{\rho}^m$, according to Eq. (6.17), results in

$$\begin{aligned} {}^j\boldsymbol{\rho}^m &= {}^j\boldsymbol{\rho}^k + {}^k\boldsymbol{\rho}^m + 2{}^j\boldsymbol{\omega}^k \times {}^k\boldsymbol{\alpha}^m + {}^j\boldsymbol{\alpha}^k \times {}^k\boldsymbol{\omega}^m + {}^j\boldsymbol{\omega}^k \times \left({}^j\boldsymbol{\omega}^k \times {}^k\boldsymbol{\omega}^m\right) \\ &= {}^j\boldsymbol{\rho}^k = \dddot{\theta}\hat{\boldsymbol{k}} = 3\hat{\boldsymbol{k}}. \end{aligned} \tag{6.51}$$

On the other hand, the linear jerk of the gripped body P as measured from body j is given [see Eq. (6.18)] by

$$ {}^j\boldsymbol{\rho}_P^m = {}^j\boldsymbol{\rho}_P^k + {}^k\boldsymbol{\rho}_P^m + 3{}^j\boldsymbol{\alpha}^k \times {}^k\boldsymbol{v}_P^m + 3{}^j\boldsymbol{\omega}^k \times {}^k\boldsymbol{a}_P^m + 3{}^j\boldsymbol{\omega}^k \times \left({}^j\boldsymbol{\omega}^k \times {}^k\boldsymbol{v}_P^m\right). \tag{6.52}$$

Considering that $\hat{\boldsymbol{u}}_{P/O}$ is a unit vector pointed from O to P, for example,

$$\hat{\boldsymbol{u}}_{P/O} = \cos(\theta)\hat{\boldsymbol{i}} + \sin(\theta)\hat{\boldsymbol{j}} = 0.866\hat{\boldsymbol{i}} + 0.5\hat{\boldsymbol{j}}, \tag{6.53}$$

the velocity vector ${}^k\boldsymbol{v}_P^m$ is computed as

$${}^k\boldsymbol{v}_P^m = \dot{l}\hat{\boldsymbol{u}}_{P/O} = 0.216\hat{\boldsymbol{i}} + 0.125\hat{\boldsymbol{j}}. \tag{6.54}$$

Similarly, the relative acceleration and jerk between bodies m and k are obtained, respectively, as

$${}^k\boldsymbol{a}_P^m = \ddot{l}\hat{\boldsymbol{u}}_{P/O} = 0.0866\hat{\boldsymbol{i}} + 0.050\hat{\boldsymbol{j}} \tag{6.55}$$

and

$${}^k\boldsymbol{\rho}_P^m = \dddot{l}\hat{\boldsymbol{u}}_{P/O} = 0.129\hat{\boldsymbol{i}} + 0.075\hat{\boldsymbol{j}}. \tag{6.56}$$

Furthermore, according to Eq. (6.7), the jerk of point P, fixed in body k, as observed from body j is given by

$$\begin{aligned} {}^j\boldsymbol{\rho}_P^k = {} & {}^j\boldsymbol{\rho}^k \times \boldsymbol{r}_{P/O} + 2{}^j\boldsymbol{\alpha}^k \times \left({}^j\boldsymbol{\omega}^k \times \boldsymbol{r}_{P/O}\right) + \\ & {}^j\boldsymbol{\omega}^k \times \left({}^j\boldsymbol{\alpha}^k \times \boldsymbol{r}_{P/O}\right) + {}^j\boldsymbol{\omega}^k \times \left[{}^j\boldsymbol{\omega}^k \times \left({}^j\boldsymbol{\omega}^k \times \boldsymbol{r}_{P/O}\right)\right]. \end{aligned} \tag{6.57}$$

Finally, since ${}^j\boldsymbol{\rho}_O^k = \boldsymbol{0}$, we obtain ${}^j\boldsymbol{\rho}_P^k = 163.43\hat{\boldsymbol{i}} - 379.070\hat{\boldsymbol{j}} + 552.399\hat{\boldsymbol{k}}$ or ${}_j\rho_{Pk} = 689.6\ \mathrm{m/s^3}$. Finally, by substituting these results into Eq. (6.52), we obtain the jerk of the gripped body P as ${}^j\boldsymbol{\rho}_P^m = 128.435\hat{\boldsymbol{i}} - 382.657\hat{\boldsymbol{j}} + 559.841\hat{\boldsymbol{k}}$ (m/s^3) or ${}_j\rho_{Pm} = 690.181\ \mathrm{m/s^3}$.

Proposition 6.3. *Let j, $j+1$, …, $m-1$, and m be rigid bodies of a multibody mechanical system (see Fig. 4.6). The reduced jerk state of body m as observed from body j may be expressed through the intermediate jerkors ${}^j\boldsymbol{J}_O^{j+1}$, ${}^{j+1}\boldsymbol{J}_O^{j+2}$, …, ${}^{m-2}\boldsymbol{J}_O^{m-1}$, and ${}^{m-1}\boldsymbol{J}_O^m$ as follows:*

$$\begin{aligned} {}^j\boldsymbol{J}_O^m = {} & {}^j\boldsymbol{J}_O^{j+1} + {}^{j+1}\boldsymbol{J}_O^{j+2} + \ldots + {}^{m-2}\boldsymbol{J}_O^{m-1} + {}^{m-1}\boldsymbol{J}_O^m + \\ & 2\left[{}^j\boldsymbol{V}_O^{j+1} \quad {}^{j+1}\boldsymbol{A}_O^m\right] + 2\left[{}^{j+1}\boldsymbol{V}_O^{j+2} \quad {}^{j+2}\boldsymbol{A}_O^m\right] + \ldots + 2\left[{}^{m-2}\boldsymbol{V}_O^{m-1} \quad {}^{m-1}\boldsymbol{A}_O^m\right] \\ & + \left[{}^j\boldsymbol{A}_O^{j+1} \quad {}^{j+1}\boldsymbol{V}_O^m\right] + \left[{}^{j+1}\boldsymbol{A}_O^{j+2} \quad {}^{j+2}\boldsymbol{V}_O^m\right] + \ldots + \left[{}^{m-2}\boldsymbol{A}_O^{m-1} \quad {}^{m-1}\boldsymbol{V}_O^m\right] \\ & + \left[{}^j\boldsymbol{V}_O^{j+1}\left[{}^j\boldsymbol{V}_O^{j+1} \quad {}^{j+1}\boldsymbol{V}_O^m\right]\right] + \left[{}^{j+1}\boldsymbol{V}_O^{j+2}\left[{}^{j+1}\boldsymbol{V}_O^{j+2} \quad {}^{j+2}\boldsymbol{V}_O^m\right]\right] \\ & + \ldots + \left[{}^{m-2}\boldsymbol{V}_O^{m-1}\left[{}^{m-2}\boldsymbol{V}_O^{m-1} \quad {}^{m-1}\boldsymbol{V}_O^m\right]\right]. \end{aligned} \tag{6.58}$$

Proof. Substituting $j = m - 2$ and $k = m - 1$ into Eq. (6.44) leads to

$$
\begin{aligned}
{}^{m-2}\boldsymbol{J}_O^m = {}^{m-2}\boldsymbol{J}_O^{m-1} + {}^{m-1}\boldsymbol{J}_O^m + 2\left[{}^{m-2}\boldsymbol{V}_O^{m-1} \quad {}^{m-1}\boldsymbol{A}_O^m\right] \\
+\left[{}^{m-2}\boldsymbol{A}_O^{m-1} \quad {}^{m-1}\boldsymbol{V}_O^m\right] + \left[{}^{m-2}\boldsymbol{V}_O^{m-1} \quad \left[{}^{m-2}\boldsymbol{V}_O^{m-1} \quad {}^{m-1}\boldsymbol{V}_O^m\right]\right].
\end{aligned} \tag{6.59}
$$

On the other hand, by the induction hypothesis, based on expression (6.58) it is possible to write

$$
\begin{aligned}
{}^{j+1}\boldsymbol{J}_O^m = {}^{j+1}\boldsymbol{J}_O^{j+2} + {}^{j+2}\boldsymbol{J}_O^{j+3} + \ldots + {}^{m-2}\boldsymbol{J}_O^{m-1} + {}^{m-1}\boldsymbol{J}_O^m + \\
2\left[{}^{j+1}\boldsymbol{V}_O^{j+2} \quad {}^{j+2}\boldsymbol{A}_O^m\right] + 2\left[{}^{j+2}\boldsymbol{V}_O^{j+3} \quad {}^{j+2}\boldsymbol{A}_O^m\right] + \ldots + 2\left[{}^{m-2}\boldsymbol{V}_O^{m-1} \quad {}^{m-1}\boldsymbol{A}_O^m\right] \\
+\left[{}^{j+1}\boldsymbol{A}_O^{j+2} \quad {}^{j+2}\boldsymbol{V}_O^m\right] + \left[{}^{j+2}\boldsymbol{A}_O^{j+3} \quad {}^{j+3}\boldsymbol{V}_O^m\right] + \ldots + \left[{}^{m-2}\boldsymbol{A}_O^{m-1} \quad {}^{m-1}\boldsymbol{V}_O^m\right] \\
+\left[{}^{j+1}\boldsymbol{V}_O^{j+2} \quad \left[{}^{j+1}\boldsymbol{V}_O^{j+2} \quad {}^{j+2}\boldsymbol{V}_O^m\right]\right] + \left[{}^{j+2}\boldsymbol{V}_O^{j+3} \quad \left[{}^{j+2}\boldsymbol{V}_O^{j+3} \quad {}^{j+3}\boldsymbol{V}_O^m\right]\right] \\
+ \ldots + \left[{}^{m-2}\boldsymbol{V}_O^{m-1} \quad \left[{}^{m-2}\boldsymbol{V}_O^{m-1} \quad {}^{m-1}\boldsymbol{V}_O^m\right]\right].
\end{aligned} \tag{6.60}
$$

Furthermore, with the substitution of $k = j + 1$ into Eq. (6.2), we obtain the result of the proposition. ◇

The velocity state of a rigid body is a twist on a screw. The same was proved for the reduced acceleration. Hence, at this point the natural next step is to prove that the reduced jerk state is also a twist on a screw.

Proposition 6.4. *With reference to Fig. 4.2, let k and l be two rigid bodies connected to each other through a helical pair ${}^k\$^l = ({}^k\hat{\boldsymbol{s}}^l, {}^k\boldsymbol{s}_O^l)$, where ${}^k\hat{\boldsymbol{s}}^l$ denotes the direction of the line while ${}^k\boldsymbol{s}_O^l$ could be the moment produced by ${}^k\hat{\boldsymbol{s}}^l$ about a point O embedded in body l. Furthermore, let $\boldsymbol{q}$ be a vector bounded on body l directed from an arbitrary point B^l of body l located on the axis of the helical pair ${}^k\$^l$ to point O^l. Then the reduced jerk state of body l as observed from body k taking O^l as the reference pole may be expressed as*

$$
{}^k\boldsymbol{J}_O^l = {}_k\rho_l\,{}^k\$^l, \tag{6.61}
$$

where ${}_k\rho_l \equiv \frac{d}{dt}{}_k\alpha_l$. Meanwhile,

$$
{}^k\$^l = \begin{bmatrix} {}^k\hat{\boldsymbol{s}}^l \\ {}^k\boldsymbol{s}_O^l \end{bmatrix} = \begin{bmatrix} {}^k\hat{\boldsymbol{s}}^l \\ {}_kh_l\,{}^k\hat{\boldsymbol{s}}^l + {}^k\hat{\boldsymbol{s}}^l \times \boldsymbol{q} \end{bmatrix}.
$$

Proof. Consider that since bodies k and l share the unit vector ${}^k\hat{\boldsymbol{s}}^l$, the angular jerk of body l with respect to body k, the three-dimensional vector ${}^k\boldsymbol{\rho}^l$ is given by

$$
{}^k\boldsymbol{\rho}^l = {}_k\rho_l\,{}^k\hat{\boldsymbol{s}}^l. \tag{6.62}
$$

On the other hand, the jerk of point O^l may be expressed as

$$ {}^k\boldsymbol{\rho}^l_O = \frac{{}^kd}{dt}({}^k\boldsymbol{a}^l_O), \tag{6.63}$$

where the acceleration vector ${}^k\boldsymbol{a}^l_O$ is given by

$$ {}^k\boldsymbol{a}^l_O = {}_kh_l{}^k\boldsymbol{\alpha}^l + {}^k\boldsymbol{\alpha}^l \times \boldsymbol{q} + {}^k\boldsymbol{\omega}^l \times \left({}^k\boldsymbol{\omega}^l \times \boldsymbol{q}\right). \tag{6.64}$$

Hence,

$$ {}^k\boldsymbol{\rho}^l_O = \frac{{}^kd}{dt}({}_kh_l{}^k\boldsymbol{\alpha}^l) + \frac{{}^kd}{dt}({}^k\boldsymbol{\alpha}^l \times \boldsymbol{q}) + \frac{{}^kd}{dt}[{}^k\boldsymbol{\omega}^l \times ({}^k\boldsymbol{\omega}^l \times \boldsymbol{q})]. \tag{6.65}$$

The time derivatives provided in Eq. (6.65) are displayed below:

$$ \left.\begin{array}{l} \frac{{}^kd}{dt}({}_kh_l{}^k\boldsymbol{\alpha}^l) = {}_kh_l{}^k\boldsymbol{\rho}^l, \\ \frac{{}^kd}{dt}({}^k\boldsymbol{\alpha}^l \times \boldsymbol{q}) = {}^k\boldsymbol{\rho}^l \times \boldsymbol{q} + {}^k\boldsymbol{\alpha}^l \times ({}^k\boldsymbol{\omega}^l \times \boldsymbol{q}), \\ \frac{{}^kd}{dt}[{}^k\boldsymbol{\omega}^l \times ({}^k\boldsymbol{\omega}^l \times \boldsymbol{q})] = {}^k\boldsymbol{\alpha}^l \times ({}^k\boldsymbol{\omega}^l \times \boldsymbol{q}) + {}^k\boldsymbol{\omega}^l \times [{}^k\boldsymbol{\alpha}^l \times \boldsymbol{q} + {}^k\boldsymbol{\omega}^l \times ({}^k\boldsymbol{\omega}^l \times \boldsymbol{q})] \end{array}\right\}. \tag{6.66}$$

Substituting Eqs. (6.66) into Eq. (6.65) leads to

$$ \begin{aligned} {}^k\boldsymbol{\rho}^l_O = {}_kh_l{}^k\boldsymbol{\rho}^l + {}^k\boldsymbol{\rho}^l \times \boldsymbol{q} + {}^k\boldsymbol{\alpha}^l \times ({}^k\boldsymbol{\omega}^l \times \boldsymbol{q}) + \\ {}^k\boldsymbol{\alpha}^l \times ({}^k\boldsymbol{\omega}^l \times \boldsymbol{q}) + {}^k\boldsymbol{\omega}^l \times [{}^k\boldsymbol{\alpha}^l \times \boldsymbol{q} + {}^k\boldsymbol{\omega}^l \times ({}^k\boldsymbol{\omega}^l \times \boldsymbol{q})]. \end{aligned} \tag{6.67}$$

However, taking into account that

$$ {}^k\boldsymbol{\rho}^l \times \boldsymbol{q} = {}^k\boldsymbol{v}^l_O \tag{6.68}$$

and

$$ {}^k\boldsymbol{\alpha}^l \times \boldsymbol{q} + {}^k\boldsymbol{\omega}^l \times ({}^k\boldsymbol{\omega}^l \times \boldsymbol{q}) = {}^k\boldsymbol{a}^l_O, \tag{6.69}$$

we can rewrite Eq. (6.67) as

$$ {}^k\boldsymbol{\rho}^l_O - 2{}^k\boldsymbol{\alpha}^l \times {}^k\boldsymbol{v}^l_O - {}^k\boldsymbol{\omega}^l \times {}^k\boldsymbol{a}^l_O = {}_kh_l{}^k\boldsymbol{\rho}^l + {}^k\boldsymbol{\rho}^l \times \boldsymbol{q} \tag{6.70}$$

or

$$ {}^k\boldsymbol{\rho}^l_O - 2{}^k\boldsymbol{\alpha}^l \times {}^k\boldsymbol{v}^l_O - {}^k\boldsymbol{\omega}^l \times {}^k\boldsymbol{a}^l_O = {}_k\rho_l\left({}_kh_l{}^k\hat{\boldsymbol{s}}^l + {}^k\hat{\boldsymbol{s}}^l \times \boldsymbol{q}\right). \tag{6.71}$$

Therefore,

$$ {}^k\boldsymbol{\rho}^l_O - 2{}^k\boldsymbol{\alpha}^l \times {}^k\boldsymbol{v}^l_O - {}^k\boldsymbol{\omega}^l \times {}^k\boldsymbol{a}^l_O = {}_k\rho_l{}^k\boldsymbol{s}^l_O. \tag{6.72}$$

On the other hand, it is evident that the jerkor of body l with respect to body k is given by

$$ {}^k\boldsymbol{J}^l_O = \begin{bmatrix} {}^k\boldsymbol{\rho}^l \\ {}^k\boldsymbol{\rho}^l_O - 2\,{}^k\boldsymbol{\alpha}^l \times {}^k\boldsymbol{v}^l_O - {}^k\boldsymbol{\omega}^l \times {}^k\boldsymbol{a}^l_O \end{bmatrix}. \tag{6.73} $$

Thus, substituting Eqs. (6.62) and (6.72) into Eq. (6.73) leads to

$$ {}^k\boldsymbol{J}^l_O = \begin{bmatrix} {}_k\rho_l\,{}^k\hat{\boldsymbol{s}}^l \\ {}_k\rho_l\,{}^k\boldsymbol{s}^l_O \end{bmatrix}, \tag{6.74} $$

or

$$ {}^k\boldsymbol{J}^l_O = {}_k\rho_l \begin{bmatrix} {}^k\hat{\boldsymbol{s}}^l \\ {}^k\boldsymbol{s}^l_O \end{bmatrix} = {}_k\rho_l\,{}^k\$^l. \quad \diamond \tag{6.75} $$

The next proposition illustrates how the jerk analysis of an open kinematic chain is expressed in screw form.

Proposition 6.5. *Consider a serially connected kinematic chain formed with* $j, j+1, j+2, \ldots, m-2, m-1$, *and* m *rigid bodies. Furthermore, consider that adjacent bodies are connected by helical pairs. Then the reduced jerk state of body* m *as observed from body* j *may be expressed as follows:*

$$
\begin{aligned}
{}^j\boldsymbol{J}^m_O &= {}_j\rho_{j+1}{}^j\$^{j+1} + {}_{j+1}\rho_{j+2}{}^{j+1}\$^{j+2} + \ldots + {}_{m-1}\rho_m{}^{m-1}\$^m \\
+2\Big[{}_j\omega_{j+1}{}^j\$^{j+1} \quad & {}_{j+1}\alpha_{j+2}{}^{j+1}\$^{j+2} + \ldots + {}_{m-2}\alpha_{m-1}{}^{m-2}\$^{m-1} + {}_{m-1}\alpha_m{}^{m-1}\$^m \\
&+ \left[{}_{j+1}\omega_{j+2}{}^{j+1}\$^{j+2} \quad {}_{j+2}\omega_{j+3}{}^{j+2}\$^{j+3} + \ldots + {}_{m-1}\omega_m{}^{m-1}\$^m \right] \\
&+ \left[{}_{j+2}\omega_{j+3}{}^{j+2}\$^{j+3} \quad {}_{j+3}\omega_{j+4}{}^{j+3}\$^{j+4} + \ldots + {}_{m-1}\omega_m{}^{m-1}\$^m \right] \\
&+ \ldots + \left[{}_{m-2}\omega_{m-1}{}^{m-2}\$^{m-1} \quad {}_{m-1}\omega_m{}^{m-1}\$^m \right] \Big] \\
+2\Big[{}_{j+1}\omega_{j+2}{}^{j+1}\$^{j+2} \quad & {}_{j+2}\alpha_{j+3}{}^{j+2}\$^{j+3} + \ldots + {}_{m-1}\alpha_m{}^{m-1}\$^m \\
&+ \left[{}_{j+2}\omega_{j+3}{}^{j+2}\$^{j+3} \quad {}_{j+3}\omega_{j+4}{}^{j+3}\$^{j+4} + \ldots + {}_{m-1}\omega_m{}^{m-1}\$^m \right] \\
&+ \left[{}_{j+3}\omega_{j+4}{}^{j+3}\$^{j+4} \quad {}_{j+4}\omega_{j+5}{}^{j+4}\$^{j+5} + \ldots + {}_{m-1}\omega_m{}^{m-1}\$^m \right] \\
&+ \ldots + \left[{}_{m-2}\omega_{m-1}{}^{m-2}\$^{m-1} \quad {}_{m-1}\omega_m{}^{m-1}\$^m \right] \Big] \\
&+ \ldots + 2\left[{}_{m-2}\omega_{m-1}{}^{m-2}\$^{m-1} \quad {}_{m-1}\alpha_m{}^{m-1}\$^m \right] \\
+ \left[{}_j\alpha_{j+1}{}^j\$^{j+1} \quad {}_{j+1}\omega_{j+2}{}^{j+1}\$^{j+2} + \ldots + {}_{m-2}\omega_{m-1}{}^{m-2}\$^{m-1} + {}_{m-1}\omega_m{}^{m-1}\$^m \right] &
\end{aligned}
$$

$$
\begin{aligned}
&+\left[{}_{j+1}\alpha_{j+2}{}^{j+1}\$^{j+2} \quad {}_{j+2}\omega_{j+3}{}^{j+2}\$^{j+3}+\ldots+{}_{m-1}\omega_m{}^{m-1}\$^m\right]\\
&+\ldots+\left[{}_{m-2}\alpha_{m-1}{}^{m-2}\$^{m-1} \quad {}_{m-1}\omega_m{}^{m-1}\$^m\right]\\
&+\left[{}_j\omega_{j+1}{}^j\$^{j+1}\left[{}_j\omega_{j+1}{}^j\$^{j+1} \quad {}_{j+1}\omega_{j+2}{}^{j+1}\$^{j+2}+\ldots+{}_{m-1}\omega_m{}^{m-1}\$^m\right]\right]\\
&+\left[{}_{j+1}\omega_{j+2}{}^{j+1}\$^{j+2}\left[{}_{j+1}\omega_{j+2}{}^{j+1}\$^{j+2} \quad {}_{j+2}\omega_{j+3}{}^{j+2}\$^{j+3}+\ldots+{}_{m-1}\omega_m{}^{m-1}\$^m\right]\right]\\
&+\ldots+\left[{}_{m-2}\omega_{m-1}{}^{m-2}\$^{m-1}\left[{}_{m-2}\omega_{m-1}{}^{m-2}\$^{m-1} \quad {}_{m-1}\omega_m{}^{m-1}\$^m\right]\right].
\end{aligned}
\tag{6.76}
$$

Proof. The proof is immediate by a direct application of Propositions 4.6, 5.4, and 6.4 to expression (6.58). ◇

For simplicity Eq. (6.76) is rewritten as

$$
{}^j\boldsymbol{J}_O^m = {}_j\rho_{j+1}{}^j\$^{j+1} + {}_{j+1}\rho_{j+2}{}^{j+1}\$^{j+2} + \ldots + {}_{m-1}\rho_m{}^{m-1}\$^m + {}^j\mathcal{J}_O^m, \tag{6.77}
$$

where

$$
\begin{aligned}
{}^j\mathcal{J}_O^m = 2\Big[&{}_j\omega_{j+1}{}^j\$^{j+1} \quad {}_{j+1}\alpha_{j+2}{}^{j+1}\$^{j+2}+\ldots+{}_{m-1}\alpha_m{}^{m-1}\$^m\\
&+\left[{}_{j+1}\omega_{j+2}{}^{j+1}\$^{j+2} \quad {}_{j+2}\omega_{j+3}{}^{j+2}\$^{j+3}+\ldots+{}_{m-1}\omega_m{}^{m-1}\$^m\right]\\
&+\left[{}_{j+2}\omega_{j+3}{}^{j+2}\$^{j+3} \quad {}_{j+3}\omega_{j+4}{}^{j+3}\$^{j+4}+\ldots+{}_{m-1}\omega_m{}^{m-1}\$^m\right]\\
&+\ldots+\left[{}_{m-2}\omega_{m-1}{}^{m-2}\$^{m-1} \quad {}_{m-1}\omega_m{}^{m-1}\$^m\right]\Big]\\
+2\Big[&{}_{j+1}\omega_{j+2}{}^{j+1}\$^{j+2} \quad {}_{j+2}\alpha_{j+3}{}^{j+2}\$^{j+3}+\ldots+{}_{m-1}\alpha_m{}^{m-1}\$^m\\
&+\left[{}_{j+2}\omega_{j+3}{}^{j+2}\$^{j+3} \quad {}_{j+3}\omega_{j+4}{}^{j+3}\$^{j+4}+\ldots+{}_{m-1}\omega_m{}^{m-1}\$^m\right]\\
&+\left[{}_{j+3}\omega_{j+4}{}^{j+3}\$^{j+4} \quad {}_{j+4}\omega_{j+5}{}^{j+4}\$^{j+5}+\ldots+{}_{m-1}\omega_m{}^{m-1}\$^m\right]\\
&+\ldots+\left[{}_{m-2}\omega_{m-1}{}^{m-2}\$^{m-1} \quad {}_{m-1}\omega_m{}^{m-1}\$^m\right]\Big]\\
&+\ldots+2\left[{}_{m-2}\omega_{m-1}{}^{m-2}\$^{m-1} \quad {}_{m-1}\alpha_m{}^{m-1}\$^m\right]\\
+\Big[{}_j\alpha_{j+1}{}^j\$^{j+1} \quad &{}_{j+1}\omega_{j+2}{}^{j+1}\$^{j+2}+\ldots+{}_{m-2}\omega_{m-1}{}^{m-2}\$^{m-1}+{}_{m-1}\omega_m{}^{m-1}\$^m\Big]\\
&+\left[{}_{j+1}\alpha_{j+2}{}^{j+1}\$^{j+2} \quad {}_{j+2}\omega_{j+3}{}^{j+2}\$^{j+3}+\ldots+{}_{m-1}\omega_m{}^{m-1}\$^m\right]\\
&+\ldots+\left[{}_{m-2}\alpha_{m-1}{}^{m-2}\$^{m-1} \quad {}_{m-1}\omega_m{}^{m-1}\$^m\right]\\
&+\left[{}_j\omega_{j+1}{}^j\$^{j+1}\left[{}_j\omega_{j+1}{}^j\$^{j+1} \quad {}_{j+1}\omega_{j+2}{}^{j+1}\$^{j+2}+\ldots+{}_{m-1}\omega_m{}^{m-1}\$^m\right]\right]
\end{aligned}
$$

$$+ \left[{}_{j+1}\omega_{j+2}{}^{j+1}\$^{j+2}\left[{}_{j+1}\omega_{j+2}{}^{j+1}\$^{j+2} \quad {}_{j+2}\omega_{j+3}{}^{j+2}\$^{j+3} + \ldots + {}_{m-1}\omega_m{}^{m-1}\$^m\right]\right]$$
$$+ \ldots + \left[{}_{m-2}\omega_{m-1}{}^{m-2}\$^{m-1}\left[{}_{m-2}\omega_{m-1}{}^{m-2}\$^{m-1} \quad {}_{m-1}\omega_m{}^{m-1}\$^m\right]\right] \tag{6.78}$$

is termed the *Lie screw of jerk* of body m as observed from body j.

Example 6.4. Figure 6.4 shows a planar two-degree-of-freedom parallel manipulator where the position of point P is controlled by two generalized coordinates notated as q_1 and q_2. The exercise consists of finding the input–output equations of velocity, acceleration, and jerk of the robot related with the control point P. Note that the primal parts of the velocity state, the accelerator and the jerkor, vanish. Thus, the dimension of the Lie algebra must be properly taken into account.

Solution. Using the geometry of the mechanism, it is straightforward to demonstrate that

$$\begin{aligned} b &= \frac{q_1^2 - q_2^2 + a^2}{2a}, \\ h &= \sqrt{q_1^2 - b^2}, \\ \beta &= \tan^{-1}(h/b), \\ \gamma &= \tan^{-1}\left(\frac{h}{a-b}\right). \end{aligned} \tag{6.79}$$

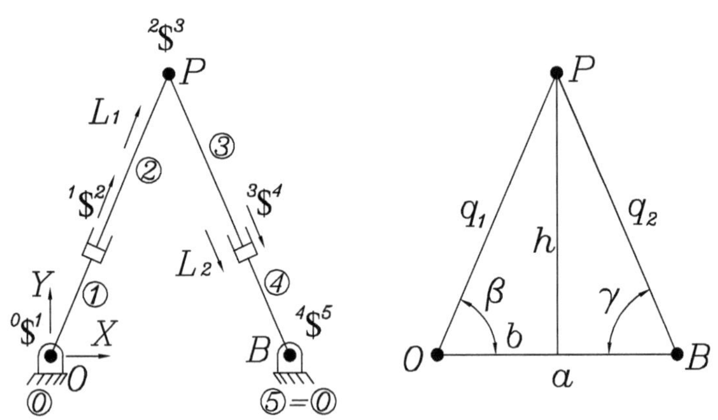

Fig. 6.4 Planar two-degree-of-freedom parallel manipulator

Next, the Plücker coordinates of the screws of the robot manipulator are computed as shown below:

$$
{}^{0}\$^{1} = \begin{bmatrix} 0 \\ 0 \\ 1 \\ -h \\ b \\ 0 \end{bmatrix}, \quad {}^{1}\$^{2} = \begin{bmatrix} 0 \\ 0 \\ 0 \\ \cos\beta \\ \sin\beta \\ 0 \end{bmatrix}, \quad {}^{2}\$^{3} = \begin{bmatrix} 0 \\ 0 \\ 1 \\ 0 \\ 0 \\ 0 \end{bmatrix},
$$

$$
{}^{3}\$^{4} = \begin{bmatrix} 0 \\ 0 \\ 0 \\ \cos\gamma \\ -\sin\gamma \\ 0 \end{bmatrix}, \quad {}^{4}\$^{5} = \begin{bmatrix} 0 \\ 0 \\ 1 \\ -h \\ b-a \\ 0 \end{bmatrix}. \tag{6.80}
$$

Let us consider $\boldsymbol{v}_P$ as the velocity of point P. The velocity state ${}^{0}\boldsymbol{V}_P^2$ of point P as observed from body 0, the vector

$$
{}^{0}\boldsymbol{V}_P^2 = \begin{bmatrix} \mathbf{0} \\ \boldsymbol{v}_P \end{bmatrix},
$$

may be expressed in screw form through the left limb of the planar manipulator as follows:

$$
{}_{0}\omega_{1}{}^{0}\$^{1} + {}_{1}\omega_{2}{}^{1}\$^{2} = {}^{0}\boldsymbol{V}_P^2, \tag{6.81}
$$

where ${}_{1}\omega_{2} = \dot{q}_1$. Furthermore, according to the closed kinematic we write

$$
\begin{aligned}
&{}_{0}\omega_{1}{}^{0}\$^{1} + {}_{1}\omega_{2}{}^{1}\$^{2} + {}_{2}\omega_{3}{}^{2}\$^{3} + {}_{3}\omega_{4}{}^{3}\$^{4} + {}_{4}\omega_{5}{}^{4}\$^{5} = \begin{bmatrix} \mathbf{0} \\ \mathbf{0} \end{bmatrix} \\
&\rightarrow {}_{0}\omega_{1}{}^{0}\$^{1} + {}_{1}\omega_{2}{}^{1}\$^{2} = -{}_{2}\omega_{3}{}^{2}\$^{3} - {}_{3}\omega_{4}{}^{3}\$^{4} - {}_{4}\omega_{5}{}^{4}\$^{5},
\end{aligned} \tag{6.82}
$$

where ${}_{3}\omega_{4} = \dot{q}_2$. Hence, it follows that

$$
-{}_{2}\omega_{3}{}^{2}\$^{3} - {}_{3}\omega_{4}{}^{3}\$^{4} - {}_{4}\omega_{5}{}^{4}\$^{5} = {}^{0}\boldsymbol{V}_P^2. \tag{6.83}
$$

Of course, Eq. (6.83) can also be obtained through the right limb of the manipulator.

On the other hand, consider two lines $\boldsymbol{L}_1$ and $\boldsymbol{L}_2$. The first line is directed from point O to the control point P, while the second line is directed from point P to point B. The corresponding Plücker coordinates of $\boldsymbol{L}_1$ and $\boldsymbol{L}_2$ are given by

$$L_1 = \begin{bmatrix} \cos\beta \\ \sin\beta \\ 0; \\ 0 \\ 0 \\ 0 \end{bmatrix}, \quad L_2 = \begin{bmatrix} \cos\gamma \\ -\sin\gamma \\ 0; \\ 0 \\ 0 \\ 0 \end{bmatrix}. \tag{6.84}$$

Applying the Klein form of the line L_1 to both sides of Eq. (6.81) and reducing terms, we have

$$\dot{q}_1 = \left\{ L_1; {}^0V_P^2 \right\}. \tag{6.85}$$

Similarly, applying the Klein form of the line L_2 to both sides of Eq. (6.83) gives us

$$-\dot{q}_2 = \left\{ L_2; {}^0V_P^2 \right\}. \tag{6.86}$$

Casting Eqs. (6.85) and (6.86) into matrix-vector form, the input–output equation of velocity of the robot results in

$$\mathbf{A}\boldsymbol{v}_P = \mathbf{B}\dot{\mathbf{q}}, \tag{6.87}$$

where

$$\mathbf{A} = \begin{bmatrix} \cos\beta & \sin\beta \\ \cos\gamma & -\sin\gamma \end{bmatrix}. \tag{6.88}$$

Meanwhile, $\mathbf{B}$ is the identity matrix of order 2, whereas

$$\dot{\mathbf{q}} = \begin{bmatrix} \dot{q}_1 & -\dot{q}_2 \end{bmatrix}^T \tag{6.89}$$

is the first-order driver matrix of the parallel manipulator.

According to expression (6.87), the robot is free of singularities concerning the inverse velocity analysis, that is, $\det \mathbf{B} \neq 0$. However, dealing with the forward velocity analysis, the robot is at a singular configuration when $\det \mathbf{A} = 0$, that is, when $\sin(\beta + \gamma) = 0$. Direct singularities occur when the movable bars are aligned.

In what follows the acceleration analysis of the robot is addressed. The reduced acceleration state of body 2 as observed from body 0 taking point P as the reference pole, the six-dimensional vector

$${}^0\boldsymbol{A}_P^2 = \begin{bmatrix} \mathbf{0} \\ \boldsymbol{a}_P - {}^0\boldsymbol{\alpha}^2 \times \boldsymbol{v}_P \end{bmatrix},$$

may be expressed in screw form as follows:

$${}_0\ddot{q}_1{}^0\$^1 + {}_1\ddot{q}_2{}^1\$^2 + {}^0\mathcal{A}_P^2 = {}^0\boldsymbol{A}_P^2. \tag{6.90}$$

Here ${}_1\alpha_2 = \ddot{q}_1$, whereas the Lie screw of acceleration ${}^0\mathcal{A}_P^2$ is given by

$$ {}^0\mathcal{A}_P^2 = \left[{}_0\omega_1{}^0\$^1 \quad {}_1\omega_2{}^1\$^2 \right]. \tag{6.91}$$

On the other hand, from the closed kinematic chain we write

$$ {}_0\alpha_1{}^0\$^1 + {}_1\alpha_2{}^1\$^2 + {}_2\alpha_3{}^2\$^3 + {}_3\alpha_4{}^3\$^4 + {}_4\alpha_5{}^4\$^5 + {}^0\mathcal{A}_P^5 = \begin{bmatrix} \mathbf{0} \\ \mathbf{0} \end{bmatrix}$$

$$ \rightarrow {}_0\alpha_1{}^0\$^1 + {}_1\alpha_2{}^1\$^2 = -{}_2\alpha_3{}^2\$^3 - {}_3\alpha_4{}^3\$^4 - {}_4\alpha_5{}^4\$^5 - {}^0\mathcal{A}_P^5, \tag{6.92}$$

where the Lie screw of acceleration ${}^0\mathcal{A}_P^5$ is computed as

$$ {}^0\mathcal{A}_P^5 = \left[{}_0\omega_1{}^0\$^1 \quad {}_1\omega_2{}^1\$^2 + {}_2\omega_3{}^2\$^3 + {}_3\omega_4{}^3\$^4 + {}_4\omega_5{}^4\$^5 \right] + \left[{}_1\omega_2{}^1\$^2 \quad {}_2\omega_3{}^2\$^3 + {}_3\omega_4{}^3\$^4 + {}_4\omega_5{}^4\$^5 \right] + \left[{}_2\omega_3{}^2\$^3 \quad {}_3\omega_4{}^3\$^4 + {}_4\omega_5{}^4\$^5 \right] + \left[{}_3\omega_4{}^3\$^4 \quad {}_4\omega_5{}^4\$^5 \right]. \tag{6.93}$$

Therefore, the reduced acceleration state ${}^0\boldsymbol{A}_P^2$ may be rewritten as follows:

$$ -{}_2\alpha_3{}^2\$^3 - {}_3\alpha_4{}^3\$^4 - {}_4\alpha_5{}^4\$^5 + \mathcal{A}_P = {}^0\boldsymbol{A}_P^2, \tag{6.94}$$

where $\mathcal{A}_P = {}^0\mathcal{A}_P^2 - {}^0\mathcal{A}_P^5$. Finally, following the trend of the velocity analysis, by resorting to reciprocal screw theory, we obtain the input–output equation of acceleration of the robot as

$$ \mathbf{A}\left[\boldsymbol{a}_P - \omega \times \boldsymbol{v}_P \right] = \mathbf{B}\ddot{\mathbf{q}} + \mathbf{M}_A. \tag{6.95}$$

Here

$$ \ddot{\mathbf{q}} = \left[\dot{q}_1 \quad -\dot{q}_2 \right]^T \tag{6.96}$$

is the second-order driver matrix of the robot. Meanwhile,

$$ \mathbf{M}_A = \left[\{\boldsymbol{L}_1; {}^0\mathcal{A}_P^5\} \quad \{\boldsymbol{L}_2; \mathcal{A}_P\} \right]^T \tag{6.97}$$

is the complementary matrix of acceleration.

The example finishes with the jerk analysis. The reduced jerk state of body 2 as observed from body 0 taking point P as the reference pole, the six-dimensional vector

$$ {}^0\boldsymbol{J}_P^2 = \begin{bmatrix} \mathbf{0} \\ {}^0\boldsymbol{\rho}_O^2 - 2{}^0\boldsymbol{\alpha}^2 \times {}^0\boldsymbol{v}_O^2 - {}^0\boldsymbol{\omega}^2 \times {}^0\boldsymbol{a}_O^2 \end{bmatrix}.$$

may be expressed in screw form as follows:

$$ {}_0\rho_1{}^0\$^1 + {}_1\rho_2{}^1\$^2 + {}^0\mathcal{J}_P^2 = {}^0\boldsymbol{J}_P^2. \tag{6.98}$$

Here ${}_1\alpha_2 = \dddot{q}_1$, whereas the Lie screw of jerk ${}^0\mathcal{J}_P^2$ is computed according to Eq. (6.78). On the other hand, from the closed kinematic chain it follows that

$$ {}_0\rho_1{}^0\$^1 + {}_1\rho_2{}^1\$^2 + {}_2\rho_3{}^2\$^3 + {}_3\rho_4{}^3\$^4 + {}_4\rho_5{}^4\$^5 + {}^0\mathcal{J}_P^5 = \begin{bmatrix} \mathbf{0} \\ \mathbf{0} \end{bmatrix}$$

$$ \rightarrow {}_0\rho_1{}^0\$^1 + {}_1\rho_2{}^1\$^2 = -{}_2\rho_3{}^2\$^3 - {}_3\rho_4{}^3\$^4 - {}_4\rho_5{}^4\$^5 - {}^0\mathcal{J}_P^5, \tag{6.99}$$

where the Lie screw of jerk is calculated according to Eq. (6.78). Later, the reduced jerk state ${}^0\boldsymbol{J}_P^2$ may be conveniently rewritten as follows:

$$ -{}_2\rho_3{}^2\$^3 - {}_3\rho_4{}^3\$^4 - {}_4\rho_5{}^4\$^5 + \mathcal{J}_P = {}^0\boldsymbol{J}_P^2, \tag{6.100}$$

where $\mathcal{J}_P = {}^0\mathcal{J}_P^2 - {}^0\mathcal{J}_P^5$. Finally, following the trend of the velocity and acceleration analyses, by resorting to reciprocal screw theory we obtain the input–output equation of jerk as

$$ \mathbf{A}\left[{}^0\boldsymbol{\rho}_P^2 - 2{}^0\boldsymbol{\alpha}^2 \times {}^0\boldsymbol{v}_P^2 - {}^0\boldsymbol{\omega}^2 \times {}^0\boldsymbol{a}_P^2\right] = \mathbf{B}\dddot{\mathbf{q}} + \mathbf{M}_J, \tag{6.101}$$

where

$$ \dddot{\mathbf{q}} = \begin{bmatrix} \dddot{q}_1 & -\dddot{q}_2 \end{bmatrix}^T \tag{6.102}$$

is the third-order driver matrix of the two-degree-of-freedom parallel manipulator. Meanwhile,

$$ \mathbf{M}_J = \begin{bmatrix} \{\boldsymbol{L}_1; {}^0\mathcal{J}_P^5\} & \{\boldsymbol{L}_2; \mathcal{J}_P\} \end{bmatrix}^T \tag{6.103}$$

is the complementary jerk matrix.

6.4 Exercises

1. The nozzle shown in Fig. 6.5 rotates with constant angular velocity ω about a fixed horizontal axis passing through point O. The velocity of the water relative to the nozzle is v and is constant. Determine the velocity, acceleration, and jerk of a water particle as it passes point A.
2. The slider P can be moved by the string S while the bar OA rotates around the pivot O (see Fig. 6.6). The orientation of the bar is given according to the expression $\theta = 0.25 + 0.1t + 0.05t^2 + 0.075t^3$, where the angle θ is given in

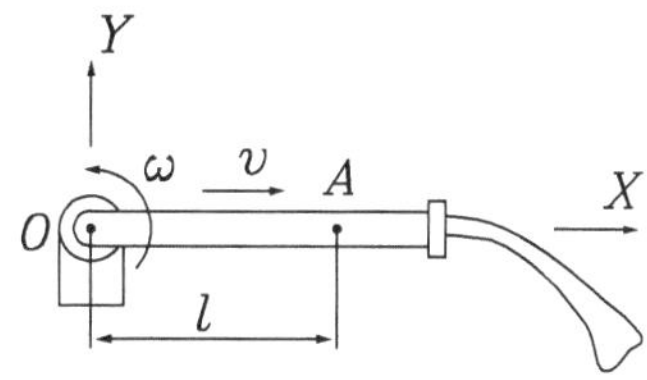

Fig. 6.5 Jerk: problem 1

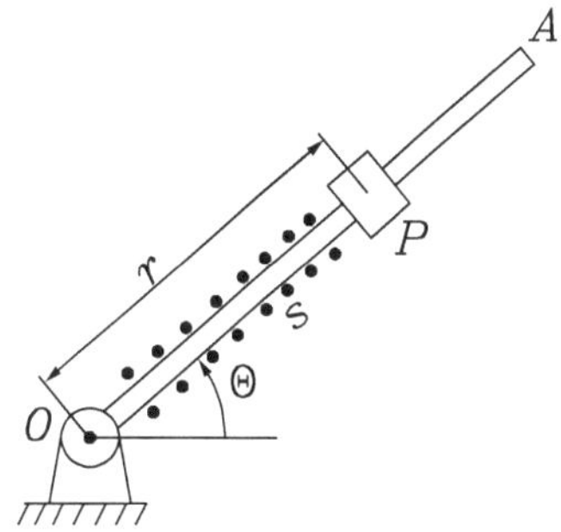

Fig. 6.6 Jerk: problem 2

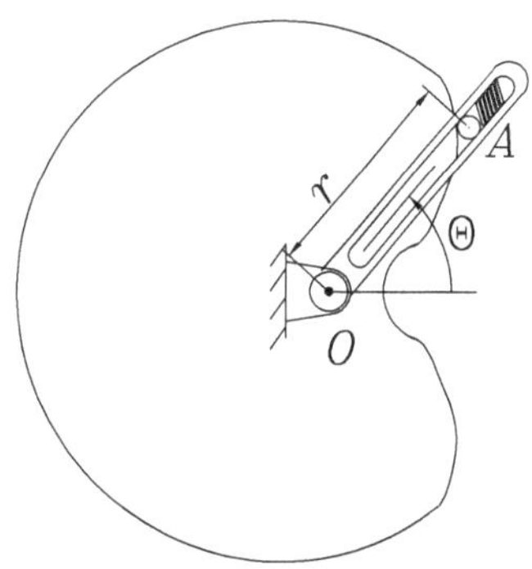

Fig. 6.7 Jerk: problem 3

radians and the time t is in seconds. The position of the slider is commanded to follow the expression $r = 0.6 - 0.25t - 0.025t^2 + 0.001t^3$, where r is given in meters and t in seconds. Determine the velocity, acceleration, and jerk of the slider at time $t = 3\,\text{s}$.

3. In the development of cam profiles due to tribological implications and the ability of the actuated body to follow the cam profile without chatter, the jerk must be taken into proper account. The cam shown in Fig. 6.7 is designed in such a way that the position of the center of the roller A must follow the function $r = b - c\cos(\theta)$, where $b > c$. Considering that the cam does not rotate, determine the magnitude of the jerk of A in terms of θ if the slotted arm revolves with a constant counterclockwise angular rate $\dot{\theta} = \omega$.
4. The disk shown in Fig. 6.8 rotates about a fixed axis passing through point O with angular velocity $\omega = 5\,\text{rad/s}$, angular acceleration $\alpha = 2\,\text{rad/s}^2$, and angular jerk $\rho = 3\,\text{rad/s}^3$. The small sphere A moves in the circular slot in such a way

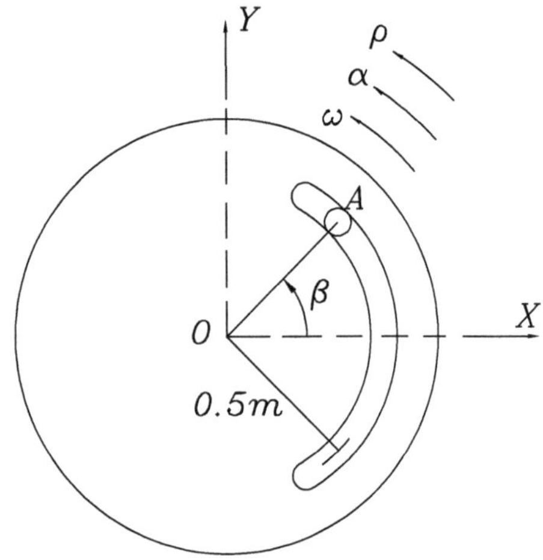

Fig. 6.8 Jerk: problem 4

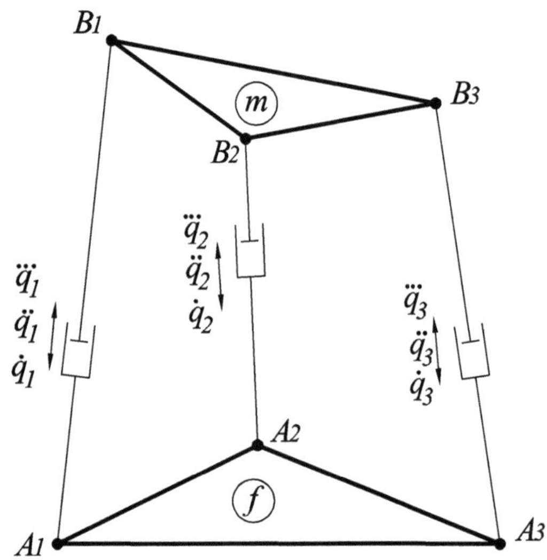

Fig. 6.9 Jerk: problem 5

that in the same instant we have $\beta = 30°$, $\dot{\beta} = 2\,\mathrm{rad/s}$, $\ddot{\beta} = -4\,\mathrm{rad/s}^2$, and $\dddot{\beta} = 4\,\mathrm{rad/s}^3$. Determine the velocity, acceleration, and jerk of A.

5. Determine the input–output equation of jerk of the 3-RPR parallel manipulator shown in Fig. 6.9 by resorting to reciprocal screw theory.

References

Cheng, W.-T. (2002). Synthesis of universal motion curves in generalized model. *ASME Journal of Mechanical Design, 124*(2), 284–293.

Crossman, E. R. F. W., & Goodeve, P. J. (1983). Feedback control of hand movements and Fitts' law. *Quarterly Journal of Experimental Psychology Section A—Human Experimental Psychology, A35*, 251–278.

Erkorkmaz, K., & Altintas, Y. (2001). High speed CNC system design, part I. Jerk limited trajectory generation and quintic spline interpolation. *International Journal of Machine Tools and Manufacture, 41*(9), 1323–1345.

Flash, T., & Hogan, N. (1985). The coordination of arm movements: an experimentally confirmed mathematical model. *Journal of Neuroscience, 5*(7), 1688–1703.

Gielen, C. C., Vrijenhock, E. J., & Neggers, S. F. (1997). Arm position constraints during pointing and reaching in 3-D space. *Journal of Neurophysiology, 78*(2), 660–673.

Goldvasser, D., McGibbon, C. A., & Krebs, D. E. (2001). High curvature and jerk analyses of arm ataxia. *Biological Cybernetics, 84*(2), 85–90.

Howard, R. A., Rosvold, J. M., Darcy, S. P., Corr, D. T., Shrive, N. G., Tapper, J. E., et al. (2007). Reproduction of in vivo motion using a parallel robot. *ASME Journal of Biomechanical Engineering, 129*(5), 743 (7 pp).

Ishii, T., Mukai, Y., Hosono, N., Sakaura, H., Fujii, R., Nakajima, Y., et al. (2006). Kinematics of the cervical spine in lateral bending: in vivo three-dimensional analysis. *Spine, 31*(2), 155–160.

Jones, M., Holt, C., & Franyuti, D. (2008). Developing a methodology for the analysis of infant spine kinematics for the investigation of the shaken baby syndrome. *Journal of Biomechanics, 41*, S355.

Konz, R. J., Fatone, S., Stine, R. L., Ganju, A., Gard, S. A., & Ondra, S. L. (2006). A kinematic model to assess spinal motion during walking. *Spine, 31*, E898–E906.

Lee, T. S., & Lin, Y. J. (1999). An improved sculptured part surface design method with jerk continuity consideration for smooth machining. *International Journal of Advanced Manufacturing Technology, 15*, 640–648.

Lessard, S., Bigras, P., & Bonev, I. A. (2007). A new medical parallel robot and its static balancing optimization. *ASME Journal of Medical Devices, 1*(4), 272–278.

Levin, S. M. (1995). The importance of soft tissue for structural support of the body. In T. A. Dorman (Ed.), *Prolotherapy in the lumbar spine and pelvis. Spine: state of the art reviews, 9*(2), 357.

Li, Y., & Xu, Q. (2007). Design and development of a medical parallel robot for cardiopulmonary resuscitation. *IEEE/ASME Transactions on Mechatronics, 12*(3), 265–273.

Macfarlane, S., & Croft, E. A. (2003). Jerk-bounded manipulator trajectory planning: design for real-time applications. *IEEE Transactions on Robotics and Automation, 19*(1), 42–52.

Mandal, M., & Nazcar, T. K. (2009). Introduction of control points in splines for synthesis of optimized cam motion program. *Mechanism and Machine Theory, 44*(1), 255–271.

Morasso, P. (1981). Spatial control arm movements. *Experimental Brain Research, 42*(2), 223–227.

Nguyen, V.-U., & Kim, D. (2007). Cam profile smoothing by modified spline curves. In *Proceedings 12th IFToMM World Congress.*

Novàk, K. E., Miller, L. E., & Houk, J. C. (2000). Kinematic properties of rapid hand movements in a knob turning task. *Experimental Brain Research, 132*(4), 419–433.

Panjabi, M. M., & White, A. A. (1980). Basic biomechanics of the spine. *Neurosurgery, 7*(1), 76–93.

Rico, J. M., Gallardo, J., & Duffy, J. (1999). Screw theory and higher order kinematic analysis of open serial and closed chains. *Mechanism and Machine Theory, 34*(4), 559–586.

Schot, S. H. (1978). Jerk: the time rate of change acceleration. *American Journal of Physics, 46*(11), 1090–1094.

Sparis, P. D., & Mouroutsos, S. G. (1984). A new matrix method for the kinematic analysis and motion simulation of planar mechanisms with lower pairs. *ASME Journal of Mechanisms, Transmissions and Automation in Design, 106*(4), 429–436.

Suh, C. H. (1971). Higher order analysis of spatial coupler curves. *Journal of Mechanisms, 6*(1), 81–95.

Tsoi, Y. H., & Xie, S. Q. (2010). Design and control of a parallel robot for ankle rehabilitation. *International Journal of Intelligent Systems Technologies and Applications, 8*(1/2/3/4), 100–113.

Uno, Y., Kawato, M., & Suzuki, R. (1989). Formation and control of optimal trajectory in human multijoint arm movements. *Biological Cybernetics, 61*(2), 89–101.

Urbinati, F., & Pennestri, E. (1998). A tool for third order kinematic analysis of spatial mechanisms. In *Proceedings 25th Biennial Mechanisms Conference, Atlanta*. CD Rom, Paper DETC/MECH-5901.

Viviani, P., & Flash, T. (1995). Minimum-jerk, two-thirds power law, and isochrony: converging approaches to movement planning. *Journal of Experimental Psychology: Human Perception and Performance, 21*(1), 32–53.

Viviani, P., & Schneider, R. (1991). A development study of the relationship between geometry and kinematics in drawing movements. *Journal of Experimental Psychology: Human Perception and Performance, 17*(1), 198–218.

Willems, J. M., Jull, G. A., & Ng, J. K. F. (1996). An in vivo study of the primary and coupled rotations of the thoracic spine. *Clinical Biomechanics, 11*, 311–316.

Zahraee, M. A., & Neff, G. P. (1996). Computer-aided analysis of cam mechanisms: a reverse engineering approach. In *Proceedings Second International Mechanical Engineering Conference, Shiraz* (pp. 905–912).

Žefran, M., Kumar, V., & Croke, C. (1995). On the generation of smooth three-dimensional rigid body motions. *IEEE Transactions on Robotics and Automation, 14*(4), 579–589.

Zhu, S. J., Huang, Z., & Zhao, M. Y. (2008). Feasible human-spine motion simulators based on parallel manipulators. In Wu, H. (Ed.), *Parallel manipulators, towards new applications*. Vienna, Austria: I-Tech Education and Publishing.

Ziddiqui, M., Karadimas, E., Nicol, M., Smith, F. W., & Wardlaw, D. (2006). Effects of X-stop device on sagittal lumbar spine kinematics in spinal stenosis. *Journal of Spinal Disorder Technology, 19*, 328–333.

Chapter 7
Hyper-Jerk Analysis

7.1 Introduction

The hyper-jerk, also known as snap or jounce, is the time rate of the jerk and is more than an academic pursuit; for example, Wohlhart (2010) showed that any planar parallel manipulator at a singular configuration could be architecturally mobile according to hyper-jerk analysis. In this chapter the hyper-jerk analysis of rigid bodies is approached using screw theory. A few decades ago the theory of screws introduced by Ball (1900) seemed to be an "old-fashioned" mathematical tool limited to elementary kinematic analyses, such as the displacement and velocity analyses of rigid bodies. Although the robustness of the definition of reduced acceleration state of a rigid body was introduced by Brand (1947), the representation in screw form of the acceleration analysis of kinematic chains was formally published half a century later by Rico and Duffy (1996). Before that contribution, most kinematicians commonly accepted that screw theory could not handle higher-order analyses of mechanisms, which today we recognize to be a wrong assumption. The escape from singular configurations of serial manipulators, the characterization of singularities of closed chains, as well as the acceleration analysis of parallel manipulators have been the most recurrent subjects since the formulation introduced by Rico and Duffy (1996) was proved successfully; see, for instance, Rico et al. (1995), Rico and Gallardo (1996), Gallardo-Alvarado et al. (2007, 2010), Gallardo et al. (2008). Furthermore, the elucidation of the acceleration analysis in screw form opened the possibility to extend it to the jerk analysis (Rico et al. 1999; Gallardo-Alvarado and Rico-Martínez 2001), which was applied in the higher-order kinematic analyses of parallel manipulators (Gallardo-Alvarado 2003, 2012; Gallardo-Alvarado and Camarillo-Gómez 2011). In that concern, as shown by Lipkin (2005), screw theory is not limited by the order of the kinematic analysis. Without a doubt, the higher-order analysis is a subject of growing interest covering topics like path-planning control (Kyriakopoulos and Saridis 1994), improvement of maneuvers of mobile robots (Oyadiji et al. 2005), high-speed machining (Zhang

J. Gallardo-Alvarado, *Kinematic Analysis of Parallel Manipulators by Algebraic Screw Theory*, DOI 10.1007/978-3-319-31126-5_7

et al. 2011), jerk dynamics (Maccari 2011), smooth path planning for conventional vehicles (Villagra et al. 2012), optimal time-jerk algorithms for trajectory planning (Gasparetto et al. 2012), and so forth.

The material exposed in this chapter presents a little effort to ensure that screw theory is a confident and trusted mathematical tool that can be used successfully in the higher-order kinematic analyses of robot manipulators. The reader is invited to find in this chapter important clues to approach not only the hyper-jerk analysis of a rigid body but also the kinematic analysis of optional order of robot manipulators.

7.2 Fundamental Hyper-Jerk Equations

Definition 7.1. Let k and l be two rigid bodies or reference frames. The angular hyper-jerk of body l with respect to body k, the vector ${}^k\boldsymbol{\sigma}^l$, is defined as follows:

$$ {}^k\boldsymbol{\sigma}^l \equiv \frac{{}^kd}{dt}\left({}^k\boldsymbol{\rho}^l\right), \tag{7.1}$$

where ${}^k\boldsymbol{\rho}^l$ is the angular jerk vector of body l as measured from body k. Furthermore, if O^l is a point of body l, then the linear hyper-jerk of point O^l as observed from body j, the vector ${}^k\boldsymbol{\sigma}^l_O$, is defined as follows:

$$ {}^k\boldsymbol{\sigma}^l_O \equiv \frac{{}^kd}{dt}\left({}^k\boldsymbol{\rho}^l_O\right), \tag{7.2}$$

where ${}^k\boldsymbol{\rho}^l_O$ is the linear jerk vector of point O^l embedded in body l as observed from body k.

It is straightforward to demonstrate that the angular hyper-jerk vector may be expressed as

$$ {}^k\boldsymbol{\sigma}^l = {}_k\sigma_l\,{}^k\hat{\boldsymbol{s}}^l, \tag{7.3}$$

where ${}_k\sigma_l \equiv \frac{{}^kd}{dt}({}_k\rho_l)$ is the magnitude of vector ${}^k\boldsymbol{\sigma}^l$, also named the joint-hyper-jerk rate between bodies k and l.

Definition 7.1 allows us to find a relationship involving the linear jerks of two points, notated as P and Q, embedded in the same body. To this end, Eq. (6.7) is rewritten here:

$$ \begin{aligned} {}^j\boldsymbol{\rho}^m_P &= {}^j\boldsymbol{\rho}^m_Q + {}^j\boldsymbol{\rho}^m \times \boldsymbol{r}_{P/Q} + 2{}^j\boldsymbol{\alpha}^m \times \left({}^j\boldsymbol{\omega}^m \times \boldsymbol{r}_{P/Q}\right) + \\ &{}^j\boldsymbol{\omega}^m \times \left({}^j\boldsymbol{\alpha}^m \times \boldsymbol{r}_{P/Q}\right) + {}^j\boldsymbol{\omega}^m \times \left[{}^j\boldsymbol{\omega}^m \times \left({}^j\boldsymbol{\omega}^m \times \boldsymbol{r}_{P/Q}\right)\right]. \end{aligned} \tag{7.4}$$

Hence, the linear hyper-jerk ${}^j\boldsymbol{\sigma}^m_P$ is obtained as

$$\begin{aligned}
{}^j\boldsymbol{\sigma}_P^m &= \frac{{}^jd}{dt}({}^j\boldsymbol{\rho}_P^m) \\
&= \frac{{}^jd}{dt}({}^j\boldsymbol{\rho}_Q^m) + \frac{{}^jd}{dt}({}^j\boldsymbol{\rho}^m \times \boldsymbol{r}_{P/Q}) + 2\frac{{}^jd}{dt}\left[{}^j\boldsymbol{\alpha}^m \times \left({}^j\boldsymbol{\omega}^m \times \boldsymbol{r}_{P/Q}\right)\right] + \\
&\quad \frac{{}^jd}{dt}\left[{}^j\boldsymbol{\omega}^m \times \left({}^j\boldsymbol{\alpha}^m \times \boldsymbol{r}_{P/Q}\right)\right] + \frac{{}^jd}{dt}\left\{{}^j\boldsymbol{\omega}^m \times \left[{}^j\boldsymbol{\omega}^m \times \left({}^j\boldsymbol{\omega}^m \times \boldsymbol{r}_{P/Q}\right)\right]\right\}, \quad (7.5)
\end{aligned}$$

where the time derivatives of Eq. (7.5) are displayed below:

$$\left.\begin{array}{l}
\frac{{}^jd}{dt}({}^j\boldsymbol{\rho}_Q^m) = {}^j\boldsymbol{\sigma}_Q^m \\
\frac{{}^jd}{dt}({}^j\boldsymbol{\rho}^m \times \boldsymbol{r}_{P/Q}) = {}^j\boldsymbol{\sigma}^m \times \boldsymbol{r}_{P/Q} + {}^j\boldsymbol{\rho}^m \times ({}^j\boldsymbol{\omega}^m \times \boldsymbol{r}_{P/Q}) \\
\frac{{}^jd}{dt}\left[2{}^j\boldsymbol{\alpha}^m \times \left({}^j\boldsymbol{\omega}^m \times \boldsymbol{r}_{P/Q}\right)\right] = 2{}^j\boldsymbol{\rho}^m \times \left({}^j\boldsymbol{\omega}^m \times \boldsymbol{r}_{P/Q}\right) + \\
2{}^j\boldsymbol{\alpha}^m \times \left[{}^j\boldsymbol{\alpha}^m \times \boldsymbol{r}_{P/Q} + {}^j\boldsymbol{\omega}^m \times \left({}^j\boldsymbol{\omega}^m \times \boldsymbol{r}_{P/Q}\right)\right] \\
\frac{{}^jd}{dt}\left[{}^j\boldsymbol{\omega}^m \times \left({}^j\boldsymbol{\alpha}^m \times \boldsymbol{r}_{P/Q}\right)\right] = {}^j\boldsymbol{\alpha}^m \times \left({}^j\boldsymbol{\alpha}^m \times \boldsymbol{r}_{P/Q}\right) + \\
{}^j\boldsymbol{\omega}^m \times \left[{}^j\boldsymbol{\rho}^m \times \boldsymbol{r}_{P/Q} + {}^j\boldsymbol{\alpha}^m \times \left({}^j\boldsymbol{\omega}^m \times \boldsymbol{r}_{P/Q}\right)\right] \\
\frac{{}^jd}{dt}\left\{{}^j\boldsymbol{\omega}^m \times \left[{}^j\boldsymbol{\omega}^m \times \left({}^j\boldsymbol{\omega}^m \times \boldsymbol{r}_{P/Q}\right)\right]\right\} = {}^j\boldsymbol{\alpha}^m \times \left[{}^j\boldsymbol{\omega}^m \times \left({}^j\boldsymbol{\omega}^m \times \boldsymbol{r}_{P/Q}\right)\right] + \\
{}^j\boldsymbol{\omega}^m \times \left\{{}^j\boldsymbol{\alpha}^m \times \left({}^j\boldsymbol{\omega}^m \times \boldsymbol{r}_{P/Q}\right) + {}^j\boldsymbol{\omega}^m \times \left[{}^j\boldsymbol{\alpha}^m \times \boldsymbol{r}_{P/Q} + {}^j\boldsymbol{\omega}^m \times \left({}^j\boldsymbol{\omega}^m \times \boldsymbol{r}_{P/Q}\right)\right]\right\}
\end{array}\right\}. \quad (7.6)$$

Substituting Eqs. (7.6) into Eq. (7.5) and reducing terms, one obtains

$$\begin{aligned}
{}^j\boldsymbol{\sigma}_P^m = {}&{}^j\boldsymbol{\sigma}_Q^m + {}^j\boldsymbol{\sigma}^m \times \boldsymbol{r}_{P/Q} + 3{}^j\boldsymbol{\rho}^m \times \left({}^j\boldsymbol{\omega}^m \times \boldsymbol{r}_{P/Q}\right) + 3{}^j\boldsymbol{\alpha}^m \times \left({}^j\boldsymbol{\alpha}^m \times \boldsymbol{r}_{P/Q}\right) \\
&+ 3{}^j\boldsymbol{\alpha}^m \times \left[{}^j\boldsymbol{\omega}^m \times \left({}^j\boldsymbol{\omega}^m \times \boldsymbol{r}_{P/Q}\right)\right] + 2{}^j\boldsymbol{\omega}^m \times \left[{}^j\boldsymbol{\alpha}^m \times \left({}^j\boldsymbol{\omega}^m \times \boldsymbol{r}_{P/Q}\right)\right] \\
&+ {}^j\boldsymbol{\omega}^m \times \left({}^j\boldsymbol{\rho}^m \times \boldsymbol{r}_{P/Q}\right) + {}^j\boldsymbol{\omega}^m \times \left[{}^j\boldsymbol{\omega}^m \times \left({}^j\boldsymbol{\alpha}^m \times \boldsymbol{r}_{P/Q}\right)\right] \\
&+ {}^j\boldsymbol{\omega}^m \times \left\{{}^j\boldsymbol{\omega}^m \times \left[{}^j\boldsymbol{\omega}^m \times \left({}^j\boldsymbol{\omega}^m \times \boldsymbol{r}_{P/Q}\right)\right]\right\}. \quad (7.7)
\end{aligned}$$

Expression (7.7) provides the means to study the hyper-jerk of three bodies in relative motion.

Proposition 7.1. *With reference to Fig. 4.5, let j, k, and m be three rigid bodies or reference frames. Furthermore, let O^m and O^k be two points, the first one fixed to body m and the second one attached to body k, where point O^m is instantaneously coincident with point O^k. The angular and linear hyper-jerks, the vectors ${}^j\boldsymbol{\sigma}^m$ and ${}^j\boldsymbol{\sigma}_O^m$, of body m with respect to body j are given, respectively, by*

$$\begin{aligned}
{}^j\boldsymbol{\sigma}^m = {}&{}^j\boldsymbol{\sigma}^k + {}^k\boldsymbol{\sigma}^m + 3{}^j\boldsymbol{\omega}^k \times {}^k\boldsymbol{\rho}^m + 3{}^j\boldsymbol{\alpha}^m \times {}^k\boldsymbol{\alpha}^m + \\
&3{}^j\boldsymbol{\omega}^m \times \left({}^j\boldsymbol{\omega}^k \times {}^k\boldsymbol{\alpha}^m\right) + 2{}^j\boldsymbol{\alpha}^k \times \left({}^j\boldsymbol{\omega}^k \times {}^k\boldsymbol{\omega}^m\right) +
\end{aligned}$$

$$ {}^{j}\boldsymbol{\rho}^{k} \times {}^{k}\boldsymbol{\omega}^{m} + {}^{j}\boldsymbol{\omega}^{k} \times \left({}^{j}\boldsymbol{\alpha}^{k} \times {}^{k}\boldsymbol{\omega}^{m}\right) + {}^{j}\boldsymbol{\omega}^{k} \times \left[{}^{j}\boldsymbol{\omega}^{k} \times \left({}^{j}\boldsymbol{\omega}^{k} \times {}^{k}\boldsymbol{\omega}^{m}\right)\right] \tag{7.8} $$

and

$$ \begin{aligned} {}^{j}\boldsymbol{\sigma}_{O}^{m} = {}^{j}\boldsymbol{\sigma}_{O}^{k} + {}^{k}\boldsymbol{\sigma}_{O}^{m} + 4{}^{j}\boldsymbol{\omega}^{k} \times {}^{k}\boldsymbol{\rho}_{O}^{m} + 4{}^{j}\boldsymbol{\rho}^{k} \times {}^{k}\boldsymbol{v}_{O}^{m} + \\ 8{}^{j}\boldsymbol{\alpha}^{k} \times \left({}^{j}\boldsymbol{\omega}^{k} \times {}^{k}\boldsymbol{v}_{O}^{m}\right) + 6{}^{j}\boldsymbol{\alpha}^{k} \times {}^{k}\boldsymbol{a}_{O}^{m} + 6{}^{j}\boldsymbol{\omega}^{k} \times \left({}^{j}\boldsymbol{\omega}^{k} \times {}^{k}\boldsymbol{a}_{O}^{m}\right) + \\ 4{}^{j}\boldsymbol{\omega}^{k} \times \left({}^{j}\boldsymbol{\alpha}^{k} \times {}^{k}\boldsymbol{v}_{O}^{m}\right) + 4{}^{j}\boldsymbol{\omega}^{k} \times \left[{}^{j}\boldsymbol{\omega}^{k} \times \left({}^{j}\boldsymbol{\omega}^{k} \times {}^{k}\boldsymbol{v}_{O}^{m}\right)\right], \end{aligned} \tag{7.9} $$

where

$$ \begin{aligned} {}^{j}\boldsymbol{\sigma}_{O}^{k} = {}^{j}\boldsymbol{\sigma}_{B}^{k} + {}^{j}\boldsymbol{\sigma}^{k} \times \boldsymbol{q} + 3{}^{j}\boldsymbol{\rho}^{k} \times ({}^{j}\boldsymbol{\omega}^{k} \times \boldsymbol{q}) + 3{}^{j}\boldsymbol{\alpha}^{k} \times ({}^{j}\boldsymbol{\alpha}^{k} \times \boldsymbol{q}) \\ +3{}^{j}\boldsymbol{\alpha}^{k} \times \left[{}^{j}\boldsymbol{\omega}^{k} \times ({}^{j}\boldsymbol{\omega}^{k} \times \boldsymbol{q})\right] + 2{}^{j}\boldsymbol{\omega}^{k} \times ({}^{j}\boldsymbol{\rho}^{k} \times \boldsymbol{q}) + {}^{j}\boldsymbol{\omega}^{k} \times ({}^{j}\boldsymbol{\rho}^{k} \times \boldsymbol{q}) \\ +{}^{j}\boldsymbol{\omega}^{k} \times \left[{}^{j}\boldsymbol{\omega}^{k} \times ({}^{j}\boldsymbol{\alpha}^{k} \times \boldsymbol{q})\right] + {}^{j}\boldsymbol{\omega}^{k} \times \left\{{}^{j}\boldsymbol{\omega}^{k} \times [{}^{j}\boldsymbol{\omega}^{k} \times ({}^{j}\boldsymbol{\omega}^{k} \times \boldsymbol{q})]\right\}. \end{aligned} \tag{7.10} $$

Proof. Based on Eq. (6.17) we have

$$ \begin{aligned} {}^{j}\boldsymbol{\sigma}^{m} &= \frac{{}^{j}d}{dt}({}^{j}\boldsymbol{\rho}^{m}) \\ &= \frac{{}^{j}d}{dt}({}^{j}\boldsymbol{\rho}^{k} + {}^{k}\boldsymbol{\rho}^{m} + 2{}^{j}\boldsymbol{\omega}^{k} \times {}^{k}\boldsymbol{\alpha}^{m} + {}^{j}\boldsymbol{\alpha}^{k} \times {}^{k}\boldsymbol{\omega}^{m} + {}^{j}\boldsymbol{\omega}^{k} \times ({}^{j}\boldsymbol{\omega}^{k} \times {}^{k}\boldsymbol{\omega}^{m}) \\ &= \frac{{}^{j}d}{dt}({}^{j}\boldsymbol{\rho}^{k}) + \frac{{}^{j}d}{dt}({}^{k}\boldsymbol{\rho}^{m}) + \frac{{}^{j}d}{dt}(2{}^{j}\boldsymbol{\omega}^{k} \times {}^{k}\boldsymbol{\alpha}^{m}) + \frac{{}^{j}d}{dt}({}^{j}\boldsymbol{\alpha}^{k} \times {}^{k}\boldsymbol{\omega}^{m}) + \\ &\quad \frac{{}^{j}d}{dt}\left[{}^{j}\boldsymbol{\omega}^{k} \times ({}^{j}\boldsymbol{\omega}^{k} \times {}^{k}\boldsymbol{\omega}^{m})\right], \end{aligned} \tag{7.11} $$

where the time derivatives in Eq. (7.11) are displayed below:

$$ \left.\begin{aligned} &\tfrac{{}^{j}d}{dt}({}^{j}\boldsymbol{\rho}^{k}) = {}^{j}\boldsymbol{\sigma}^{k}, \\ &\tfrac{{}^{j}d}{dt}({}^{k}\boldsymbol{\rho}^{m}) = {}^{j}\boldsymbol{\omega}^{k} \times {}^{k}\boldsymbol{\rho}^{m} + {}^{k}\boldsymbol{\sigma}^{m}, \\ &2\tfrac{{}^{j}d}{dt}({}^{j}\boldsymbol{\omega}^{k} \times {}^{k}\boldsymbol{\alpha}^{m}) = 2[{}^{j}\boldsymbol{\alpha}^{k} \times {}^{k}\boldsymbol{\alpha}^{m} + {}^{j}\boldsymbol{\omega}^{k} \times ({}^{j}\boldsymbol{\omega}^{k} \times {}^{k}\boldsymbol{\alpha}^{m} + {}^{k}\boldsymbol{\rho}^{m})], \\ &\tfrac{{}^{j}d}{dt}({}^{j}\boldsymbol{\alpha}^{k} \times {}^{k}\boldsymbol{\omega}^{m}) = {}^{j}\boldsymbol{\rho}^{k} \times {}^{k}\boldsymbol{\omega}^{m} + {}^{j}\boldsymbol{\alpha}^{k} \times ({}^{j}\boldsymbol{\omega}^{k} \times {}^{k}\boldsymbol{\omega}^{m} + {}^{k}\boldsymbol{\alpha}^{m}), \\ &\tfrac{{}^{j}d}{dt}[{}^{j}\boldsymbol{\omega}^{k} \times ({}^{j}\boldsymbol{\omega}^{k} \times {}^{k}\boldsymbol{\omega}^{m})] = {}^{j}\boldsymbol{\alpha}^{k} \times ({}^{j}\boldsymbol{\omega}^{k} \times {}^{k}\boldsymbol{\omega}^{m}) \\ &+{}^{j}\boldsymbol{\omega}^{k} \times [{}^{j}\boldsymbol{\alpha}^{k} \times {}^{k}\boldsymbol{\omega}^{m} + {}^{j}\boldsymbol{\omega}^{k} \times ({}^{j}\boldsymbol{\omega}^{k} \times {}^{k}\boldsymbol{\omega}^{m} + {}^{k}\boldsymbol{\alpha}^{m})]. \end{aligned}\right\} \tag{7.12} $$

Substituting Eqs. (7.12) into Eq. (7.11) yields Eq. (7.8).

To prove the second part of the proposition based on Eq. (6.18), it is possible to write

$$
\begin{aligned}
{}^j\boldsymbol{\sigma}_O^m &= \frac{{}^jd}{dt}({}^j\boldsymbol{\rho}_O^m) \\
&= \frac{{}^jd}{dt}\Big({}^j\boldsymbol{\rho}_B^k + {}^j\boldsymbol{\rho}^k \times \boldsymbol{q} + 2{}^j\boldsymbol{\alpha}^k \times ({}^j\boldsymbol{\omega}^k \times \boldsymbol{q}) + {}^j\boldsymbol{\omega}^k \times ({}^j\boldsymbol{\alpha}^k \times \boldsymbol{q}) \\
&\quad + {}^j\boldsymbol{\omega}^k \times \left[{}^j\boldsymbol{\omega}^k \times ({}^j\boldsymbol{\omega}^k \times \boldsymbol{q})\right] + {}^k\boldsymbol{\rho}_O^m + 3{}^j\boldsymbol{\alpha}^k \times {}^k\boldsymbol{v}_O^m + 3{}^j\boldsymbol{\omega}^k \times {}^k\boldsymbol{a}_O^m \\
&\quad + 3{}^j\boldsymbol{\omega}^k \times ({}^j\boldsymbol{\omega}^k \times {}^k\boldsymbol{v}_O^m)\Big) \\
&= \frac{{}^jd}{dt}({}^j\boldsymbol{\rho}_B^k) + \frac{{}^jd}{dt}({}^j\boldsymbol{\rho}^k \times \boldsymbol{q}) + 2\frac{{}^jd}{dt}[{}^j\boldsymbol{\alpha}^k \times ({}^j\boldsymbol{\omega}^k \times \boldsymbol{q})] + \\
&\quad \frac{{}^jd}{dt}[{}^j\boldsymbol{\omega}^k \times ({}^j\boldsymbol{\alpha}^k \times \boldsymbol{q})] + \frac{{}^jd}{dt}\left\{{}^j\boldsymbol{\omega}^k \times \left[{}^j\boldsymbol{\omega}^k \times ({}^j\boldsymbol{\omega}^k \times \boldsymbol{q})\right]\right\} + \frac{{}^jd}{dt}({}^k\boldsymbol{\rho}_O^m) + \\
&\quad 3\frac{{}^jd}{dt}({}^j\boldsymbol{\alpha}^k \times {}^k\boldsymbol{v}_O^m) + 3\frac{{}^jd}{dt}({}^j\boldsymbol{\omega}^k \times {}^k\boldsymbol{a}_O^m) + 3\frac{{}^jd}{dt}[{}^j\boldsymbol{\omega}^k \times ({}^j\boldsymbol{\omega}^k \times {}^k\boldsymbol{v}_O^m)],
\end{aligned}
\tag{7.13}
$$

where the time derivatives in Eq. (7.13) are detailed below:

$$
\left.
\begin{aligned}
&\tfrac{{}^jd}{dt}({}^j\boldsymbol{\rho}_B^k) = {}^j\boldsymbol{\sigma}_B^k, \\
&\tfrac{{}^jd}{dt}({}^j\boldsymbol{\rho}^k \times \boldsymbol{q}) = {}^j\boldsymbol{\sigma}^k \times \boldsymbol{q} + {}^j\boldsymbol{\rho}^k \times [{}^j\boldsymbol{\omega}^k \times \boldsymbol{q} + {}^k\boldsymbol{v}_O^m], \\
&2\tfrac{{}^jd}{dt}[{}^j\boldsymbol{\alpha}^k \times ({}^j\boldsymbol{\omega}^k \times \boldsymbol{q})] = 2{}^j\boldsymbol{\rho}^k \times ({}^j\boldsymbol{\omega}^k \times \boldsymbol{q}) + 2{}^j\boldsymbol{\alpha}^k \times ({}^j\boldsymbol{\alpha}^k \times \boldsymbol{q} + {}^j\boldsymbol{\omega}^k \times {}^k\boldsymbol{v}_O^m), \\
&\tfrac{{}^jd}{dt}[{}^j\boldsymbol{\omega}^k \times ({}^j\boldsymbol{\alpha}^k \times \boldsymbol{q})] = {}^j\boldsymbol{\alpha}^k \times ({}^j\boldsymbol{\alpha}^k \times \boldsymbol{q}) + \\
&{}^j\boldsymbol{\omega}^k \times [{}^j\boldsymbol{\rho}^k \times \boldsymbol{q} + {}^j\boldsymbol{\alpha}^k \times ({}^j\omega^k \times \boldsymbol{q} + {}^k\boldsymbol{v}_O^m)], \\
&\tfrac{{}^jd}{dt}\left\{{}^j\boldsymbol{\omega}^k \times \left[{}^j\boldsymbol{\omega}^k \times ({}^j\boldsymbol{\omega}^k \times \boldsymbol{q})\right]\right\} = {}^j\boldsymbol{\alpha}^k \times \left[{}^j\boldsymbol{\omega}^k \times ({}^j\boldsymbol{\omega}^k \times \boldsymbol{q})\right] + \\
&{}^j\boldsymbol{\omega}^k \times \left\{{}^j\boldsymbol{\alpha}^k \times ({}^j\boldsymbol{\omega}^k \times \boldsymbol{q}) + {}^j\boldsymbol{\omega}^k \times \left[{}^j\boldsymbol{\alpha}^k \times \boldsymbol{q} + {}^j\boldsymbol{\omega}^k \times ({}^j\boldsymbol{\omega}^k \times \boldsymbol{q} + {}^k\boldsymbol{v}_O^m)\right]\right\}, \\
&\tfrac{{}^jd}{dt}({}^k\boldsymbol{\rho}_O^m) = {}^j\boldsymbol{\omega}^k \times {}^k\boldsymbol{\rho}_O^m + {}^k\boldsymbol{\sigma}_O^m, \\
&3\tfrac{{}^jd}{dt}({}^j\boldsymbol{\alpha}^k \times {}^k\boldsymbol{v}_O^m) = 3{}^j\boldsymbol{\rho}^k \times {}^k\boldsymbol{v}_O^m + 3{}^j\boldsymbol{\alpha}^k \times [{}^j\boldsymbol{\omega}^k \times {}^k\boldsymbol{v}_O^m + {}^k\boldsymbol{a}_O^m], \\
&3\tfrac{{}^jd}{dt}({}^j\boldsymbol{\omega}^k \times {}^k\boldsymbol{a}_O^m) = 3{}^j\boldsymbol{\alpha}^k \times {}^k\boldsymbol{a}_O^m + 3{}^j\boldsymbol{\omega}^k \times ({}^j\boldsymbol{\omega}^k \times {}^k\boldsymbol{a}_O^m + {}^k\boldsymbol{\rho}_O^m), \\
&3\tfrac{{}^jd}{dt}[{}^j\boldsymbol{\omega}^k \times ({}^j\boldsymbol{\omega}^k \times {}^k\boldsymbol{v}_O^m)] = 3{}^j\boldsymbol{\alpha}^k \times ({}^j\boldsymbol{\omega}^k \times {}^k\boldsymbol{v}_O^m) + \\
&3{}^j\boldsymbol{\omega}^k \times [{}^j\boldsymbol{\alpha}^k \times {}^k\boldsymbol{v}_O^m + {}^j\boldsymbol{\omega}^k \times ({}^j\boldsymbol{\omega}^k \times {}^k\boldsymbol{v}_O^m + {}^k\boldsymbol{a}_O^m)].
\end{aligned}
\right\}
\tag{7.14}
$$

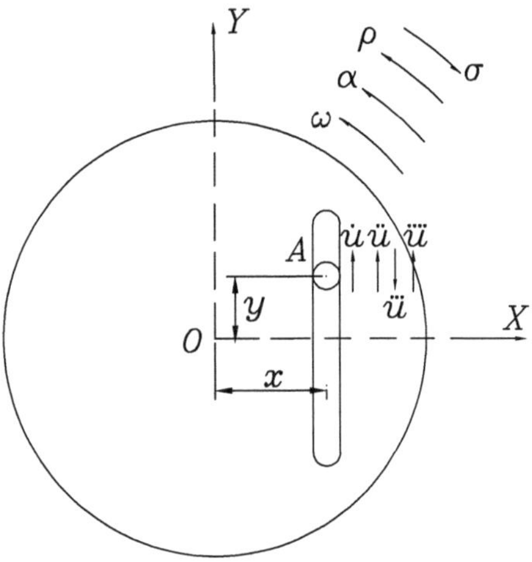

Fig. 7.1 Example 7.1. Rotating disk

Finally, substituting Eqs. (7.14) into Eq. (7.13) and reducing terms, we obtain Eq. (7.9). ◇

Postulate 3. *Some terms of Eq. (7.9) may be grouped as follows:*

$$\begin{aligned} \boldsymbol{hj}_{cor} = \; & 4\,{}^j\boldsymbol{\omega}^k \times {}^k\boldsymbol{\rho}_O^m + 4\,{}^j\boldsymbol{\rho}^k \times {}^k\boldsymbol{v}_O^m + \\ & 8\,{}^j\boldsymbol{\alpha}^k \times ({}^j\boldsymbol{\omega}^k \times {}^k\boldsymbol{v}_O^m) + 6\,{}^j\boldsymbol{\alpha}^k \times {}^k\boldsymbol{a}_O^m + 6\,{}^j\boldsymbol{\omega}^k \times ({}^j\boldsymbol{\omega}^k \times {}^k\boldsymbol{a}_O^m) + \\ & 4\,{}^j\boldsymbol{\omega}^k \times ({}^j\boldsymbol{\alpha}^k \times {}^k\boldsymbol{v}_O^m) + 4\,{}^j\boldsymbol{\omega}^k \times [{}^j\boldsymbol{\omega}^k \times ({}^j\boldsymbol{\omega}^k \times {}^k\boldsymbol{v}_O^m)], \end{aligned} \tag{7.15}$$

where $\boldsymbol{hj}_{cor}$ *is termed the Coriolis hyper-jerk.*

Example 7.1. The disk shown in Fig. 7.1 rotates about a fixed axis passing through point O with angular velocity $\omega = 5\,\mathrm{rad/s}$, angular acceleration $\alpha = 2\,\mathrm{rad/s^2}$, angular jerk $\rho = 3\,\mathrm{rad/s^3}$, and angular hyper-jerk $\sigma = 2\,\mathrm{rad/s^4}$ in the indicated directions. On the other hand, the small sphere A moves in the slot undergoing relative motions with respect to the disk as follows: $\dot{u} = 2\,\mathrm{m/s}$, $\ddot{u} = 1.5\,\mathrm{m/s^2}$, $\dddot{u} = 0.5\,\mathrm{m/s^3}$, and $\ddddot{u} = 3\,\mathrm{m/s^4}$ in the indicated directions. We must compute the total instantaneous velocity, acceleration, jerk, and hyper-jerk of sphere A considering that the instantaneous coordinates of A are given by $x = 0.75\,\mathrm{m}$ and $y = 0.5\,\mathrm{m}$.

Solution. Let XYZ be a fixed reference frame with associated unit vectors $\hat{\boldsymbol{i}}\hat{\boldsymbol{j}}\hat{\boldsymbol{k}}$ whose origin is located at point O. The total velocity of the sphere A, the vector $\boldsymbol{v}_A = {}^j\boldsymbol{v}_A^m$, can be obtained as

$$\boldsymbol{v}_A = {}^j\boldsymbol{v}_A^m = {}^j\boldsymbol{v}_A^k + {}^k\boldsymbol{v}_A^m, \tag{7.16}$$

where j denotes the fixed reference frame. Meanwhile, k denotes the disk, whereas body m is the proper sphere A.

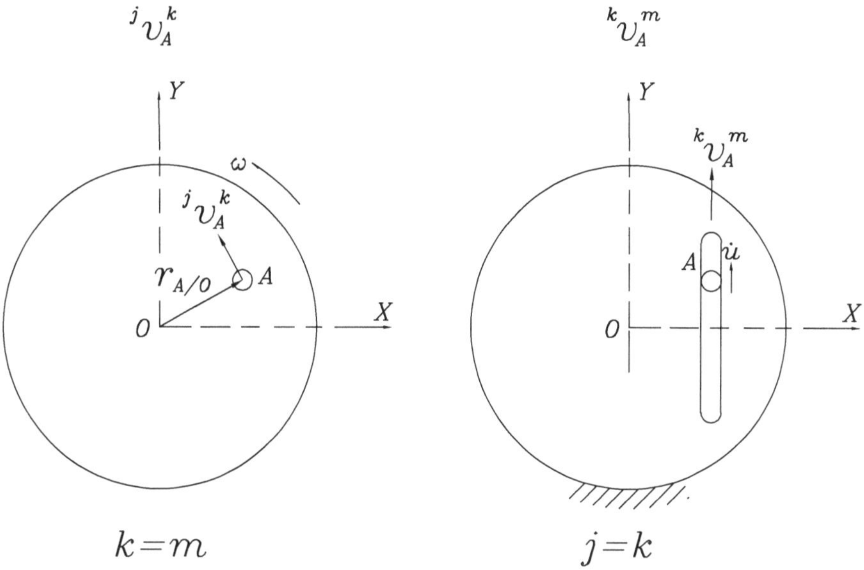

Fig. 7.2 Example 7.1. Decomposition of the velocity analysis

With reference to Fig. 7.2, evidently the velocity of A fixed in body k as observed from j is obtained as ${}^j\boldsymbol{v}_A^k = \boldsymbol{\omega} \times \boldsymbol{r}_{A/O}$, where $\boldsymbol{r}_{A/O} = x\hat{\boldsymbol{i}} + y\hat{\boldsymbol{j}}$ is the position vector of A with respect to O. Furthermore, the velocity of A fixed in body m as observed from k is given by ${}^k\boldsymbol{v}_A^m = \dot{u}\hat{\boldsymbol{j}}$. Hence, we have

$$ {}^j\boldsymbol{v}_A^m = \boldsymbol{\omega} \times (x\hat{\boldsymbol{i}} + y\hat{\boldsymbol{j}}) + \dot{u}\hat{\boldsymbol{j}} = -\omega y\hat{\boldsymbol{i}} + (\omega x + \dot{u})\hat{\boldsymbol{j}}, \tag{7.17} $$

or ${}^j\boldsymbol{v}_A^m = -2.5\hat{\boldsymbol{i}} + 5.75\hat{\boldsymbol{j}} \rightarrow v_A = 6.27\,\text{m/s}$.

On the other hand, the total acceleration of sphere A as measured from body j, the vector $\boldsymbol{a}_A = {}^j\boldsymbol{a}_A^m$, can be computed as

$$ \boldsymbol{a}_A = {}^j\boldsymbol{a}_A^m = {}^j\boldsymbol{a}_A^k + {}^k\boldsymbol{a}_A^m + 2{}^j\boldsymbol{\omega}^k \times {}^k\boldsymbol{v}_A^m. \tag{7.18} $$

With reference to Fig. 7.3, it follows that

$$ {}^j\boldsymbol{a}_A^k = \boldsymbol{a}_{At} + \boldsymbol{a}_{An} = \boldsymbol{\alpha} \times \boldsymbol{r}_{A/O} + \boldsymbol{\omega} \times (\boldsymbol{\omega} \times \boldsymbol{r}_{A/O}), \tag{7.19} $$

while

$$ {}^k\boldsymbol{a}_A^m = \ddot{u}\hat{\boldsymbol{j}}. \tag{7.20} $$

Furthermore, the Coriolis acceleration is computed as

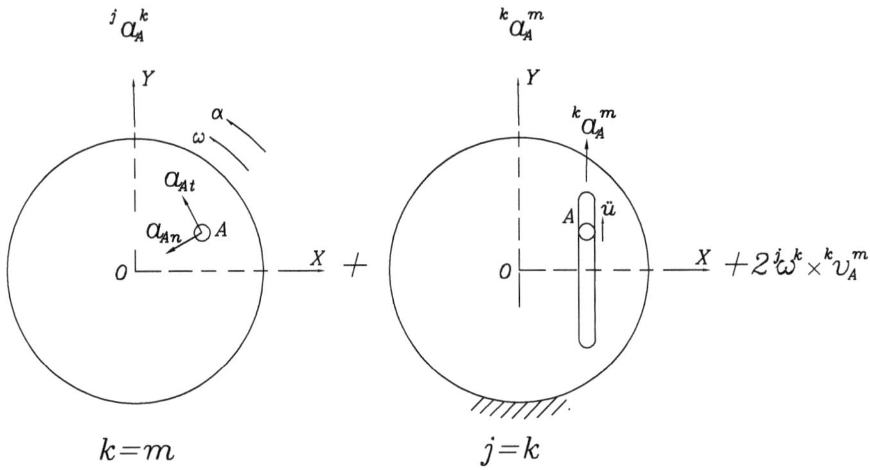

Fig. 7.3 Example 7.1. Decomposition of the acceleration analysis

$$2{}^j\boldsymbol{\omega}^k \times {}^k\boldsymbol{v}_A^m = 2\boldsymbol{\omega} \times \dot{u}\hat{\boldsymbol{j}}. \tag{7.21}$$

Therefore, one obtains

$$\begin{aligned}\boldsymbol{a}_A &= \boldsymbol{\alpha} \times \boldsymbol{r}_{A/O} + \boldsymbol{\omega} \times (\boldsymbol{\omega} \times \boldsymbol{r}_{A/O}) + \ddot{u}\hat{\boldsymbol{j}} + 2\boldsymbol{\omega} \times \dot{u}\hat{\boldsymbol{j}} \\ &= -(\alpha y + \omega^2 x + 2\omega\dot{u})\hat{\boldsymbol{\imath}} + (\alpha x - \omega^2 y + \ddot{u})\hat{\boldsymbol{j}},\end{aligned} \tag{7.22}$$

or $\boldsymbol{a}_A = -39.75\hat{\boldsymbol{\imath}} - 9.50\hat{\boldsymbol{j}} \rightarrow a_A = 40.869\,\text{m/s}^2$.

On the other hand, the jerk of the sphere A, the vector $\boldsymbol{\rho}_A = {}^j\boldsymbol{\rho}_A^m$, is computed as

$${}^j\boldsymbol{\rho}_A^m = {}^j\boldsymbol{\rho}_A^k + {}^k\boldsymbol{\rho}_A^m + 3{}^j\boldsymbol{\alpha}^k \times {}^j\boldsymbol{v}_A^m + 3{}^j\boldsymbol{\omega}^k \times {}^k\boldsymbol{a}_A^m + 3{}^j\boldsymbol{\omega}^k \times ({}^j\boldsymbol{\omega}^k \times {}^k\boldsymbol{v}_A^m), \tag{7.23}$$

where the jerk of A fixed in k as observed from body j is computed as

$$\begin{aligned}{}^j\boldsymbol{\rho}_A^k &= {}^j\boldsymbol{\rho}^k \times \boldsymbol{r}_{A/O} + 2{}^j\boldsymbol{\alpha}^k \times ({}^j\boldsymbol{\omega}^k \times \boldsymbol{r}_{A/O}) + {}^j\boldsymbol{\omega}^k \times ({}^j\boldsymbol{\alpha}^k \times \boldsymbol{r}_{A/O}) \\ &\quad + {}^j\boldsymbol{\omega}^k \times \left[{}^j\boldsymbol{\omega}^k \times ({}^j\boldsymbol{\omega}^k \times \boldsymbol{r}_{A/O})\right] \\ &= (-\rho y - 3\alpha\omega x + \omega^3 y)\hat{\boldsymbol{\imath}} + (\rho x - 3\alpha\omega y - \omega^3 x)\hat{\boldsymbol{j}}.\end{aligned} \tag{7.24}$$

Meanwhile,

$${}^k\boldsymbol{\rho}_A^m = \dddot{u}\hat{\boldsymbol{j}}. \tag{7.25}$$

Hence,

$${}^j\boldsymbol{\rho}_A^m = \left[-\rho y - 3\alpha\omega x + \omega^3 y - 3\alpha(\omega x + \dot{u}) - 3\omega\ddot{u}\right]\hat{\boldsymbol{\imath}} +$$

$$(\rho x - 6\alpha\omega y - \omega^3 x + \dddot{u} - 3\omega^2\dot{u})\hat{\boldsymbol{j}}, \tag{7.26}$$

or ${}^j\boldsymbol{\rho}_A^m = -18.5\hat{\boldsymbol{\imath}} - 272\hat{\boldsymbol{\jmath}} \rightarrow \rho_A = 272.628\,\mathrm{m/s^3}$.

Finally, the total hyper-jerk of the sphere A as measured from body j, the vector $\boldsymbol{\sigma}_A = {}^j\boldsymbol{\sigma}_A^m$, is computed according to Eq. (7.9) as

$$\begin{aligned}{}^j\boldsymbol{\sigma}_A^m &= {}^j\boldsymbol{\sigma}_A^k + {}^k\boldsymbol{\sigma}_A^m + 4{}^j\boldsymbol{\omega}^k \times {}^k\boldsymbol{\rho}_A^m + 4{}^j\boldsymbol{\rho}^k \times {}^k\boldsymbol{v}_A^m + \\ &8{}^j\boldsymbol{\alpha}^k \times \left({}^j\boldsymbol{\omega}^k \times {}^k\boldsymbol{v}_A^m\right) + 6{}^j\boldsymbol{\alpha}^k \times {}^k\boldsymbol{a}_A^m + 6{}^j\boldsymbol{\omega}^k \times \left({}^j\boldsymbol{\omega}^k \times {}^k\boldsymbol{a}_A^m\right) + \\ &4{}^j\boldsymbol{\omega}^k \times \left({}^j\boldsymbol{\alpha}^k \times {}^k\boldsymbol{v}_A^m\right) + 4{}^j\boldsymbol{\omega}^k \times \left[{}^j\boldsymbol{\omega}^k \times \left({}^j\boldsymbol{\omega}^k \times {}^k\boldsymbol{v}_A^m\right)\right],\end{aligned} \tag{7.27}$$

where [see Eq. (7.7)] the hyper-jerk of A fixed in body k as measured from body j, the vector ${}^j\boldsymbol{\sigma}_A^k$, is given by

$$\begin{aligned}{}^j\boldsymbol{\sigma}_A^k &= {}^j\boldsymbol{\sigma}^k \times \boldsymbol{r}_{A/O} + 3{}^j\boldsymbol{\rho}^k \times \left({}^j\boldsymbol{\omega}^k \times \boldsymbol{r}_{A/O}\right) + 3{}^j\boldsymbol{\alpha}^k \times \left({}^j\boldsymbol{\alpha}^k \times \boldsymbol{r}_{A/O}\right) + \\ &3{}^j\boldsymbol{\alpha}^k \times \left[{}^j\boldsymbol{\omega}^k \times \left({}^j\boldsymbol{\omega}^k \times \boldsymbol{r}_{A/O}\right)\right] + 2{}^j\boldsymbol{\omega}^k \times \left[{}^j\boldsymbol{\alpha}^k \times \left({}^j\boldsymbol{\omega}^k \times \boldsymbol{r}_{A/O}\right)\right] + \\ &{}^j\boldsymbol{\omega}^k \times \left({}^j\boldsymbol{\rho}^k \times \boldsymbol{r}_{A/O}\right) + {}^j\boldsymbol{\omega}^k \times \left[{}^j\boldsymbol{\omega}^k \times \left({}^j\boldsymbol{\alpha}^k \times \boldsymbol{r}_{A/O}\right)\right] + \\ &{}^j\boldsymbol{\omega}^k \times \left\{{}^j\boldsymbol{\omega}^k \times \left[{}^j\boldsymbol{\omega}^k \times \left({}^j\boldsymbol{\omega}^k \times \boldsymbol{r}_{A/O}\right)\right]\right\},\end{aligned} \tag{7.28}$$

while

$$ {}^k\boldsymbol{\sigma}_A^m = \ddddot{u}\hat{\boldsymbol{j}}. \tag{7.29}$$

Hence, after a few computations it follows that

$$\begin{aligned}{}^j\boldsymbol{\sigma}_A^m &= (-\sigma y - 4\rho\omega x - 3\alpha^2 x + 6\alpha\omega^2 y + \omega^4 x - 4\omega\dddot{u} - 4\rho\dot{u} - 6\alpha\ddot{u} + 4\omega^3\dot{u})\hat{\boldsymbol{\imath}} \\ &+(\sigma x - 4\rho\omega y - 3\alpha^2 y - 6\alpha\omega^2 x + \omega^4 y + \ddddot{u} - 12\alpha\omega\dot{u} - 6\omega^2\ddot{u})\hat{\boldsymbol{j}},\end{aligned} \tag{7.30}$$

or ${}^j\boldsymbol{\sigma}_A^m = 1533.75\hat{\boldsymbol{\imath}} - 412\hat{\boldsymbol{\jmath}} \rightarrow \sigma_A = 1588.12\ \mathrm{m/s^4}$.

7.3 Hyper-Jerk Equations in Screw Form

This section begins with a formal introduction of the concept of reduced hyper-jerk state of a rigid body as observed from another body or reference frame.

Proposition 7.2. *Let j and m be two rigid bodies. The reduced hyper-jerk state of body m with respect to body j, the six-dimensional vector ${}^j\boldsymbol{H}_O^m$, is given by*

$$ {}^j\boldsymbol{H}_O^m = \begin{bmatrix} {}^j\boldsymbol{\sigma}^m \\ {}^j\boldsymbol{\sigma}_O^m - 3{}^j\boldsymbol{\rho}^m \times {}^j\boldsymbol{v}_O^m - 3{}^j\boldsymbol{\alpha}^m \times {}^j\boldsymbol{a}_O^m - {}^j\boldsymbol{\omega}^m \times {}^j\boldsymbol{\rho}_O^m, \end{bmatrix} \tag{7.31}$$

where point O^m embedded in body m plays the role of reference pole.

Proof. Let P and Q be two points of body m where $P \neq Q$; then it is possible to write two representations of the hyper-jerk reduced state as follows:

$$^j\boldsymbol{H}_P^m = \begin{bmatrix} ^j\boldsymbol{\sigma}^m \\ ^j\boldsymbol{\sigma}_P^m + \boldsymbol{F}_P \end{bmatrix} \tag{7.32}$$

and

$$^j\boldsymbol{H}_Q^m = \begin{bmatrix} ^j\boldsymbol{\sigma}^m \\ ^j\boldsymbol{\sigma}_Q^m + \boldsymbol{F}_Q \end{bmatrix}, \tag{7.33}$$

where $^j\boldsymbol{\sigma}^m$ is the vector of the helicoidal field and $\boldsymbol{F}$ is an unknown entity. In accordance with the properties of helicoidal vector fields, we have

$$^j\boldsymbol{\sigma}_P^m + \boldsymbol{F}_P = {^j\boldsymbol{\sigma}_Q^m} + \boldsymbol{F}_Q + {^j\boldsymbol{\sigma}^m} \times \boldsymbol{r}_{P/Q}. \tag{7.34}$$

Substituting Eq. (7.7) into Eq. (7.34) leads to

$$\begin{aligned} &^j\boldsymbol{\sigma}_Q^m + {^j\boldsymbol{\sigma}^m} \times \boldsymbol{r}_{P/Q} + 3\,{^j\boldsymbol{\rho}^m} \times \left({^j\boldsymbol{\omega}^m} \times \boldsymbol{r}_{P/Q}\right) + 3\,{^j\boldsymbol{\alpha}^m} \times \left({^j\boldsymbol{\alpha}^m} \times \boldsymbol{r}_{P/Q}\right) \\ &+3\,{^j\boldsymbol{\alpha}^m} \times \left[{^j\boldsymbol{\omega}^m} \times \left({^j\boldsymbol{\omega}^m} \times \boldsymbol{r}_{P/Q}\right)\right] + 2\,{^j\boldsymbol{\omega}^m} \times \left[{^j\boldsymbol{\alpha}^m} \times \left({^j\boldsymbol{\omega}^m} \times \boldsymbol{r}_{P/Q}\right)\right] \\ &+{^j\boldsymbol{\omega}^m} \times \left({^j\boldsymbol{\rho}^m} \times \boldsymbol{r}_{P/Q}\right) + {^j\boldsymbol{\omega}^m} \times \left[{^j\boldsymbol{\omega}^m} \times \left({^j\boldsymbol{\alpha}^m} \times \boldsymbol{r}_{P/Q}\right)\right] \\ &+{^j\boldsymbol{\omega}^m} \times \left\{{^j\boldsymbol{\omega}^m} \times \left[{^j\boldsymbol{\omega}^m} \times \left({^j\boldsymbol{\omega}^m} \times \boldsymbol{r}_{P/Q}\right)\right]\right\} + \boldsymbol{F}_P = {^j\boldsymbol{\sigma}_Q^m} + \boldsymbol{F}_Q + {^j\boldsymbol{\sigma}^m} \times \boldsymbol{r}_{P/Q}. \end{aligned} \tag{7.35}$$

On the other hand, upon Eq. (6.7) it follows that

$$\begin{aligned} &^j\boldsymbol{\rho}^m \times \boldsymbol{r}_{P/Q} = {^j\boldsymbol{\rho}_P^m} - {^j\boldsymbol{\rho}_Q^m} - 2\,{^j\boldsymbol{\alpha}^m} \times \left({^j\boldsymbol{\omega}^m} \times \boldsymbol{r}_{P/Q}\right) - \\ &{^j\boldsymbol{\omega}^m} \times \left({^j\boldsymbol{\alpha}^m} \times \boldsymbol{r}_{P/Q}\right) - {^j\boldsymbol{\omega}^m} \times \left[{^j\boldsymbol{\omega}^m} \times \left({^j\boldsymbol{\omega}^m} \times \boldsymbol{r}_{P/Q}\right)\right]. \end{aligned} \tag{7.36}$$

Premultiplying Eq. (7.36) by $^j\boldsymbol{\omega}^m$, one obtains

$$\begin{aligned} &^j\boldsymbol{\omega}^m \times \left({^j\boldsymbol{\rho}^m} \times \boldsymbol{r}_{P/Q}\right) = {^j\boldsymbol{\omega}^m} \times \left({^j\boldsymbol{\rho}_P^m} - {^j\boldsymbol{\rho}_Q^m}\right) - \\ &2\,{^j\boldsymbol{\omega}^m} \times \left[{^j\boldsymbol{\alpha}^m} \times \left({^j\boldsymbol{\omega}^m} \times \boldsymbol{r}_{P/Q}\right)\right] - {^j\boldsymbol{\omega}^m} \times \left[{^j\boldsymbol{\omega}^m} \times \left({^j\boldsymbol{\alpha}^m} \times \boldsymbol{r}_{P/Q}\right)\right] \\ &-{^j\boldsymbol{\omega}^m} \times \left\{{^j\boldsymbol{\omega}^m} \times \left[{^j\boldsymbol{\omega}^m} \times \left({^j\boldsymbol{\alpha}^m} \times \boldsymbol{r}_{P/Q}\right)\right]\right\}. \end{aligned} \tag{7.37}$$

Substituting Eq. (7.37) into Eq. (7.35) and reducing terms, we get

$$3\,{^j\boldsymbol{\rho}^m} \times \left({^j\boldsymbol{\omega}^m} \times \boldsymbol{r}_{P/Q}\right) + 3\,{^j\boldsymbol{\alpha}^m} \times \left({^j\boldsymbol{\alpha}^m} \times \boldsymbol{r}_{P/Q}\right) +$$

$$3{}^j\boldsymbol{\alpha}^m \times \left[{}^j\boldsymbol{\omega}^m \times \left({}^j\boldsymbol{\omega}^m \times \boldsymbol{r}_{P/Q}\right)\right] + {}^j\boldsymbol{\omega}^m \times \left({}^j\boldsymbol{\rho}_P^m - {}^j\boldsymbol{\rho}_Q^m\right) + \boldsymbol{F}_P = \boldsymbol{F}_Q. \tag{7.38}$$

Furthermore, taking into account that

$${}^j\boldsymbol{\alpha}^m \times \boldsymbol{r}_{P/Q} = {}^j\boldsymbol{a}_P^m - {}^j\boldsymbol{a}_Q^m - {}^j\boldsymbol{\omega}^m \times \left({}^j\boldsymbol{\omega}^m \times \boldsymbol{r}_{P/Q}\right) \tag{7.39}$$

and

$${}^j\boldsymbol{\omega}^m \times \boldsymbol{r}_{P/Q} = {}^j\boldsymbol{v}_P^m - {}^j\boldsymbol{v}_Q^m, \tag{7.40}$$

then the substitution of Eqs. (7.39) and (7.40) into Eq. (7.38) leads to

$$\begin{aligned} &\boldsymbol{F}_P + 3{}^j\boldsymbol{\rho}^m \times {}^j\boldsymbol{v}_P^m + 3{}^j\boldsymbol{\alpha}^m \times {}^j\boldsymbol{a}_P^m + {}^j\boldsymbol{\omega}^m \times {}^j\boldsymbol{\rho}_P^m = \\ &\quad \boldsymbol{F}_Q + 3{}^j\boldsymbol{\rho}^m \times {}^j\boldsymbol{v}_Q^m + 3{}^j\boldsymbol{\alpha}^m \times {}^j\boldsymbol{a}_Q^m + {}^j\boldsymbol{\omega}^m \times {}^j\boldsymbol{\rho}_Q^m. \end{aligned} \tag{7.41}$$

Hence, following the trend of the acceleration and jerk analyses, Eq. (7.41) leads to

$$\boldsymbol{F}_P = -3{}^j\boldsymbol{\rho}^m \times {}^j\boldsymbol{v}_P^m - 3{}^j\boldsymbol{\alpha}^m \times {}^j\boldsymbol{a}_P^m - {}^j\boldsymbol{\omega}^m \times {}^j\boldsymbol{\rho}_P^m \tag{7.42}$$

and

$$\boldsymbol{F}_Q = -3{}^j\boldsymbol{\rho}^m \times {}^j\boldsymbol{v}_Q^m - 3{}^j\boldsymbol{\alpha}^m \times {}^j\boldsymbol{a}_Q^m - {}^j\boldsymbol{\omega}^m \times {}^j\boldsymbol{\rho}_Q^m. \tag{7.43}$$

Finally, substituting Eq. (7.42) into Eq. (7.32) and Eq. (7.43) into Eq. (7.33), and removing points P and Q, yields expression (7.31). ◇

It is interesting to note that the reduced hyper-jerk state of a rigid body can support, like the acceleration and reduced jerk states, more than one representation; for instance, the left and right sides of Eq. (7.41) can equate separately to an arbitrary vector $\boldsymbol{C}$ without affecting the meaning of Eq. (7.41). For example, we have

$$\begin{aligned} &\boldsymbol{F}_P + 3{}^j\boldsymbol{\rho}^m \times {}^j\boldsymbol{v}_P^m + 3{}^j\boldsymbol{\alpha}^m \times {}^j\boldsymbol{a}_P^m + {}^j\boldsymbol{\omega}^m \times {}^j\boldsymbol{\rho}_P^m = \\ &\boldsymbol{F}_Q + 3{}^j\boldsymbol{\rho}^m \times {}^j\boldsymbol{v}_Q^m + 3{}^j\boldsymbol{\alpha}^m \times {}^j\boldsymbol{a}_Q^m + {}^j\boldsymbol{\omega}^m \times {}^j\boldsymbol{\rho}_Q^m = \boldsymbol{C}, \end{aligned} \tag{7.44}$$

where in general $\boldsymbol{C} \neq \mathbf{0}$. However, the simplest representation of the reduced hyper-jerk state arises precisely when $\boldsymbol{C} = \mathbf{0}$. The correctness of the representation of the hyper-jerk state of a rigid body through Eq. (7.31) has been proven; now let us consider three rigid bodies j, k, and m in relative motion (see Fig. 4.5). The reduced hyper-jerk state of body m as measured from body j, based on Eqs. (4.63), (4.43), (4.55), (5.5), (5.6), (5.30), (5.36), (6.17), (6.18), (6.43), (6.44), (7.8), and (7.9), may be determined as follows:

$$
\begin{aligned}
{}^j\boldsymbol{H}_O^m = & \begin{bmatrix} {}^j\boldsymbol{\sigma}^m \\ {}^j\boldsymbol{\sigma}_O^m - 3{}^j\boldsymbol{\rho}^m \times {}^j\boldsymbol{v}_O^m - 3{}^j\boldsymbol{\alpha}^m \times {}^j\boldsymbol{a}_O^m - {}^j\boldsymbol{\omega}^m \times {}^j\boldsymbol{\rho}_O^m \end{bmatrix} \\
= & \begin{bmatrix} {}^j\boldsymbol{\sigma}^k \\ {}^j\boldsymbol{\sigma}_O^k - 3{}^j\boldsymbol{\rho}^k \times {}^j\boldsymbol{v}_Q^k - 3{}^j\boldsymbol{\alpha}^k \times {}^j\boldsymbol{a}_O^k - {}^j\boldsymbol{\omega}^k \times {}^j\boldsymbol{\rho}_O^k \end{bmatrix} \\
& + \begin{bmatrix} {}^k\boldsymbol{\sigma}^m \\ {}^k\boldsymbol{\sigma}_O^m - 3{}^k\boldsymbol{\rho}^m \times {}^k\boldsymbol{v}_O^m - 3{}^k\boldsymbol{\alpha}^m \times {}^k\boldsymbol{a}_O^m - {}^k\boldsymbol{\omega}^m \times {}^k\boldsymbol{\rho}_O^m \end{bmatrix} \\
& +3 \begin{bmatrix} {}^j\boldsymbol{\omega}^k \times {}^k\boldsymbol{\rho}^m \\ {}^j\boldsymbol{\omega}^k \times ({}^k\boldsymbol{\rho}_O^m - 2{}^k\boldsymbol{\alpha}^m \times {}^k\boldsymbol{v}_O^m - {}^k\boldsymbol{\omega}^m \times {}^k\boldsymbol{a}_O^m) - {}^k\boldsymbol{\rho}^m \times {}^j\boldsymbol{v}_O^k \end{bmatrix} \\
& +3 \begin{bmatrix} {}^j\boldsymbol{\alpha}^k \times {}^k\boldsymbol{\alpha}^m \\ {}^j\boldsymbol{\alpha}^k \times ({}^k\boldsymbol{a}_O^m - {}^k\boldsymbol{\omega}^m \times {}^k\boldsymbol{v}_O^m) - {}^k\boldsymbol{\alpha}^m \times ({}^j\boldsymbol{a}_O^k - {}^j\boldsymbol{\omega}^k \times {}^j\boldsymbol{v}_O^k) \end{bmatrix} \\
& +3 \begin{bmatrix} {}^j\boldsymbol{\omega}^k \times ({}^j\boldsymbol{\omega}^k \times {}^k\boldsymbol{\alpha}^m) \\ {}^j\boldsymbol{\omega}^k \times \begin{pmatrix} {}^j\boldsymbol{\omega}^k \times {}^k\boldsymbol{a}_O^m - {}^j\boldsymbol{\omega}^k \times ({}^k\boldsymbol{\omega}^m \times {}^k\boldsymbol{v}_O^m) \\ -{}^k\boldsymbol{\alpha}^m \times {}^j\boldsymbol{v}_O^k \end{pmatrix} - ({}^j\boldsymbol{\omega}^k \times {}^k\boldsymbol{\alpha}^m) \times {}^j\boldsymbol{v}_O^k \end{bmatrix} \\
& +2 \begin{bmatrix} {}^j\boldsymbol{\alpha}^k \times ({}^j\boldsymbol{\omega}^k \times {}^k\boldsymbol{\omega}^m) \\ {}^j\boldsymbol{\alpha}^k \times ({}^j\boldsymbol{\omega}^k \times {}^k\boldsymbol{v}_O^m - {}^k\boldsymbol{\omega}^m \times {}^j\boldsymbol{v}_O^k) - ({}^j\boldsymbol{\omega}^k \times {}^k\boldsymbol{\omega}^m) \times {}^j\boldsymbol{a}_O^k - {}^j\boldsymbol{\omega}^k \times {}^j\boldsymbol{v}_O^k) \end{bmatrix} \\
& + \begin{bmatrix} {}^j\boldsymbol{\rho}^k \times {}^k\boldsymbol{\omega}^m \\ {}^j\boldsymbol{\rho}^k \times {}^k\boldsymbol{v}_O^m - {}^k\boldsymbol{\omega}^m \times ({}^j\boldsymbol{\rho}_O^k - 2{}^j\boldsymbol{\alpha}^k \times {}^j\boldsymbol{v}_O^k - {}^j\boldsymbol{\omega}^k \times {}^j\boldsymbol{a}_O^k) \end{bmatrix} \\
& + \begin{bmatrix} {}^j\boldsymbol{\omega}^k \times ({}^j\boldsymbol{\alpha}^k \times {}^k\boldsymbol{\omega}^m) \\ \\ {}^j\boldsymbol{\omega}^k \times ({}^j\boldsymbol{\alpha}^k \times {}^k\boldsymbol{v}_O^m) - {}^j\boldsymbol{\omega}^k \times ({}^k\boldsymbol{\omega}^m \times {}^j\boldsymbol{a}_O^k) \\ +{}^j\boldsymbol{\omega}^k \times ({}^k\boldsymbol{\omega}^m \times ({}^j\boldsymbol{\omega}^k \times {}^j\boldsymbol{v}_O^k)) - ({}^j\boldsymbol{\alpha}^k \times {}^k\boldsymbol{\omega}^m) \times {}^j\boldsymbol{v}_O^k \end{bmatrix} \\
& + \begin{bmatrix} {}^j\boldsymbol{\omega}^k \times ({}^j\boldsymbol{\omega}^k \times ({}^j\boldsymbol{\omega}^k \times {}^k\boldsymbol{\omega}^m)) \\ \\ {}^j\boldsymbol{\omega}^k \times ({}^j\boldsymbol{\omega}^k \times ({}^j\boldsymbol{\omega}^k \times {}^k\boldsymbol{v}_O^m)) - {}^j\boldsymbol{\omega}^k \times ({}^j\boldsymbol{\omega}^k \times ({}^k\boldsymbol{\omega}^m \times {}^j\boldsymbol{v}_O^k)) \\ -{}^j\boldsymbol{\omega}^k \times (({}^j\boldsymbol{\omega}^k \times {}^k\boldsymbol{\omega}^m) \times {}^j\boldsymbol{v}_O^k) - ({}^j\boldsymbol{\omega}^k \times ({}^j\boldsymbol{\omega}^k \times {}^k\boldsymbol{\omega}^m)) \times {}^j\boldsymbol{v}_O^k) \end{bmatrix} .
\end{aligned}
\tag{7.45}
$$

Hence, one obtains

$$
\begin{aligned}
{}^j\boldsymbol{H}_O^m = {}^j\boldsymbol{H}_O^k + {}^k\boldsymbol{H}_O^m + 3\left[{}^j\boldsymbol{V}_O^k \quad {}^k\boldsymbol{J}_O^m\right] + 3\left[{}^j\boldsymbol{A}_O^k \quad {}^k\boldsymbol{A}_O^m\right] + \\
3\left[{}^j\boldsymbol{V}_O^k \quad \left[{}^j\boldsymbol{V}_O^k \quad {}^k\boldsymbol{A}_O^m\right]\right] + 2\left[{}^j\boldsymbol{A}_O^k \quad \left[{}^j\boldsymbol{V}_O^k \quad {}^k\boldsymbol{V}_O^m\right]\right] + \\
\left[{}^j\boldsymbol{J}_O^k \quad {}^k\boldsymbol{V}_O^m\right] + \left[{}^j\boldsymbol{V}_O^k\left[{}^j\boldsymbol{A}_O^k \quad {}^k\boldsymbol{V}_O^m\right]\right] + \left[{}^j\boldsymbol{V}_O^k\left[{}^j\boldsymbol{V}_O^k\left[{}^j\boldsymbol{V}_O^k \quad {}^k\boldsymbol{V}_O^m\right]\right]\right].
\end{aligned}
\tag{7.46}
$$

The next proposition is a generalization of Eq. (7.46). Indeed, in what follows the connection of the reduced hyper-jerk states of physically disconnected rigid bodies is elucidated systematically mainly based on the induction hypothesis. Although the notation seems to be rather involved, it is necessary to clarify this issue.

Proposition 7.3. *Let j, $j+1$, …, m be a set of rigid bodies. Then their reduced hyper-jerk states notated as* ${}^{j}\boldsymbol{H}_O^{j+1}$, ${}^{j+1}\boldsymbol{H}_O^{j+2}$, …, ${}^{m-1}\boldsymbol{H}_O^{m}$ *are related by*

$$\begin{aligned}
{}^{j}\boldsymbol{H}_O^{m} = {}&{}^{j}\boldsymbol{H}_O^{j+1} + {}^{j+1}\boldsymbol{H}_O^{j+2} + \ldots + {}^{m-1}\boldsymbol{H}_O^{m} + 3\left[{}^{j}\boldsymbol{V}_O^{j+1}\;\; {}^{j+1}\boldsymbol{J}_O^{j+2}\right] + \\
&3\left[{}^{j+1}\boldsymbol{V}_O^{j+2}\;\; {}^{j+2}\boldsymbol{J}_O^{m}\right] + \ldots + 3\left[{}^{m-2}\boldsymbol{V}_O^{m-1}\;\; {}^{m-1}\boldsymbol{J}_O^{m}\right] + \\
&3\left[{}^{j}\boldsymbol{A}_O^{j+1}\;\; {}^{j+1}\boldsymbol{A}_O^{m}\right] + 3\left[{}^{j+1}\boldsymbol{A}_O^{j+2}\;\; {}^{j+2}\boldsymbol{A}_O^{m}\right] + \ldots + 3\left[{}^{m-2}\boldsymbol{A}_O^{m-1}\;\; {}^{m-1}\boldsymbol{A}_O^{m}\right] \\
&+3\left[{}^{j}\boldsymbol{V}_O^{j+1}\left[{}^{j}\boldsymbol{V}_O^{j+1}\;\; {}^{j+1}\boldsymbol{A}_O^{m}\right]\right] + 3\left[{}^{j+1}\boldsymbol{V}_O^{j+2}\left[{}^{j+1}\boldsymbol{V}_O^{j+2}\;\; {}^{j+2}\boldsymbol{A}_O^{m}\right]\right] \\
&+\ldots + 3\left[{}^{m-2}\boldsymbol{V}_O^{m-1}\left[{}^{m-2}\boldsymbol{V}_O^{m-1}\;\; {}^{m-1}\boldsymbol{A}_O^{m}\right]\right] + \\
&2\left[{}^{j}\boldsymbol{A}_O^{j+1}\left[{}^{j}\boldsymbol{V}_O^{j+1}\;\; {}^{j+1}\boldsymbol{V}_O^{m}\right]\right] + 2\left[{}^{j+1}\boldsymbol{A}_O^{j+2}\left[{}^{j+1}\boldsymbol{V}_O^{j+2}\;\; {}^{j+2}\boldsymbol{V}_O^{m}\right]\right] \\
&+\ldots + 2\left[{}^{m-2}\boldsymbol{A}_O^{m-1}\left[{}^{m-2}\boldsymbol{V}_O^{m-1}\;\; {}^{m-1}\boldsymbol{V}_O^{m}\right]\right] + \\
&\left[{}^{j}\boldsymbol{J}_O^{j+1}\;\; {}^{j+1}\boldsymbol{V}_O^{m}\right] + \left[{}^{j+1}\boldsymbol{J}_O^{j+2}\;\; {}^{j+2}\boldsymbol{V}_O^{m}\right] + \ldots + \left[{}^{m-2}\boldsymbol{J}_O^{m-1}\;\; {}^{m-1}\boldsymbol{V}_O^{m}\right] + \\
&\left[{}^{j}\boldsymbol{V}_O^{j+1}\left[{}^{j}\boldsymbol{A}_O^{j+1}\;\; {}^{j+1}\boldsymbol{V}_O^{m}\right]\right] + \left[{}^{j+1}\boldsymbol{V}_O^{j+2}\left[{}^{j+1}\boldsymbol{A}_O^{j+2}\;\; {}^{j+2}\boldsymbol{V}_O^{m}\right]\right] \\
&+\ldots + \left[{}^{m-2}\boldsymbol{V}_O^{m-1}\left[{}^{m-2}\boldsymbol{A}_O^{m-1}\;\; {}^{m-1}\boldsymbol{V}_O^{m}\right]\right] + \\
&\left[{}^{j}\boldsymbol{V}_O^{j+1}\left[{}^{j}\boldsymbol{V}_O^{j+1}\left[{}^{j}\boldsymbol{V}_O^{j+1}\;\; {}^{j+1}\boldsymbol{V}_O^{m}\right]\right]\right] + \\
&\left[{}^{j+1}\boldsymbol{V}_O^{j+2}\left[{}^{j+1}\boldsymbol{V}_O^{j+2}\left[{}^{j+1}\boldsymbol{V}_O^{j+2}\;\; {}^{j+2}\boldsymbol{V}_O^{m}\right]\right]\right] \\
&+\ldots + \left[{}^{m-2}\boldsymbol{V}_O^{m-1}\left[{}^{m-2}\boldsymbol{V}_O^{m-1}\left[{}^{m-2}\boldsymbol{V}_O^{m-1}\;\; {}^{m-1}\boldsymbol{V}_O^{m}\right]\right]\right].
\end{aligned} \tag{7.47}$$

Proof. Let us consider three bodies $m-2$, $m-1$, and m. According to expression (7.46), it is possible to write

$$\begin{aligned}
{}^{m-2}\boldsymbol{H}_O^{m} = {}&{}^{m-2}\boldsymbol{H}_O^{m-1} + {}^{m-1}\boldsymbol{H}_O^{m} + 3\left[{}^{m-2}\boldsymbol{V}_O^{m-1}\;\; {}^{m-1}\boldsymbol{J}_O^{m}\right] + \\
&3\left[{}^{m-2}\boldsymbol{A}_O^{m-1}\;\; {}^{m-1}\boldsymbol{A}_O^{m}\right] + 3\left[{}^{m-2}\boldsymbol{V}_O^{m-1}\left[{}^{m-2}\boldsymbol{V}_O^{m-1}\;\; {}^{m-1}\boldsymbol{A}_O^{m}\right]\right] + \\
&2\left[{}^{m-2}\boldsymbol{A}_O^{m-1}\left[{}^{m-2}\boldsymbol{V}_O^{m-1}\;\; {}^{m-1}\boldsymbol{V}_O^{m}\right]\right] + \left[{}^{m-2}\boldsymbol{J}_O^{m-1}\;\; {}^{m-1}\boldsymbol{V}_O^{m}\right] + \\
&\left[{}^{m-2}\boldsymbol{V}_O^{m-1}\left[{}^{m-2}\boldsymbol{A}_O^{m-1}\;\; {}^{m-1}\boldsymbol{V}_O^{m}\right]\right] +
\end{aligned}$$

$$\left[{}^{m-2}\boldsymbol{V}_O^{m-1}\left[{}^{m-2}\boldsymbol{V}_O^{m-1}\left[{}^{m-2}\boldsymbol{V}_O^{m-1}\quad {}^{m-1}\boldsymbol{V}_O^{m}\right]\right]\right]. \tag{7.48}$$

Furthermore, by the induction hypothesis, assume that

$$\begin{aligned}
{}^{j+1}\boldsymbol{H}_O^m = {}^{j+1}\boldsymbol{H}_O^{j+2} + {}^{j+2}\boldsymbol{H}_O^{j+3} + \ldots + {}^{m-1}\boldsymbol{H}_O^m \\
+3\left[{}^{j+1}\boldsymbol{V}_O^{j+2}\quad {}^{j+2}\boldsymbol{J}_O^m\right] + 3\left[{}^{j+2}\boldsymbol{V}_O^{j+3}\quad {}^{j+3}\boldsymbol{J}_O^m\right] + \ldots + \\
3\left[{}^{m-2}\boldsymbol{V}_O^{m-1}\quad {}^{m-1}\boldsymbol{J}_O^m\right] + 3\left[{}^{j+1}\boldsymbol{A}_O^{j+2}\quad {}^{j+2}\boldsymbol{A}_O^m\right] + \\
3\left[{}^{j+2}\boldsymbol{A}_O^{j+3}\quad {}^{j+3}\boldsymbol{A}_O^m\right] + \ldots + 3\left[{}^{m-2}\boldsymbol{A}_O^{m-1}\quad {}^{m-1}\boldsymbol{A}_O^m\right] + \\
3\left[{}^{j+1}\boldsymbol{V}_O^{j+2}\left[{}^{j+1}\boldsymbol{V}_O^{j+2}\quad {}^{j+2}\boldsymbol{A}_O^m\right]\right] + 3\left[{}^{j+2}\boldsymbol{V}_O^{j+3}\left[{}^{j+2}\boldsymbol{V}_O^{j+3}\quad {}^{j+3}\boldsymbol{A}_O^m\right]\right] \\
+ \ldots + 3\left[{}^{m-2}\boldsymbol{V}_O^{m-1}\quad \left[{}^{m-2}\boldsymbol{V}_O^{m-1}\quad {}^{m-1}\boldsymbol{A}_O^m\right]\right] + \\
2\left[{}^{j+1}\boldsymbol{A}_O^{j+2}\left[{}^{j+1}\boldsymbol{V}_O^{j+2}\quad {}^{j+2}\boldsymbol{V}_O^m\right]\right] + 2\left[{}^{j+2}\boldsymbol{A}_O^{j+3}\left[{}^{j+2}\boldsymbol{V}_O^{j+3}\quad {}^{j+3}\boldsymbol{V}_O^m\right]\right] \\
+ \ldots + 2\left[{}^{m-2}\boldsymbol{A}_O^{m-1}\left[{}^{m-2}\boldsymbol{V}_O^{m-1}\quad {}^{m-1}\boldsymbol{V}_O^m\right]\right] + \\
\left[{}^{j+1}\boldsymbol{J}_O^{j+2}\quad {}^{j+2}\boldsymbol{V}_O^m\right] + \left[{}^{j+2}\boldsymbol{J}_O^{j+3}\quad {}^{j+3}\boldsymbol{V}_O^m\right] \\
+ \ldots + \left[{}^{m-2}\boldsymbol{J}_O^{m-1}\quad {}^{m-1}\boldsymbol{V}_O^m\right] \\
+\left[{}^{j+1}\boldsymbol{V}_O^{j+2}\left[{}^{j+1}\boldsymbol{A}_O^{j+2}\quad {}^{j+2}\boldsymbol{V}_O^m\right]\right] + \left[{}^{j+2}\boldsymbol{V}_O^{j+3}\left[{}^{j+2}\boldsymbol{A}_O^{j+3}\quad {}^{j+3}\boldsymbol{V}_O^m\right]\right] \\
+ \ldots + \left[{}^{m-2}\boldsymbol{V}_O^{m-1}\left[{}^{m-2}\boldsymbol{A}_O^{m-1}\quad {}^{m-1}\boldsymbol{V}_O^m\right]\right] \\
+\left[{}^{j+1}\boldsymbol{V}_O^{j+2}\left[{}^{j+1}\boldsymbol{V}_O^{j+2}\left[{}^{j+1}\boldsymbol{V}_O^{j+2}\quad {}^{j+2}\boldsymbol{V}_O^m\right]\right]\right] \\
+\left[{}^{j+2}\boldsymbol{V}_O^{j+3}\left[{}^{j+2}\boldsymbol{V}_O^{j+3}\left[{}^{j+2}\boldsymbol{V}_O^{j+3}\quad {}^{j+3}\boldsymbol{V}_O^m\right]\right]\right] \\
+ \ldots + \left[{}^{m-2}\boldsymbol{V}_O^{m-1}\left[{}^{m-2}\boldsymbol{V}_O^{m-1}\left[{}^{m-2}\boldsymbol{V}_O^{m-1}\quad {}^{m-1}\boldsymbol{V}_O^m\right]\right]\right].
\end{aligned} \tag{7.49}$$

Finally, considering Eqs. (7.48) and (7.49), if we substitute k by $j+1$ into Eq. (7.46), we obtain expression (7.47). ◇

In what follows it will be proved that the reduced hyper-jerk state of a rigid body is, in reality, a twist on a screw.

Proposition 7.4. *Let k and l be two rigid bodies (see Fig. 4.2) connected to each other by a helical pair ${}^k\$^l$. Furthermore, let P be an arbitrary point embedded in body l playing the role of pole. The reduced hyper-jerk state can be considered a twist on a screw given by*

$${}^k\boldsymbol{H}_P^l = {}_k\sigma_l{}^k\$^l, \tag{7.50}$$

where ${}_k\sigma_l$ *is the magnitude of the vector* ${}^k\boldsymbol{\sigma}^l$.

Proof. Since the unit vector ${}^k\hat{\boldsymbol{s}}^l$ is the direction of the helical pair, the angular hyper-jerk of body l as observed from body k is obtained upon the angular jerk ${}^k\boldsymbol{\rho}^l = {}_k\rho_l{}^k\hat{\boldsymbol{s}}^l$ as follows:

$$ {}^k\boldsymbol{H}^l = \frac{{}^kd}{dt}\left({}^k\boldsymbol{\rho}^l\right) = \frac{{}^kd}{dt}\left({}_k\rho_l{}^k\hat{\boldsymbol{s}}^l\right) = {}_k\sigma_l{}^k\hat{\boldsymbol{s}}^l, \tag{7.51} $$

where ${}_k\rho_l$ is the magnitude of vector ${}^k\boldsymbol{\rho}^l$. On the other hand, the velocity of point P as observed from body k can be determined as

$$ {}^k\boldsymbol{v}_P^l \equiv \frac{{}^kd}{dt}\boldsymbol{p} = \frac{{}^kd}{dt}(\boldsymbol{q}+\boldsymbol{r}), \tag{7.52} $$

where

- $\boldsymbol{p}$ is a vector directed from an arbitrary point $\tilde{A}$ of body l to point P;
- $\boldsymbol{r}$ is the position vector of an arbitrary point $\tilde{B}$ of the axis of the helical pair with respect to point $\tilde{A}$;
- $\boldsymbol{q}$ is a vector beginning in $\tilde{B}$ and ending in P.

Furthermore, according to the pitch z of the helical pair, it is evident that $\frac{{}^kd}{dt}\boldsymbol{r} = z{}^k\boldsymbol{\omega}^l$ while $\frac{{}^kd}{dt}\boldsymbol{q} = {}^k\boldsymbol{\omega}^l \times \boldsymbol{q}$. Therefore,

$$ {}^k\boldsymbol{v}_P^l = z{}^k\boldsymbol{\omega}^l + {}^k\boldsymbol{\omega}^l \times \boldsymbol{q}. \tag{7.53} $$

Hence, considering that ${}^k\boldsymbol{a}_P^l \equiv \frac{{}^kd}{dt}({}^k\boldsymbol{v}_P^l)$, it follows that

$$ {}^k\boldsymbol{a}_P^l = \frac{{}^kd}{dt}(z{}^k\boldsymbol{\omega}^l + {}^k\boldsymbol{\omega}^l \times \boldsymbol{q}) = z{}^k\boldsymbol{\alpha}^l + {}^k\boldsymbol{\alpha}^l \times \boldsymbol{q} + {}^k\boldsymbol{\omega}^l \times ({}^k\boldsymbol{\omega}^l \times \boldsymbol{q}). \tag{7.54} $$

On the other hand, taking into account that ${}^k\boldsymbol{\rho}_P^l \equiv \frac{{}^kd}{dt}({}^k\boldsymbol{a}_P^l)$, we get

$$ \begin{aligned} {}^k\boldsymbol{\rho}_P^l = \frac{{}^kd}{dt}[z{}^k\boldsymbol{\alpha}^l + {}^k\boldsymbol{\alpha}^l \times \boldsymbol{q} + {}^k\boldsymbol{\omega}^l \times ({}^k\boldsymbol{\omega}^l \times \boldsymbol{q})] = \\ z{}^k\boldsymbol{\rho}^l + {}^k\boldsymbol{\rho}^l \times \boldsymbol{q} + 2{}^k\boldsymbol{\alpha}^l \times {}^k\boldsymbol{v}_P^l + {}^k\boldsymbol{\omega}^l \times {}^k\boldsymbol{a}_P^l. \end{aligned} \tag{7.55} $$

Furthermore, the hyper-jerk ${}^k\boldsymbol{\sigma}_P^l \equiv \frac{{}^kd}{dt}({}^k\boldsymbol{\rho}_P^l)$ is obtained as

$$ \begin{aligned} {}^k\boldsymbol{\sigma}_P^l &= \frac{{}^kd}{dt}\left(z{}^k\boldsymbol{\rho}^l + {}^k\boldsymbol{\rho}^l \times \boldsymbol{q} + 2{}^k\boldsymbol{\alpha}^l \times {}^k\boldsymbol{v}_P^l + {}^k\boldsymbol{\omega}^l \times {}^k\boldsymbol{a}_P^l\right) \\ &= z{}^k\boldsymbol{\sigma}^l + {}^k\boldsymbol{\sigma}^l \times \boldsymbol{q} + {}^k\boldsymbol{\rho}^l \times \left({}^k\boldsymbol{\omega}^l \times \boldsymbol{q}\right) + 2{}^k\boldsymbol{\rho}^l \times {}^k\boldsymbol{v}_P^l + 2{}^k\boldsymbol{\alpha}^l \times {}^k\boldsymbol{a}_P^l \\ &\quad + {}^k\boldsymbol{\alpha}^l \times {}^k\boldsymbol{a}_P^l + {}^k\boldsymbol{\omega}^l \times {}^k\boldsymbol{\rho}_P^l. \end{aligned} \tag{7.56} $$

Hence,

$$ {}^k\boldsymbol{\sigma}^l_P - 3{}^k\boldsymbol{\rho}^l \times {}^k\boldsymbol{v}^l_P - 3{}^k\boldsymbol{\alpha}^l \times {}^k\boldsymbol{a}^l_P - {}^k\boldsymbol{\omega}^l \times {}^k\boldsymbol{\rho}^l_P = z{}^k\boldsymbol{\sigma}^l + {}^k\boldsymbol{\sigma}^l \times \boldsymbol{q}. \tag{7.57} $$

However,

$$ z{}^k\boldsymbol{\sigma}^l + {}^k\boldsymbol{\sigma}^l \times \boldsymbol{q} = {}_k\sigma_l\left(z{}^k\hat{\boldsymbol{s}}^l + {}^k\hat{\boldsymbol{s}}^l \times \boldsymbol{q}\right) = {}_k\sigma_l{}^k\boldsymbol{s}^l_P. \tag{7.58} $$

Next, according to expression (7.31) it follows that

$$ {}^k\boldsymbol{\sigma}^l_P \equiv \begin{bmatrix} {}^k\boldsymbol{\sigma}^l \\ {}^k\boldsymbol{\sigma}^l_P - 3{}^k\boldsymbol{\rho}^l \times {}^k\boldsymbol{v}^l_P - 3{}^k\boldsymbol{\alpha}^l \times {}^k\boldsymbol{a}^l_P - {}^k\boldsymbol{\omega}^l \times {}^k\boldsymbol{\rho}^l_P \end{bmatrix}. \tag{7.59} $$

Gathering these results, one obtains

$$ {}^k\boldsymbol{\sigma}^l_P \equiv \begin{bmatrix} {}^k\boldsymbol{\sigma}^l \\ {}_k\sigma_l{}^k\boldsymbol{s}^l_P \end{bmatrix} = \begin{bmatrix} {}_k\sigma_l{}^k\hat{\boldsymbol{s}}^l \\ {}_k\sigma_l{}^k\boldsymbol{s}^l_P \end{bmatrix} = {}_k\sigma_l \begin{bmatrix} {}^k\hat{\boldsymbol{s}}^l \\ {}^k\boldsymbol{s}^l_P \end{bmatrix} = {}_k\sigma_l\,{}^k\$^l. \diamond \tag{7.60} $$

Proposition 7.5. *Let us consider an open kinematic chain where adjacent bodies are interconnected by helical pairs. The reduced hyper-jerk state in screw form of body m as measured from body j is given by*

$$ {}^j\boldsymbol{H}^m_O = {}_j\sigma_{j+1}{}^j\$^{j+1} + {}_{j+1}\sigma_{j+2}{}^{j+1}\$^{j+2} + \ldots + {}_{m-1}\sigma_m{}^{m-1}\$^m + {}^j\mathcal{H}^m, \tag{7.61} $$

where here ${}^j\mathcal{H}^m$ *is named the Lie screw of hyper-jerk, which is given by*

$$ \begin{aligned} {}^j\mathcal{H}^m = {} & 3\left[{}_j\omega_{j+1}{}^j\$^{j+1} \quad {}_{j+1}\rho_{j+2}{}^{j+1}\$^{j+2} + \ldots + {}_{m-1}\rho_m{}^{m-1}\$^m + {}^j\$p^m\right] \\ & +3\left[{}_{j+1}\omega_{j+2}{}^{j+1}\$^{j+2} \quad {}_{j+2}\rho_{j+3}{}^{j+2}\$^{j+3} + \ldots + {}_{m-1}\rho_m{}^{m-1}\$^m + {}^{j+1}\$p^m\right] \\ & + \ldots + 3\left[{}_{m-2}\omega_{m-1}{}^{m-2}\$^{m-1} \quad {}_{m-1}\rho_m{}^{m-1}\$^m\right] \\ & +3\left[{}_j\alpha_{j+1}{}^j\$^{j+1} \quad {}_{j+1}\alpha_{j+2}{}^{j+1}\$^{j+2} + \ldots + {}_{m-1}\alpha_m{}^{m-1}\$^m + {}^j\$a^m\right] \\ & +3\left[{}_{j+1}\alpha_{j+2}{}^{j+1}\$^{j+2} \quad {}_{j+2}\alpha_{j+3}{}^{j+2}\$^{j+3} + \ldots + {}_{m-1}\alpha_m{}^{m-1}\$^m + {}^{j+1}\$a^m\right] \\ & + \ldots + 3\left[{}_{m-2}\alpha_{m-1}{}^{m-2}\$^{m-1} \quad {}_{m-1}\alpha_m{}^{m-1}\$^m\right] \\ & +3\left[{}_j\omega_{j+1}{}^j\$^{j+1}\left[{}_j\omega_{j+1}{}^j\$^{j+1} \quad {}_{j+1}\alpha_{j+2}{}^{j+1}\$^{j+2} + \ldots + {}_{m-1}\alpha_m{}^{m-1}\$^m + {}^j\$a^m\right]\right] \\ & +3\left[{}_{j+1}\omega_{j+2}{}^{j+1}\$^{j+2}\left[{}_{j+1}\omega_{j+2}{}^{j+1}\$^{j+2} \quad {}_{j+2}\alpha_{j+3}{}^{j+2}\$^{j+3} + \ldots + {}_{m-1}\alpha_m{}^{m-1}\$^m + {}^{j+1}\$a^m\right]\right] \\ & + \ldots + 3\left[{}_{m-2}\omega_{m-1}{}^{m-2}\$^{m-1}\left[{}_{m-2}\omega_{m-1}{}^{m-2}\$^{m-1} \quad {}_{m-1}\alpha_m{}^{m-1}\$^m\right]\right] \\ & +2\left[{}_j\alpha_{j+1}{}^j\$^{j+1}\left[{}_j\omega_{j+1}{}^j\$^{j+1} \quad {}_{j+1}\omega_{j+2}{}^{j+1}\$^{j+2} + \ldots + {}_{m-1}\omega_m{}^{m-1}\$^m\right]\right] \end{aligned} $$

$$
\begin{aligned}
&+2\left[{}_{j+1}\alpha_{j+2}{}^{j+1}\$^{j+2}\left[{}_{j+1}\omega_{j+2}{}^{j+1}\$^{j+2}\quad {}_{j+2}\omega_{j+3}{}^{j+2}\$^{j+3}+\ldots+{}_{m-1}\omega_m{}^{m-1}\$^m\right]\right]\\
&\quad+\ldots+2\left[{}_{m-2}\alpha_{m-1}{}^{m-2}\$^{m-1}\left[{}_{m-2}\omega_{m-1}{}^{m-2}\$^{m-1}\quad {}_{m-1}\omega_m{}^{m-1}\$^m\right]\right]\\
&\quad+\left[{}_{j}\rho_{j+1}{}^{j}\$^{j+1}\quad {}_{j+1}\omega_{j+2}{}^{j+1}\$^{j+2}+\ldots+{}_{m-1}\omega_m{}^{m-1}\$^m\right]\\
&\quad+\left[{}_{j+1}\rho_{j+2}{}^{j+1}\$^{j+2}\quad {}_{j+2}\omega_{j+3}{}^{j+2}\$^{j+3}+\ldots+{}_{m-1}\omega_m{}^{m-1}\$^m\right]\\
&\quad+\ldots+\left[{}_{m-2}\rho_{m-1}{}^{m-2}\$^{m-1}\quad {}_{m-1}\omega_m{}^{m-1}\$^m\right]\\
&\quad+\left[{}_{j}\omega_{j+1}{}^{j}\$^{j+1}\left[{}_{j}\alpha_{j+1}{}^{j}\$^{j+1}\quad {}_{j+1}\omega_{j+2}{}^{j+1}\$^{j+2}+\ldots+{}_{m-1}\omega_m{}^{m-1}\$^m\right]\right]\\
&+\left[{}_{j+1}\omega_{j+2}{}^{j+1}\$^{j+2}\left[{}_{j+1}\alpha_{j+2}{}^{j+1}\$^{j+2}\quad {}_{j+2}\omega_{j+3}{}^{j+1}\$^{j+2}+\ldots+{}_{m-1}\omega_m{}^{m-1}\$^m\right]\right]\\
&\quad+\ldots+\left[{}_{m-2}\omega_{m-1}{}^{m-2}\$^{m-1}\left[{}_{m-2}\alpha_{m-1}{}^{m-2}\$^{m-1}\quad {}_{m-1}\omega_m{}^{m-1}\$^m\right]\right]\\
&+\left[{}_{j}\omega_{j+1}{}^{j}\$^{j+1}\left[{}_{j}\omega_{j+1}{}^{j}\$^{j+1}\left[{}_{j}\omega_{j+1}{}^{j}\$^{j+1}\ {}_{j+1}\omega_{j+2}{}^{j+1}\$^{j+2}+\ldots+{}_{m-1}\omega_m{}^{m-1}\$^m\right]\right]\right]\\
&+\ldots+\left[{}_{m-2}\omega_{m-1}{}^{m-2}\$^{m-1}\left[{}_{m-2}\omega_{m-1}{}^{m-2}\$^{m-1}\left[{}_{m-2}\omega_{m-1}{}^{m-2}\$^{m-1}\ {}_{m-1}\omega_m{}^{m-1}\$^m\right]\right]\right].
\end{aligned}
\tag{7.62}
$$

Proof. The proof is immediate by a recursive application of Eqs. (4.78), (5.52), (6.76), and (7.50) into Eq. (7.47). ◇

Example 7.2. Curved surfaces A of a production line are painted by means of a paint-spraying robot (see Fig. 7.4). The length of the telescoping arm is controlled according to the periodic function $l = (1/4)\sin(\pi t/2)$ where the extendable length l is given in meters while the time t is in seconds. Simultaneously, the arm is programmed to rotate according to $\theta = \pi/3 + (\pi/6)\sin(\pi t/2)$, where θ is in radians. Determine (a) the magnitude of the velocity, acceleration, jerk, and hyper-jerk of the nozzle N for $t = 2$ s, and (b) the symbolic expressions for the reduced velocity, acceleration, jerk, and hyper-jerk states of the nozzle N as measured from the fixed link.

Solution. Let XYZ be a reference frame with associated unit vectors $\hat{\boldsymbol{i}}\hat{\boldsymbol{j}}\hat{\boldsymbol{k}}$ attached to the fixed body whose origin is located at point O. The velocity of N as measured from body j, the vector $\boldsymbol{v}_N = {}^j\boldsymbol{v}_N^m$, is computed as

$$
{}^j\boldsymbol{v}_N^m = {}^j\boldsymbol{v}_N^k + {}^k\boldsymbol{v}_N^m, \tag{7.63}
$$

where ${}^j\boldsymbol{v}_N^k$ is the velocity vector of N considering the body m is joined to body k. Furthermore, ${}^k\boldsymbol{v}_N^m$ is the velocity vector of N assuming that j and k are the same rigid body.

With reference to Fig. 7.5, it is evident that

$$
{}^j\boldsymbol{v}_N^k = \dot{\theta}\hat{\boldsymbol{k}} \times \boldsymbol{r}_{N/O}, \tag{7.64}
$$

Fig. 7.4 Example 7.2. Paint-spraying robot

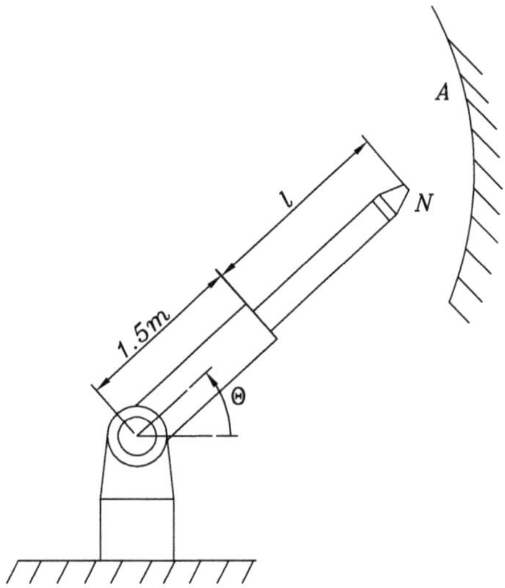

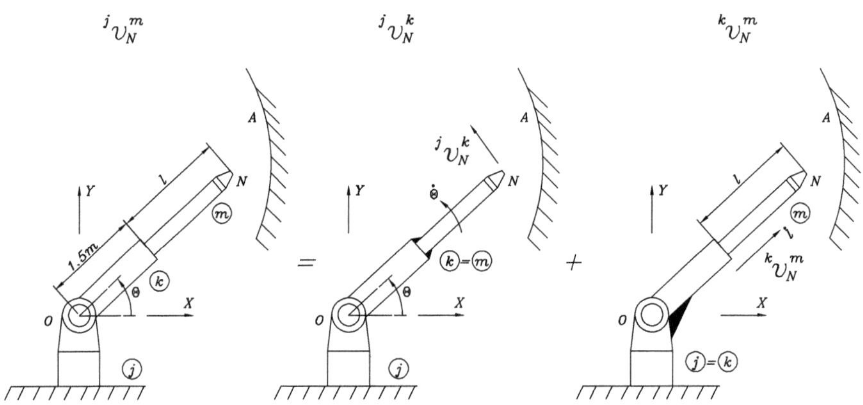

Fig. 7.5 Example 7.2. Decomposition of the velocity analysis

where

$$\boldsymbol{r}_{N/O} = (1.5 + l)\left[\cos(\theta)\hat{\boldsymbol{i}} + \sin(\theta)\hat{\boldsymbol{j}}\right] \tag{7.65}$$

is a vector pointed from O to N and

$$\dot{\theta} = \frac{d}{dt}\left[\pi/3 + (\pi/6)\sin(\pi t/2)\right] = (\pi^2/12)\cos(\pi t/2). \tag{7.66}$$

Furthermore, the relative velocity vector ${}^k\boldsymbol{v}_N^m$ is given by

$${}^k\boldsymbol{v}_N^m = \dot{l}\hat{\boldsymbol{u}}_{N/O}, \tag{7.67}$$

where

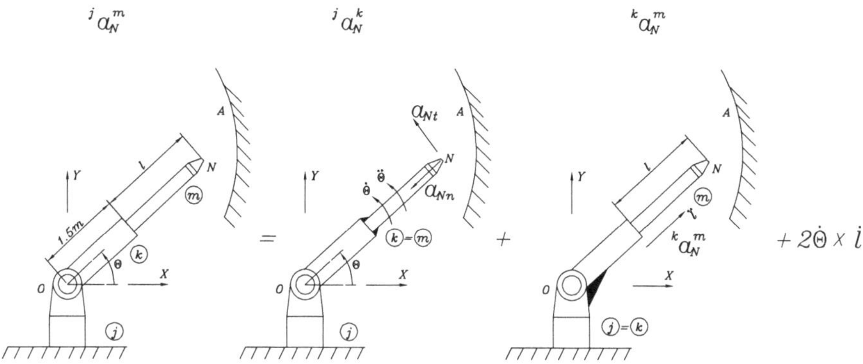

Fig. 7.6 Example 7.2. Decomposition of the acceleration analysis

$$\hat{\boldsymbol{u}}_{N/O} = \cos(\theta)\hat{\boldsymbol{i}} + \sin(\theta)\hat{\boldsymbol{j}} \tag{7.68}$$

is a unit vector along the arm of the robot. Meanwhile,

$$\dot{l} = \frac{d}{dt}\left[(1/4)\sin(\pi t/2)\right] = (\pi/8)\cos(\pi t/2). \tag{7.69}$$

Hence,

$$\begin{aligned} {}^j\boldsymbol{v}_N^m &= \dot{\theta}\hat{\boldsymbol{k}} \times (1.5 + l)\left[\cos(\theta)\hat{\boldsymbol{i}} + \sin(\theta)\hat{\boldsymbol{i}}\right] + \dot{l}\left[\cos\theta\hat{\boldsymbol{i}} + \sin(\theta)\hat{\boldsymbol{j}}\right] \\ &= \left[-\dot{\theta}(1.5 + l)\sin(\theta) + \dot{l}\cos(\theta)\right]\hat{\boldsymbol{i}} + \left[\dot{\theta}(1.5 + l)\cos(\theta) + \dot{l}\sin(\theta)\right]\hat{\boldsymbol{j}}, \end{aligned} \tag{7.70}$$

or ${}^j\boldsymbol{v}_N^m = 0.872\hat{\boldsymbol{i}} - 0.956\hat{\boldsymbol{j}} \rightarrow v_N = 1.294\,\text{m/s}$.

In order to approach the acceleration analysis, consider Fig. 7.6.

The acceleration of the nozzle N as measured from point O, the vector $\boldsymbol{a}_N = {}^j\boldsymbol{a}_N^m$, is computed as

$$ {}^j\boldsymbol{a}_N^m = {}^j\boldsymbol{a}_N^k + {}^k\boldsymbol{a}_N^m + 2\,{}^j\boldsymbol{\omega}^k \times {}^k\boldsymbol{v}_N^m, \tag{7.71}$$

where, according to Fig. 7.6,

$$ {}^j\boldsymbol{a}_N^k = \boldsymbol{a}_{Nt} + \boldsymbol{a}_{Nt} = \ddot{\theta}\hat{\boldsymbol{k}} \times \boldsymbol{r}_{N/O} + \dot{\theta}\hat{\boldsymbol{k}} \times (\dot{\theta}\hat{\boldsymbol{k}} \times \boldsymbol{r}_{N/O}). \tag{7.72}$$

Meanwhile,

$$ {}^k\boldsymbol{a}_N^m = \ddot{l}\hat{\boldsymbol{u}}_{N/O}, \tag{7.73}$$

where

$$\ddot{\theta} = \frac{d}{dt}(\dot{\theta}) = -(\pi^3/24)\sin(\pi t/2) \tag{7.74}$$

and

$$\ddot{l} = \frac{d}{dt}(\dot{l}) = -(\pi^2/16)\sin(\pi t/2). \tag{7.75}$$

Furthermore, the Coriolis acceleration is computed as

$$2{}^j\boldsymbol{\omega}^k \times {}^k\boldsymbol{v}_N^m = 2\dot{\theta}\hat{\boldsymbol{k}} \times \dot{l}\hat{\boldsymbol{u}}_{N/O}. \tag{7.76}$$

Hence,

$$\begin{aligned} {}^j\boldsymbol{a}_N^m &= \left[-\ddot{\theta}(1.5+l)\sin(\theta) - \dot{\theta}^2(1.5+l)\cos(\theta) + \ddot{l}\cos(\theta) - 2\dot{\theta}\dot{l}\sin(\theta)\right]\hat{\boldsymbol{i}} \\ &+ \left[\ddot{\theta}(1.5+l)\cos(\theta) - \dot{\theta}^2(1.5+l)\sin(\theta) + \ddot{l}\sin(\theta) + 2\dot{\theta}\dot{l}\cos(\theta)\right]\hat{\boldsymbol{j}}, \end{aligned} \tag{7.77}$$

or ${}^j\boldsymbol{a}_N^m = -1.0667\hat{\boldsymbol{i}} - 0.5557\hat{\boldsymbol{j}} \rightarrow a_N = 1.2027\,\text{m/s}^2$.

On the other hand, the jerk of the nozzle N as measured from the fixed body j, the vector $\boldsymbol{\rho}_N = {}^j\boldsymbol{\rho}_N^m$, is computed as

$${}^j\boldsymbol{\rho}_N^m = {}^j\boldsymbol{\rho}_N^k + {}^k\boldsymbol{\rho}_N^m + 3{}^j\boldsymbol{\alpha}^k \times {}^j\boldsymbol{v}_N^m + 3{}^j\boldsymbol{\omega}^k \times {}^k\boldsymbol{a}_N^m + 3{}^j\boldsymbol{\omega}^k \times ({}^j\boldsymbol{\omega}^k \times {}^k\boldsymbol{v}_N^m), \tag{7.78}$$

where the jerk of N fixed in k as observed from body j is computed as

$$\begin{aligned} {}^j\boldsymbol{\rho}_N^k = {}^j\boldsymbol{\rho}^k \times \boldsymbol{r}_{N/O} + 2{}^j\boldsymbol{\alpha}^k \times ({}^j\boldsymbol{\omega}^k \times \boldsymbol{r}_{N/O}) + {}^j\boldsymbol{\omega}^k \times ({}^j\boldsymbol{\alpha}^k \times \boldsymbol{r}_{N/O}) \\ + {}^j\boldsymbol{\omega}^k \times \left[{}^j\boldsymbol{\omega}^k \times ({}^j\boldsymbol{\omega}^k \times \boldsymbol{r}_{N/O})\right]. \end{aligned} \tag{7.79}$$

Meanwhile,

$${}^k\boldsymbol{\rho}_N^m = \dddot{l}\hat{\boldsymbol{u}}_{N/O}. \tag{7.80}$$

Hence, one obtains

$$\begin{aligned} {}^j\boldsymbol{\rho}_N^m = \big[&-\dddot{\theta}(1.5+l)\sin(\theta) - 3\ddot{\theta}\dot{\theta}(1.5+l)\cos(\theta) + \dot{\theta}^3(1.5+l)\sin(\theta) + \\ &\dddot{l}\cos(\theta) - 3\ddot{\theta}(\dot{\theta}(1.5+l)\cos(\theta) + \dot{l}\sin(\theta)) - 3\dot{\theta}\ddot{l}\sin(\theta) - 3\dot{\theta}^2\dot{l}\cos(\theta)\big]\hat{\boldsymbol{i}} + \\ &\big[\dddot{\theta}(1.5+l)\cos(\theta) - 3\ddot{\theta}\dot{\theta}(1.5+l)\sin(\theta) - \dot{\theta}^3(1.5+l)\cos(\theta) + \\ &\dddot{l}\sin(\theta) + 3\ddot{\theta}(-\dot{\theta}(1.5+l)\sin(\theta) + \dot{l}\cos(\theta)) + 3\dot{\theta}\ddot{l}\cos(\theta) - 3\dot{\theta}^2\dot{l}\sin(\theta)\big]\hat{\boldsymbol{j}}, \end{aligned} \tag{7.81}$$

or ${}^j\boldsymbol{\rho}_N^m = -2.476\hat{\boldsymbol{i}} + 3.468\hat{\boldsymbol{j}} \rightarrow \rho_N = 4.261\ \text{m/s}^3$.

Finally, the total hyper-jerk of the nozzle N as measured from body j, the vector $\boldsymbol{\sigma}_N = {}^j\boldsymbol{\sigma}_N^m$, is computed according to Eq. (7.9) as

$$
\begin{aligned}
{}^j\boldsymbol{\sigma}_N^m = {}^j\boldsymbol{\sigma}_N^k + {}^k\boldsymbol{\sigma}_N^m + 4{}^j\boldsymbol{\omega}^k \times {}^k\boldsymbol{\rho}_N^m + 4{}^j\boldsymbol{\rho}^k \times {}^k\boldsymbol{v}_N^m + \\
8{}^j\boldsymbol{\alpha}^k \times \left({}^j\boldsymbol{\omega}^k \times {}^k\boldsymbol{v}_N^m\right) + 6{}^j\boldsymbol{\alpha}^k \times {}^k\boldsymbol{a}_N^m + 6{}^j\boldsymbol{\omega}^k \times \left({}^j\boldsymbol{\omega}^k \times {}^k\boldsymbol{a}_N^m\right) + \\
4{}^j\boldsymbol{\omega}^k \times \left({}^j\boldsymbol{\alpha}^k \times {}^k\boldsymbol{v}_N^m\right) + 4{}^j\boldsymbol{\omega}^k \times \left[{}^j\boldsymbol{\omega}^k \times \left({}^j\boldsymbol{\omega}^k \times {}^k\boldsymbol{v}_N^m\right)\right],
\end{aligned}
\tag{7.82}
$$

where according to Eq. (7.7), the hyper-jerk of N fixed in body k as measured from body j, the vector ${}^j\boldsymbol{\sigma}_N^k$, is given by

$$
\begin{aligned}
{}^j\boldsymbol{\sigma}_N^k = {}^j\boldsymbol{\sigma}^k \times \boldsymbol{r}_{A/O} + 3{}^j\boldsymbol{\rho}^k \times \left({}^j\boldsymbol{\omega}^k \times \boldsymbol{r}_{A/O}\right) + 3{}^j\boldsymbol{\alpha}^k \times \left({}^j\boldsymbol{\alpha}^k \times \boldsymbol{r}_{A/O}\right) + \\
3{}^j\boldsymbol{\alpha}^k \times \left[{}^j\boldsymbol{\omega}^k \times \left({}^j\boldsymbol{\omega}^k \times \boldsymbol{r}_{A/O}\right)\right] + 2{}^j\boldsymbol{\omega}^k \times \left[{}^j\boldsymbol{\alpha}^k \times \left({}^j\boldsymbol{\omega}^k \times \boldsymbol{r}_{A/O}\right)\right] + \\
{}^j\boldsymbol{\omega}^k \times \left({}^j\boldsymbol{\rho}^k \times \boldsymbol{r}_{A/O}\right) + {}^j\boldsymbol{\omega}^k \times \left[{}^j\boldsymbol{\omega}^k \times \left({}^j\boldsymbol{\alpha}^k \times \boldsymbol{r}_{A/O}\right)\right] + \\
{}^j\boldsymbol{\omega}^k \times \left\{{}^j\boldsymbol{\omega}^k \times \left[{}^j\boldsymbol{\omega}^k \times \left({}^j\boldsymbol{\omega}^k \times \boldsymbol{r}_{A/O}\right)\right]\right\}.
\end{aligned}
\tag{7.83}
$$

Meanwhile,

$$
{}^k\boldsymbol{\sigma}_N^m = \ddddot{l}\hat{\boldsymbol{u}}_{N/O}. \tag{7.84}
$$

Hence, we have

$$
\begin{aligned}
{}^j\boldsymbol{\sigma}_N^m = \big[-\ddddot{\theta}(1.5+l)\sin(\theta) - 4\dddot{\theta}\dot{\theta}(1.5+l)\cos\theta - 3\ddot{\theta}^2(1.5+l)\cos\theta + \\
6\ddot{\theta}\dot{\theta}^2(1.5+l)\sin(\theta) + \dot{\theta}^4(1.5+l)\cos\theta + \ddddot{l}\cos\theta - 4\dot{\theta}\dddot{l}\sin(\theta) - \\
4\dddot{\theta}\dot{l}\sin(\theta) - 12\ddot{\theta}\dot{\theta}\dot{l}\cos\theta - 6\ddot{\theta}\ddot{l}\sin(\theta) - 6\dot{\theta}^2\ddot{l}\cos\theta + 4\dot{\theta}^3\dot{l}\sin(\theta)\big]\hat{\boldsymbol{i}} \\
\big[\ddddot{\theta}(1.5+l)\cos\theta - 4\dddot{\theta}\dot{\theta}(1.5+l)\sin(\theta) - 3\ddot{\theta}^2(1.5+l)\sin(\theta) - \\
6\ddot{\theta}\dot{\theta}^2(1.5+l)\cos\theta + \dot{\theta}^4(1.5+l)\sin(\theta) + \ddddot{l}\sin(\theta) + 4\dot{\theta}\dddot{l}\cos\theta + \\
4\dddot{\theta}\dot{l}\cos\theta - 12\ddot{\theta}\dot{\theta}\dot{l}\sin(\theta) + 6\ddot{\theta}\ddot{l}\cos\theta - 6\dot{\theta}^2\ddot{l}\sin(\theta) - 4\dot{\theta}^3\dot{l}\cos\theta\big]\hat{\boldsymbol{j}},
\end{aligned}
\tag{7.85}
$$

or ${}^j\boldsymbol{\sigma}_N^m = 11.6285\hat{\boldsymbol{i}} + 5.6245\hat{\boldsymbol{j}} \rightarrow \sigma_N = 12.925\ \mathrm{m/s^4}$.

This part of the example does not require the inclusion of the infinitesimal screws associated to the kinematic pairs of the robot. To obtain the kinematic states of the end-effector of the serial manipulator, in what follows the kinematics of the paint-spraying robot is approached by incorporating the corresponding infinitesimal screws. To this end, the model of the screws is depicted in Fig. 7.7.

With reference to Fig. 7.7, the revolute joint connecting body k to body j is represented by the screw ${}^j\$^k$, while ${}^k\$^m$ is the screw associated to the prismatic joint connecting body m to body k. Hence, considering point N as the reference pole, the infinitesimal screws are computed as

$$ {}^{j}\$^{k} = \begin{bmatrix} 0 \\ 0 \\ 1 \\ -(1.5+l)\sin(\theta) \\ (1.5+l)\cos(\theta) \\ 0 \end{bmatrix}, \quad {}^{k}\$^{m} = \begin{bmatrix} 0 \\ 0 \\ 0 \\ \cos(\theta) \\ \sin(\theta) \\ 0 \end{bmatrix}. \tag{7.86} $$

Next, the velocity state of body m as measured from body j, the vector ${}^{j}\boldsymbol{V}_{N}^{m}$, is given by

$$ {}^{j}\boldsymbol{V}_{N}^{m} = {}^{j}\boldsymbol{V}_{N}^{k} + {}^{k}\boldsymbol{V}_{N}^{m}, \tag{7.87} $$

where

$$ {}^{j}\boldsymbol{V}_{N}^{k} = \dot{\theta}\,{}^{j}\$^{k} = \dot{\theta} \begin{bmatrix} 0 \\ 0 \\ 1 \\ -(1.5+l)\sin(\theta) \\ (1.5+l)\cos(\theta) \\ 0 \end{bmatrix}. \tag{7.88} $$

Meanwhile,

$$ {}^{k}\boldsymbol{V}_{N}^{m} = \dot{l}\,{}^{k}\$^{m} = \dot{l} \begin{bmatrix} 0 \\ 0 \\ 0 \\ \cos(\theta) \\ \sin(\theta) \\ 0 \end{bmatrix}. \tag{7.89} $$

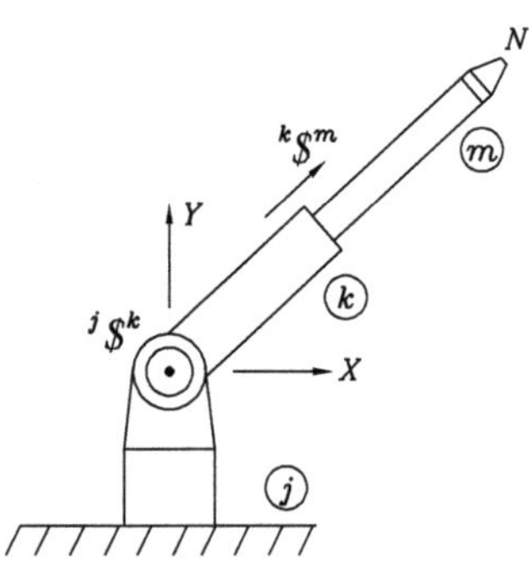

Fig. 7.7 Example 7.2. Model of the screws

Thus,

$$
{}^{j}\boldsymbol{V}_N^m = \dot{\theta}\begin{bmatrix} 0 \\ 0 \\ 1 \\ -(1.5+l)\sin(\theta) \\ (1.5+l)\cos(\theta) \\ 0 \end{bmatrix} + \dot{l}\begin{bmatrix} 0 \\ 0 \\ 0 \\ \cos(\theta) \\ \sin(\theta) \\ 0 \end{bmatrix}
= \begin{bmatrix} 0 \\ 0 \\ \dot{\theta} \\ -\dot{\theta}(1.5+l)\sin(\theta) + \dot{l}\cos(\theta) \\ \dot{\theta}(1.5+l)\cos(\theta) + \dot{l}\sin(\theta) \\ 0 \end{bmatrix}. \tag{7.90}
$$

Furthermore, the reduced acceleration state of body m as measured from body j, the vector ${}^{j}\boldsymbol{A}_N^m$, may be computed as

$$
{}^{j}\boldsymbol{A}_N^m = {}^{j}\boldsymbol{A}_N^k + {}^{k}\boldsymbol{A}_N^m + \left[{}^{j}\boldsymbol{V}_N^k \quad {}^{k}\boldsymbol{V}_N^m\right], \tag{7.91}
$$

where

$$
{}^{j}\boldsymbol{A}_N^k = \ddot{\theta}\,{}^{j}\$^k \tag{7.92}
$$

and

$$
{}^{k}\boldsymbol{A}_N^m = \ddot{l}\,{}^{k}\$^m. \tag{7.93}
$$

Therefore,

$$
{}^{j}\boldsymbol{A}_N^m = \ddot{\theta}\,{}^{j}\$^k + \ddot{l}\,{}^{k}\$^m + \left[\dot{\theta}\,{}^{j}\$^k \quad \dot{l}\,{}^{k}\$^m\right]
= \begin{bmatrix} 0 \\ 0 \\ \ddot{\theta} \\ -\ddot{\theta}(1.5+l)\sin(\theta) + \ddot{l}\cos(\theta) - \dot{\theta}\dot{l}\sin(\theta) \\ \ddot{\theta}(1.5+l)\cos(\theta) + \ddot{l}\sin(\theta) + \dot{\theta}\dot{l}\cos(\theta) \\ 0 \end{bmatrix}. \tag{7.94}
$$

On the other hand, the reduced jerk state of body m as measured from body j, the vector ${}^{j}\boldsymbol{J}_N^m$, may be computed as

$$
{}^{j}\boldsymbol{J}_N^m = {}^{j}\boldsymbol{J}_N^k + {}^{k}\boldsymbol{J}_N^m + 2\left[{}^{j}\boldsymbol{V}_N^k \quad {}^{k}\boldsymbol{A}_N^m\right] + \left[{}^{j}\boldsymbol{A}_N^k \quad {}^{k}\boldsymbol{V}_N^m\right] + \left[{}^{j}\boldsymbol{V}_N^k\left[{}^{j}\boldsymbol{V}_N^k \quad {}^{k}\boldsymbol{V}_N^m\right]\right], \tag{7.95}
$$

where

$$ {}^{j}\boldsymbol{J}_{N}^{k} = \dddot{\theta}{}^{j}\$^{k} \tag{7.96} $$

and

$$ {}^{k}\boldsymbol{J}_{N}^{m} = \dddot{l}{}^{k}\$^{m}. \tag{7.97} $$

Hence,

$$ \begin{aligned} {}^{j}\boldsymbol{J}_{N}^{m} &= \dddot{\theta}{}^{j}\$^{k} + \dddot{l}{}^{k}\$^{m} + 2\left[\dot{\theta}{}^{j}\$^{k} \quad \ddot{l}{}^{k}\$^{m}\right] + \left[\ddot{\theta}{}^{j}\$^{k} \quad \dot{l}{}^{k}\$^{m}\right] + \left[\dot{\theta}{}^{j}\$^{k}\left[\dot{\theta}{}^{j}\$^{k} \quad \dot{l}{}^{k}\$^{m}\right]\right] \\ &= \begin{bmatrix} 0 \\ 0 \\ \dddot{\theta} \\ -\dddot{\theta}(1.5+l)\sin(\theta) + \dddot{l}\cos(\theta) - 2\dot{\theta}\ddot{l}\sin(\theta) - \ddot{\theta}\dot{l}\sin(\theta) - \dot{\theta}^{2}\dot{l}\cos(\theta) \\ \dddot{\theta}(1.5+l)\cos(\theta) + \dddot{l}\sin(\theta) + 2\dot{\theta}\ddot{l}\cos(\theta) + \ddot{\theta}\dot{l}\cos(\theta) - \dot{\theta}^{2}\dot{l}\sin(\theta) \\ 0 \end{bmatrix}. \end{aligned} \tag{7.98} $$

Finally, the reduced hyper-jerk state of body m as measured from body j, the vector ${}^{j}\boldsymbol{H}_{N}^{m}$, is obtained as

$$ \begin{aligned} {}^{j}\boldsymbol{H}_{N}^{m} = {}& {}^{j}\boldsymbol{H}_{N}^{k} + {}^{k}\boldsymbol{H}_{N}^{m} + 3\left[{}^{j}\boldsymbol{V}_{N}^{k} \quad {}^{k}\boldsymbol{J}_{N}^{m}\right] + 3\left[{}^{j}\boldsymbol{A}_{N}^{k} \quad {}^{k}\boldsymbol{A}_{N}^{m}\right] + \\ & 3\left[{}^{j}\boldsymbol{V}_{N}^{k} \quad \left[{}^{j}\boldsymbol{V}_{N}^{k} \quad {}^{k}\boldsymbol{A}_{N}^{m}\right]\right] + 2\left[{}^{j}\boldsymbol{A}_{N}^{k} \quad \left[{}^{j}\boldsymbol{V}_{N}^{k} \quad {}^{k}\boldsymbol{V}_{N}^{m}\right]\right] + \\ & \left[{}^{j}\boldsymbol{J}_{N}^{k} \quad {}^{k}\boldsymbol{V}_{N}^{m}\right] + \left[{}^{j}\boldsymbol{V}_{N}^{k}\left[{}^{j}\boldsymbol{A}_{N}^{k} \quad {}^{k}\boldsymbol{V}_{N}^{m}\right]\right] + \left[{}^{j}\boldsymbol{V}_{N}^{k}\left[{}^{j}\boldsymbol{V}_{N}^{k}\left[{}^{j}\boldsymbol{V}_{N}^{k} \quad {}^{k}\boldsymbol{V}_{N}^{m}\right]\right]\right], \end{aligned} \tag{7.99} $$

where

$$ {}^{j}\boldsymbol{H}_{N}^{k} = \ddddot{\theta}{}^{j}\$^{k} \tag{7.100} $$

and

$$ {}^{k}\boldsymbol{H}_{N}^{m} = \ddddot{l}{}^{k}\$^{m}. \tag{7.101} $$

Therefore,

$$ \begin{aligned} {}^{j}\boldsymbol{H}_{N}^{m} = {}& \ddddot{\theta}{}^{j}\$^{k} + \ddddot{l}{}^{k}\$^{m} + 3\left[\dot{\theta}{}^{j}\$^{k} \quad \dddot{l}{}^{k}\$^{m}\right] + 3\left[\ddot{\theta}{}^{j}\$^{k} \quad \ddot{l}{}^{k}\$^{m}\right] + \\ & 3\left[\dot{\theta}{}^{j}\$^{k} \quad \left[\dot{\theta}{}^{j}\$^{k} \quad \ddot{l}{}^{k}\$^{m}\right]\right] + 2\left[\ddot{\theta}{}^{j}\$^{k} \quad \left[\dot{\theta}{}^{j}\$^{k} \quad \dot{l}{}^{k}\$^{m}\right]\right] + \end{aligned} $$

$$\left[\dddot{\theta}^j\$^k \quad \dot{l}^k\$^m\right]+\left[\dot{\theta}^j\$^k\left[\ddot{\theta}^j\$^k \quad \dot{l}^k\$^m\right]\right]+\left[\dot{\theta}^j\$^k\left[\dot{\theta}^j\$^k\left[\dot{\theta}^j\$^k \quad \dot{l}^k\$^m\right]\right]\right]$$

$$=\begin{bmatrix} 0 \\ 0 \\ \ddddot{\theta} \\ -\ddddot{\theta}(1.5+l)\sin(\theta)+\ddddot{l}\cos(\theta)-3\dot{\theta}\dddot{l}\sin(\theta)-3\dddot{\theta}\dot{l}\sin(\theta) \\ -3\dot{\theta}^2\ddot{l}\cos(\theta)-3\ddot{\theta}\dot{\theta}\dot{l}\cos(\theta)-\dddot{\theta}\dot{l}\sin(\theta)+\dot{\theta}^3\dot{l}\sin(\theta) \\ \ddddot{\theta}(1.5+l)\cos(\theta)+\ddddot{l}\sin(\theta)+3\dot{\theta}\dddot{l}\cos(\theta)+3\dddot{\theta}\dot{l}\cos(\theta) \\ -3\dot{\theta}^2\ddot{l}\sin(\theta)-3\ddot{\theta}\dot{\theta}\dot{l}\sin(\theta)+\dddot{\theta}\dot{l}\cos(\theta)-\dot{\theta}^3\dot{l}\cos(\theta) \\ 0 \end{bmatrix}. \quad (7.102)$$

7.4 Exercises

1. The disk shown in Fig. 7.8 rotates about a fixed axis passing through point O with angular velocity $\omega = 5\,\text{rad/s}$, angular acceleration $\alpha = 2\,\text{rad/s}^2$, angular jerk $\rho = 3\,\text{rad/s}^3$, and angular hyper-jerk $\sigma = 2\,\text{rad/s}^4$ in the indicated directions. On the other hand, the small sphere A moves in the slot undergoing relative motions with respect to the slot as follows: $\dot{u} = 2\,\text{m/s}$, $\ddot{u} = 1.5\,\text{m/s}^2$, $\dddot{u} = 0.5\,\text{m/s}^3$, and $\ddddot{u} = 3\,\text{m/s}^4$ in the indicated directions. We must compute the instantaneous velocity, acceleration, jerk, and hyper-jerk of sphere A considering $x = 0.75\,\text{m}$ and $y = 0.5\,\text{m}$.
2. Figure 7.9 shows an actuating mechanism for a telescoping antenna on a spacecraft. The supporting shaft j rotates about the fixed Y-axis while the orientation of the telescoping antenna is determined according to the angles $\theta(t) = \frac{\pi}{9} + \frac{\pi}{6}\sin(t)$ and $\phi(t) = \frac{\pi}{6} + \frac{\pi}{9}\sin(t)$. At the same time, the antenna extends according to the variable length $l(t) = 1 + \frac{\pi}{18}\sin(t)$ while body m rotates freely with respect to body k along the unit axis $\hat{\boldsymbol{u}}_{P/O}$ according to the angle $\beta(t) = \frac{\pi}{4} + \frac{\pi}{12}\sin(t)$. Determine the velocity, acceleration, jerk, and hyper-jerk of P as measured from the fixed body at time $t = 4$.
3. Determine the input–output equation of hyper-jerk of the two-degree-of-freedom parallel manipulator shown in Fig. 7.10 that was introduced by Nam and Park (2006).

Fig. 7.8 Problem 1. Hyper-jerk analysis

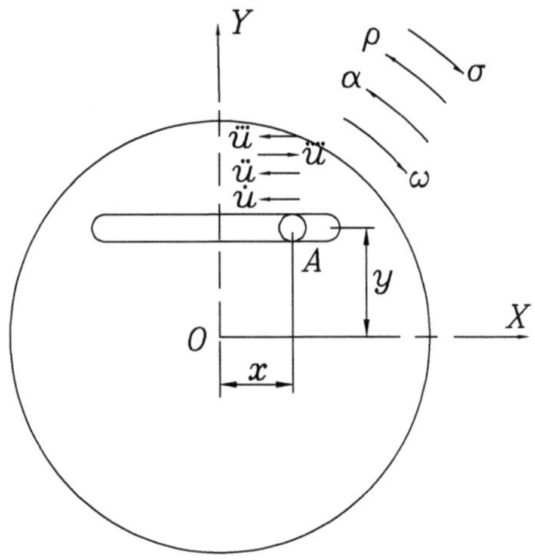

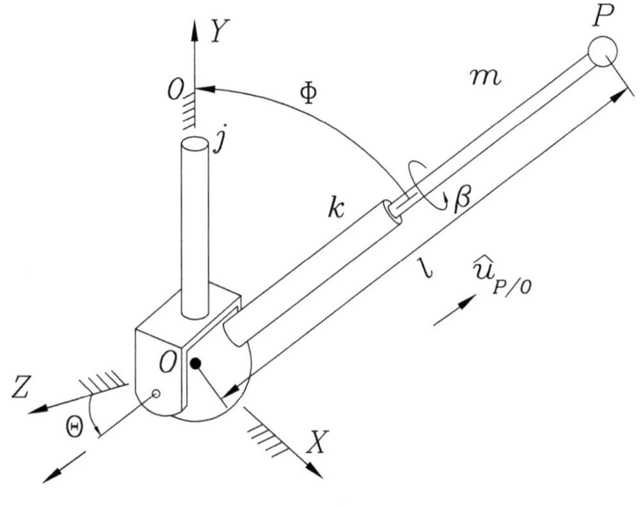

Fig. 7.9 Problem 2. Hyper-jerk analysis

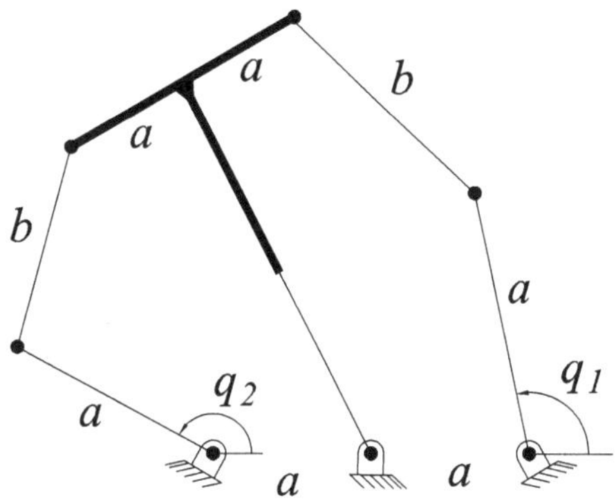

Fig. 7.10 Problem 3. Hyper-jerk analysis

References

Ball, R. S. (1900). *A treatise on the theory of screws*. Cambridge: Cambridge University Press (reprinted 1998).

Brand, L. (1947). *Vector and tensor analysis*. New York: Wiley.

Gallardo, J., Orozco, H., & Rico, J. M. (2008). Kinematics of 3-RPS parallel manipulators by means of screw theory. *International Journal of Advanced Manufacturing Technology, 36*(5), 598–605.

Gallardo-Alvarado, J. (2003). Jerk distribution of a 6–3 Gough-Stewart platform. *Proceedings Institution of Mechanical Engineers Part K: Journal of Multi-Body Dynamics, 217*(1), 77–84.

Gallardo-Alvarado, J. (2012). Jerk analysis of a six-degrees-of-freedom three-legged parallel manipulator. *Robotics and Computer-Integrated Manufacturing, 28*, 220–226.

Gallardo-Alvarado, J., & Camarillo-Gómez, K. A. (2011). Inverse jerk analysis of symmetric zero-torsion parallel manipulators. *Robotics and Autonomous Systems, 59*(11), 859–866.

Gallardo-Alvarado, J., Alici, G., & Pérez-González, L. (2010). A new family of constrained redundant parallel manipulators. *Multibody System Dynamics, 23*(1), 57–75.

Gallardo-Alvarado, J., Orozco-Mendoza, H., Rodríguez-Castro, R., & Rico-Martínez, J. M. (2007). Kinematics of a class of parallel manipulators which generates structures with three limbs. *Multibody System Dynamics, 17*(1), 27–46.

Gallardo-Alvarado, J., & Rico-Martínez, J. M. (2001). Jerk influence coefficients, via screw theory, of closed chains. *Meccanica, 36*(2), 213–228.

Gasparetto, A., Lanzutti, A., Vidoni, R., & Zanotto, V. (2012). Experimental validation and comparative analysis of optimal time-jerk algorithms for trajectory planning. *Robotics and Computer-Integrated Manufacturing, 28*(2), 164–181.

Kyriakopoulos, K. J., & Saridis, G. N. (1994). Minimum jerk for trajectory planning and control. *Robotica, 12*(2), 109–113.

Lipkin, H. (2005). Time derivatives of screws with applications to dynamic and stiffness. *Mechanism and Machine Theory, 4*, 259–273.

Maccari, A. (2011). Vibration control by nonlocal feedback and jerk dynamics. *Nonlinear Dynamics, 63*(1), 159–169.

Nam, Y.-J., & Park, M.-K. (2006). Kinematics and optimization of 2-dof parallel manipulator with revolute actuators and a passive leg. *Journal of Mechanical Science and Technology, 20*(6), 828–839.

Oyadiji, S. O., Ayalew, A., & Lai, E. (2005). Analytical framework for the smooth manoeuvre of wheeled mobile robots traversing obstacles. *Proceedings Royal Society A, 461*(2054), 481–507.

Rico, J. M., & Duffy, J. (1996). An application of screw algebra to the acceleration analysis of serial chains. *Mechanism and Machine Theory, 31*(4), 445–457.

Rico, J. M., & Gallardo, J. (1996). Acceleration analysis, via screw theory and characterization of singularities of closed chains. In J. Lenarčič & V. Parenti Castelli (Eds.), *Recent advances in robot kinematics* (pp. 139–148). Dordrecht: Kluwer.

Rico, J. M., Gallardo, J., & Duffy, J. (1995). A determination of singular configurations of serial non-redundant manipulators and their escapement from singularities using lie products. In J. P. Merlet & B. Ravani (Eds.), *Computational Kinematics '95* (pp. 143–152). Dordrecht: Kluwer.

Rico, J. M., Gallardo, J., & Duffy, J. (1999). Screw theory and higher order kinematic analysis of open serial and closed chains. *Mechanism and Machine Theory, 34*(4), 559–586.

Villagra, J., Milanés, V., Pérez, J., & Godoy, J. (2012). Smooth path and speed planning for an automated public transport vehicle. *Robotics and Autonomous Systems, 60*(2), 252–265.

Wohlhart, K. (2010). From higher degrees of shakiness to mobility. *Mechanism and Machine Theory, 45*(3), 467–476.

Zhang, L. B., You, Y. P., He, J., & Yang, X. F. (2011). The transition algorithm based on parametric spline curve for high-speed machining of continuous short line segments. *International Journal of Advanced Manufacturing Technology, 52*(1), 245–254.

Part IV
Kinematics of Parallel Manipulators by Means of Screw Theory Exemplified

Chapter 8
3R2T Parallel Manipulator

8.1 Introduction

A parallel manipulator with fully decoupled motions is a robot where the orientation and position of the moving platform with respect to the fixed platform can be computed separately, which can increase the robot's dexterity and workspace. This option was investigated by Hunt and Primrose (1993) for fully parallel manipulators and was applied by Zlatanov et al. (1992) in the design of a six-degree-of-freedom (DOF), three-legged parallel manipulator. Jin et al. (2006) proposed a versatile three-legged parallel manipulator to perform 6 DOF, in which by changing the scheme of actuation it is possible to generate decoupled and hybrid motions over the moving platform. On the other hand, a spherical parallel wrist can be connected at the moving platform of a Tricept, yielding a 6-DOF mechanism (Innocenti and Wenger 2004; Carbone and Ceccarelli 2005a,b). Further, taking into account that if the spherical wrist is an open chain, to improve the stiffness of the composed robot, the serial manipulator would be substituted by a parallel manipulator, producing a series-parallel manipulator; see, for instance, Tanev (2000), Zheng et al. (2004), Gallardo-Alvarado (2005), Lu and Hu (2006). Furthermore, taking into account that decoupled parallel manipulators allow for closed-form solutions, Gallardo-Alvarado et al. (2010) introduced a family of non-overconstrained redundantly actuated parallel manipulators. This possibility is also available for parallel manipulators having fewer than 6 DOF (Gallardo-Alvarado et al. 2006, 2007, 2008).

This chapter is motivated by the good acceptance of limited-DOF parallel manipulators in an industrial context; see, for instance, Clavel (1990, 1991). In fact, the so-called defective parallel manipulators have been successfully proposed as machine tools (Weck and Staimer 2002; Zoppi et al. 2004; Zhao et al. 2002; Song and Mu 2005; Gao et al. 2006; Lu and Xu 2008; Li et al. 2004, 2010), hyper-redundant manipulators (Chirikjian 1992), haptic devices based on parallel mechanisms (Angerilli et al. 2001; Birglen et al. 2002), micro robotics (Merlet 2001; Wu et al. 2005; Ng et al. 2006; Peirs et al. 1999), and so forth. In this area, 5-DOF

J. Gallardo-Alvarado, *Kinematic Analysis of Parallel Manipulators by Algebraic Screw Theory*, DOI 10.1007/978-3-319-31126-5_8

parallel manipulators represent an excellent option for the development of multiaxis machine tools that will possibly replace traditional Cartesian machine tools in the near future. On the other hand, in this area it is noticeable how most researchers are focused on the study of the so-called two rotation (R) + three translation (T) parallel mechanisms, written as 2R3T for brevity. Certainly, the restriction of one rotation, usually normal to the moving platform, is the main attraction of such robots. In that vein, interesting applications using 3R2T parallel mechanisms include the kinematic simulation of the human spinal column, which has recently attracted the attention of kinematicians Zhu et al. (2008, 2009).

In this chapter, the kinematic analysis of a 5-DOF decoupled parallel manipulator, three rotations + two translations, is addressed using screw theory.

8.2 Description of the 3R2T Parallel Manipulator

With reference to Fig. 8.1, the parallel manipulator under analysis consists of a moving platform and a fixed platform connected to each other by three UPS-type kinematic chains (where U, P, and S respectively stand for universal, prismatic, and spherical joints) whose role is to control the orientation of the moving platform and one central 2-DOF PPS-type kinematic chain whose role is to control the position of the moving platform. As expected, according to a revised version of Kutzbach–Grübler's formula, the robot manipulator at hand possesses 5 DOF, where the prismatic pairs have the privilege of being selected as the active joints of the robot manipulator. Furthermore, the robot's topology allows for decoupled translational and rotational motions of the moving platform as observed from the fixed platform.

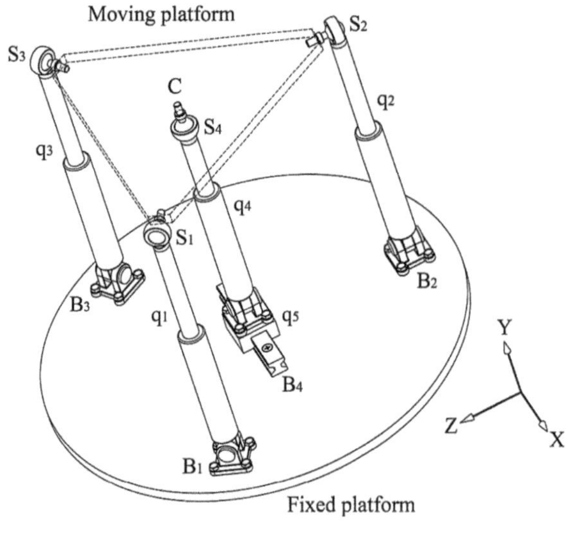

Fig. 8.1 3R2T parallel manipulator and its geometric scheme

In what follows we explain the notation used to approach the displacement analysis of the 3R2T parallel manipulator. Let XYZ be a reference frame attached to the center B_4 of the fixed platform, with associated unit vectors $\boldsymbol{ijk}$, where the Y-axis is normal to the plane of the fixed platform. The points $B_i(i = 1,2,3)$, located by vectors $\boldsymbol{b}_i(i = 1,2,3)$, shape an equilateral triangle over the fixed platform and denote the nominal positions of the universal joints. On the other hand, the points $S_i(i = 1,2,3,4)$ denote the centers of the spherical joints and are located, respectively, by vectors $\boldsymbol{s}_i(i = 1,2,3,4)$. Similar to the fixed platform, points $S_i(i = 1,2,3)$ shape an equilateral triangle $\Delta S_1S_2S_3$ of side a inscribed in a circle of radius r named the radius of the moving platform. Furthermore, the center $C = S_4$ of the triangle $\Delta S_1S_2S_3$ is named the control point of the parallel manipulator. Finally, $q_i(i = 1,2,\ldots,5)$ are the generalized coordinates of the mechanism and denote the expandable lengths of the parallel manipulator.

8.3 Finite Kinematics of the 3R2T Robot

This section presents the inverse and forward displacement analyses of the parallel manipulator under study.

8.3.1 *Forward Displacement Analysis of the 3R2T Robot*

The forward position analysis of the parallel manipulator consists of finding the pose of the moving platform with respect to the fixed platform for a prescribed set $\{q_1, q_2, \ldots, q_5\}$ of generalized coordinates. In other words, we must compute the coordinates of points $S_i = (X_i, Y_i, Z_i)$ given the robot's limb lengths. The coordinates of point S_4 are immediately computed because of the decoupled motion of the parallel manipulator. In fact, the vector $\boldsymbol{s}_4$ is directly obtained as

$$\boldsymbol{s}_4 = q_5\hat{\boldsymbol{i}} + q_4\hat{\boldsymbol{j}} \rightarrow S_4 = (q_5, q_4, 0). \tag{8.1}$$

On the other hand, to compute the coordinates of the remaining points S_i, let us consider that because $\Delta S_1S_2S_3$ is an equilateral triangle whose geometric center is located by vector $\boldsymbol{S}_4$, it follows that

$$(\boldsymbol{s}_1 + \boldsymbol{s}_2 + \boldsymbol{s}_3)/3 = \boldsymbol{s}_4, \tag{8.2}$$

which yields three linear equations. Furthermore, points S_i and B_i must satisfy the following closure equations:

$$(\boldsymbol{s}_i - \boldsymbol{b}_i) \bullet (\boldsymbol{s}_i - \boldsymbol{b}_i) = q_i^2, \quad i = 1,2,3. \tag{8.3}$$

Finally, according to the radius r of the moving platform, three compatibility kinematic constraint equations may be written as

$$(\boldsymbol{s}_i - \boldsymbol{s}_4) \bullet (\boldsymbol{s}_i - \boldsymbol{s}_4) = r^2, \quad i = 1, 2, 3. \tag{8.4}$$

After a few computations, from Eqs. (8.3) and (8.4) we get

$$(\boldsymbol{s}_i - \boldsymbol{b}_i) \bullet (\boldsymbol{s}_i - \boldsymbol{b}_i) - (\boldsymbol{s}_i - \boldsymbol{s}_4) \bullet (\boldsymbol{s}_i - \boldsymbol{s}_4) - q_i^2 + r^2 = 0. \tag{8.5}$$

Equations (8.2) and (8.5) represent a linear system of six equations in the nine unknown coordinates of points $S_i (i = 1, 2, 3)$. This linear system may be solved in terms of the variables X_1, Y_2, and Z_3. On the other hand, from the sides of the equilateral triangle $\Delta S_1 S_2 S_3$ it is possible to write three closure equations as follows:

$$\left.\begin{array}{l} (\boldsymbol{s}_1 - \boldsymbol{s}_2) \bullet (\boldsymbol{s}_1 - \boldsymbol{s}_2) = a^2 \\ (\boldsymbol{s}_2 - \boldsymbol{s}_3) \bullet (\boldsymbol{s}_2 - \boldsymbol{s}_3) = a^2 \\ (\boldsymbol{s}_1 - \boldsymbol{s}_3) \bullet (\boldsymbol{s}_1 - \boldsymbol{s}_3) = a^2 \end{array}\right\}. \tag{8.6}$$

Afterward, a nonlinear system of three equations in the unknowns X_1, Y_2, and Z_3 is obtained as

$$\left.\begin{array}{l} K_1'Y_2^2 + K_2'Z_3^2 + K_3'Y_2^2Z_3 + K_4'Y_2Z_3^2 + K_5'Y_2Z_3 + K_6'Y_2 + \\ K_7'Z_3 + K_8' = 0 \\ K_1''X_1^2 + K_2''Z_3^2 + K_3''X_1^2Z_3 + K_4''X_1Z_3^2 + K_5''Y_2Z_3 + K_6''X_1 + \\ K_7''Z_3 + K_8'' = 0 \\ K_1'''X_1^2 + K_2'''Y_2^2 + K_3'''X_1^2Y_2 + K_4'''X_1Y_2^2 + K_5'''X_1Y_2 + K_6'''X_1 + \\ K_7'''Y_2 + K_8''' = 0 \end{array}\right\}, \tag{8.7}$$

where K_*', K_*'', and K_*''' are coefficients computed according to the parameters and generalized coordinates of the parallel manipulator. Equation (8.7) are solved using the Sylvester dialytic method of elimination.

Once the coordinates of the centers of the spherical joints are computed, the pose of the moving platform with respect to the fixed platform may be interpreted as the 4×4 homogeneous coordinate transformation matrix $\mathbf{T}$ given by

$$\mathbf{T} = \begin{bmatrix} \mathbf{R} & \boldsymbol{s}_4 \\ 0_{1\times 3} & 1 \end{bmatrix}, \tag{8.8}$$

where the rotation matrix $\mathbf{R}$ is computed based on the coordinates of points $S_i (i = 1, 2, 3)$ (see Chap. 16).

Finally, in order to disregard spurious solutions, the numerical solutions of the forward displacement analysis can be verified by Eqs. (8.2)–(8.4).

8.3.2 Inverse Displacement Analysis of the 3R2T Robot

The inverse position analysis of the robot at hand consists of finding its limb lengths $q_i(i = 1, 2, \ldots, 5)$ given the pose of the moving platform as measured from the fixed platform. To this end, consider that xyz is a reference frame attached to the moving platform with the y-axis normal to the plane of the moving platform.

Clearly, given the pose of the moving platform, the coordinates of the centers of the spherical joints, the points $S_i = (X_i, Y_i, Z_i)$, are computed directly by the relationship

$$\begin{bmatrix} X_i \\ Y_i \\ Z_i \\ 1 \end{bmatrix} = \begin{bmatrix} \mathbf{R} & \boldsymbol{\tau} \\ 0_{1\times 3} & 1 \end{bmatrix} \begin{bmatrix} x_i \\ y_i \\ z_i \\ 1 \end{bmatrix}, \quad i = 1, \ldots, 4, \tag{8.9}$$

where $\boldsymbol{\tau}$ is the position vector of point S_4 while x_i, y_i, and z_i are the coordinates of the centers of the spherical joints expressed in the moving reference frame xyz. Once the points S_i are computed, the generalized coordinates $q_i(i = 1, 2, 3)$ are obtained as

$$q_i^2 = (\boldsymbol{s}_i - \boldsymbol{b}_i) \bullet (\boldsymbol{s}_i - \boldsymbol{b}_i), \quad i = 1, 2, 3, \tag{8.10}$$

whereas, taking into account that $S_4 = (X_4, Y_4, Z_4)$, we calculate the generalized coordinates q_4 and q_5 directly as $q_4 = Y_4$ and $q_5 = X_4$.

8.4 Infinitesimal Kinematics of the 3R2T Robot

In this section the velocity and acceleration analyses of the 3R2T parallel manipulator are addressed using screw theory. The model of the infinitesimal screws is depicted in Fig. 8.2.

8.4.1 Velocity Analysis of the 3R2T Parallel Manipulator

Let

$$\boldsymbol{V}_C = \begin{bmatrix} \boldsymbol{\omega} \\ \boldsymbol{v}_C \end{bmatrix}$$

be the velocity state of the moving platform, with respect to the fixed platform, where the primal part, $P(\boldsymbol{V}_C) = \boldsymbol{\omega}$, is the angular velocity vector of the moving

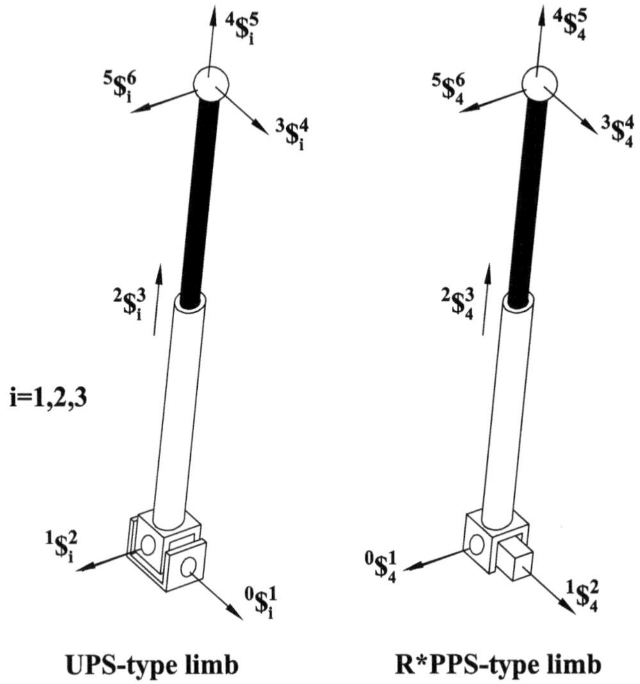

Fig. 8.2 Infinitesimal screws of the 3R2T parallel manipulator

platform with respect to the fixed platform, and the dual part, $D(\boldsymbol{V}_C) = \boldsymbol{v}_C$, is the linear velocity of point $C = S_4$ embedded in the moving platform as measured from the fixed platform. The velocity state of the moving platform with respect to the fixed platform, the six-dimensional vector $\boldsymbol{V}_C$, may be expressed in screw form through any of the UPS-type limbs as follows:

$$
{}_0\omega_1^i\,{}^0\$_i^1 + {}_1\omega_2^i\,{}^1\$_i^2 + {}_2\omega_3^i\,{}^2\$_i^3 + {}_3\omega_4^i\,{}^3\$_i^4 + {}_4\omega_5^i\,{}^4\$_i^5 + {}_5\omega_6^i\,{}^5\$_i^6 = \boldsymbol{V}_C
$$

$$
\rightarrow \mathbf{J}_i \Omega_i = \boldsymbol{V}_C, \quad i = 1, 2, 3, \qquad (8.11)
$$

where

$$
\mathbf{J}_i = [{}^0\$_i^1 \quad {}^1\$_i^2 \quad {}^2\$_i^3 \quad {}^3\$_i^4 \quad {}^4\$_i^5 \quad {}^5\$_i^6]
$$

is the screw-coordinate Jacobian of the ith limb. Meanwhile,

$$
\Omega_i = [{}_0\omega_1^i \quad {}_1\omega_2^i \quad {}_2\omega_3^i \quad {}_3\omega_4^i \quad {}_4\omega_5^i \quad {}_5\omega_6^i]^T
$$

is a matrix containing the joint-velocity rates of the ith limb. It should be noted that three of the five generalized speeds of the robot are notated as ${}_2\omega_3^i = \dot{q}_i$.

On the other hand, in terms of the central limb, in order to satisfy an algebraic requirement, the PPS-type limb is modeled as a R*PPS-type limb where the extra revolute joint R* is a fictitious kinematic pair; for example, its corresponding joint-velocity rate is zero. With this consideration in mind, the velocity state of the moving platform as measured from the fixed platform may be expressed in screw form through the R*PPS-type limb as follows:

$$ {}_0\omega_1^{4\,0}\$_4^1 + {}_1\omega_2^{4\,1}\$_4^2 + {}_2\omega_3^{4\,2}\$_4^3 + {}_3\omega_4^{4\,3}\$_4^4 + {}_4\omega_5^{4\,4}\$_4^5 + {}_5\omega_6^{4\,5}\$_4^6 = \boldsymbol{V}_C \tag{8.12} $$

or

$$ \mathbf{J}_4 \Omega_4 = \boldsymbol{V}_C, \tag{8.13} $$

where

$$ \mathbf{J}_4 = [{}^0\$_4^1 \quad {}^1\$_4^2 \quad {}^2\$_4^3 \quad {}^3\$_4^4 \quad {}^4\$_4^5 \quad {}^5\$_4^6] $$

is the screw-coordinate Jacobian matrix of the central limb. Meanwhile,

$$ \Omega_4 = [{}_0\omega_1^4 \quad {}_1\omega_2^4 \quad {}_2\omega_3^4 \quad {}_3\omega_4^4 \quad {}_4\omega_5^4 \quad {}_5\omega_6^4]^T $$

is a matrix containing the joint-velocity rates of the central limb. It should be noted that ${}_1\omega_2^4 = \dot{q}_4$ and ${}_2\omega_3^4 = \dot{q}_5$ are the generalized speeds of the central limb, while ${}_0\omega_1^4 = 0$ is the joint-velocity rate of the virtual kinematic pair. In what follows, the input–output equation of velocity of the parallel manipulator is systematically obtained by resorting to reciprocal screw theory.

Applying the Klein form of the screw ${}^4\$_i^5$ to both sides of Eq. (8.11), we have

$$ \begin{aligned} &\{{}^0\$_i^1; {}^4\$_i^5\}{}_0\omega_1^i + \{{}^1\$_i^2; {}^4\$_i^5\}{}_1\omega_2^i + \{{}^2\$_i^3; {}^4\$_i^5\}\dot{q}_i + \\ &\{{}^3\$_i^4; {}^4\$_i^5\}{}_3\omega_4^i + \{{}^4\$_i^5; {}^4\$_i^5\}{}_4\omega_5^i + \{{}^5\$_i^6; {}^4\$_i^5\}{}_5\omega_6^i = \{{}^4\$_i^5; \boldsymbol{V}_C\}. \end{aligned} \tag{8.14} $$

Taking into account that the screw ${}^4\$_i^5$ is reciprocal to all the screws in the same limb, except the screw associated to the prismatic joint, we can reduce Eq. (8.14) to

$$ \{{}^4\$_i^5; \boldsymbol{V}_C\} = \dot{q}_i, \quad i = 1, 2, 3. \tag{8.15} $$

To complete the equations required to obtain the input–output equation of velocity of the robot, from the central limb we get three equations as follows. Considering that ${}^0\$_4^1$ is a screw related to a fictitious revolute joint, then it is evident that ${}_0\omega_1^4 = 0$. Hence, applying the Klein form of the screw ${}^3\$_4^4$ to both sides of Eq. (8.13) and reducing terms, we have

$$ \{{}^3\$_4^4; \boldsymbol{V}_C\} = \dot{q}_4. \tag{8.16} $$

Similarly, applying the Klein form of the screw ${}^4\$^5_4$ to both sides of Eq. (8.13) leads to

$$\{{}^4\$^5_4; \boldsymbol{V}_C\} = \dot{q}_5. \tag{8.17}$$

Furthermore, since the screw ${}^5\$^6_4$ is reciprocal to all the screws of the central kinematic chain, it follows that

$$\{{}^5\$^6_4; \boldsymbol{V}_C\} = 0. \tag{8.18}$$

Casting Eqs. (8.15)–(8.18) into matrix-vector form, we find that the input–output equation of velocity of the 3R2T parallel manipulator results in

$$\mathbf{A}\boldsymbol{V}_C = \dot{\mathbf{q}}, \tag{8.19}$$

where $\mathbf{A} = \mathbf{J}^T\ \boldsymbol{\Delta}$ so that

$$\mathbf{J} = \begin{bmatrix} {}^4\$^5_1 & {}^4\$^5_2 & {}^4\$^5_3 & {}^3\$^4_4 & {}^4\$^5_4 & {}^5\$^6_4 \end{bmatrix}$$

is the screw-coordinate Jacobian matrix of the parallel manipulator. Meanwhile,

$$\boldsymbol{\Delta} = \begin{bmatrix} \mathbf{0} & \mathbf{I} \\ \mathbf{I} & \mathbf{0} \end{bmatrix}$$

is an operator of polarity formed with the zero matrix $\mathbf{0}$ and the identity matrix $\mathbf{I}$. Furthermore,

$$\dot{\mathbf{q}} = \begin{bmatrix} \dot{q}_1 & \dot{q}_2 & \dot{q}_3 & \dot{q}_4 & \dot{q}_5 & 0 \end{bmatrix}^T$$

is the first-order driver matrix of the parallel manipulator.

Concluding the velocity analysis, the inverse velocity analysis consists of finding matrix $\dot{\mathbf{q}}$ given the velocity state $\boldsymbol{V}_C$, while the forward velocity analysis consists of finding the velocity state $\boldsymbol{V}_C$ given matrix $\dot{\mathbf{q}}$. To achieve this, the matrix $\mathbf{J}$ must be invertible; otherwise, the parallel manipulator is at a singular configuration. Determining singular configurations of the robot is left as an exercise.

8.4.2 Acceleration Analysis of the 3R2T Parallel Manipulator

Let

$$\boldsymbol{A}_C = \begin{bmatrix} \boldsymbol{\alpha} \\ \boldsymbol{a}_C - \boldsymbol{\omega} \times \boldsymbol{v}_C \end{bmatrix}$$

be the reduced acceleration state of the moving platform with respect to the fixed platform where $\boldsymbol{\alpha}$ is the angular acceleration vector of the moving platform with respect to the fixed platform, whereas $\boldsymbol{a}_C$ is the linear acceleration of the center C of the moving platform. The accelerator $\boldsymbol{A}_C$ may be expressed in screw form through any of the UPS-type limbs of the parallel manipulator as follows:

$$ {}_0\dot{\omega}_1^{i}\,{}^0\$_i^1 + {}_1\dot{\omega}_2^{i}\,{}^1\$_i^2 + {}_2\dot{\omega}_3^{i}\,{}^2\$_i^3 + {}_3\dot{\omega}_4^{i}\,{}^3\$_i^4 + {}_4\dot{\omega}_5^{i}\,{}^4\$_i^5 + {}_5\dot{\omega}_6^{i}\,{}^5\$_i^6 + \mathcal{A}_i = \boldsymbol{A}_C, \quad i = 1,2,3, \tag{8.20}$$

or

$$\mathbf{J}_i \dot{\Omega}_i + \mathcal{A}_i = \boldsymbol{A}_C, \quad i = 1,2,3, \tag{8.21}$$

where

$$\dot{\Omega}_i = \begin{bmatrix} {}_0\dot{\omega}_1^i & {}_1\dot{\omega}_2^i & \ddot{q}_i & {}_3\dot{\omega}_4^i & {}_4\dot{\omega}_5^i & {}_5\dot{\omega}_6^i \end{bmatrix}^T$$

is a matrix containing the joint-acceleration rates of the ith limb. Meanwhile, $\mathcal{A}_i$ is the Lie screw of acceleration of the ith limb. Indeed,

$$\mathcal{A}_i = \begin{bmatrix} {}_0\omega_1^{i}\,{}^0\$_i^1 & {}_1\omega_2^{i}\,{}^1\$_i^2 + \ldots + {}_5\omega_6^{i}\,{}^5\$_i^6 \end{bmatrix} + \begin{bmatrix} {}_1\omega_2^{i}\,{}^1\$_i^2 & {}_2\omega_3^{i}\,{}^2\$_i^3 + \ldots + {}_5\omega_6^{i}\,{}^5\$_i^6 \end{bmatrix} + \ldots + \begin{bmatrix} {}_4\omega_5^{i}\,{}^4\$_i^5 & {}_5\omega_6^{i}\,{}^5\$_i^6 \end{bmatrix}. \tag{8.22}$$

Furthermore, the reduced acceleration state may be expressed in screw form through the R*PPS-type limb as follows:

$$ {}_0\dot{\omega}_1^{4}\,{}^0\$_4^1 + {}_1\dot{\omega}_2^{4}\,{}^1\$_4^2 + {}_2\dot{\omega}_3^{4}\,{}^2\$_4^3 + {}_3\dot{\omega}_4^{4}\,{}^3\$_4^4 + {}_4\dot{\omega}_5^{4}\,{}^4\$_4^5 + {}_5\dot{\omega}_6^{4}\,{}^5\$_4^6 + \mathcal{A}_4 = \boldsymbol{A}_C$$

or

$$\mathbf{J}_4 \dot{\Omega}_4 + \mathcal{A}_4 = \boldsymbol{A}_C, \tag{8.23}$$

where

$$\dot{\Omega}_4 = \begin{bmatrix} {}_0\dot{\omega}_1^4 & \ddot{q}_4 & \ddot{q}_5 & {}_3\dot{\omega}_4^4 & {}_4\dot{\omega}_5^4 & {}_5\dot{\omega}_6^4 \end{bmatrix}^T$$

is a matrix containing the joint-acceleration rates of the R*PPS-type limb. Meanwhile,

$$\mathcal{A}_4 = \begin{bmatrix} {}_0\omega_1^{4}\,{}^0\$_4^1 & {}_1\omega_2^{4}\,{}^1\$_4^2 + \ldots + {}_5\omega_6^{4}\,{}^5\$_4^6 \end{bmatrix} + \begin{bmatrix} {}_1\omega_2^{4}\,{}^1\$_4^2 & {}_2\omega_3^{4}\,{}^2\$_4^3 + \ldots + {}_5\omega_6^{4}\,{}^5\$_4^6 \end{bmatrix} + \ldots + \begin{bmatrix} {}_4\omega_5^{4}\,{}^4\$_4^5 & {}_5\omega_6^{4}\,{}^5\$_4^6 \end{bmatrix} \tag{8.24}$$

is the Lie screw of acceleration of the central limb.

Following the trend in the velocity analysis, the input–output equation of acceleration of the robot manipulator results in

$$\mathbf{A}\boldsymbol{A}_C = \ddot{\mathbf{q}} + \mathbf{C}, \tag{8.25}$$

where

$$\ddot{\mathbf{q}} = \begin{bmatrix} \ddot{q}_1 & \ddot{q}_2 & \ddot{q}_3 & \ddot{q}_4 & \ddot{q}_5 & 0 \end{bmatrix}^T$$

is the second-order driver matrix of the parallel manipulator, whereas

$$\mathbf{C} = \begin{bmatrix} \{{}^4\$_1^5; \mathcal{A}_1\} \\ \{{}^4\$_2^5; \mathcal{A}_2\} \\ \{{}^4\$_3^5; \mathcal{A}_3\} \\ \{{}^3\$_4^4; \mathcal{A}_4\} \\ \{{}^4\$_4^5; \mathcal{A}_4\} \\ \{{}^5\$_4^6; \mathcal{A}_4\} \end{bmatrix} \tag{8.26}$$

is the complementary matrix of acceleration of the 3R2T parallel manipulator. The inverse acceleration analysis consists of finding matrix $\ddot{\mathbf{q}}$ given the six-dimensional vector $\boldsymbol{A}_C$. On the other hand, the forward acceleration analysis consists of computing the reduced acceleration state $\boldsymbol{A}_C$ given matrix $\ddot{\mathbf{q}}$.

Finally, as pointed out by Zhu et al. (2009), because of their short history the kinematics of 3R2T parallel manipulators has not been well studied. Therefore, in this chapter the reader can find an option to systematically approach the position, velocity, and acceleration analyses of this class of limited-DOF parallel manipulators.

Example 8.1. To illustrate the application of the kinematic analysis method developed for the 3R2T parallel manipulator, we've included a numerical example here. To this end, consider that the parameters of the moving platform are given by $r = 0.25$ and $a = 0.433$. Meanwhile, the coordinates of points B_i are selected as

$$B_1 = (-0.25, 0, 0.433), \quad B_2 = (0.5, 0, 0),$$
$$B_3 = (-0.25, 0, -0.433), \quad B_4 = (0, 0, 0)$$

Furthermore, the generalized coordinates are commanded to follow periodic functions given by

$$\begin{aligned} q_1 &= 0.559 + 0.1\sin(t), \\ q_2 &= 0.579 + 0.025\sin(t), \\ q_3 &= 0.559 + 0.05\sin^2(t), \\ q_4 &= 0.075\sin(t), \\ q_5 &= 0.5 - 0.1\sin^2(t), \end{aligned}$$

where the time t is confined to the interval $0 \le t \le 2\pi$.

With these data it is necessary to determine the following kinematic properties of the parallel manipulator: (1) all the real feasible locations of the moving platform, with respect to the fixed platform, at the beginning of the motion; (2) select the home position of the parallel manipulator; according to the periodic functions assigned to the generalized coordinates q_i, compute the time history of the linear and angular infinitesimal kinematic quantities of the moving platform as measured from the fixed platform.

Solution. According to Eq. (8.7), a nonlinear system of three equations in the unknowns X_1, Y_2, and Z_3 is obtained at the beginning of the motion, that is, when time t is equal to zero, as follows:

$$\left.\begin{aligned} &1.25X_1^2 + 19.250Y_2^2 + 9.749Z_3^2 + 0.5X_1Y_2 - 25.115Y_2Z_3 - \\ &0.866X_1Z_3 - 0.84X_1 - 25.66Y_2 + 18.147Z_3 + 8.696 = 0 \\ \\ &1.25X_1^2 + 7.250Y_2^2 + 9.749Z_3^2 + 3.5X_1Y_2 - 9.526Y_2Z_3 - \\ &0.866X_1Z_3 - 0.977X_1 - 6.841Y_2 + 7.676Z_3 + 1.844 = 0 \\ \\ &5.X_1^2 + 8.Y_2^2 + 2.999Z_3^2 + 8.X_1Y_2 - 6.928Y_2Z_3 - 3.464X_1Z_3 - \\ &3.636X_1 - 10.182Y_2 + 4.408Z_3 + 3.228 = 0 \end{aligned}\right\} . \tag{8.27}$$

Equation (8.27) were solved using the Sylvester dialytic method of elimination. The solutions are listed in Table 8.1. Ignoring complex and spurious solutions, there are only two real solutions available for the forward position analysis. One solution yields the following coordinates for the centers of the spherical joints:

$$\begin{aligned} S_1 &= (-0.069, 0.492, 0.240), \\ S_2 &= (0.242, 0.514, -0.060), \\ S_3 &= (-0.173, 0.492, -0.180), \\ S_4 &= (0, 0.5, 0), \end{aligned}$$

Table 8.1 Example 8.1. Reachable values for X_1, Y_2, and Z_3

X_1	Y_2	Z_3
0.145e−1+0.358i	0.243−0.504e−10i	−0.445+0.319e−1i
0.145e−1−0.358i	0.243+0.504e−10i	−0.445−0.319e−1i
−0.069	0.514	−0.180
−0.173	0.514	−0.240
−0.375+0.203i	0.638−0.138i	0.728e−2+0.389e−1i
$-0.375-0.203i$	0.638+0.138i	0.728c−2 − 0.389e−1i
0.928e−2−0.649e−1i	0.638−0.138i	−0.212−0.278i
0.928e−2+0.649e−1i	0.638+0.138i	−0.212+0.278i

whereas the second solution indicates that

$$\begin{aligned}
S_1 &= (-0.173, 0.492, 0.180),\\
S_2 &= (0.242, 0.514, 0.060),\\
S_3 &= (-0.069, 0.492, -0.240),\\
S_4 &= (0, 0.5, 0).
\end{aligned}$$

Because of the decoupled motion of the robot, the center S_4 of the moving platform depends on the generalized coordinate q_4, representing the lower prismatic joint, and the generalized coordinate q_5, associated to the adjustable length of the central limb. Hence, in reality there is only one solution of the forward displacement analysis for point S_4. On the other hand, because of the presence of multiple solutions, the centers of the remaining spherical joints can reach two different positions. Thus, determining the actual configuration of the mechanism is a mandatory task to approach the infinitesimal kinematics of the parallel manipulator. This task can be achieved using sensors. In that way it is opportune to emphasize that this chapter reports on a simple academic numerical example; therefore, there are no particular reasons to select one of the two computed configurations of the 5-DOF parallel manipulator as the actual configuration of the robot. With this consideration in mind, take the first solution of the forward displacement analysis as the home position or reference configuration of the parallel manipulator and solve the forward displacement analysis by applying a numerical technique such as the Newton–Raphson method. The relevant results of the angular and linear velocities and accelerations of the center of the moving platform as measured from the fixed platform computed via the theory of screws are provided, respectively, in Figs. 8.3 and 8.4.

As was expected, the linear velocity and acceleration of the center of the moving platform of the robot manipulator vanishes along the Z-axis.

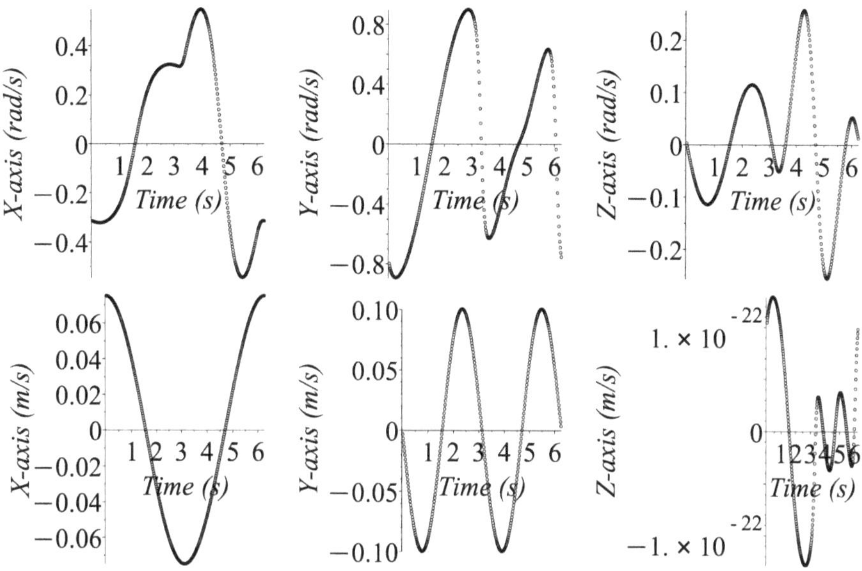

Fig. 8.3 Example 8.1. Time history of the angular and linear velocities of the center of the moving platform

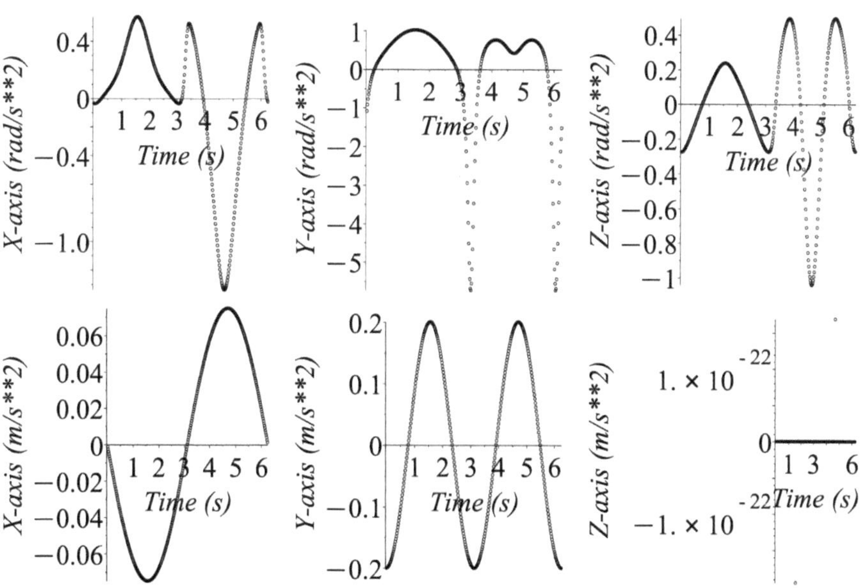

Fig. 8.4 Example 8.1. Time history of the angular and linear accelerations of the center of the moving platform

8.5 Exercises

1. Because of the lack of closed-form solutions for the forward displacement analysis of the 3R2T parallel manipulator, its singularity analysis in loci form is a rather complicated task. Based on the input–output equation of velocity, determine two singular configurations of such a robot.
2. Explain why the 3R2T parallel manipulator is free of singularities concerned with the inverse velocity analysis.

References

Angerilli, M., Frisoli, A., Salsedo, F., Marcheschi, S., & Bergamasco, M. (2001). Haptic simulation of an automotive manual gearshift. In *Proceedings Romancy 2001, International Workshop on Robot-Human Communication, Bordeaux, France.*

Birglen, L., Gosselin, C. M., Pouliot, N., Monsarrat, B., & Lalibertè, T. (2002). SHaDe, a new 3-DOF haptic device. *IEEE Transactions on Robotics and Automation, 18*(2), 166–175.

Carbone, G., & Ceccarelli, M. (2005a). A serial-parallel robotic architecture for surgical tasks. *Robotica, 23*(3), 345–354.

Carbone, G., & Ceccarelli, M. (2005b). A stiffness analysis for a hybrid parallel-serial manipulator. *Robotica, 22*(5), 567–576.

Chirikjian, G. S. (1992). *Theory and applications of hyper-redundant robotic manipulators*. Ph.D. thesis, Department of Applied Mechanics, Division of Engineering and Applied Science, California Institute of Technology.

Clavel, R. (1990). Device for the movement and positioning of an element in space. U.S. Patent No. 4,976,582, December 11.

Clavel, R. (1991). *Conception d'un robot parallèle rapide à 4 degrés de liberté*. Ph.D. thesis, Ecole Polytechique Fédérale de Lausanne (EPFL).

Gallardo-Alvarado, J. (2005). Kinematics of a hybrid manipulator by means of screw theory. *Multibody System Dynamics, 14*(3/4), 345–366.

Gallardo-Alvarado, J., Alici, G., & Pérez-González, L. (2010). A new family of constrained redundant parallel manipulators. *Multibody System Dynamics, 23*(1), 57–75.

Gallardo-Alvarado, J., Orozco-Mendoza, H., & Maeda-Sánchez, A. (2007). Acceleration and singularity analyses of a parallel manipulator with a particular topology. *Meccanica, 42*(3), 223–238.

Gallardo-Alvarado, J., Rico-Martínez, J. M., & Alici, G. (2006). Kinematics and singularity analyses of a 4-dof parallel manipulator using screw theory. *Mechanism and Machine Theory, 41*(9), 1048–1061.

Gallardo-Alvarado, J., Rodríguez-Castro, R., Caudillo-Ramírez, M., & Rico-Martínez, J. M. (2008). A family of spherical parallel manipulators with two legs. *Mechanism and Machine Theory, 43*(2), 201–216.

Gao, F., Feng, B., Zhao, H., & Li, W. (2006). A novel 5-DOF fully parallel kinematic machine tool. *International Journal of Advanced Manufacturing Technology, 31*(1), 201–207.

Hunt, K. H., & Primrose, E. J. F. (1993). Assembly configurations of some in-parallel-actuated manipulators. *Mechanism and Machine Theory, 28*(1), 31–42.

Innocenti, C., & Wenger, P. (2004). Position analysis of the RRP-3(SS) multiloop spatial structure. In *Proceedings DETC'04 ASME 2004 Design Engineering Technical Conferences and Computers and Information in Engineering Conference, Salt Lake City, UT.* Paper DETC2004-57382 CD-ROM format.

Jin, Y., Chen, I. M., & Yang, G. (2006). Kinematic design of a 6-DOF parallel manipulator with decoupled translation and rotation. *IEEE Transactions on Robotics, 22*(3), 545–551.

Li, Q. C., Huang, Z., & Hervé, J. M. (2004). Type synthesis of 3R2T 5-DoF parallel manipulators using the Lie group of displacements. *IEEE Transactions on Robotics and Automation, 20*(2), 173–180.

Li, Y. G., Liu, H. T., Zhao, X. M., Huang, T., & Chetwynd, D. (2010). Design of a 3-DOF PKM module for large structural component machining. *Mechanism and Machine Theory, 45*(6), 941–954.

Lu, Y., & Hu, B. (2006). Solving driving forces of 2(3-SPR) serial-parallel manipulator by CAD variation geometry approach. *ASME Journal of Mechanical Design, 128*(6), 1349–1351.

Lu, Y., & Xu, J. Y. (2008). Simulation of three-dimensional free-form surface normal machining by 3SPS+RRPU and 2SPS+RRPRR parallel machine tools. *Institution of Mechanical Engineers Part B: Journal of Engineering Manufacture, 222*(4), 485–494.

Merlet, J. P. (2001). Micro parallel robot MIPS for medical applications. In *IEEE International Conference on Emerging Technologies and Factory Automation*, Antibes.

Ng, C. C., Ong, S. K., & Nee, A. Y. C. (2006). Design and development of 3-DOF modular micro parallel kinematic manipulator. *International Journal of Advanced Manufacturing Technology, 31*(1), 188–200.

Peirs, J., Reynaerts, D., & Van Brussel, H. (1999). A miniature hydraulic parallel manipulator for integration in a self-propelling endoscope. In *EUROSENSORS XIII The 13th European Conference on Solid-State Transducers* (pp. 753–756), The Hague, 22C4 Medical applications.

Song, J., & Mu, J. (2005). A near-optimal part setup algorithm for 5-axis machining using a parallel kinematic machine. *International Journal of Advanced Manufacturing Technology, 25*(1), 130–139.

Tanev, T. K. (2000). Kinematics of a hybrid (parallel-serial) robot manipulator. *Mechanism and Machine Theory, 35*(9), 1183–1196.

Weck, M., & Staimer, M. (2002). Parallel kinematic machine tools - current state and future potentials. *Annals of the CIRP, 51*(2), 671–683.

Wu, G., Li, J., Fei, R., Wang, X., & Liu, D. (2005). Analysis and design of a novel micro-dissection manipulator based on ultrasonic vibration. In *Proceedings 2005 IEEE ICRA, Barcelona* (pp. 468–473), April 2005.

Zhao, J. W., Fan, K. C., Chang, T. H., & Li, Z. (2002). Error analysis of a serial-parallel type machine tool. *International Journal of Advanced Manufacturing Technology, 19*(3), 174–179.

Zheng, X. Z., Bin, H. Z., & Luo, Y. G. (2004). Kinematic analysis of a hybrid serial-parallel manipulator. *The International Journal of Advanced Manufacturing Technology, 23*(11–12), 925–930.

Zhu, S. J., Huang, Z., & Zhao, M. Y. (2007). Kinematics of a 5-DOF 5-$\underline{R}$RR(RR) prototype. In *Proceedings ASME 2007 International Design Engineering Technical Conferences and Computers and Information in Engineering Conference IDETC/CIE 2007, Las Vegas, NV.* Paper DETC2007-35916 CD-ROM format.

Zhu, S. J., Huang, Z., & Zhao, M. Y. (2009). Singularity analysis for six practicable 5-DOF fully-symmetrical parallel manipulators. *Mechanism and Machine Theory, 44*(4), 710–725.

Zlatanov, D., Dai, M. Q., Fenton, R. G., & Benhabib, B. (1992). Mechanical design and kinematic analysis of a three-legged six-degree-of-freedom parallel manipulator. In *Proceedings 22nd Biennial Mechanisms Conference* (pp. 529–536).

Zoppi, M., Bruzzone, L. E., & Molfino, R. M. (2004). A novel 5-dof interconnected-chains PKM for manufacturing revolute surfaces. In *Proceedings Fourth Chemnitzer Parallel Kinematic Seminar, Chemnitz.*

Chapter 9
Two-Degree-of-Freedom Parallel Wrist

9.1 Introduction

A spherical mechanism has the virtue that all of its moving points describe paths forming concentric spherical surfaces (Chiang 1988). The study of spherical motions achieved with parallel manipulators is, in reality, an old kinematic topic. In fact, it dates back to 1897 to the pioneering contribution of Bricard (1897). Undoubtedly, the most celebrated parallel manipulator possessing this property is credited to Gosselin and Angeles (1989), who developed a 3-RRR (where R represents a resolute joint) parallel wrist named *Agile Eye* in which all motors are mounted on the fixed platform. Screws with axes intersecting at a common point form the main characteristic of the Agile Eye. Some applications require only two of the three degrees of freedom (DOF) of a general parallel wrist. A 2-DOF parallel wrist is a parallel manipulator capable of orienting a line in space and can be used in antennas, telescopes, biomechanics, telesurgery, haptic devices (joysticks), car gearshifts, cameras, and other applications. On the other hand, it is well known that parallel manipulators with coupled motions exhibit a complex-shaped workspace, singularities, and higher nonlinear forward kinematics. A decoupled parallel manipulator is a mechanism in which 1 DOF of the moving platform is affected by only one active joint of the mechanism; therefore, linear input–output kinematic equations are available for such mechanisms, which diminishes the drawbacks of parallel manipulators cited in this paragraph. In this field Carricato and Parenti-Castelli (2004) introduced a novel pointing fully decoupled 2-DOF parallel wrist with linear actuators. Vertechy and Parenti-Castelli (2006) investigated 2-DOF US/UPS (where U, S, and P respectively represent universal, spherical, and prismatic joints) spherical parallel manipulators with linear actuators. Li et al. (2007) proposed a fully decoupled 2-DOF wrist called Hemisphere that is composed of prismatic and revolute active joints. In Gallardo et al. (2003) the kinematics of

J. Gallardo-Alvarado, *Kinematic Analysis of Parallel Manipulators by Algebraic Screw Theory*, DOI 10.1007/978-3-319-31126-5_9

a 2-DOF wrist is carried out by combining screw theory and the so-called first-order coefficients. None of the above-mentioned contributions approached the jerk analysis.

This chapter reports on the position, velocity, acceleration, and jerk analyses of a simple 2-DOF parallel manipulator.

9.2 Description of the Parallel Wrist

The parallel wrist considered for analysis is depicted in Fig. 9.1.

The spatial mechanism consists of a moving platform, named the knob, and a fixed platform, named the base, connected to each other by two distinct kinematic chains, R̲R and R̲RR, where the active revolute joints or actuators are conveniently mounted on the fixed platform and are indicated with underlines. All the axes of the revolute joints are concurrent, and the linear infinitesimal kinematic properties of the intersecting point therefore vanish, yielding spherical motions of the knob as observed from the base. Furthermore, one primary feature of the wrist at hand is that the axes associated to adjacent revolute joints are orthogonal.

The mobility of the mechanism can be easily explained if we take into account that because of the chosen architecture, while the R̲RR-type limb is a 3-DOF spherical serial manipulator, the connection of the knob though the R̲R-type limb is such that the knob cannot rotate along an axis perpendicular to the axes of the revolute joints of this leg. Therefore, the wrist at hand is limited to 2 DOF.

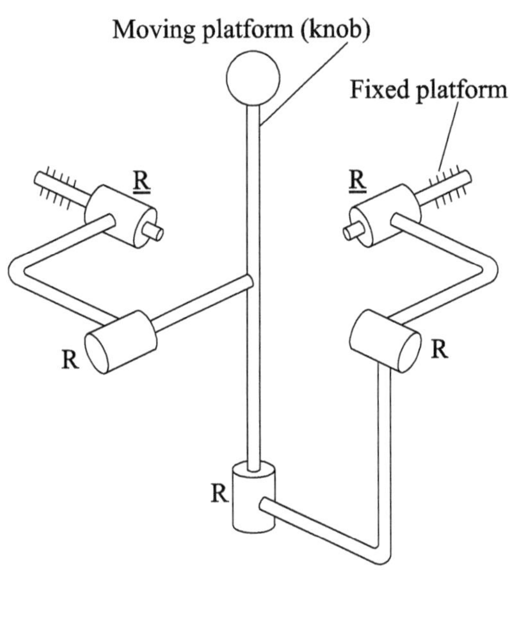

Fig. 9.1 The R̲R + R̲RR spherical parallel manipulator

9.3 Finite Kinematics of the Parallel Wrist

In this section the displacement analysis of the limited-DOF spherical parallel manipulator is presented. Because of the robot's simple architecture, closed-form solutions are easily derived. The parameters and generalized coordinates of the mechanism are outlined in Fig. 9.2.

9.3.1 Forward Displacement Analysis of the Parallel Wrist

The forward position analysis of the parallel wrist is formulated as follows. Given the generalized coordinates q_1 and q_2, we must compute the rotation matrix $\mathbf{R}$ of the knob with respect to the base. To accomplish this, consider that *XYZ* is a reference frame attached to the base whose origin *O* is instantaneously coincident with the point of zero linear kinematic properties of the moving knob. Next, it immediately emerges that the coordinates of points P_1 and P_3 are obtained as

$$\left.\begin{array}{l} P_1 = (0, -r\sin q_1, r\cos q_1) \\ P_3 = (r\cos q_2, r\sin q_2, 0) \end{array}\right\} . \qquad (9.1)$$

Furthermore, to compute the coordinates of point $P_2 = (X_2, Y_2, Z_2)$, let us consider the following closure equations:

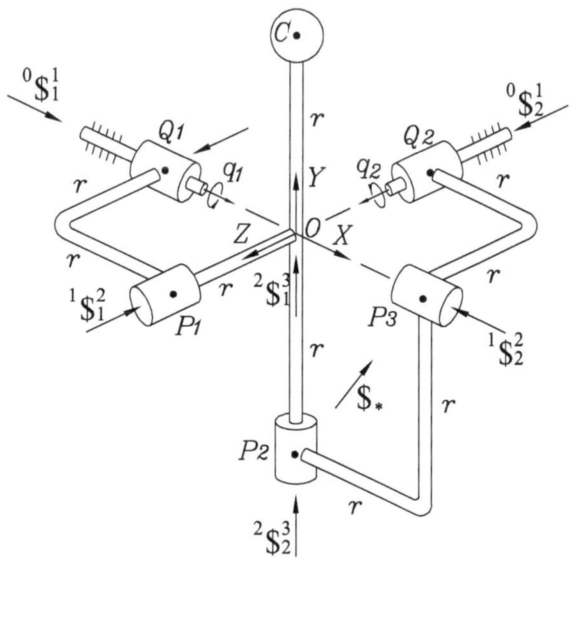

Fig. 9.2 Geometric scheme of the RR + RRR parallel wrist

$$\left.\begin{aligned} (\boldsymbol{p}_2 - \boldsymbol{p}_1) \bullet (\boldsymbol{p}_2 - \boldsymbol{p}_1) &= 2r^2 \\ (\boldsymbol{p}_2 - \boldsymbol{p}_3) \bullet (\boldsymbol{p}_2 - \boldsymbol{p}_3) &= 2r^2 \\ \boldsymbol{p}_2 \bullet \boldsymbol{p}_2 &= r^2 \end{aligned}\right\}, \tag{9.2}$$

where $\boldsymbol{p}_i (i = 1, 2, 3)$ are the position vectors of points $P_i (i = 1, 2, 3)$ expressed in the reference frame XYZ. After a few computations it follows that

$$P_2 = \lambda(\tan q_2, -1, -\tan q_1), \tag{9.3}$$

where

$$\lambda = r / \sqrt{1 + \tan^2 q_1 + \tan^2 q_2}. \tag{9.4}$$

Furthermore, the coordinates of point C are obtained as

$$C = \lambda(-\tan q_2, 1, \tan q_1). \tag{9.5}$$

Finally, the rotation matrix $\mathbf{R}$ is determined using the method introduced by Gallardo-Alvarado et al. (2008) (see Chap. 16) as follows:

$$\mathbf{R} = \left[\frac{\boldsymbol{c} \times \boldsymbol{p}_1}{r^2} \; \frac{\boldsymbol{c}}{r} \; \frac{\boldsymbol{p}_1}{r} \right]. \tag{9.6}$$

Example 9.1. Given the generalized coordinates $q_1 = \pi/6$ and $q_2 = \pi/3$, we are required to compute the rotation matrix. To fulfill this aim, consider that the radius of the parallel wrist is given by $r = 0.5$.

Solution. The coordinates of point P_1 are obtained immediately as

$$P_1 = (0, -r \sin q_1, r \cos q_1) = (0, -0.25, 0.433).$$

Meanwhile, the coordinates of point P_3 are computed as

$$P_3 = (r \cos q_2, r \sin q_2, 0) = (0.25, 0.433, 0).$$

Furthermore, the scalar λ results in

$$\lambda = r / \sqrt{1 + \tan^2 q_1 + \tan^2 q_2} = 0.24.$$

Thus, the coordinates of point P_2 are given by

$$P_2 = \lambda(\tan q_2, -1, -\tan q_1) = (0.416, -0.24, -0.138),$$

while the coordinates of the control point C result in

$$C = \lambda(-\tan q_2, 1, \tan q_1) = (-0.416, 0.24, 0.138).$$

Finally, the rotation matrix **R** is obtained as

$$\mathbf{R} = \begin{bmatrix} \frac{c \times p_1}{r^2} & \frac{c}{r} & \frac{p_1}{r} \end{bmatrix} = \begin{bmatrix} 0.554 & -0.832 & 0 \\ 0.720 & 0.480 & -0.5 \\ 0.416 & 0.277 & 0.866 \end{bmatrix},$$

where evidently $\det(\mathbf{R}) = 1$.

9.3.2 *Inverse Displacement Analysis of the Parallel Wrist*

The inverse position analysis of the parallel wrist consists of finding the generalized coordinates q_1 and q_2 given a prescribed trajectory of the point C of the knob with respect to the base. The inverse displacement analysis is a simple task and is included here only for the sake of completeness.

As reported by Gallardo et al. (2003), the inverse position analysis is immediately solved, taking advantage of the decoupled motions of the knob. In fact, let

$$C = (C_X, C_Y, C_Z)$$

be the coordinates of point C where $C_X^2 + C_Y^2 + C_Z^2 = r^2$. The generalized coordinates q_1 and q_2 are immediately obtained as

$$q_1 = -\arctan\left(\frac{C_Y}{C_Z}\right) \tag{9.7}$$

and

$$q_2 = \arctan\left(\frac{C_X}{C_Z}\right), \tag{9.8}$$

respectively.

9.4 Infinitesimal Kinematics of the Parallel Wrist

In this section the velocity, acceleration, and jerk analyses of the parallel wrist are addressed by screw theory. To reach this goal, the infinitesimal screws representing the kinematic pairs of the mechanism are depicted in Fig. 9.2.

9.4.1 Velocity Analysis of the Parallel Wrist

Let $\boldsymbol{\omega}$ be the angular velocity of the knob. Furthermore, let

$$\boldsymbol{V}_O = \begin{bmatrix} \boldsymbol{\omega} \\ \boldsymbol{v}_O \end{bmatrix}$$

be the velocity state of the knob as measured from the base where, to simplify the analysis of the robot, point O is chosen as the reference pole; that is, the velocity $\boldsymbol{v}_O$ of point O vanishes ($\boldsymbol{v}_O = \boldsymbol{0}$). The *six-dimensional vector* $\boldsymbol{V}_O$ may be expressed in screw form through any of the two limbs of the manipulator as follows:

$$\dot{q}_i{}^0\$_i^1 + {}_1\omega_2^i{}^1\$_i^2 + {}_2\omega_3^i{}^2\$_i^3 = \boldsymbol{V}_O, \quad i = 1, 2, \tag{9.9}$$

where ${}^2\$_1^3$ is an auxiliary screw representing a fictitious revolute joint introduced to satisfy an algebraic requirement, that is, ${}_2\omega_3^1 = 0$. Furthermore, based on Eq. (9.9) we have

$$\dot{q}_1{}^0\$_1^1 + {}_1\omega_2^{1}{}^1\$_1^2 + {}_2\omega_3^{1}{}^2\$_1^3 - (\dot{q}_2{}^0\$_2^1 + {}_1\omega_2^{2}{}^1\$_2^2 + {}_2\omega_3^{2}{}^2\$_2^3) = \boldsymbol{0}. \tag{9.10}$$

On the other hand, let

$$\$_* = \begin{bmatrix} \hat{\boldsymbol{s}}_* \\ \boldsymbol{s}_{O*} \end{bmatrix} \tag{9.11}$$

be a line in Plücker coordinates directed from point P_2 to point P_3, where the primal and dual parts of the line are given by

$$\hat{\boldsymbol{s}}_* = \frac{\boldsymbol{p}_3 - \boldsymbol{p}_2}{||\,\boldsymbol{p}_3 - \boldsymbol{p}_2\,||} \tag{9.12}$$

and

$$\boldsymbol{s}_{O*} = \boldsymbol{r}_3 \times \hat{\boldsymbol{s}}_*, \tag{9.13}$$

respectively.

Applying the Klein form of the reciprocal line $\$_*$ to both sides of Eq. (9.10) and reducing terms allows us to compute the passive joint-velocity rate ${}_1\omega_2^1$. Next, substituting ${}_1\omega_2^1$ into the first equation of (9.9) leads to

$$\boldsymbol{V}_O = \dot{q}_1{}^0\$_i^1 - \frac{\{(\dot{q}_1{}^0\$_1^1 - \dot{q}_2{}^0\$_2^1); \$_*\}}{\{{}^1\$_1^2; \$_*\}}{}^1\$_1^2. \tag{9.14}$$

Equation (9.14) would be considered the input–output equation of velocity of the robot. The inverse velocity analysis of the parallel wrist consists of finding the generalized speeds $\dot{q}_1$ and $\dot{q}_2$ given a prescribed velocity state $\boldsymbol{V}_O$. On the other hand, the forward velocity analysis consists of computing the velocity state $\boldsymbol{V}_O$ given the generalized speeds $\dot{q}_1$ and $\dot{q}_2$.

Example 9.2. The input–output equation of velocity of parallel manipulators may be obtained using different sets of reciprocal screws. This example shows how the input–output equation of velocity of the parallel wrist at hand may be obtained by using a different strategy than that used at the beginning of this subsection. For clarity, most of the notations used in Fig. 9.2 are repeated here (see Fig. 9.3).

Solution. The procedure avoids the inclusion of virtual screws. The velocity state of the knob as measured from the base may be written in screw form through the 2-DOF kinematic chain as follows:

$$\dot{q}_1\,{}^0\$_1^1 + {}_1\omega_2^1\,{}^1\$_1^2 = \boldsymbol{V}_O. \tag{9.15}$$

Let $\boldsymbol{L}_1$ be a line of Plücker coordinates passing through point P_1 parallel to the OC line. Applying the Klein form of this line to both sides of Eq. (9.15) and canceling terms leads to

$$\dot{q}_1\left\{{}^0\$_1^1;\boldsymbol{L}_1\right\} = \left\{\boldsymbol{V}_O;\boldsymbol{L}_1\right\}. \tag{9.16}$$

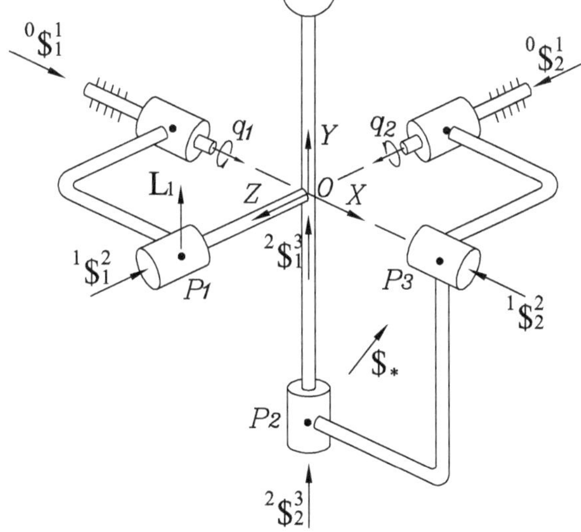

Fig. 9.3 Infinitesimal screws of the RR + RRR parallel wrist

On the other hand, based on the 3-DOF kinematic chain, the velocity state $\boldsymbol{V}_O$ may be written in screw form as follows:

$$\dot{q}_2{}^0\$_2^1 + {}_1\omega_2^{21}\$_2^2 + {}_2\omega_3^{22}\$_2^3 = \boldsymbol{V}_O. \tag{9.17}$$

Hence, applying the Klein form of the line $\$_*$ to both sides of Eq. (9.17) and reducing terms, we get

$$\dot{q}_2\left\{{}^0\$_2^1; \$_*\right\} = \left\{\boldsymbol{V}_O; \$_*\right\}. \tag{9.18}$$

To complete the velocity analysis, consider that the 2-DOF kinematic chain, that is, the <u>R</u>R limb, constrains the spherical motion of the knob such that

$$\boldsymbol{\omega} \bullet \hat{\boldsymbol{u}}_{C/O} = 0, \tag{9.19}$$

where $\hat{\boldsymbol{u}}_{C/O} = u_X\hat{\boldsymbol{i}} + u_Y\hat{\boldsymbol{j}} + u_Z\hat{\boldsymbol{k}}$ is a unit vector directed from point O to point C. Meanwhile, $\boldsymbol{\omega} = \omega_X\hat{\boldsymbol{i}} + \omega_Y\hat{\boldsymbol{j}} + \omega_Z\hat{\boldsymbol{k}}$ is the angular velocity vector of the knob expressed in the fixed reference frame XYZ. In other words, the knob cannot rotate about an axis along the $\hat{\boldsymbol{u}}_{C/O}$ vector.

Casting Eqs. (9.16)–(9.19)into matrix-vector form, we find that the input–output equation of velocity of the parallel wrist results in

$$\mathbf{A}\boldsymbol{\omega} = \mathbf{B}\dot{\mathbf{q}}, \tag{9.20}$$

where

- $\mathbf{A} = \mathbf{J}^T\boldsymbol{\Delta}$, so that

$$\mathbf{J} = \left[\boldsymbol{L}_1\ \$_*\ \boldsymbol{u}\right]^T$$

  is the active Jacobian matrix of the manipulator in which

$$\boldsymbol{u} = \left[\hat{\boldsymbol{u}}_{C/O}\ \mathbf{0}\right]^T$$

  and

$$\boldsymbol{\Delta} = \left[\mathbf{0}\ \mathbf{I}\right]^T$$

  is a nonsquare matrix whose role is to act as an operator of polarity.
- $\mathbf{B}$ is the first-order coefficient matrix computed as

$$\mathbf{B} = \begin{bmatrix} \left\{{}^0\$_1^1; \boldsymbol{L}_1\right\} & 0 & 0 \\ 0 & \left\{{}^0\$_1^2; \$_*\right\} & 1 \\ 0 & 0 & 1 \end{bmatrix}.$$

- $\dot{\mathbf{q}} = \left[\dot{q}_1 \; \dot{q}_2 \; 0\right]^T$ is the first-order driver matrix of the robot.

At a glance, the procedure to obtain the input–output equation of velocity of the parallel wrist provided in this example appears to be a very involved method. Its advantages may be noted in the higher-order kinematic analyses. Thus, the acceleration analysis is left as an exercise.

9.4.2 *Acceleration Analysis of the Parallel Wrist*

Let $\boldsymbol{\alpha}$ be the angular acceleration of the knob. Furthermore, let

$$\boldsymbol{A}_O = \begin{bmatrix} \boldsymbol{\alpha} \\ \boldsymbol{a}_O - \boldsymbol{\omega} \times \boldsymbol{v}_O \end{bmatrix}$$

be the reduced acceleration state of the knob as observed from the base, where $\boldsymbol{a}_O$ is the acceleration of point O. It is evident that the dual part of $\boldsymbol{A}_O$ vanishes; that is, $\boldsymbol{a}_O - \boldsymbol{\omega} \times \boldsymbol{v}_O = \mathbf{0}$. The *six-dimensional vector* $\boldsymbol{A}_O$ may be written in screw form through any of the two limbs of the manipulator as follows:

$$\ddot{q}_i\,{}^0\$_i^1 + {}_1\alpha_2^i\,{}^1\$_i^2 + {}_2\alpha_3^i\,{}^2\$_i^3 + \mathcal{A}_i = \boldsymbol{A}_O, \quad i = 1, 2, \tag{9.21}$$

where the ith Lie screw of acceleration is given by

$$\mathcal{A}_i = [\dot{q}_i\,{}^0\$_i^1 \quad {}_1\omega_2^i\,{}^1\$_i^2 + {}_2\omega_3^i\,{}^2\$_i^3] + [{}_1\omega_2^i\,{}^1\$_i^2 \quad {}_2\omega_3^i\,{}^2\$_i^3]. \tag{9.22}$$

Based on Eq. (9.21), we have

$$\ddot{q}_1\,{}^0\$_1^1 + {}_1\alpha_2^1\,{}^1\$_1^2 + {}_2\alpha_3^1\,{}^2\$_1^3 + \mathcal{A}_1 - (\ddot{q}_2\,{}^0\$_2^1 + {}_1\alpha_2^2\,{}^1\$_2^2 + {}_2\alpha_3^2\,{}^2\$_2^3 + \mathcal{A}_2) = \mathbf{0}. \tag{9.23}$$

Following the trend in the velocity analysis, the input–output equation of acceleration of the parallel wrist results in

$$\boldsymbol{A}_O = \ddot{q}_i\,{}^0\$_i^1 - \frac{\{(\ddot{q}_1\,{}^0\$_1^1 + \mathcal{A}_1 - \ddot{q}_2\,{}^0\$_2^1 - \mathcal{A}_2); \$_*\}}{\{{}^1\$_1^2; \$_*\}}\,{}^1\$_1^2 + \mathcal{A}_1. \tag{9.24}$$

The inverse acceleration analysis of the parallel wrist consists of finding the joint-rate accelerations $\ddot{q}_1$ and $\ddot{q}_2$ given the reduced acceleration state $\boldsymbol{A}_O$, while the forward acceleration analysis consists of computing the reduced acceleration state $\boldsymbol{A}_O$ given $\ddot{q}_1$ and $\ddot{q}_2$.

9.4.3 Jerk Analysis of the Parallel Wrist

Let $\boldsymbol{\rho}$ be the angular jerk of the knob. Furthermore, let

$$\boldsymbol{J}_O = \begin{bmatrix} \boldsymbol{\rho} \\ \boldsymbol{j}_O - 2\boldsymbol{\alpha} \times \boldsymbol{v}_O - \boldsymbol{\omega} \times \boldsymbol{a}_O \end{bmatrix}$$

be the reduced jerk state of the knob as observed from the base where $\boldsymbol{j}_O$ is the jerk of point O. It is evident that the dual part of $\boldsymbol{J}_O$ vanishes; that is, $\boldsymbol{j}_O - 2\boldsymbol{\alpha} \times \boldsymbol{v}_O - \boldsymbol{\omega} \times \boldsymbol{a}_O = \mathbf{0}$. The *six-dimensional vector* $\boldsymbol{J}_O$ may be written in screw form through any of the two limbs of the manipulator as follows:

$$\dddot{q}_i {}^0\$_i^1 + {}_1\rho_2^{i\,1}\$_i^2 + {}_2\rho_3^{i\,2}\$_i^3 + \mathcal{J}_i = \boldsymbol{J}_O, \quad i = 1, 2, \tag{9.25}$$

where the ith Lie screw of jerk is computed as

$$\begin{aligned} \mathcal{J}_i = {} & 2[\dot{q}_i {}^0\$_i^1 \quad {}_1\alpha_2^{i\,1}\$_i^2 + {}_2\alpha_3^{i\,2}\$_i^3 + [{}_1\omega_2^{i\,1}\$_i^2 \quad {}_2\omega_3^{i\,2}\$_i^3]] \\ & +2[{}_1\omega_2^{i\,1}\$_i^2 \quad {}_2\alpha_3^{i\,2}\$_i^3] + [\ddot{q}_i {}^0\$_i^1 \quad {}_1\omega_2^{i\,1}\$_i^2 + {}_2\omega_3^{i\,2}\$_i^3] + \\ & [{}_1\alpha_2^{i\,1}\$_i^2 \quad {}_2\omega_3^{i\,2}\$_i^3] + [\dot{q}_i {}^0\$_i^1 [\dot{q}_i {}^0\$_i^1 \quad {}_1\omega_2^{i\,1}\$_i^2 + {}_2\omega_3^{i\,2}\$_i^3]] \\ & +[{}_1\omega_2^{i\,1}\$_i^2 \quad [{}_1\omega_2^{i\,1}\$_i^2 \quad {}_2\omega_3^{i\,2}\$_i^3]]. \end{aligned} \tag{9.26}$$

Hence, based on Eq. (9.25) one obtains

$$\dddot{q}_1 {}^0\$_1^1 + {}_1\rho_2^{1\,1}\$_1^2 + {}_2\rho_3^{1\,2}\$_1^3 + \mathcal{J}_1 - (\dddot{q}_2 {}^0\$_2^1 + {}_1\rho_2^{2\,1}\$_2^2 + {}_2\rho_3^{2\,2}\$_2^3 + \mathcal{J}_2) = \mathbf{0}. \tag{9.27}$$

Following the trend in the velocity analysis, the input–output equation of jerk of the parallel wrist is obtained as

$$\boldsymbol{J}_O = \dddot{q}_i {}^0\$_i^1 - \frac{\{(\dddot{q}_1 {}^0\$_1^1 + \mathcal{J}_1 - \dddot{q}_2 {}^0\$_2^1 - \mathcal{J}_2); \$_*\}}{\{{}^1\$_1^2; \$_*\}} {}^1\$_1^2 + \mathcal{J}_1. \tag{9.28}$$

The inverse jerk analysis of the mechanism consists of finding the joint-rate jerks $\dddot{q}_1$ and $\dddot{q}_2$ given the reduced jerk state $\boldsymbol{J}_O$. On the other hand, the forward jerk analysis consists of computing the reduced jerk state given the joint-rate jerks $\dddot{q}_1$ and $\dddot{q}_2$.

Expressions (9.14), (9.24), and (9.28) are simple, linear, and compact; to the authors' knowledge, they have not been derived in screw form in previous works.

9.4.4 Singularity Analysis of the Parallel Wrist

A singularity occurs when the moving platform gains or loses degrees of freedom. The first type of singularity to be analyzed is related to the finite kinematics of the wrist, usually referred to as structural singularities. On this topic, note that according to Eq. (9.4), the forward position analysis is indefinite if

$$1 + \tan^2 q_1 + \tan^2 q_2 = 0. \tag{9.29}$$

Of course, it is an unrealistic possibility and must be disregarded immediately from the analysis. On the other hand, the input–output equation of position for the wrist may be expressed as follows:

$$P = \mathbf{R}p, \tag{9.30}$$

where P are the coordinates of a point of the knob expressed in the reference frame XYZ, p are the coordinates of the same point but expressed in a moving reference frame attached to the knob, and $\mathbf{R}$ is the rotation matrix in accordance with Eq. (9.6). It is straightforward to show that the inverse position analysis is indefinite when $\det(\mathbf{R})$ vanishes. Therefore, this singularity occurs when

$$(\cos^2 q_2 \cos^2 q_1 - \cos^2 q_1 - \cos^2 q_2) \cos q_2 \cos q_1 = 0. \tag{9.31}$$

The second type of singularity is related with the infinitesimal kinematics of the wrist. Here a linear combination of the screws in the same limb may be expressed as

$$\lambda_{1i}{}^0\$_i^1 + \lambda_{2i}{}^1\$_i^2 + \lambda_{3i}{}^2\$_i^3 = \mathbf{0}. \tag{9.32}$$

In the inverse velocity analysis, a singularity is present when the screws in expression (9.32) are linearly dependent. Because of the physical orthogonality imposed on adjacent revolute joints in the same limb, none of the screws in Eq. (9.32) can be obtained as a linear combination of the remaining screws. Therefore, the wrist is free of this type of singularity. Finally, note that according to Eq. (9.14), the forward velocity analysis is indefinite when $\{{}^1\$_1^2; \$_*\} = 0$—in other words, when the screws ${}^1\$_1^2$ and $\$_*$ are reciprocal to each other, for instance, when both screws intersect and the equation governing such a situation, considering that $r \neq 0$, is given by

$$\sin q_1 \cos q_2 \tan q_1 = 0. \tag{9.33}$$

Example 9.3. The numerical example is devoted to the forward kinematics of the parallel wrist. Of course, we are required also to solve the inverse, velocity, and acceleration analyses of the robot as intermediate steps. The parameters of the numerical example are chosen as follows: $r = 0.1$, $Q_1 = (-0.1, 0, 0)$, and

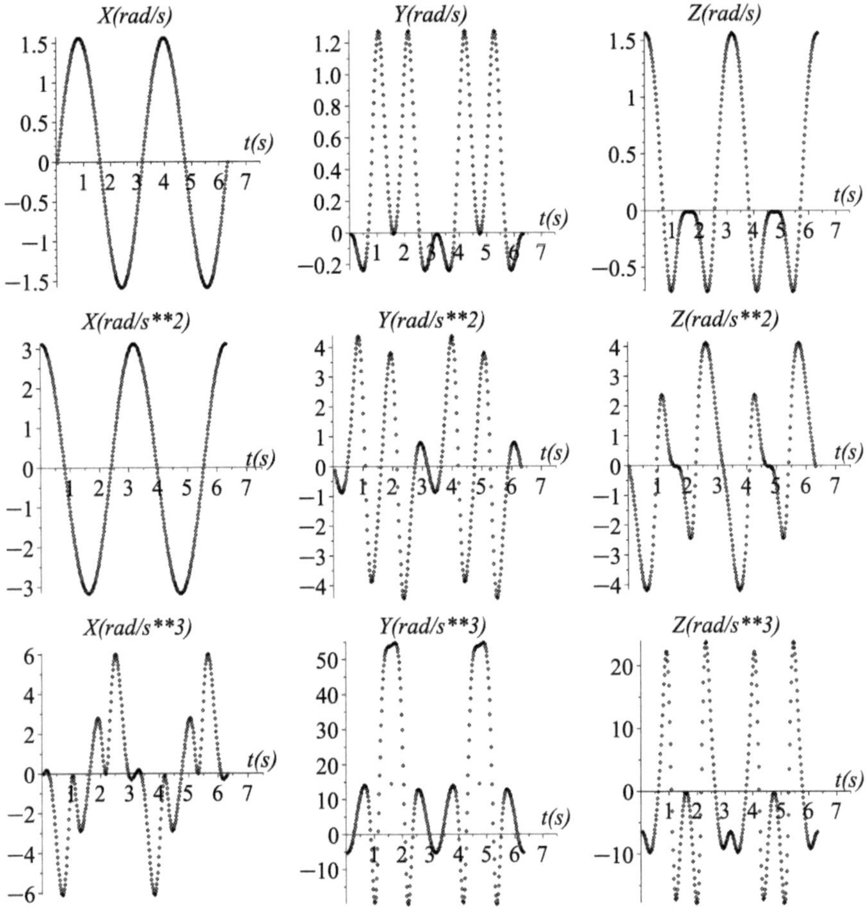

Fig. 9.4 Example 9.3. Time history of the angular quantities of the knob as measured from the base

$Q_2 = (0, 0, -0.1)$. Meanwhile, the generalized coordinates are commanded to follow periodic functions given by $q_1 = 0.5\pi \sin^2 t$ and $q_2 = 0.5\pi \sin t \cos t$, where the time t is chosen in the interval $0 \le t \le 2\pi$.

Solution. The numerical results obtained with screw theory are reported in Fig. 9.4.

9.5 Exercises

1. Following the trend of Example 9.2, obtain the input–output equation of acceleration of the parallel wrist.
2. Determine the singularity analysis of the parallel wrist based on Eq. (9.20).

References

Bricard, R. (1897). Mémoire sur la théorie de lóctahèdre articulé. *Journal de Mathématiques Pures et Appliqués, 3*, 113–150.

Carricato, M., & Parenti-Castelli, V. (2004). A novel fully decoupled two-degrees-of-freedom parallel wrist. *International Journal of Robotics Research, 23*(6), 661–667.

Chiang, C. H. (1988). *Kinematics of spherical mechanisms*. Cambridge: Cambridge University Press.

Gallardo, J., Rico, J. M., Frisoli, A., Checcacci, D., & Bergamasco, M. (2003). Dynamics of parallel manipulators by means of screw theory. *Mechanism and Machine Theory, 38*(11), 1113–1131.

Gallardo-Alvarado, J., Aguilar-Nájera, C. R., Casique-Rosas, L., Pérez-González, L., & Rico-Martínez, J. M. (2008). Solving the kinematics and dynamics of a modular spatial hyper-redundant manipulator by means of screw theory. *Multibody System Dynamics, 20*(4), 307–325.

Gosselin, C. M., & Angeles, J. (1989). The optimum kinematic design of a spherical three-degree-of-freedom parallel manipulator. *ASME Journal of Mechanisms, Transmissions and Automation in Design, 111*(2), 202–207.

Li, W., He, K., Qu, Y., Zhang, J., & Du, R. (2007). Hemisphere, a fully decoupled parallel 2-DOF spherical mechanism. In *Proceedings 7th WSEAS International Conference on Robotics, Control and Manufacturing Technology, Hangzhou* (pp. 301–306).

Vertechy, R., & Parenti-Castelli, V. (2006). Synthesis of 2-DOF spherical US/U$\underline{P}$S parallel mechanisms. In J. Lenarčič & B. Roth (Eds.), *Advances in robot kinematics: mechanisms and motion* (pp. 385–394). New York: Springer.

Chapter 10
3-RRPS Parallel Manipulator

10.1 Introduction

The weaknesses and strengths of the six-legged in-parallel manipulator have been extensively discussed, sometimes bravely, in the kinematician community. The forward finite kinematics of the general parallel manipulator is affected by a higher nonlinear and complex coupled finite kinematics, which, among its other disadvantages, makes the commercialization of such robots difficult because of the necessity to include expensive and sophisticated control systems. Remarkable efforts have been devoted to solving the forward displacement analysis of the general parallel manipulator, commonly known as the Gough–Stewart platform; see, for instance, Innocenti and Parenti-Castelli (1991), Raghavan (1993), Husty (1996), Innocenti (2001), Rolland (2005). Freudenstein called this issue "the Everest of modern kinematics" (Merlet 2009). Elegant solutions in the form of univariate polynomial equations capable of computing all the reachable poses of the moving platform as measured from the fixed platform have been recurrently reported in the literature. However, closed-form solutions of the forward displacement analysis of the general parallel manipulator are not available at this time. Certainly, a linear input–output relationship involving the actuated joint displacement vector and the task-space displacement vector seems to be an unrealistic task. On the other hand, it is straightforward to show that as a result of the closed-loop nature of the mechanism, the number of limbs is a crucial factor limiting the motions of the moving platform and creating complex singularities in its workspace. With these considerations in mind and taking into account that a rigid body is statically balanced when it is supported on three different points, by reducing the number of limbs of the Gough–Stewart platform one can increase the workspace, improve the manipulability, and diminish the presence of singularities of the manipulator. The development of three-legged parallel manipulators, preserving the six degrees of freedom (DOF) of the moving platform, is a viable option to alleviate the drawbacks of the Gough–Stewart platform.

J. Gallardo-Alvarado, *Kinematic Analysis of Parallel Manipulators by Algebraic Screw Theory*, DOI 10.1007/978-3-319-31126-5_10

Guilin et al. (2001) and Angeles et al. (2001) carried out the singularity analysis of the 3-RRRS (where R and S represented revolute and spherical joints, respectively) 6-DOF parallel manipulator. The singularity condition of three-legged nonredundant parallel manipulators was extensively investigated by Ben-Horin and Shoham (2006) by resorting to Grassmann–Cayley algebra. Three-legged parallel manipulators with PRPS- and PPSP-type limbs (where P represents prismatic joints) were introduced, respectively, by Behi (1988) and Byun and Cho (1997). The workspace of the 3-RRPS parallel manipulator was investigated by Qi et al. (2011) considering the constant orientation of the moving platform. Applications of the 3-RRPS parallel manipulator were investigated by Ji et al. (2006), overcoming the parasitic motions observed in the 3-RPS limited-DOF parallel manipulator. The forward kinematics of parallel manipulators may be simplified by using three-dimensional moving platforms. In that way, García-Murillo et al. (2013) approached the dynamics of a 2-(3RRPS) parallel manipulator using screw theory. It is evident that in general simpler forward kinematics and control are some benefits when parallel manipulators with fewer than six limbs are used; see, for instance, Husain and Waldron (1992), Zlatanov et al. (1992), Geng and Haynes (1994), Notash and Podhorodeski (1995), Portman and Sandler (2002), Zabalza et al. (2002), Goldsmith (2003), Jin et al. (2009).

In this chapter the velocity, acceleration, and jerk analyses of the 3-RRPS parallel manipulator are carried out with screw theory. As an intermediate step, the displacement analysis of the three-legged 6-DOF parallel manipulator is also addressed.

10.2 Description of the 3-RRPS Parallel Manipulator

The topology of the 3-RRPS parallel manipulator is outlined in Fig. 10.1. Unless otherwise specified, in the remainder of the chapter let us consider $i = 1, 2, 3$.

The parallel manipulator chosen for the analysis is composed of a moving platform (labeled body m), a fixed platform (labeled body 0), and three RRPS-type limbs. Each kinematic chain of the manipulator is connected to the fixed platform by an active revolute joint q_i whose nominal position A_i is located by the vector $\boldsymbol{a}_i$. On the other hand, the arm is connected to the forearm in the same limb by a passive revolute joint whose nominal position is denoted by B_i and located by vector $\boldsymbol{b}_i$. The unit vector $\hat{\boldsymbol{u}}_i$ represents the axis of the passive revolute joint. The length of the forearm is controlled by an active prismatic joint $\overline{q}_i$. Furthermore, the limbs are connected to the moving platform via passive spherical joints whose geometric centers are denoted as points C_i located by vectors $\boldsymbol{c}_i$. The axes of the revolute joints in the same limb are perpendicular to each other, and the spherical joints therefore move on concentric spheres about the corresponding points B_i. The centers of the passive revolute joints form an equilateral triangle $\Delta C_1C_2C_3$ of side e whose geometric center C is given by

$$C = (C_1 + C_2 + C_3)/3. \tag{10.1}$$

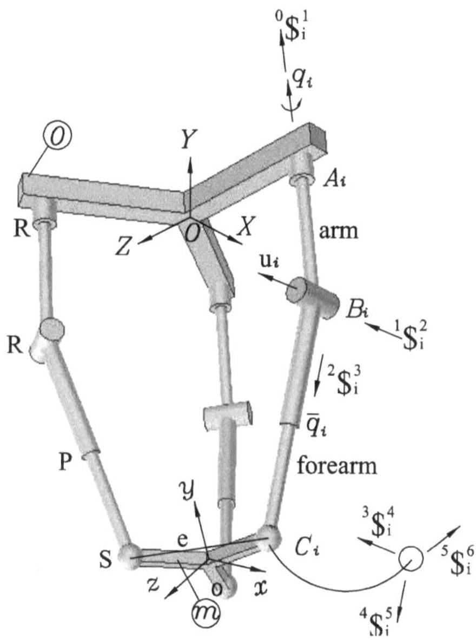

Fig. 10.1 The 3-RRPS parallel manipulator

10.3 Finite Kinematics of the 3-RRPS Manipulator

In this section the displacement analysis of the nonredundant parallel manipulator is presented.

10.3.1 Forward Displacement Analysis of the 3-RRPS Manipulator

Let XYZ be a reference frame attached at the center of the fixed platform with associated unit vectors $\hat{i}\hat{j}\hat{k}$, and let xyz be a reference frame attached at the center of the moving platform. The forward position analysis is formulated as follows: Given a set of generalized coordinates $\{\overline{q}_i, q_i\}$, we need to compute the pose, position, and orientation of the moving platform as measured from the fixed platform. Clearly, this analysis is equivalent to the computation of points C_i upon the generalized coordinates of the robot manipulator.

As an initial step to solve the forward displacement analysis, consider that the passive revolute joints constrain the motion of the spherical joints such that

$$(\boldsymbol{c}_i - \boldsymbol{b}_i) \bullet \hat{\boldsymbol{u}}_i = 0, \tag{10.2}$$

where the unit vectors $\hat{\boldsymbol{u}}_i$ are obtained directly according to the generalized coordinates q_i (see Fig. 10.2). Indeed, $\hat{\boldsymbol{u}}_i = \sin(q_i)\hat{\boldsymbol{i}} + \cos(q_i)\hat{\boldsymbol{k}}$.

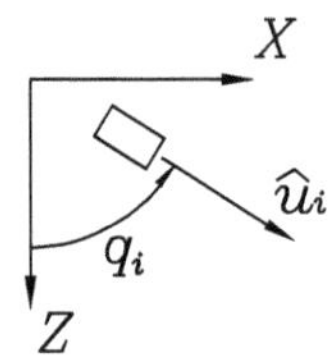

Fig. 10.2 3-RRPS parallel manipulator. Active revolute joint

On the other hand, three closure equations may be written based on the generalized coordinates $\overline{q}_i$ as follows:

$$(\boldsymbol{c}_i - \boldsymbol{b}_i) \bullet (\boldsymbol{c}_i - \boldsymbol{b}_i) = \overline{q}_i^2. \tag{10.3}$$

Furthermore, from the equilateral triangle $\Delta C_1 C_2 C_3$ we can write three compatibility equations as

$$(\boldsymbol{c}_i - \boldsymbol{c}_j) \bullet (\boldsymbol{c}_i - \boldsymbol{c}_j) = e^2 \quad i, j = 1, 2, 3 \bmod(3). \tag{10.4}$$

Expressions (10.2)–(10.4) can be reduced systematically into a 16th-order univariate polynomial equation by applying the Sylvester dialytic method of elimination (Merlet 1989; Tsai 2007; Gallardo-Alvarado et al. 2008b). Such a polynomial equation leads to reflected solutions. Therefore, there are at most 16 reachable poses of the moving platform. In fact, the center of a spherical joint admits at most two distinct positions. Once the coordinates of the spherical joints are computed, the 4×4 homogeneous matrix ${}^0\mathbf{T}^m$ is obtained as

$${}^0\mathbf{T}^m = \begin{bmatrix} {}^0\mathbf{R}^m & \boldsymbol{r}_{o/O} \\ \mathbf{0} & 1 \end{bmatrix}, \tag{10.5}$$

where ${}^0\mathbf{R}^m$ is the rotation matrix between the moving and fixed platforms, whereas $\boldsymbol{r}_{o/O} = (\boldsymbol{c}_1 + \boldsymbol{c}_2 + \boldsymbol{c}_2)/3$ is the position vector of the center o of the moving platform with respect to the center O of the fixed platform. Matrix ${}^0\mathbf{R}^m$ may be calculated using the method introduced by Gallardo-Alvarado et al. (2008a); see Chap. 16.

It is worth noting how easy it is to solve the forward position analysis of the manipulator at hand compared to solving the forward displacement analysis of the general parallel manipulator.

Example 10.1. The example deals with the forward displacement analysis of the 3-RRPS parallel manipulator. We use SI units in the numerical example; the nominal positions of the revolute joints of the parallel manipulator are given by $A_1 = (0.25, 0, 0)$, $A_2 = (-0.125, 0, -0.2165)$, $A_3 = (-0.125, 0, 0.2165)$, $B_1 = (0.25, -0.25, 0)$, $B_2 = (-0.125, -0.25, -0.2165)$, and $B_3 = (-0.125, -0.25, 0.2165)$. On the other hand, the active revolute joints are chosen as $q_1 = 0.03$, $q_2 = 2.2$, and $q_3 = 3.6$, while the active prismatic joints are given by $\overline{q}_1 = 0.25$, $\overline{q}_2 = 0.275$, and $\overline{q}_3 = 0.3$. With these data it is necessary to compute all the reachable centers $\boldsymbol{r}_{o/O}$ of the moving platform.

Solution. According to the active revolute joints, the unit vectors $\hat{\boldsymbol{u}}_i$ result in $\hat{\boldsymbol{u}}_1 = 0.0299\hat{\boldsymbol{i}} + 0.9995\hat{\boldsymbol{k}}$, $\hat{\boldsymbol{u}}_2 = 0.8084\hat{\boldsymbol{i}} - 0.5885\hat{\boldsymbol{k}}$, and $\hat{\boldsymbol{u}}_3 = -0.4425\hat{\boldsymbol{i}} - 0.8967\hat{\boldsymbol{k}}$. Consider that the coordinates of the centers of the spherical joints are given by $C_1 = (w_1, w_2, w_3)$, $C_2 = (w_4, w_5, w_6)$, and $C_3 = (w_7, w_8, w_9)$. Applying Eq. (10.2) yields

$$\left.\begin{array}{l} 0.02999w_1 - 0.007498 + 0.9995500337w_3 = 0 \\ 0.80849w_4 - 0.026348 - 0.5885011173w_6 = 0 \\ -0.4425204433w_7 + 0.1388331417 - 0.8967584163w_9 = 0 \end{array}\right\}. \tag{10.6}$$

Meanwhile, according to Eq. (10.3) it follows that

$$\left.\begin{array}{l} w_1^2 - 0.5w_1 + 0.0625 + w_2^2 + 0.5w_2 + w_3^2 = 0 \\ w_4^2 + 0.25w_4 + 0.049372 + w_5^2 + 0.5w_5 + w_6^2 + 0.433w_6 = 0 \\ w_7^2 + 0.25w_7 + 0.034997 + w_8^2 + 0.5w_8 + w_9^2 - 0.433w_9 = 0 \end{array}\right\}. \tag{10.7}$$

Furthermore, applying Eq. (10.4) leads to

$$\left.\begin{array}{l} w_1^2 - 2w_1w_4 + w_4^2 + w_2^2 - 2w_2w_5 + w_5^2 + w_3^2 - 2w_3w_6 + w_6^2 - 0.04 = 0 \\ w_1^2 - 2w_1w_7 + w_7^2 + w_2^2 - 2w_2w_8 + w_8^2 + w_3^2 - 2w_3w_9 + w_9^2 - 0.04 = 0 \\ w_7^2 - 2w_7w_4 + w_4^2 + w_8^2 - 2w_8w_5 + w_5^2 + w_9^2 - 2w_9w_6 + w_6^2 - 0.04 = 0 \end{array}\right\}. \tag{10.8}$$

Next, the solutions of Eqs. (10.6)–(10.8) are listed in Table 10.1.

The reachable coordinates of the center o of the moving platform are listed in Table 10.2.

10.3.2 *Inverse Displacement Analysis of the 3-RRPS Manipulator*

The inverse position analysis consists of finding the generalized coordinates $\{\overline{q}_i, q_i\}$ given the pose, matrix ${}^0\mathbf{R}^m$, and vector $\boldsymbol{r}_{o/O}$ of the moving platform with respect to the fixed platform.

It is straightforward to demonstrate that the coordinates of points C_i may be computed as

$$\boldsymbol{c}_i = {}^0\mathbf{R}^m\boldsymbol{d}_i + \boldsymbol{r}_{o/O}, \tag{10.9}$$

where $\boldsymbol{d}_i$ is the position vector of the center of the ith spherical joint expressed in the moving reference frame xyz. Next, the generalized coordinates $\overline{q}_i$ are computed [see Eq. (10.3)] as

$$\overline{q}_i = \sqrt{(\boldsymbol{c}_i - \boldsymbol{b}_i) \bullet (\boldsymbol{c}_i - \boldsymbol{b}_i)}. \tag{10.10}$$

Table 10.1 Example 10.1. Solutions of the unknowns $w_i (i = 1, 2, \ldots, 9)$

Sol.	w_1	w_2	w_3	w_4
w_5	w_6	w_7	w_8	w_9
1	0.00024	−0.24189	0.00749	−0.00429
−0.43319	−0.05067	0.11906	−0.37619	0.09606
2	0.00024	−0.25810	0.00749	−0.00429
−0.06680	−0.05067	0.11906	−0.12380	0.09606
3	0.17816	−0.01055	0.00215	0.021735
−0.13400	−0.01491	0.03226	−0.00659	0.138895
4	0.17816	−0.48944	0.00215	0.021735
−0.36599	−0.01491	0.03226	−0.49340	0.13889
5	0.18102	−0.00971	0.00206	−0.00294
−0.06941	−0.04881	0.03424	−0.00820	0.13791
6	0.18102	−0.49028	0.00206	−0.00294
−0.43058	−0.04881	0.03424	−0.49179	0.13791
7	0.18547	−0.00847	0.00193	0.007351
−0.09173	−0.03467	0.13720	−0.18288	0.08710
8	0.18547	−0.49152	0.00193	0.007351
−0.40826	−0.03467	0.13720	−0.31711	0.08710
9	0.56082	−0.25 + 0.184 I	−0.00932	0.24957
−0.25 + 0.574 I	0.29809	0.29544	−0.25 + 0.36 I	0.00902
10	0.56082	−0.25 − 0.18 I	−0.00932	0.24957
−0.25 − 0.57 I	0.29809	0.2954	−0.25 − 0.36 I	0.00902
11	−0.01529 + 0.001 I	−0.24409 + 0.089 I	0.00796 − 0.00005 I	−0.00465 − 0.045 I
−0.03741 + 0.073 I	−0.05116− 0.062 I	0.17159 − 0.049 I	−0.14058 + 0.168 I	0.07014 + 0.024 I
12	−0.01529 + 0.001 I	−0.2559 − 0.089 I	0.00796 − 0.00005 I	−0.00465 − 0.045 I
−0.46258 − 0.073 I	−0.05116 − 0.062 I	0.17159 − 0.049 I	−0.35941 − 0.168 I	0.07014 + 0.024 I
13	−0.07494	−0.25 + 0.207 I	0.00975	0.12373

−0.25 + 0.32 I	0.12522	−0.66208	−0.25 − 0.518 I	0.48153
14	−0.07494	−0.25 − 0.207 I	0.00975	0.12373
−0.25 − 0.32 I	0.12522	−0.66208	−0.25 + 0.518 I	0.48153
15	−0.01529 − 0.001 I	−0.24409 − 0.089 I	0.00796 + 0.00005 I	−0.00465 + 0.045 I
−0.03741 − 0.073 I	−0.05116 + 0.062 I	0.17159 + 0.049 I	−0.14058 − 0.168 I	0.07014 − 0.024 I
16	−0.01529 − 0.001 I	−0.2559 + 0.089 I	0.00796 + 0.00005 I	−0.00465 + 0.04525 I
−0.46258 + 0.073 I	−0.05116 + 0.062 I	0.17159 + 0.049 I	−0.35941 + 0.168 I	0.07014 − 0.024 I

Table 10.2 Example 10.1. Available coordinates for the center of the moving platform

Sol.	X	Y	Z
1	0.3833820868e−1	−0.3504250117	0.1762599537e−1
2	0.3833820868e−1	−0.1495749883	0.1762599537e−1
3	0.7738689211e−1	−0.5038366289e−1	0.4204693347e−1
4	0.7738689211e−1	−0.4496163371	0.4204693347e−1
5	0.7077655395e−1	−0.2911097777e−1	0.3038878376e−1
6	0.7077655395e−1	−0.4708890222	0.3038878376e−1
7	0.1100112237	−0.9436583230e−1	0.1812449112e−1
8	0.1100112237	−0.4056341677	0.1812449112e−1
9	0.3686133367	−0.2499999999+0.3730805859*I	0.9926289064e−1
10	0.3686133367	−0.2499999999−0.3730805859*I	0.9926289064e−1
11	0.5054887468e−1−0.3105075748e−1*I	−0.1406996009+0.1104745972*I	0.898053633e−2−0.1253998598e−1*I
12	0.5054887468e−1−0.3105075748e−1*I	−0.3593003991−0.1104745972*I	0.898053633e−2−0.1253998598e−1*I
13	−0.2044283563	−0.2499999999+0.34679818e−2*I	0.2055014157
14	−0.2044283563	−0.2499999999−0.34679818e−2*I	0.2055014157
15	0.5054887468e−1+0.3105075748e−1*I	−0.1406996009−0.1104745972*I	0.898053633e−2+0.1253998598e−1*I
16	0.5054887468e−1+0.3105075748e−1*I	−0.3593003991+0.1104745972*I	0.898053633e−2+0.1253998598e−1*I

Finally, according to Eq. (10.2) the generalized coordinates q_i are computed as

$$q_i = \arctan \frac{(\boldsymbol{c}_i - \boldsymbol{b}_i) \bullet \hat{k}}{(\boldsymbol{c}_i - \boldsymbol{b}_i) \bullet \hat{i}}. \tag{10.11}$$

10.4 Infinitesimal Kinematics of the 3-RRPS Manipulator

To approach the infinitesimal kinematics of the 3-RRPS parallel manipulator, the corresponding infinitesimal screws are outlined in Fig. 10.1.

10.4.1 Velocity Analysis of the 3-RRPS Manipulator

Let

$$\boldsymbol{V}_C = \begin{bmatrix} \boldsymbol{\omega} \\ \boldsymbol{v}_C \end{bmatrix}$$

be the velocity state of the center of the moving platform, the point C, as measured from the fixed platform, where $\boldsymbol{\omega}$ and $\boldsymbol{v}_C$ are the angular and linear velocities of the center C of the moving platform. The velocity state of the moving platform, with respect to the fixed platform, may be expressed in screw form through any of the connector chains as

$${}_0\omega_1^i\,{}^0\$_i^1 + {}_1\omega_2^i\,{}^1\$_i^2 + {}_2\omega_3^i\,{}^2\$_i^3 + {}_3\omega_4^i\,{}^3\$_i^4 + {}_4\omega_5^i\,{}^4\$_i^5 + {}_5\omega_6^i\,{}^5\$_i^6 = \boldsymbol{V}_C. \tag{10.12}$$

To obtain the input–output equation of velocity of the robot, note that the screw ${}^4\$_i^5$ is reciprocal to all the screws representing revolute joints in the same limb. Therefore, applying the Klein form of this reciprocal screw to both sides of expression (10.12) and reducing terms, we find that

$$\{{}^4\$_i^5; \boldsymbol{V}_C\} = \dot{\bar{q}}_i. \tag{10.13}$$

Similarly, since the screw ${}^3\$_i^4$ is reciprocal to all the screws in the same limb—excluding the screw ${}^0\$_i^1$—by applying this screw to both sides of Eq. (10.12) and reducing terms, one obtains

$$\{{}^3\$_i^4; \boldsymbol{V}_C\} = \dot{q}_i\{{}^3\$_i^4; {}^0\$_i^1\}. \tag{10.14}$$

Casting expressions (10.13) and (10.14) into matrix-vector form, we see that the input–output equation of velocity of the manipulator results in

$$\mathbf{A}\boldsymbol{V}_C = \mathbf{B}\dot{\mathbf{q}}, \tag{10.15}$$

where

$$\mathbf{A} = \mathbf{J}^T \, \boldsymbol{\Delta}$$

is the first-order coefficient matrix of the manipulator, so that

$$\mathbf{J} = \begin{bmatrix} {}^4\$_1^5 & {}^4\$_2^5 & {}^4\$_3^5 & {}^3\$_1^4 & {}^3\$_2^4 & {}^3\$_3^4 \end{bmatrix}$$

is the screw-coordinate Jacobian matrix of the parallel manipulator. Meanwhile,

$$\boldsymbol{\Delta} = \begin{bmatrix} \mathbf{0} & \mathbf{I} \\ \mathbf{I} & \mathbf{0} \end{bmatrix}$$

is an operator of polarity. Furthermore, matrix **B** is given by

$$\mathbf{B} = diag \begin{bmatrix} 1 & 1 & 1 & \{{}^3\$_1^4; {}^0\$_1^1\} & \{{}^3\$_2^4; {}^0\$_2^1\} & \{{}^3\$_3^4; {}^0\$_3^1\} \end{bmatrix}. \tag{10.16}$$

Meanwhile,

$$\dot{\mathbf{q}} = \begin{bmatrix} \dot{\bar{q}}_1 & \dot{\bar{q}}_2 & \dot{\bar{q}}_3 & \dot{q}_1 & \dot{q}_2 & \dot{q}_3 \end{bmatrix}^T \tag{10.17}$$

is the first-order driver matrix of the robot manipulator.

The inverse velocity analysis consists of finding the generalized speeds given the velocity state $\boldsymbol{V}_C$, while the forward velocity analysis consists of finding the velocity state $\boldsymbol{V}_C$ meeting a set of generalized speeds. Clearly, to perform the inverse velocity analysis, matrix **B** must be invertible, whereas to perform the forward velocity analysis matrix **A** must be invertible. The singularity analysis of the 3-RRPS parallel manipulator is left as an exercise.

10.4.2 Acceleration Analysis of the 3-RRPS Manipulator

Let

$$\boldsymbol{A}_C = \begin{bmatrix} \boldsymbol{\alpha} \\ \boldsymbol{a}_C - \boldsymbol{\omega} \times \boldsymbol{v}_C) \end{bmatrix}$$

be the reduced acceleration state of the moving platform as measured from the fixed platform where $\boldsymbol{\alpha}$ and $\boldsymbol{a}_C$ are the angular and linear accelerations of the center C of the moving platform. The reduced acceleration state $\boldsymbol{A}_C$ may be expressed in screw form through any of the limbs of the manipulator as follows:

$${}_0\dot{\omega}_1^i \, {}^0\$_i^1 + {}_1\dot{\omega}_2^i \, {}^1\$_i^2 + {}_2\dot{\omega}_3^i \, {}^2\$_i^3 + {}_3\dot{\omega}_4^i \, {}^3\$_i^4 + {}_4\dot{\omega}_5^i \, {}^4\$_i^5 + {}_5\dot{\omega}_6^i \, {}^5\$_i^6 + \mathcal{A}_i = \boldsymbol{A}_C, \tag{10.18}$$

where $\mathcal{A}_i$ is the ith Lie screw of acceleration, which is calculated as

$$\begin{aligned}\mathcal{A}_i = & \left[{}_0\omega_1^i \, {}^0\$_i^1 \quad {}_1\omega_2^i \, {}^1\$_i^2 + \ldots + {}_5\omega_6^i \, {}^5\$_i^6\right] + \\ & \left[{}_1\omega_2^i \, {}^1\$_i^2 \quad {}_2\omega_3^i \, {}^2\$_i^3 + \ldots + {}_5\omega_6^i \, {}^5\$_i^6\right] + \\ & \left[{}_2\omega_3^i \, {}^2\$_i^3 \quad {}_3\omega_4^i \, {}^3\$_i^4 + \ldots + {}_5\omega_6^i \, {}^5\$_i^6\right] + \\ \left[{}_3\omega_4^i \, {}^3\$_i^4 \quad {}_4\omega_5^i \, {}^4\$_i^5 + {}_5\omega_6^i \, {}^5\$_i^6\right] + & \left[{}_4\omega_5^i \, {}^4\$_i^5 \quad {}_5\omega_6^i \, {}^4\$_i^5\right].\end{aligned} \tag{10.19}$$

To compute the six-dimensional vector $\mathcal{A}_i$, we calculate the joint-velocity rates in the same limb according to Eq. (10.12) as follows:

$$\Omega_i = \mathbf{J}_i^{-1} \, V_C, \tag{10.20}$$

where

$$\Omega_i = \left[{}_0\omega_1^i \; {}_1\omega_2^i \; {}_2\omega_3^i \; {}_3\omega_4^i \; {}_4\omega_5^i \; {}_5\omega_6^i\right]^T$$

is a matrix containing the joint-velocity rates of the ith limb. Meanwhile,

$$\mathbf{J}_i = \left[{}^0\$_i^1 \; {}^1\$_i^2 \; {}^2\$_i^3 \; {}^3\$_i^4 \; {}^4\$_i^5 \; {}^5\$_i^6\right]$$

is the screw-coordinate Jacobian matrix of the ith limb. Following the trend of the velocity analysis, the application of reciprocal screw theory yields the following input–output equation of acceleration:

$$\mathbf{A}\boldsymbol{A}_C = \mathbf{B}\ddot{\mathbf{q}} + \mathbf{H}, \tag{10.21}$$

where

$$\ddot{\mathbf{q}} = \left[\ddot{\bar{q}}_1 \; \ddot{\bar{q}}_2 \; \ddot{\bar{q}}_3 \; \ddot{q}_1 \; \ddot{q}_2 \; \ddot{q}_3\right]^T \tag{10.22}$$

is the second-order driver matrix of the robot manipulator, whereas

$$\mathbf{H} = \begin{bmatrix} \{{}^4\$_1^5; \mathcal{A}_1\} \\ \{{}^4\$_2^5; \mathcal{A}_2\} \\ \{{}^4\$_3^5; \mathcal{A}_3\} \\ \{{}^3\$_1^4; \mathcal{A}_1\} \\ \{{}^3\$_2^4; \mathcal{A}_2\} \\ \{{}^3\$_3^4; \mathcal{A}_3\} \end{bmatrix} \tag{10.23}$$

is the complementary matrix of acceleration of the 3-RRPS parallel manipulator.

10.4.3 Jerk Analysis of the 3-RRPS Manipulator

Let

$$ {}^{0}\boldsymbol{J}_{C}^{m} = \begin{bmatrix} {}^{0}\ddot{\boldsymbol{\omega}}^{m} \\ {}^{0}\dot{\boldsymbol{j}}_{C}^{m} - 2\boldsymbol{\alpha} \times \boldsymbol{v}_{C} - \boldsymbol{\omega} \times \boldsymbol{a}_{C} \end{bmatrix} $$

be the reduced jerk state of the moving platform as measured from the fixed platform, where ${}^{0}\ddot{\boldsymbol{\omega}}^{m}$ and ${}^{0}\dot{\boldsymbol{j}}_{C}^{m}$ are the angular and linear jerks of the center of the moving platform. The six-dimensional vector ${}^{0}\boldsymbol{J}_{C}^{m}$ may be written in screw form through the connector limbs as follows:

$$ {}_{0}\ddot{\omega}_{1}^{i}\,{}^{0}\$_{i}^{1} + {}_{1}\ddot{\omega}_{2}^{i}\,{}^{1}\$_{i}^{2} + {}_{2}\ddot{\omega}_{3}^{i}\,{}^{2}\$_{i}^{3} + {}_{3}\ddot{\omega}_{4}^{i}\,{}^{3}\$_{i}^{4} + {}_{4}\ddot{\omega}_{5}^{i}\,{}^{4}\$_{i}^{5} + {}_{5}\ddot{\omega}_{6}^{i}\,{}^{5}\$_{i}^{6} + \mathcal{J}_{i} = {}^{0}\boldsymbol{J}_{C}^{m}, \qquad (10.24) $$

where $\mathcal{J}_{i}$ is the ith Lie screw of jerk, which is computed as

$$ \begin{aligned} \mathcal{J}_{i} = 2\Big[& {}_{0}\omega_{1}{}^{0}\$^{1} \quad {}_{1}\dot{\omega}_{2}{}^{1}\$^{2} + \ldots + {}_{4}\dot{\omega}_{5}{}^{4}\$^{5} + {}_{5}\dot{\omega}_{6}{}^{5}\$^{6} \\ & + \left[{}_{1}\omega_{2}{}^{1}\$^{2} \quad {}_{2}\omega_{3}{}^{2}\$^{3} + \ldots + {}_{4}\omega_{5}{}^{4}\$^{5} + {}_{5}\omega_{6}{}^{5}\$^{6}\right] \\ & + \left[{}_{2}\omega_{3}{}^{2}\$^{3} \quad {}_{3}\omega_{4}{}^{3}\$^{4} + \ldots + {}_{4}\omega_{5}{}^{4}\$^{5} + {}_{5}\omega_{6}{}^{5}\$^{6}\right] \\ & + \ldots + \left[{}_{4}\omega_{5}{}^{4}\$^{5} \quad {}_{5}\omega_{6}{}^{5}\$^{6}\right]\Big] \\ + 2\Big[& {}_{1}\omega_{2}{}^{1}\$^{2} \quad {}_{2}\dot{\omega}_{3}{}^{2}\$^{3} + \ldots + {}_{4}\dot{\omega}_{5}{}^{4}\$^{5} + {}_{5}\dot{\omega}_{6}{}^{5}\$^{6} \\ & + \left[{}_{2}\omega_{3}{}^{2}\$^{3} \quad {}_{3}\omega_{4}{}^{3}\$^{4} + \ldots + {}_{4}\omega_{5}{}^{4}\$^{5} + {}_{5}\omega_{6}{}^{5}\$^{6}\right] \\ & + \left[{}_{3}\omega_{4}{}^{3}\$^{4} \quad {}_{4}\omega_{5}{}^{4}\$^{5} + \ldots + {}_{4}\omega_{5}{}^{4}\$^{5} + {}_{5}\omega_{6}{}^{5}\$^{6}\right] \\ & + \ldots + \left[{}_{4}\omega_{5}{}^{4}\$^{5} \quad {}_{5}\omega_{6}{}^{5}\$^{6}\right]\Big] \\ & + \ldots + 2\left[{}_{4}\omega_{5}{}^{4}\$^{5} \quad {}_{5}\dot{\omega}_{6}{}^{5}\$^{6}\right] \\ & + \left[{}_{0}\dot{\omega}_{1}{}^{0}\$^{1} \quad {}_{1}\omega_{2}{}^{1}\$^{2} + \ldots + {}_{4}\omega_{5}{}^{4}\$^{5} + {}_{5}\omega_{6}{}^{5}\$^{6}\right] \\ & + \left[{}_{1}\dot{\omega}_{2}{}^{1}\$^{2} \quad {}_{2}\omega_{3}{}^{2}\$^{3} + \ldots + {}_{4}\omega_{5}{}^{4}\$^{5} + {}_{5}\omega_{6}{}^{5}\$^{6}\right] \\ & + \ldots + \left[{}_{4}\dot{\omega}_{5}{}^{4}\$^{5} \quad {}_{5}\omega_{6}{}^{5}\$^{6}\right] \\ + \Big[& {}_{0}\omega_{1}{}^{0}\$^{1} \left[{}_{0}\omega_{1}{}^{0}\$^{1} \quad {}_{1}\omega_{2}{}^{1}\$^{2} + \ldots + {}_{4}\omega_{5}{}^{4}\$^{5} + {}_{5}\omega_{6}{}^{5}\$^{6}\right]\Big] \\ & + \Big[{}_{1}\omega_{2}{}^{1}\$^{2} \left[{}_{1}\omega_{2}{}^{1}\$^{2} \quad {}_{2}\omega_{3}{}^{2}\$^{3} + \ldots + {}_{5}\omega_{6}{}^{5}\$^{6}\right]\Big] \\ & + \ldots + \Big[{}_{4}\omega_{5}{}^{4}\$^{5} \left[{}_{4}\omega_{5}{}^{4}\$^{5} \quad {}_{5}\omega_{6}{}^{5}\$^{6}\right]\Big]. \end{aligned} \qquad (10.25) $$

In this case, the joint-acceleration rates are computed according to Eq. (10.18) as follows:

$$\dot{\Omega}_i = \mathbf{J}_i^{-1}({}^0\boldsymbol{A}_C^m - \mathcal{A}_i), \tag{10.26}$$

where

$$\dot{\Omega}_i = \left[{}_0\dot{\omega}_1^i \; {}_1\dot{\omega}_2^i \; {}_2\dot{\omega}_3^i \; {}_3\dot{\omega}_4^i \; {}_4\dot{\omega}_5^i \; {}_5\dot{\omega}_6^i\right]^T$$

is a matrix containing the joint-jerk rates of the ith limb. Following the trend of the velocity analysis, by applying reciprocal screw theory we find that the input–output equation of jerk of the robot manipulator results in

$$\mathbf{A}\boldsymbol{J}_C = \mathbf{B}\dddot{\mathbf{q}} + \mathbf{W}, \tag{10.27}$$

where

$$\dddot{\mathbf{q}} = \left[\dddot{\overline{q}}_1 \; \dddot{\overline{q}}_2 \; \dddot{\overline{q}}_3 \; \dddot{q}_1 \; \dddot{q}_2 \; \dddot{q}_3\right]^T \tag{10.28}$$

is the third-order driver matrix of the robot manipulator. Meanwhile,

$$\mathbf{W} = \begin{bmatrix} \{{}^4\$_1^5; \mathcal{J}_1\} \\ \{{}^4\$_2^5; \mathcal{J}_2\} \\ \{{}^4\$_3^5; \mathcal{J}_3\} \\ \{{}^3\$_1^4; \mathcal{J}_1\} \\ \{{}^3\$_2^4; \mathcal{J}_2\} \\ \{{}^3\$_3^4; \mathcal{J}_3\} \end{bmatrix} \tag{10.29}$$

is the complementary matrix of jerk of the 3-RRPS parallel manipulator.

Example 10.2. The numerical example consists of solving the forward kinematics, up to the jerk analysis, of the 3-RRPS parallel manipulator. To accomplish this, consider that when we use SI units in the numerical example, the nominal coordinates of the revolute joints A_i are given by

$$A_1 = (0.25, 0, 0), \; A_2 = (-0.125, 0, -0.2165), \; A_3 = (-0.125, 0, 0.2165).$$

On the other hand, the coordinates of points B_i are chosen as

$$B_1 = (0.25, -0.25, 0), \; B_2 = (-0.125, -0.25, -0.2165),$$
$$B_3 = (-0.125, -0.25, 0.2165).$$

Furthermore, consider $e = 0.2165$ as the side of the equilateral triangle $\Delta C_1 C_2 C_3$, while in the reference configuration of the robot manipulator the unit vectors $\hat{\boldsymbol{u}}_i$ are given by

$$\hat{\boldsymbol{u}}_1 = \hat{\boldsymbol{k}},\ \hat{\boldsymbol{u}}_2 = -0.866\hat{\boldsymbol{i}} + 0.5\hat{\boldsymbol{k}},\ \hat{\boldsymbol{u}}_3 = 0.5\hat{\boldsymbol{i}} + 0.866\hat{\boldsymbol{k}}.$$

Finally, the generalized coordinates are commanded to follow periodic functions given by

$$\begin{array}{l} q_1 = 0.5\sin(t),\ q_2 = 7\pi/6 + 0.2\sin(t),\ q_3 = \pi/3 + 0.5\sin(t), \\ \overline{q}_1 = 0.371 + 0.14\sin(t),\ \overline{q}_2 = 0.371 + 0.15\sin(t),\ \overline{q}_3 = 0.371 + 0.12\sin(t) \end{array},$$

where the interval of time t is selected as $0 \leq t \leq 2\pi$. Of course, at time $t = 0$ the parallel manipulator is considered at its home position. With this data it is necessary to compute the time history of the velocity, acceleration, and jerk of the center C of the moving platform.

Solution. The resulting temporal behavior of the angular and linear velocities, accelerations, and jerks of the center of the moving platform using screw theory is shown in Fig. 10.3.

10.5 Exercises

1. Consider the 3-RRPS parallel manipulator. Use SI units in the exercise; the nominal positions of the revolute joints of the parallel manipulator are given by $A_1 = (0.25, 0, 0)$, $A_2 = (-0.125, 0, -0.2165)$, $A_3 = (-0.125, 0, 0.2165)$, $B_1 = (0.25, -0.25, 0)$, $B_2 = (-0.125, -0.25, -0.2165)$, and $B_3 = (-0.125, -0.25, 0.2165)$. On the other hand, the active revolute joints are chosen as $q_1 = 0.05$, $q_2 = 2.0$, and $q_3 = 3.25$, while the active prismatic joints are given by $\overline{q}_1 = 0.275$, $\overline{q}_2 = 0.275$, and $\overline{q}_3 = 0.3$. With these data we are required to compute one solution of the forward displacement analysis. *Hint*: The reader can resort to using a numerical technique such as the Newton–Raphson method.
2. Determine the general conditions of singular configurations of the 3-RRPS parallel manipulator based on the input–output equation of velocity.

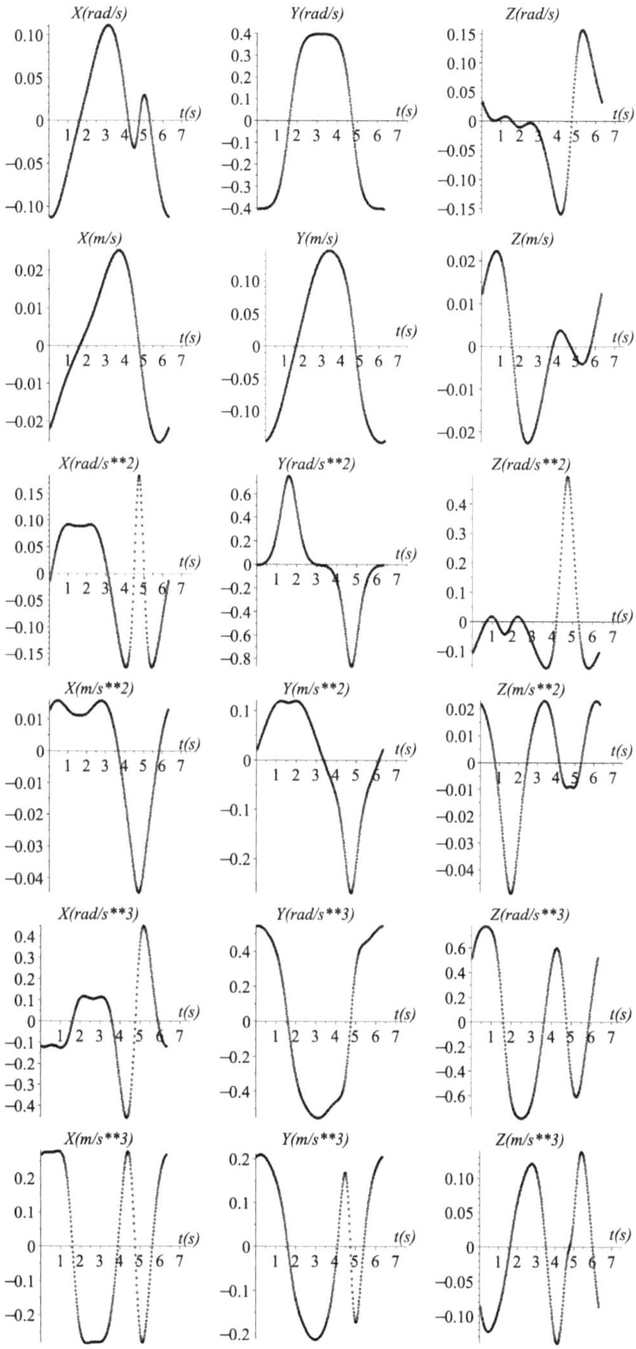

Fig. 10.3 Example 10.2. Time history of the infinitesimal kinematics of the center of the moving platform

References

Angeles, J., Yang, G., & Chen, I.-M. (2001). Singularity analysis of three-legged, six-DOF platform manipulators with RRRS legs. In *Proceedings 2001 IEEE/ASME International Conference on Advanced Intelligent Mechatronics* (pp. 32–36).

Behi, F. (1988). Kinematic analysis for a six-degree-of-freedom 3-PRPS parallel mechanism. *IEEE Journal of Robotics and Innovation, 4*(5), 561–565.

Ben-Horin, P., & Shoham, M. (2006). Singularity condition of six-degree-of-freedom three-legged parallel robots based on Grassmann & Cayley algebra. *IEEE Transactions on Robotics, 22*(4), 577–590.

Byun, Y., & Cho, H. S. (1997). Analysis of a novel 6-DOF, 3-PPSP parallel manipulator. *International Journal of Robotics Research, 16*(6), 859–872.

Gallardo-Alvarado, J., Aguilar-Nájera, C. R., Casique-Rosas, L., Pérez-González, L., & Rico-Martínez, J. M. (2008a). Solving the kinematics and dynamics of a modular spatial hyper-redundant manipulator by means of screw theory. *Multibody System Dynamics, 20*(4), 307–325.

Gallardo-Alvarado, J., Rodríguez-Castro, R., Islam, M. N. (2008b). Analytical solution of the forward position analysis of parallel manipulators that generate 3-RS structures. *Advanced Robotics, 22*, 215–234.

García-Murillo, M. A., Castillo-Castaneda, E., & Gallardo-Alvarado, J. (2013). 9th Workshop on Robot Motion and Control (RoMoCo) (pp. 270–275), Kuslin.

Geng, Z., & Haynes, L. S. (1994). A 3-2-1 kinematic configuration of a Stewart platform and its application to six-degree-of-freedom pose measurements. *Robotics and Computer-Integrated Manufacturing, 11*(1), 23–34.

Goldsmith, P. B. (2003). Design and kinematics of a three-legged parallel manipulator. *IEEE Transactions on Robotics and Automation, 19*(4), 726–731.

Guilin, Y., Chen, I.-M., Lin, W., & Angeles, J. (2001). Singularity analysis of three-legged parallel robots based on passive-joint velocities. In *IEEE International Conference on Robotics and Automation* (pp. 2407–2412).

Husain, M., & Waldron, K. J. (1992). Position kinematics of a mixed mechanism. In *Proceedings 22nd Biennial Mechanisms Conference, Scottsdale, AZ* (pp. 41–48).

Husty, M. L. (1996). An algorithm for solving the direct kinematic of general Stewart-Gough platforms. *Mechanism and Machine Theory, 31*(4), 365–380.

Innocenti, C. (2001). Forward kinematics in polynomial form of the general Stewart platform. *ASME Journal of Mechanical Design, 123*(2), 254–260.

Innocenti, C., & Parenti-Castelli, V. (1991). A new kinematic model for the closure equations of the generalized Stewart platform mechanism. *Meccanica, 26*(4), 247–252.

Ji, C.-Y., Chen, T.-C., & Lee, Y.-L. (2007). Investigation of kinematic analysis and applications for a 3-$\underline{R}\underline{R}$PS parallel manipulator. *Journal of the Chinese Society of Mechanical Engineers, 28*(6), 623–632.

Jin, Y., Chen, I.-M., & Yang, G. (2009). Kinematic design of a family of 6-DOF partially decoupled parallel manipulators. *Mechanism and Machine Theory, 44*(5), 912–922.

Merlet, J.-P. (1989). Manipulateurs parallèles, 4ème partie: mode d'assemblage et cinematique directe sous forme polynomiale. INRIA Research Report no. 1135, Centre de Sophia Antipolis, Valbonne.

Merlet, J.-P. (2009). Interval analysis for certified numerical solution of problems in robotics. *International Journal of Applied Mathematics and Computer Science, 19*(3), 399–412.

Notash, L., & Podhorodeski, R. P. (1995). On the forward displacement problem of three-branch parallel manipulators. *Mechanism and Machine Theory, 30*(3), 391–404.

Portman, T. V., & Sandler, B.-Z. (2002). Tripod robot with cylindrically actuated limbs: structure and kinematics. *Mechanism and Machine Theory, 37*(12), 1447–1463.

Qi, C. K., Zhao, X. C., Jin, Z. L., Zhang, X. X., Ma, C. X., & Gao, F. (2011). Workspace of a 3-RRPS parallel robot leg with a constant orientation. In *Proceedings 13th World Congress in Mechanism and Machine Science, Gto, Mexico* (pp. 19–25).

Raghavan, M. (1993). The Stewart platform of general geometry has 40 configurations. *ASME Journal of Mechanical Design, 115*(2), 277–282.

Rolland, L. (2005). Certified solving of the forward kinematics problem with an exact algebraic method for the general parallel manipulator. *Advanced Robotics, 19*(9), 995–1025.

Tsai, L.-W. (1999). *Robot analysis*. New York: Wiley.

Zabalza, I., Ros, J., Gil, J. J., Pintor, J. M., & Jiménez, J. M. (2002). TRI-SCOTT. A new kinematic structure for a 6-DOF decoupled parallel manipulator. In *Proceedings Workshop on Fundamental Issues and Future Research Directions for Parallel Mechanisms and Manipulators* (pp. 1–4).

Zlatanov, D., Dai, M. Q., Fenton, R. G., & Benhabib, B. (1992). Mechanical design and kinematic analysis of a three-legged six-degree-of-freedom parallel manipulator. In *Proceedings 22nd Biennial Mechanisms Conference* (pp. 529–536).

Chapter 11
3RRRS+3RRPS Parallel Manipulator

11.1 Introduction

Combined schemes of actuation, such as the combination of linear and rotary actuators, is a viable strategy to enlarge the workspace of parallel manipulators, diminishing singular regions. This chapter uses screw theory to address the kinematics of a six-degree-of-freedom (DOF) parallel manipulator equipped with three rotary and three linear actuators. For the sake of completeness the displacement analysis of the robot at hand is also investigated. The finite kinematics of the chosen parallel manipulator is precisely where the benefits of using mixed schemes of actuation are evident when compared with the Gough–Stewart platform.

11.2 Description of the 3RRRS+3RRPS Parallel Manipulator

In Fig. 11.1 a 6-DOF parallel manipulator equipped with a mixed scheme of actuation is outlined.

Referring to Fig. 11.1, the robot under study consists of a moving platform (m) and a fixed platform (0), connected to each other by three RRP*S-type and three R*RRS-type kinematic chains, where superscript asterisks denote active kinematic pairs while R, P, and S stand for revolute, prismatic, and spherical joints, respectively. Thus, the limbs of the parallel manipulator are in touch with the moving and fixed platforms through spherical and revolute joints, respectively. According to the Chebychev–Grübler–Kutzbach criterion, the mobility F of a spatial mechanism formed from N links, including the fixed link, and j kinematic pairs each one with freedom $f_i(i = 1, 2, \ldots, j)$ can be obtained through the expression

J. Gallardo-Alvarado, *Kinematic Analysis of Parallel Manipulators by Algebraic Screw Theory*, DOI 10.1007/978-3-319-31126-5_11

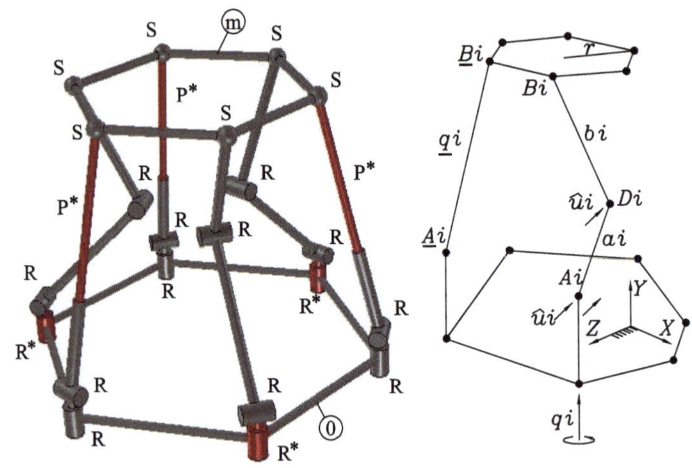

Fig. 11.1 Six-legged parallel manipulator provided with linear and rotary actuators

$$F = 6(N - 1 - j) + \sum_{i=1}^{j} f_i. \tag{11.1}$$

Hence, taking into account that for the parallel manipulator at hand $N = 20$, $j = 24$, and $\sum_{i=1}^{j} f_i = 36$, we have $F = 6$. Then the robot is capable of realizing 6 DOF. Furthermore, arbitrary poses of the moving platform with respect to the fixed platform, of course excluding singular postures, are available for this robot manipulator.

To explain the notation used in this chapter, let XYZ be a reference frame attached to the fixed platform. The axes of the revolute joints connecting the limbs to the fixed platform are along the Y-axis, whereas the orientations of the axes of the remaining revolute joints are located in the XZ-plane. Furthermore, the axes of the revolute joints in the same limb containing an active revolute joint have the same orientation, $\hat{\boldsymbol{u}}_i$, in the remainder of the chapter unless otherwise $i = 1, 2, 3$. The moving platform is a regular hexagon inscribed in a circle of radius r whose corners are points B_i and $\underline{B}_i$ denotes the centers of the spherical joints, which are located, respectively, by vectors $\boldsymbol{b}_i$ and $\underline{\boldsymbol{b}}_i$. On the other hand, points A_i and $\underline{A}_i$, located by vectors $\boldsymbol{a}_i$ and $\underline{\boldsymbol{a}}_i$, assume to be the nominal positions of the lower passive revolute joints whereas a point D_i, located by vector $\boldsymbol{d}_i$, is the instantaneous position of the revolute joint connecting the arm and forearm of a RRRS-type limb. Finally, the generalized coordinates are specified by q_i and $\underline{q}_i$.

11.3 Finite Kinematics of the 3RRRS+3RRPS Parallel Manipulator

11.3.1 Forward Displacement Analysis of the 3RRRS+3RRPS Parallel Manipulator

In what follows the finite kinematics of the parallel manipulator is presented. The forward displacement analysis consists of finding the pose (position and orientation) of the moving platform with respect to the fixed platform for a given set of generalized coordinates $\left\{\underline{q}_i, q_i\right\}$.

Taking into account that the coordinates of three points embedded in a rigid body provide all the information required to compute its pose with respect to another body or reference frame (Merlet 2004; Gallardo-Alvarado 2014), then we may formulate closure equations concerned with the displacement analysis of the robot using linear combinations of the unknown coordinates $w_i (i = 1, 2, \ldots, 9)$ of points $\underline{B}_1 = (w_1, w_2, w_3)$, $\underline{B}_2 = (w_4, w_5, w_6)$, and $\underline{B}_3 = (w_7, w_8, w_9)$ named the control points of the parallel manipulator. For example, assuming that the moving platform is a regular hexagon of side r (see Fig. 11.2), we may express point B_2 as follows:

$$B_2 = \underline{B}_3 + \left[\frac{(\underline{B}_1 + \underline{B}_2)/2 - \underline{B}_3}{3r/2}\right] 2r \tag{11.2}$$

or

$$B_2 = 2(\underline{B}_1 + \underline{B}_2)/3 - \underline{B}_3/3. \tag{11.3}$$

It is straightforward to demonstrate that similar expressions can be formulated for B_1 and B_3. Next, in terms of the limbs containing active revolute joints, three closure equations emerge immediately if we consider that the passive revolute joints constrain the motion of the spherical joint in the same limb such that

$$(\boldsymbol{b}_i - \boldsymbol{a}_i) \bullet \hat{\boldsymbol{u}}_i = 0, \tag{11.4}$$

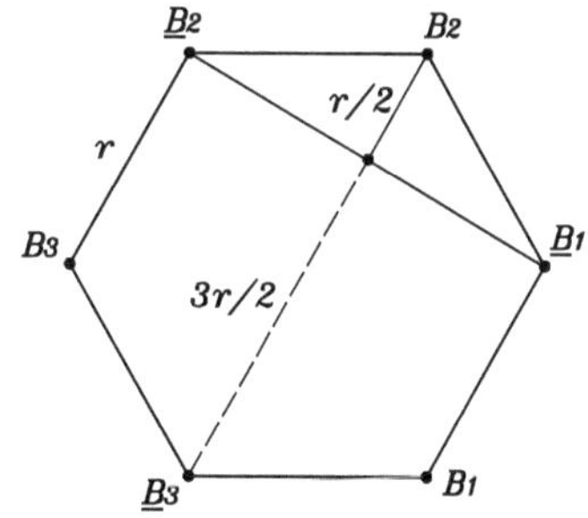

Fig. 11.2 Hexagonal moving platform of the 3RRRS+3RRPS parallel manipulator

where the ith unit vector $\hat{\boldsymbol{u}}_i$ is determined according to the orientation of the limb defined by the rotational ith generalized coordinate q_i. Furthermore, from the limbs containing the active prismatic joints it follows that

$$(\underline{\boldsymbol{b}}_i - \underline{\boldsymbol{a}}_i) \bullet (\underline{\boldsymbol{b}}_i - \underline{\boldsymbol{a}}_i) = \underline{q}_i^2. \tag{11.5}$$

Finally, three compatibility kinematic constraint equations may be written based on the symmetry of the moving platform as follows:

$$(\underline{\boldsymbol{b}}_i - \underline{\boldsymbol{b}}_j) \bullet (\underline{\boldsymbol{b}}_i - \underline{\boldsymbol{b}}_j) = 3r^2, \quad i,j = 1,2,3 \bmod(3). \tag{11.6}$$

Gathering these results, we can derive three algebraic equations in the unknowns w_1, w_4, and w_7 as follows:

$$\begin{aligned} c_{1i}w_1^4 + c_{2i}w_4^4 + c_{3i}w_7^4 + c_{4i}w_1^3 + c_{5i}w_4^3 + c_{6i}w_7^3 + \\ c_{7i}w_1^2w_4w_7 + c_{8i}w_1w_4^2w_7 + c_{9i}w_1w_4w_7^2 + c_{10i}w_1w_4w_7 + \\ c_{11i}w_1^3w_4 + c_{12i}w_1w_4^3 + c_{13i}w_1^3w_7 + c_{14i}w_13w_7^3 + \\ c_{15i}w_4^3w_7 + c_{16i}w_4w_7^3 + c_{17i}w_1^2w_4^2 + c_{18i}w_1^2w_7^2 + \\ c_{19i}w_4^2w_7^2 + c_{20i}w_1^2w_4 + c_{21i}w_1w_4^2 + c_{22i}w_4^2w_7 + \\ c_{23i}w_4w_7^2 + c_{24i}w_1^2w_7 + c_{25i}w_1w_7^2 + c_{26i}w_1w_4 + \\ c_{27i}w_4w_7 + c_{28i}w_1w_7 = 0, \end{aligned} \tag{11.7}$$

where the coefficients c_* of Eq. (11.7) are calculated according to the parameters and generalized coordinates of the parallel manipulator. Expressions (11.7) are named the *characteristic equations of the robot*. These equations are solved in a reasonable time by using special software like Maple12©. Once the variables w_1, w_4, and w_7 are determined, the unknowns w_2, w_3, w_5, w_6, w_8, and w_9 are computed recursively based on Eqs. (11.4) and (11.5). After, the center C of the moving platform, located by vector $\boldsymbol{c}$, is computed as

$$\boldsymbol{c} = (\underline{\boldsymbol{b}}_1 + \underline{\boldsymbol{b}}_2 + \underline{\boldsymbol{b}}_3)/3. \tag{11.8}$$

Meanwhile, the vectors $\boldsymbol{d}_i$ are determined according to the following closure equations:

$$\left.\begin{aligned} (\boldsymbol{d}_i - \boldsymbol{a}_i) \bullet (\boldsymbol{d}_i - \boldsymbol{a}_i) &= a^2 \\ (\boldsymbol{d}_i - \boldsymbol{b}_i) \bullet (\boldsymbol{d}_i - \boldsymbol{b}_i) &= b^2 \\ (\boldsymbol{d}_i - \boldsymbol{a}_i) \bullet \hat{\boldsymbol{u}}_i &= 0 \end{aligned}\right\}. \tag{11.9}$$

11.3.2 Closed-Form Solution of the Forward Position Analysis Using Three Rotary Sensors

Assuming that each RRP*S limb is provided with a rotary sensor capable of measuring the rotation θ_i of the revolute joint connecting it to the fixed platform, it is possible to write three auxiliary equations as

$$(\underline{b}_i - \underline{a}_i) \bullet \hat{\underline{u}}_i = 0, \tag{11.10}$$

where

$$\hat{\underline{u}}_i = \sin(\theta_i)\boldsymbol{i} + \cos(\theta_i)\boldsymbol{k}. \tag{11.11}$$

Afterward, a linear system of six equations in the unknowns w_1, w_3, w_4, w_6, w_7, and w_9 is obtained as follows:

$$\mathbf{MW} = \mathbf{N}, \tag{11.12}$$

where

$$\mathbf{M} = \begin{bmatrix} \sin(\theta_1) & \cos(\theta_1) & 0 & 0 & 0 & 0 \\ 0 & 0 & \sin(\theta_2) & \cos(\theta_2) & 0 & 0 \\ 0 & 0 & 0 & 0 & \sin(\theta_3) & \cos(\theta_3) \\ \frac{2}{3}\sin(q_1) & \frac{2}{3}\cos(q_1) & \frac{2}{3}\sin(q_1) & \frac{2}{3}\cos(q_1) & -\frac{1}{3}\sin(q_1) & -\frac{1}{3}\cos(q_1) \\ -\frac{1}{3}\sin(q_2) & -\frac{1}{3}\cos(q_2) & \frac{2}{3}\sin(q_2) & \frac{2}{3}\cos(q_2) & \frac{2}{3}\sin(q_2) & \frac{2}{3}\cos(q_2) \\ \frac{2}{3}\sin(q_3) & \frac{2}{3}\cos(q_3) & -\frac{1}{3}\sin(q_3) & -\frac{1}{3}\cos(q_3) & \frac{2}{3}\sin(q_3) & \frac{2}{3}\cos(q_3) \end{bmatrix}, \tag{11.13}$$

$$\mathbf{W} = \begin{bmatrix} w_1 & w_3 & w_4 & w_6 & w_7 & w_9 \end{bmatrix}^T, \tag{11.14}$$

and

$$\mathbf{N} = \begin{bmatrix} \sin(\theta_1)\underline{ax}_1 + \cos(\theta_1)\underline{az}_1 \\ \sin(\theta_2)\underline{ax}_2 + \cos(\theta_2)\underline{az}_2 \\ \sin(\theta_3)\underline{ax}_3 + \cos(\theta_3)\underline{az}_3 \\ \sin(q_1)ax_1 + \cos(q_1)az_1 \\ \sin(q_2)ax_2 + \cos(q_2)az_2 \\ \sin(q_3)ax_3 + \cos(q_3)az_3 \end{bmatrix}. \tag{11.15}$$

Finally, the remaining unknowns w_2, w_5, and w_8 are computed using Eq. (11.5).

11.3.3 Inverse Displacement Analysis of the 3RRRS+3RRPS Parallel Manipulator

The inverse displacement analysis consists of finding the generalized coordinates of the robot given the pose of the moving platform with respect to the fixed platform. Given the points $\underline{B}_i$ it is evident that the generalized coordinates $\underline{q}_i$ can be calculated with Eq. (11.5). Furthermore, the generalized coordinates q_i are obtained from Eq. (11.4). It is interesting to note that according to the inverse position analysis, the workspace of the robot is restricted only by its singular regions and the limits of the RRRS-type limbs; for example, the maximum limit is governed by the expression

$$a_i + b_i = \sqrt{(\boldsymbol{b}_i - \boldsymbol{a}_i) \bullet (\boldsymbol{b}_i - \boldsymbol{a}_i)}. \tag{11.16}$$

11.4 Infinitesimal Kinematics of the 3RRRS+3RRPS Parallel Manipulator

In this section the infinitesimal kinematics of the robot manipulator at hand is approached using screw theory. For this purpose, the infinitesimal screws of the parallel manipulator are outlined in Fig. 11.3.

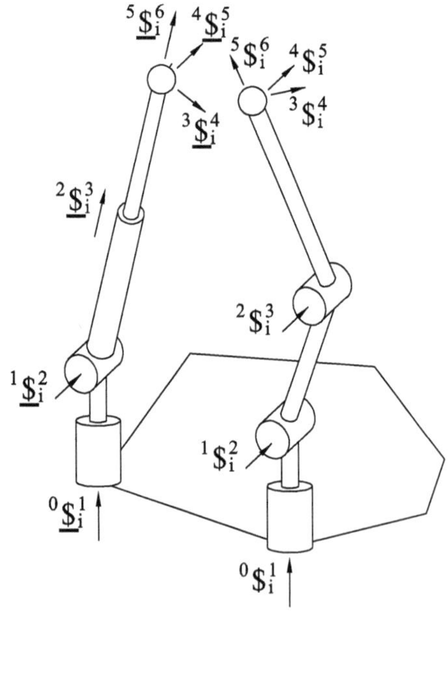

Fig. 11.3 Infinitesimal screws of two distinct limbs of the 3RRRS+3RRPS parallel manipulator

11.4.1 Velocity Analysis of the 3RRRS+3RRPS Parallel Manipulator

Let $\boldsymbol{\omega}$ and $\boldsymbol{v}_C$ be the angular and linear velocities of the center C of the moving platform as observed from the fixed reference frame. In what follows the inverse/forward equation of velocity of the robot manipulator is obtained by resorting to reciprocal screw theory. To this end, consider that the velocity state of the moving platform as measured from the fixed platform is given by

$$\boldsymbol{V}_C = \begin{bmatrix} \boldsymbol{\omega} \\ \boldsymbol{v}_C \end{bmatrix}.$$

Furthermore, the velocity state $\boldsymbol{V}_C$ may be written in screw form through the RRP*S-type limbs as follows:

$${}_0\underline{\omega}_1^i \, {}^0\underline{\$}_i^1 + {}_1\underline{\omega}_2^i \, {}^1\underline{\$}_i^2 + \underline{\dot{q}}_i \, {}^2\underline{\$}_i^3 + {}_3\underline{\omega}_4^i \, {}^3\underline{\$}_i^4 + {}_4\underline{\omega}_5^i \, {}^4\underline{\$}_i^5 + {}_5\underline{\omega}_6^i \, {}^5\underline{\$}_i^6 = \boldsymbol{V}_C. \tag{11.17}$$

According to the properties of the Klein form, it follows that

$$\left\{ {}^5\underline{\$}_i^6 ; {}^0\underline{\$}_i^1 \right\} = \left\{ {}^5\underline{\$}_i^6 ; {}^1\underline{\$}_i^2 \right\} = \left\{ {}^5\underline{\$}_i^6 ; {}^3\underline{\$}_i^4 \right\} = \left\{ {}^5\underline{\$}_i^6 ; {}^4\underline{\$}_i^5 \right\} = \left\{ {}^5\underline{\$}_i^6 ; {}^5\underline{\$}_i^6 \right\} = 0 \tag{11.18}$$

and

$$\left\{ {}^5\underline{\$}_i^6 ; {}^2\underline{\$}_i^3 \right\} = 1. \tag{11.19}$$

In other words, the screw ${}^5\underline{\$}_i^6$ is reciprocal to all the screws, excluding the screw ${}^2\underline{\$}_i^3$ associated to the active prismatic joint, in the same limb. After reducing terms in Eq. (11.17), one obtains

$$\left\{ {}^5\underline{\$}_i^6 ; \boldsymbol{V}_C \right\} = \underline{\dot{q}}_i. \tag{11.20}$$

On the other hand, the six-dimensional vector $\boldsymbol{V}_C$ can be written in screw form through the R*RRS-type limbs as follows:

$$\dot{q}_i \, {}^0\$_i^1 + {}_1\omega_2^i \, {}^1\$_i^2 + {}_2\omega_3^i \, {}^2\$_i^3 + {}_3\omega_4^i \, {}^3\$_i^4 + {}_4\omega_5^i \, {}^4\$_i^5 + {}_5\omega_6^i \, {}^5\$_i^6 = \boldsymbol{V}_C, \tag{11.21}$$

where the screw ${}^4\$_i^5$ is reciprocal to all the screws in the same limb, except the screw ${}^0\$_i^1$, that is,

$$\left\{ {}^4\$_i^5 ; {}^1\$_i^2 \right\} = \left\{ {}^4\$_i^5 ; {}^2\$_i^3 \right\} = \left\{ {}^4\$_i^5 ; {}^3\$_i^4 \right\} = \left\{ {}^4\$_i^5 ; {}^4\$_i^5 \right\} = \left\{ {}^4\$_i^5 ; {}^5\$_i^6 \right\} = 0. \tag{11.22}$$

Hence, after reducing terms in Eq. (11.21) we have

$$\left\{ {}^4\$_i^5 ; \boldsymbol{V}_C \right\} = \dot{q}_i \left\{ {}^4\$_i^5 ; {}^0\$_i^1 \right\}. \tag{11.23}$$

Casting Eqs. (11.20) and (11.23) into matrix-vector form, we see that the input–output equation of velocity of the robot results in

$$\mathbf{G}\boldsymbol{V}_C = \mathbf{H}\dot{\mathbf{q}}, \tag{11.24}$$

where

$$\dot{\mathbf{q}} = [\dot{\underline{q}}_1 \ \dot{\underline{q}}_2 \ \dot{\underline{q}}_3 \ \dot{q}_1 \ \dot{q}_2 \ \dot{q}_3]^T \tag{11.25}$$

is the first-order driver matrix of the parallel manipulator. Meanwhile,

$$\mathbf{G} = \mathbf{J}^T \, \boldsymbol{\Delta} \tag{11.26}$$

is the first-order coefficient matrix of the robot in which

$$\mathbf{J} = \left[{}^5\$^6_{\underline{1}} \ {}^5\$^6_{\underline{2}} \ {}^5\$^6_{\underline{3}} \ {}^4\$^5_1 \ {}^4\$^5_2 \ {}^4\$^5_3\right] \tag{11.27}$$

is the active screw-coordinate Jacobian matrix of the hexapod and $\boldsymbol{\Delta}$ is a 6×6 partitioned matrix, playing the role of an operator of polarity, defined according to the identity matrix $\mathbf{I}$ and the zero matrix $\mathbf{0}$ as

$$\boldsymbol{\Delta} = \begin{bmatrix} \mathbf{0} & \mathbf{I} \\ \mathbf{I} & \mathbf{0} \end{bmatrix}.$$

Finally, $\mathbf{H}$ is a diagonal matrix given by

$$\mathbf{H} = \text{diag}\left[1 \ 1 \ 1 \ \left\{{}^4\$^5_1; {}^0\$^1_1\right\} \ \left\{{}^4\$^5_2; {}^0\$^1_2\right\} \ \left\{{}^4\$^5_3; {}^0\$^1_3\right\}\right]. \tag{11.28}$$

Hence, the inverse velocity analysis consists of determining the generalized joint-velocity rates, matrix $\dot{\mathbf{q}}$, given the velocity state $\boldsymbol{V}_C$, while the forward velocity analysis consists of computing the six-dimensional vector $\boldsymbol{V}_C$ given matrix $\dot{\mathbf{q}}$.

11.4.2 Acceleration Analysis of the 3RRRS+3RRPS Parallel Manipulator

Let $\boldsymbol{\alpha}$ and $\boldsymbol{a}_C$ be the angular and linear accelerations of the center C of the moving platform with respect to the fixed platform. The reduced acceleration state between the moving and fixed platforms may be written as

$$\boldsymbol{A}_C = \begin{bmatrix} \boldsymbol{\alpha} \\ \boldsymbol{a}_C - \boldsymbol{\omega} \times \boldsymbol{v}_C \end{bmatrix}.$$

Furthermore, the accelerator $\boldsymbol{A}_C$ may be expressed through the connector chains of the parallel manipulator in screw form as follows:

$$ {}_0\underline{\alpha}_1^i\ {}^0\underline{\$}_i^1 + {}_1\underline{\alpha}_2^i\ {}^1\underline{\$}_i^2 + \ddot{\underline{q}}_i\ {}^2\underline{\$}_i^3 + {}_3\underline{\alpha}_4^i\ {}^3\underline{\$}_i^4 + {}_4\underline{\alpha}_5^i\ {}^4\underline{\$}_i^5 + {}_5\underline{\alpha}_6^i\ {}^5\underline{\$}_i^6 + \underline{\mathcal{A}}_i = \boldsymbol{A}_C \tag{11.29} $$

and

$$ \ddot{q}_i\ {}^0\$_i^1 + {}_1\alpha_2^i\ {}^1\$_i^2 + {}_2\alpha_3^i\ {}^2\$_i^3 + {}_3\alpha_4^i\ {}^3\$_i^4 + {}_4\alpha_5^i\ {}^4\$_i^5 + {}_5\alpha_6^i\ {}^5\$_i^6 + \mathcal{A}_i = \boldsymbol{A}_C, \tag{11.30} $$

where the Lie screws of acceleration $\underline{\mathcal{A}}_i$ and $\mathcal{A}_i$ are given by

$$ \begin{aligned} \underline{\mathcal{A}}_i = & \left[{}_0\underline{\omega}_1^i\ {}^0\underline{\$}_i^1 \quad {}_1\underline{\omega}_2^i\ {}^1\underline{\$}_i^2 + \ldots + {}_5\underline{\omega}_6^i\ {}^5\underline{\$}_i^6\right] + \left[{}_1\underline{\omega}_2^i\ {}^1\underline{\$}_i^2 \quad \dot{\underline{q}}_i\ {}^2\underline{\$}^3{}_i + \ldots + {}_5\underline{\omega}_6^i\ {}^5\underline{\$}_i^6\right] \\ & + \left[\dot{\underline{q}}_i\ {}^2\underline{\$}_i^3 \quad {}_3\underline{\omega}_4^i\ {}^3\underline{\$}_i^4 + {}_4\underline{\omega}_5^i\ {}^4\underline{\$}_i^5 + {}_5\underline{\omega}_6^i\ {}^5\underline{\$}_i^6\right] + \cdots + \left[{}_4\underline{\omega}_5^i\ {}^4\underline{\$}_i^5 \quad {}_5\underline{\omega}_6^i\ {}^5\underline{\$}_i^6\right] \end{aligned} \tag{11.31} $$

and

$$ \begin{aligned} \mathcal{A}_i = & \left[\dot{q}_i\ {}^0\$_i^1 \quad {}_1\omega_2^i\ {}^1\$_i^2 + \ldots + {}_5\omega_6^i\ {}^5\$_i^6\right] + \left[{}_1\omega_2^i\ {}^1\$_i^2 \quad {}_2\omega_3^i\ {}^2\$_i^3 + \ldots + {}_5\omega_6^i\ {}^5\$_i^6\right] \\ & + \left[{}_2\omega_3^i\ {}^2\$_i^3 \quad {}_3\omega_4^i\ {}^3\$_i^4 + {}_4\omega_5^i\ {}^4\$_i^5 + {}_5\omega_6^i\ {}^5\$_i^6\right] + \cdots + \left[{}_4\omega_5^i\ {}^4\$_i^5 \quad {}_5\omega_6^i\ {}^5\$_i^6\right]. \end{aligned} \tag{11.32} $$

Following the trend of the velocity analysis, the input–output equation of acceleration of the robot results in

$$ \mathbf{G}\boldsymbol{A}_C = \mathbf{H}\ddot{\mathbf{q}} + \mathbb{A}, \tag{11.33} $$

where

$$ \ddot{\mathbf{q}} = \left[\ddot{\underline{q}}_1\ \ddot{\underline{q}}_2\ \ddot{\underline{q}}_3\ \ddot{q}_1\ \ddot{q}_2\ \ddot{q}_3\right]^T \tag{11.34} $$

is the second-order driver matrix of the parallel manipulator. Meanwhile,

$$ \mathbb{A} = \begin{bmatrix} \left\{\underline{\mathcal{A}}_1;\ {}^5\underline{\$}_1^6\right\} \\ \left\{\underline{\mathcal{A}}_2;\ {}^5\underline{\$}_2^6\right\} \\ \left\{\underline{\mathcal{A}}_3;\ {}^5\underline{\$}_3^6\right\} \\ \left\{\mathcal{A}_1;\ {}^4\$_1^5\right\} \\ \left\{\mathcal{A}_2;\ {}^4\$_2^5\right\} \\ \left\{\mathcal{A}_3;\ {}^4\$_3^5\right\} \end{bmatrix} \tag{11.35} $$

is the complementary matrix of acceleration. It is worth noting that the terms concerned with the Coriolis acceleration are self-contained in matrix $\mathbb{A}$.

Given matrix $\ddot{\mathbf{q}}$, the forward acceleration analysis consists of finding the corresponding reduced acceleration state $\boldsymbol{A}_C$. On the other hand, given the six-dimensional vector $\boldsymbol{A}_C$, the inverse acceleration analysis of the robot consists of determining the corresponding generalized joint-acceleration rates contained in matrix $\ddot{\mathbf{q}}$. The inverse acceleration analysis is a fundamental step in the formulation of the dynamic equations that may be used, among traditional applications, in the path-planning trajectories of robot manipulators (Bahara 1996; Ataa 2010).

11.4.3 Jerk Analysis of the 3RRRS+3RRPS Parallel Manipulator

Let $\boldsymbol{\rho}$ and $\boldsymbol{j}_C$ be the angular and linear jerks of the center C of the moving platform as observed from the fixed platform. The reduced jerk state of the moving platform with respect to the fixed platform is given by

$$\boldsymbol{J}_C = \begin{bmatrix} \boldsymbol{\rho} \\ \boldsymbol{j}_C - \boldsymbol{\alpha} \times \boldsymbol{v}_C - \boldsymbol{\omega} \times \boldsymbol{a}_C \end{bmatrix}.$$

On the other hand, the jerkor $\boldsymbol{J}_C$ may be written in screw form through the connector chains of the robot manipulator as follows:

$${}_0\underline{\rho}_1^i\ {}^0\underline{\$}_i^1 + {}_1\underline{\rho}_2^i\ {}^1\underline{\$}_i^2 + \dddot{\underline{q}}_i\ {}^2\underline{\$}_i^3 + {}_3\underline{\rho}_4^i\ {}^3\underline{\$}_i^4 + {}_4\underline{\rho}_5^i\ {}^4\underline{\$}_i^5 + {}_5\underline{\rho}_6^i\ {}^5\underline{\$}_i^6 + \underline{\mathcal{J}}_i = \boldsymbol{J}_C \tag{11.36}$$

and

$$\dddot{q}_i\ {}^0\$_i^1 + {}_1\rho_2^i\ {}^1\$_i^2 + {}_2\rho_3^i\ {}^2\$_i^3 + {}_3\rho_4^i\ {}^3\$_i^4 + {}_4\rho_5^i\ {}^4\$_i^5 + {}_5\rho_6^i\ {}^5\$_i^6 + \mathcal{J}_i = \boldsymbol{J}_C, \tag{11.37}$$

where $\underline{\mathcal{J}}_i$ and $\mathcal{J}_i$ are the Lie screws of jerk of the limbs of the manipulator. For brevity, only the last is provided here.

$$\begin{aligned}
\mathcal{J}_i = 2\Big[& \dot{q}_i{}^0\$_i^1 \quad {}_1\alpha_2^{i\,1}\$_i^2 + \ldots + {}_4\alpha_5^{i\,4}\$_i^5 + {}_5\alpha_6^{i\,5}\$_i^6 \\
& + \left[{}_1\omega_2^{i\,1}\$_i^2 \quad {}_2\omega_3^{i\,2}\$_i^3 + \ldots + {}_4\omega_5^{i\,4}\$_i^5 + {}_5\omega_6^{i\,5}\$_i^6\right] \\
& + \left[{}_2\omega_3^{i\,2}\$_i^3 \quad {}_3\omega_4^{i\,3}\$_i^4 + \ldots + {}_4\omega_5^{i\,4}\$_i^5 + {}_5\omega_6^{i\,5}\$_i^6\right] \\
& + \ldots + \left[{}_4\omega_5^{i\,4}\$_i^5 \quad {}_5\omega_6^{i\,5}\$_i^6\right]\Big] \\
& + 2\Big[{}_1\omega_2^{i\,1}\$_i^2 \quad {}_2\alpha_3{}^2\$_i^3 + \ldots + {}_4\alpha_5^{i\,4}\$_i^5 + {}_5\alpha_6^{i\,5}\$_i^6
\end{aligned}$$

$$
\begin{aligned}
&+\left[{}_{2}\omega_{3}^{i\,2}\$_{i}^{3} \quad {}_{3}\omega_{4}^{i\,3}\$_{i}^{4}+\ldots+{}_{4}\omega_{5}^{i\,4}\$_{i}^{5}+{}_{5}\omega_{6}^{i\,5}\$_{i}^{6}\right]\\
&+\left[{}_{3}\omega_{4}^{i\,3}\$_{i}^{4} \quad {}_{4}\omega_{5}^{i\,4}\$_{i}^{5}+\ldots+{}_{4}\omega_{5}^{i\,4}\$_{i}^{5}+{}_{5}\omega_{6}^{i\,5}\$_{i}^{6}\right]\\
&\qquad\qquad +\ldots+\left[{}_{4}\omega_{5}^{i\,4}\$_{i}^{5} \quad {}_{5}\omega_{6}^{i\,5}\$_{i}^{6}\right]\Bigg]\\
&\qquad\qquad +\ldots+2\left[{}_{4}\omega_{5}^{i\,4}\$_{i}^{5} \quad {}_{5}\alpha_{6}^{i\,5}\$_{i}^{6}\right]\\
&+\left[{}_{0}\alpha_{1}{}^{0}\$_{i}^{1} \quad {}_{1}\omega_{2}^{i\,1}\$_{i}^{2}+\ldots+{}_{4}\omega_{5}^{i\,4}\$_{i}^{5}+{}_{5}\omega_{6}^{i\,5}\$_{i}^{6}\right]\\
&+\left[{}_{1}\alpha_{2}^{i\,1}\$_{i}^{2} \quad {}_{2}\omega_{3}^{i\,2}\$_{i}^{3}+\ldots+{}_{4}\omega_{5}^{i\,4}\$_{i}^{5}+{}_{5}\omega_{6}^{i\,5}\$_{i}^{6}\right]\\
&\qquad\qquad +\ldots+\left[{}_{4}\alpha_{5}^{i\,4}\$_{i}^{5} \quad {}_{5}\omega_{6}^{i\,5}\$_{i}^{6}\right]\\
&+\left[\dot{q}_{i}{}^{0}\$_{i}^{1}\left[\dot{q}_{i}{}^{0}\$_{i}^{1} \quad {}_{1}\omega_{2}^{i\,1}\$_{i}^{2}+\ldots+{}_{4}\omega_{5}^{i\,4}\$_{i}^{5}+{}_{5}\omega_{6}^{i\,5}\$_{i}^{6}\right]\right]\\
&\quad +\left[{}_{1}\omega_{2}^{i\,1}\$_{i}^{2}\left[{}_{1}\omega_{2}^{i\,1}\$_{i}^{2} \quad {}_{2}\omega_{3}^{i\,2}\$_{i}^{3}+\ldots+{}_{5}\omega_{6}^{i\,5}\$_{i}^{6}\right]\right]\\
&\qquad +\ldots+\left[{}_{4}\omega_{5}^{i\,4}\$_{i}^{5}\left[{}_{4}\omega_{5}^{i\,4}\$_{i}^{5} \quad {}_{5}\omega_{6}^{i\,5}\$_{i}^{6}\right]\right]
\end{aligned}
\tag{11.38}
$$

Following the trend of the velocity analysis, the input–output equation of jerk of the robot results in

$$
\mathbf{G}\boldsymbol{J}_{C}=\mathbf{H}\dddot{\mathbf{q}}+\mathbb{J}, \tag{11.39}
$$

where

$$
\dddot{\mathbf{q}}=\left[\dddot{\underline{q}}_{1}\ \dddot{\underline{q}}_{2}\ \dddot{\underline{q}}_{3}\ \dddot{q}_{1}\ \dddot{q}_{2}\ \dddot{q}_{3}\right]^{T} \tag{11.40}
$$

is the third-order driver matrix of the parallel manipulator. Meanwhile,

$$
\mathbb{J}=\begin{bmatrix}
\{\underline{\mathcal{J}}_{1};{}^{5}\underline{\$}_{1}^{6}\}\\
\{\underline{\mathcal{J}}_{2};{}^{5}\underline{\$}_{2}^{6}\}\\
\{\underline{\mathcal{J}}_{3};{}^{5}\underline{\$}_{3}^{6}\}\\
\{\mathcal{J}_{1};{}^{4}\$_{1}^{5}\}\\
\{\mathcal{J}_{2};{}^{4}\$_{2}^{5}\}\\
\{\mathcal{J}_{3};{}^{4}\$_{3}^{5}\}
\end{bmatrix} \tag{11.41}
$$

is the complementary matrix of jerk.

The inverse jerk analysis consists of computing matrix $\dddot{\mathbf{q}}$ given the jerkor $\boldsymbol{J}_{C}$, while the forward jerk analysis consists of determining the six-dimensional vector $\boldsymbol{J}_{C}$ given the generalized joint-jerk rates contained in matrix $\dddot{\mathbf{q}}$.

11.4.4 Hyper-Jerk Analysis of the 3RRRS+3RRPS Parallel Manipulator

Let $\boldsymbol{\sigma}$ be the angular hyper-jerk of the moving platform as measured from the fixed platform and let $\boldsymbol{h}_C$ be the linear hyper-jerk of point C. Then the reduced hyper-jerk state, or hyper-jerkor, of the moving platform as observed from the fixed platform is given by

$$\boldsymbol{H}_C = \begin{bmatrix} \boldsymbol{\sigma} \\ \boldsymbol{h}_C - 3^j\boldsymbol{\rho}^m \times {}^j\boldsymbol{v}_O^m - 3^j\boldsymbol{\alpha}^m \times {}^j\boldsymbol{a}_O^m - {}^j\boldsymbol{\omega}^m \times {}^j\boldsymbol{j}_O^m \end{bmatrix}.$$

Furthermore, the six-dimensional vector $\boldsymbol{H}_C$ may be written in screw form through the limbs of the parallel manipulator as follows:

$${}_0\underline{\sigma}_1^i\ {}^0\underline{\$}_i^1 + {}_1\underline{\sigma}_2^i\ {}^1\underline{\$}_i^2 + \dddot{\underline{q}}_i\ {}^2\underline{\$}_i^3 + {}_3\underline{\sigma}_4^i\ {}^3\underline{\$}_i^4 + {}_4\underline{\sigma}_5^i\ {}^4\underline{\$}_i^5 + {}_5\underline{\sigma}_6^i\ {}^5\underline{\$}_i^6 + \underline{\mathcal{H}}_i = \boldsymbol{H}_C \tag{11.42}$$

and

$$\dddot{q}_i\ {}^0\$_i^1 + {}_1\sigma_2^i\ {}^1\$_i^2 + {}_2\sigma_3^i\ {}^2\$_i^3 + {}_3\sigma_4^i\ {}^3\$_i^4 + {}_4\sigma_5^i\ {}^4\$_i^5 + {}_5\sigma_6^i\ {}^5\$_i^6 + \mathcal{H}_i = \boldsymbol{H}_C, \tag{11.43}$$

where $\underline{\mathcal{H}}_i$ and $\mathcal{H}_i$ are the Lie screws of hyper-jerk of the limbs of the manipulator.

Following the trend of the velocity analysis, the input–output equation of hyper-jerk of the parallel manipulator results in

$$\mathbf{G}\boldsymbol{H}_C = \mathbf{H}\ddddot{\mathbf{q}} + \mathbb{H}, \tag{11.44}$$

where

$$\ddddot{\mathbf{q}} = \begin{bmatrix} \ddddot{\underline{q}}_1 & \ddddot{\underline{q}}_2 & \ddddot{\underline{q}}_3 & \ddddot{q}_1 & \ddddot{q}_2 & \ddddot{q}_3 \end{bmatrix}^T \tag{11.45}$$

is the fourth-order driver matrix of the parallel manipulator. Meanwhile,

$$\mathbb{H} = \begin{bmatrix} \{\underline{\mathcal{H}}_1;\ {}^5\underline{\$}_1^6\} \\ \{\underline{\mathcal{H}}_2;\ {}^5\underline{\$}_2^6\} \\ \{\underline{\mathcal{H}}_3;\ {}^5\underline{\$}_3^6\} \\ \{\mathcal{H}_1;\ {}^4\$_1^5\} \\ \{\mathcal{H}_2;\ {}^4\$_2^5\} \\ \{\mathcal{H}_3;\ {}^4\$_3^5\} \end{bmatrix} \tag{11.46}$$

is the complementary matrix of hyper-jerk.

The inverse hyper-jerk analysis consists of finding the joint-hyper-jerk rates of the robot contained in matrix $\ddddot{\mathbf{q}}$ given the reduced hyper-jerk state $\boldsymbol{H}_C$, while the forward hyper-jerk analysis consists of computing the six-dimensional vector $\boldsymbol{H}_C$ for a given set of generalized joint-hyper-jerk rates.

Example 11.1. Applying screw theory in the hyper-jerk analysis of the parallel manipulator at hand appears to be a very involved procedure. In reality, the corresponding mathematical formulas may be easily translated into computer codes by symbolic software such as Maple©. In order to illustrate the application of the formulas obtained to solve the kinematics of the robot at hand, the example finishes with numerical applications related mainly to the forward kinematics of the six-legged parallel manipulator. The obtained results are presented in simple plots. Using SI units through the numerical example, the arm and forearm lengths of the limbs containing actuated revolute joints are chosen as $a = 0.4472$ and $b = 0.5656$, respectively, while the radius of the moving platform is given by $r = 0.4$. The nominal positions of passive revolute joints are given by the coordinates

$$\underline{A}_1 = (0.3, 0, -0.5196), \underline{A}_2 = (-0.6, 0, 0.0), \underline{A}_3 = (0.3, 0, 0.5196),$$
$$A_1 = (0.6, 0, 0), A_2 = (-0.3, 0, -0.5196), A_3 = (-0.3, 0, 0.5196).$$

On the other hand, in the reference configuration of the parallel manipulator consider that the coordinates of the center C of the moving platform as well as those of the points B_i, $\underline{B}_i$, and D_i are chosen as

$$C = (0, 0.8, 0),$$
$$B_1 = (0.4, 0.8, 0), B_2 = (-0.2, 0.8, -0.3464), B_3 = (-0.2, 0.8, 0.3464),$$
$$\underline{B}_1 = (0.2, 0.8, -0.3464), \underline{B}_2 = (-0.4, 0.8, 0), \underline{B}_3 = (0.2, 0.8, 0.3464),$$
$$D_1 = (0.2, 0.4, 0), D_2 = (-0.1, 0.4, -0.1732), D_3 = (-0.1, 0.4, 0.1732).$$

Furthermore, the generalized coordinates are commanded to follow periodic functions given by

$$\underline{q}_1 = 0.8246 - 0.35\sin(t)\cos(t),$$
$$\underline{q}_2 = 0.8246 + 0.1\sin(t),$$
$$\underline{q}_3 = 0.8246 + 0.1\sin(t),$$
$$q_1 = 0.5\sin(t),$$
$$q_2 = \pi/6 + 0.25\sin(t),$$
$$q_3 = \pi/3 + 0.75\sin(t),$$

where the time t is given in the interval $0 < t < 2\pi$.

Solution. The method reported here to approach the kinematic analyses of the robot was translated into a Maple16©sheet. The temporal behavior of the forward kinematics of the center C of the moving platform, up to the hyper-jerk analysis, using screw theory is shown in Figs. 11.4, 11.5, 11.6, and 11.7. For clarity the results are presented separately.

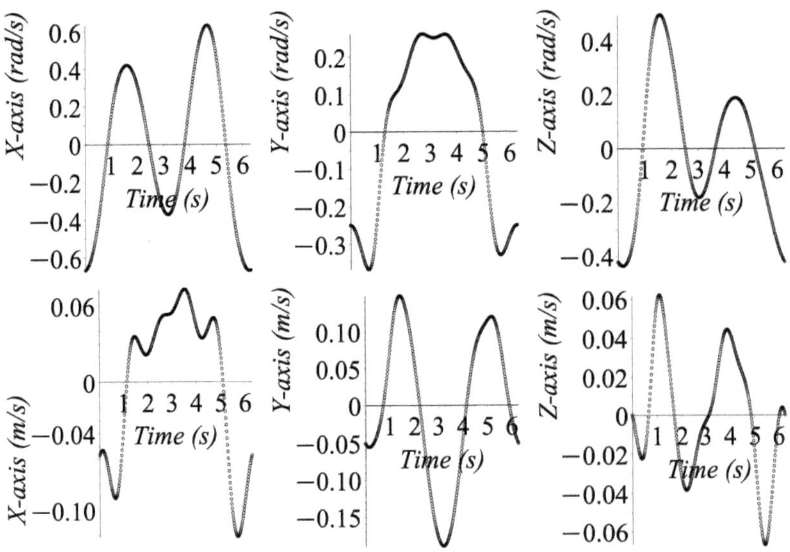

Fig. 11.4 Example 11.1. Time history of the velocity of the center of the moving platform of the 3RRRS+3RRPS parallel manipulator

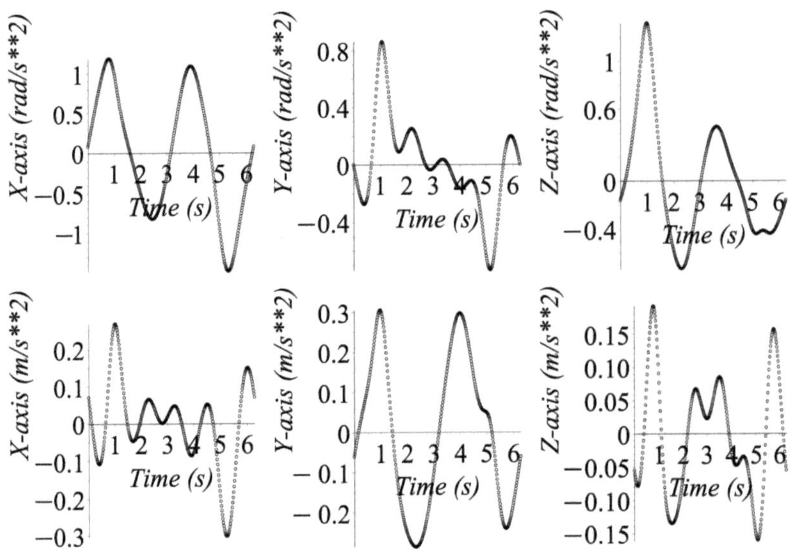

Fig. 11.5 Example 11.1. Time history of the acceleration of the center of the moving platform

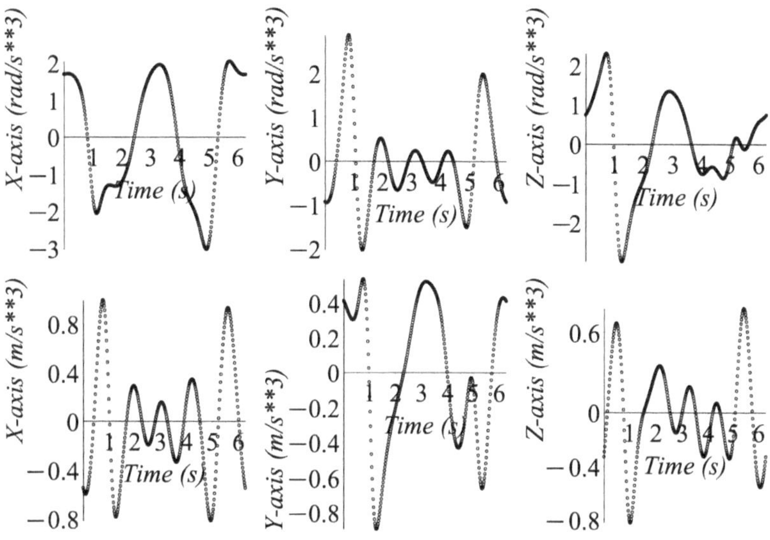

Fig. 11.6 Example 11.1. Time history of the jerk of the center of the moving platform

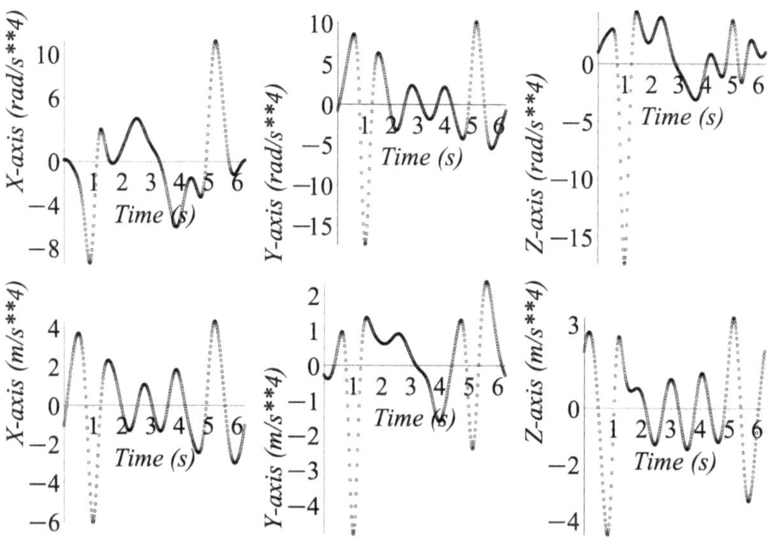

Fig. 11.7 Example 11.1. Time history of the hyper-jerk of the center of the moving platform

11.5 Exercises

1. Determine two singular configurations of the 3RRRS+3RRPS parallel manipulator associated to the forward velocity analysis of the robot.
2. Propose one alternative arrangement of the spherical joints mounted on the moving platform of the 3RRRS+3RRPS parallel manipulator.

References

Ataa, A. A. (2010). Inverse dynamic analysis and trajectory planning for flexible manipulator. *Inverse Problems in Engineering, 18*(4), 549–566.

Bahara, L. Y. (1996). An inverse problem in nonholonomic dynamics. *Inverse Problems in Engineering, 3*(4), 219–232.

Gallardo-Alvarado, J. (2014). A simple method to solve the forward displacement analysis of the general six-legged parallel manipulator. *Robotics and Computer-Integrated Manufacturing, 30*(1), 55–61.

Merlet, J.-P. (2004). Solving the forward kinematics of a Gough-type parallel manipulator with interval analysis. *International Journal of Robotics Research, 23*(3), 221–235.

Part V
Emblematic Parallel Manipulators

Chapter 12
Gough's Tyre Testing Machine

12.1 Introduction

Possibly the most celebrated application of a six-legged parallel manipulator is due to Gough, who in 1947 built an octahedral hexapod (see Fig. 12.1) to determine the properties of tyre under the action of combined loads (Gough 1957; Gough and Whitehall 1962). However, the idea of two rigid bodies connected to each other by extendable limbs dates back to the nineteenth century (Merlet 2008).

"Hexapod," from the Greek *hexapoda* means "six-footed" and may be understood as a mechanical vehicle with six legs capable of walking or as an in-parallel manipulator used in robotics applications. The latter is the concept discussed in this chapter. The first mathematical model of a hexapod system was proposed in the 1800s by Augustin-Louis Cauchy, who wrote an article about the possible motion and stiffness of an *articulated octahedron*, an early design of the modern-day six-legged parallel manipulator. Hence, the six-legged parallel manipulator could be correctly named the Cauchy platform. Two decades after the introduction of Gough's tyre testing machine, a new idea of flight simulator was proposed by Stewart in his celebrated theoretical paper entitled "A platform with six degrees of freedom" (Stewart 1965–1966, 1966). At the time, the forward displacement analyses of both manipulators were overlooked by scientists and engineers, causing the vague assumption that Stewart's platform was practically identical to Gough's machine. Curiously, mechanisms based on the six-legged parallel manipulator topology are commonly known today as Gough–Stewart platforms although Gough's tyre testing machine and Stewart's flight simulator are mechanisms with quite different kinematic characteristics; see, for instance, Fichter et al. (2009) and Bonev (2011). Hundreds of published papers describe Stewart's platform as Gough's hexapod. Along those lines, it is possible to affirm that while Gough's robot is a parallel manipulator with highly coupled kinematics, Stewart's flight simulator may be classified as a partially decoupled robot and therefore has the benefit of simpler kinematics (Gallardo-Alvarado et al. 2010). Furthermore, while there are

J. Gallardo-Alvarado, *Kinematic Analysis of Parallel Manipulators by Algebraic Screw Theory*, DOI 10.1007/978-3-319-31126-5_12

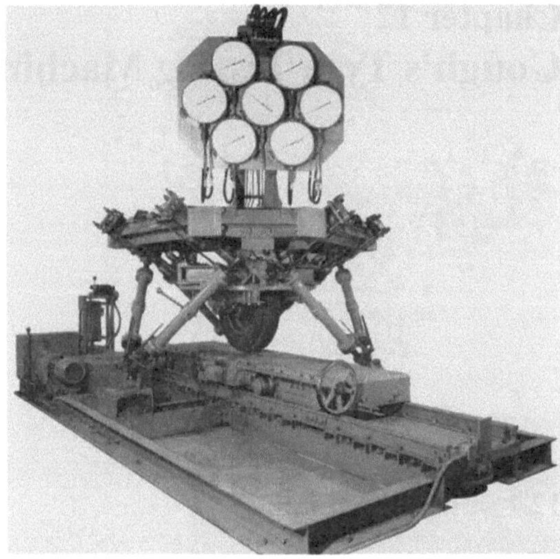

Fig. 12.1 The original Gough platform, perhaps the first octahedral hexapod

a lot of robots with topologies based on Gough's machine working around the world, to this author's best knowledge, there are no robots based on the topology of Stewart's original platform.

One of the most interesting topics dealing with the finite kinematics of the Gough platform is its forward displacement analysis. It has already been shown that the number of reachable configurations of the Gough platform is at most 40 (Lazard 1993; Mourrain 1993; Faugère and Lazard 1995; Wampler 1995) and different algorithms have been developed to compute these postures (Raghavan 1993; Husty 1996; Innocenti and Wenger 2004; Rolland 2005; Dai et al. 2006). Taking into account that at the design stage of any manipulator the forward displacement analysis plays a central role in optimizing the performance of the robot, considerable effort has been devoted to improving such algorithms; for instance, the elimination of spurious solutions as well as the reduction of computation time have been remarkable advances in the field.

This chapter reports on the finite and infinitesimal kinematics of the original Gough platform. In that regard it is worth emphasizing that the forward displacement analysis of the general six-legged in-parallel manipulator has been one of the most attractive and challenging problems of modern kinematics. Even today finding a closed-form solution to this issue seems to be an unrealistic task. In attempts to solve the forward displacement analysis, kinematicians have recurrently used combinations of numerical methods and instrumental resources like sensors to find the poses that the moving platform can reach as measured from the fixed platform for a given set of generalized coordinates of the hexapod. Usually, the nonlinear equations associated to the forward displacement analysis of the hexapod are obtained handling the rotation matrix of a rigid body, a translation vector between two reference frames, and the kinematic constraint conditions imposed on the limbs of the parallel manipulator. The most interesting topic treated in this chapter is the application of a comprehensive method to find all the reachable poses of the moving platform based on simple linear combinations of coordinates of three points embedded in it. The reader would be somewhat surprised by the simplicity with which a complicated problem such as the complete solution of the forward displacement analysis of the hexapod may be formulated based on the linear dependence of vectors describing the noncollinear centers of three spherical joints mounted on the moving platform.

12.2 Description of the General Parallel Manipulator

The Gough platform is a type of fully in-parallel manipulator incorporating six prismatic actuators, usually hydraulic jacks. For simplicity, the Gough platform is usually modeled kinematically as a 6-SPS (where S and P represent spherical and prismatic joints, respectively) fully in-parallel manipulator where redundant and uncontrollable rotations exist along the limbs.

With reference to Fig. 12.2, a 6-6 fully in-parallel manipulator of the 6-SPS type has six limbs, each one having two links connected to each other by a prismatic joint, and two spherical joints connecting the extendable limb to the moving and fixed platforms. The prismatic joints are selected as the actable-joint variables, notated as the generalized coordinates q_i, of the mechanism. The centers of the spherical joints mounted on the moving platform are denoted by points $B_i (i = 1, 2, \ldots, 6)$

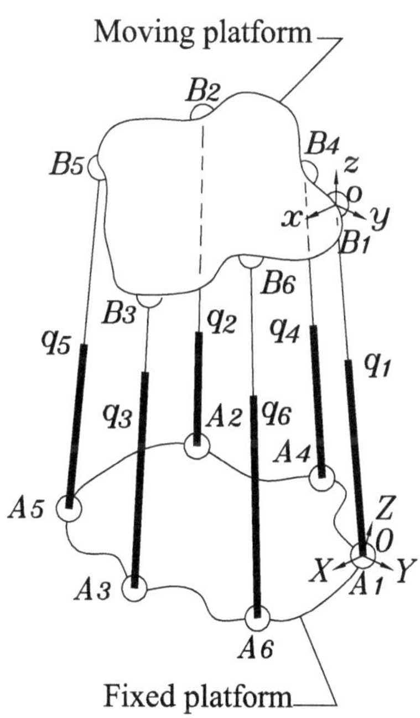

Fig. 12.2 Geometric scheme of the general six-legged in-parallel platform

located, respectively, by vectors $\boldsymbol{b}_i$. Unless otherwise specified, in the remainder of this chapter let us consider $i = 1, 2, \ldots, 6$. Meanwhile, A_i denotes the center of the ith spherical joint mounted on the fixed platform that is located by vector $\boldsymbol{a}_i$. In the general case, the parallel manipulator is equipped with three-dimensional platforms.

12.3 Finite Kinematics of the General Hexapod

In this section the displacement analyses of the hexapod are reported. For the sake of completeness, the section also includes a brief explanation of the classical method most kinematicians use to formulate the closure equations associated to the forward displacement analysis of the parallel manipulator. In that regard, the formidable effort devoted to this elusive topic by the kinematician community is noticeable. It is hoped that in the near future publications approaching the subject would give increasingly precise and efficient methods. However, the introduction of a theoretical closed-form solution seems to be an unrealistic task, and the development of efficient and precise sensors could be the most viable option to solve this problem. By the way, assuming that only one solution is required, for instance, determining the actual configuration of the parallel manipulator and then applying numerical techniques like the Newton–Raphson method, undoubtedly acquires a relevant role.

12.3.1 Forward Displacement Analysis of the Hexapod

12.3.1.1 Typical Formulation of the Forward Displacement Analysis of the Gough Platform

The forward displacement analysis of the six-legged in-parallel manipulator is one of the most challenging problems of modern kinematics. This analysis consists of finding all the reachable poses of the moving platform as observed from the fixed platform for a given set of generalized coordinates q_i. To do this, consider that *O_XYZ* is a reference frame attached to the fixed platform whose origin O is located at point A_1. Furthermore, consider that *o_xyz* is a reference frame attached to the moving platform whose origin o is coincident with the position of point B_1. Thus, it immediately emerges that the limb labeled 1 is constrained to

$$\boldsymbol{q}_1 \bullet \boldsymbol{q}_1 = q_1^2, \tag{12.1}$$

where $\boldsymbol{q}_1$ is the position vector of o with respect to O. Furthermore, the remaining limb lengths of the parallel manipulator may be determined through the equations

$$(\boldsymbol{q}_1 + \mathbf{R}\boldsymbol{b}_i^* - \boldsymbol{a}_i) \bullet (\boldsymbol{q}_1 + \mathbf{R}\boldsymbol{b}_i^* - \boldsymbol{a}_i) = q_i^2, \quad i = 2, 3, \ldots, 6, \tag{12.2}$$

where vector $\boldsymbol{b}_i^*$ is obtained according to the coordinates of the ith spherical joint expressed in the moving reference frame o_xyz, whereas $\mathbf{R}$ is the rotation matrix between the moving and fixed reference frames. On the other hand, conveniently it is possible to consider that a general representation of matrix $\mathbf{R}$ can be notated as

$$\mathbf{R} = \begin{bmatrix} r_1 & r_4 & r_7 \\ r_2 & r_5 & r_8 \\ r_3 & r_6 & r_9 \end{bmatrix}. \tag{12.3}$$

Afterward, taking into account that $\mathbf{R}$ is an orthogonal matrix, the following properties hold:

$$\left.\begin{aligned} & r_1^2 + r_2^2 + r_3^2 = 1 \\ & r_4^2 + r_5^2 + r_6^2 = 1 \\ & r_1 r_4 + r_2 r_5 + r_3 r_6 = 0 \\ & r_4 r_8 - r_5 r_7 - r_3 = 0 \\ & r_2 r_7 - r_1 r_8 - r_6 = 0 \\ & r_1 r_5 - r_2 r_4 - r_9 = 0 \end{aligned}\right\}. \tag{12.4}$$

On the other hand, in order to simplify the forward displacement analysis, according to Cayley's formula (see Bottema and Roth 1979), the rotation matrix $\mathbf{R}$ may be rewritten as follows:

$$\mathbf{R} = (\mathbf{I} - \mathbf{CH})^{-1}(\mathbf{I} + \mathbf{CH}), \tag{12.5}$$

where $\mathbf{I}$ is the identity matrix of order 3 and $\mathbf{CH}$ is an arbitrary 3×3 skew-symmetric matrix with three independent parameters whose most common representation is given by

$$\mathbf{CH} = \begin{bmatrix} 0 & -ch_3 & ch_2 \\ ch_3 & 0 & -ch_1 \\ -ch_2 & ch_1 & 0 \end{bmatrix}. \tag{12.6}$$

Gathering these results, we obtain a system composed of second-order quadratic equations. Finally, after a lengthy mathematical procedure the higher nonlinear system of equations is systematically reduced into a univariate polynomial equation; for details the reader is referred to Raghavan (1993), Husty (1996), Innocenti and Wenger (2004), Rolland (2005), and Dai et al. (2006). Although the aforementioned procedure seems to be easy to translate into computer codes, in reality it is a very involved mathematical method. The search for simpler alternatives, at least to derive the closure equations of the general hexapod, is an advisable task in order to simplify the forward displacement analysis. In what follows, the closure equations associated to the forward displacement analysis of the six-legged in-parallel manipulator are easily established based on the unknown coordinates of three points B_i of the moving platform expressed in the fixed reference frame. The method does not require the application of the rotation matrix between the moving and fixed platforms, and of course it is unnecessary to consider a reference frame attached to the moving platform, which are mandatory conditions in most conventional methods. For clarity the method is subdivided into two main types: (1) parallel manipulator with planar platforms; (2) parallel manipulator with three-dimensional platforms.

12.3.1.2 Forward Displacement Analysis of the Six-Legged Parallel Manipulator with Planar Platforms

With reference to Fig. 12.3, taking into account that the pose of any rigid body may be specified knowing the coordinates of three points attached to it, then the pose of the moving platform with respect to the fixed platform may be determined by computing the unknowns $w_i(i = 1, 2, \ldots, 9)$, which are the coordinates of three points of the moving platform expressed in the fixed reference frame *XYZ* denoted as $O = (w_1, w_2, w_3)$, $P = (w_4, w_5, w_6)$, and $R = (w_7, w_8, w_9)$. Points *O*, *P*, and *R* are named *the control points of the parallel manipulator*.

With these considerations in mind, the forward displacement analysis of the parallel manipulator consists of finding the pose of the moving platform with respect to the fixed platform upon the geometry of the moving platform and the generalized coordinates $q_i(i = 1, 2, \ldots, 6)$ of the spatial mechanism. To this end, in order to formulate the closure equations required to solve the forward displacement analysis, the coordinates of any point *Q* of the moving platform are expressed as a linear combination of points *O*, *P*, and *R*; this is the fundamental idea of the proposed method.

According to Fig. 12.3, after a few computations, the vector $\boldsymbol{q}$, which denotes the position of point *Q* with respect to the origin of the reference frame *XYZ*, may be obtained in terms of the geometry of the hexagon, for example, according to the lengths of its sides as

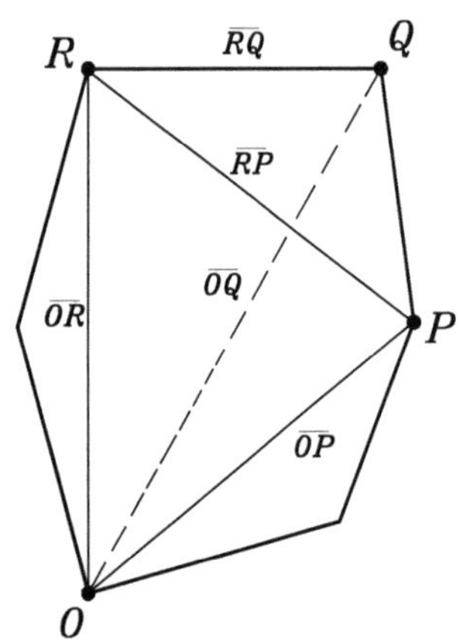

Fig. 12.3 Moving platform with three control points

$$q = o + \frac{\overline{OQ}(K_1\sqrt{1-K_2^2} + K_2\sqrt{1-K_1^2})}{\overline{OR}\sqrt{1-K_2^2}} t, \tag{12.7}$$

where vector t is given by

$$t = \frac{\overline{OR}\sqrt{1-K_1^2}}{\overline{RP}(K_1\sqrt{1-K_2^2} + K_2\sqrt{1-K_1^2})}(p - r) - o. \tag{12.8}$$

Meanwhile, the coefficients K_1 and K_2 are computed upon the geometry of the moving platform as follows:

$$K_1 = \frac{\overline{OR}^2 + \overline{OQ}^2 - \overline{RQ}^2}{2\overline{OR}\,\overline{OQ}} \tag{12.9}$$

and

$$K_2 = \frac{\overline{OR}^2 + \overline{RP}^2 - \overline{OP}^2}{2\overline{OR}\,\overline{RP}}. \tag{12.10}$$

Once the vertices of the moving platform are expressed in terms of the unknowns $w_i (i = 1, 2, \ldots, 9)$, six closure equations may be easily written as

$$(b_i - a_i) \bullet (b_i - a_i) = q_i^2, \quad i = 1, 2, \ldots, 6. \tag{12.11}$$

Furthermore, to complete the required closure equations, three compatibility equations are given by

$$\left. \begin{aligned} (o - r) \bullet (o - r) &= \overline{OR}^2 \\ (o - p) \bullet (o - p) &= \overline{OP}^2 \\ (r - p) \bullet (r - p) &= \overline{RP}^2 \end{aligned} \right\}. \tag{12.12}$$

Expressions (12.11) and (12.12) form a nonlinear system of nine equations in the unknowns $w_i (i = 1, 2, \ldots, 9)$. These equations may be easily solved by using special software like Maple©.

Example 12.1. The numerical exercise consists of solving the forward displacement analysis of a six-legged parallel manipulator provided with planar platforms. This case study was reported by Huang et al. (2007). For clarity, Fig. 12.4 shows the notation used by Huang et al. (2007). Furthermore, the corresponding input data of the numerical example are listed in Table 12.1.

Solution. According to Table 12.1, some relevant dimensions of the planar moving platform are computed as

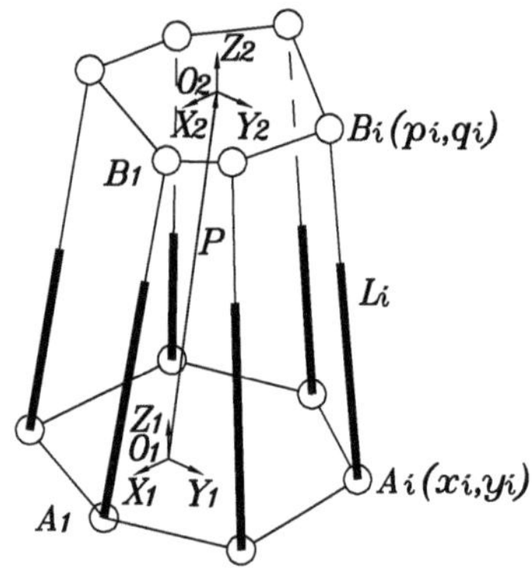

Fig. 12.4 Example 12.1. Hexapod with planar platforms

Table 12.1 Example 12.1. Input data

i	x_i	y_i	p_i	q_i	L_i
1	9	3	3	1	$\sqrt{\frac{2785}{13}}$
2	6	8	2	3	$\sqrt{\frac{11608}{65}}$
3	0	14	1	5	$\sqrt{\frac{14049}{65}}$
4	−8	13	−3	4	$\sqrt{237}$
5	−7	−6	−2	2	$\sqrt{462}$
6	−3	−5	−1	−4	$\sqrt{\frac{25848}{65}}$

$$\overline{B_1B_2} = \sqrt{5}, \quad \overline{B_2B_3} = \sqrt{5}, \quad \overline{B_3B_4} = \sqrt{17},$$
$$\overline{B_4B_5} = \sqrt{5}, \quad \overline{B_5B_6} = \sqrt{37}, \quad \overline{B_6B_1} = \sqrt{41}.$$

To obtain the closure equations of the manipulator, let us consider $B_1 = (w_1, w_2, w_3)$, $B_3 = (w_4, w_5, w_6)$, and $B_5 = (w_7, w_8, w_9)$ to be the control points of the parallel manipulator. Next, the coordinates of the remaining points B_i must be obtained as linear combinations of the coordinates of the control points. To this aim, as a preliminary step, the coordinates of point O_2, which are expressed in the reference frame $X_1Y_1Z_1$ attached to the fixed platform, are computed according to Eq. (12.7). Hence,

$$O_2 = (6w_1 - 4w_4 + 7w_7, 6w_2 - 4w_5 + 7w_8, 6w_3 - 4w_6 + 7w_9)/9. \qquad (12.13)$$

Furthermore, applying the proposed method leads to the following closure equations:

$$\left.\begin{aligned}
&\rightarrow w_1^2 + w_2^2 + w_3^2 - 18w_1 - 6w_2 = 1615/13\\
&\rightarrow w_1^2 + w_2^2 + w_3^2 + w_4^2 + w_5^2 + w_6^2 - 2(w_1w_4 + w_2w_5 + w_3w_6) = 20\\
&\rightarrow w_1^2 + w_2^2 + w_3^2 + w_7^2 + w_8^2 + w_9^2 - 2(w_1w_7 + w_2w_8 + w_3w_9) = 26\\
&\rightarrow w_4^2 + w_5^2 + w_6^2 + w_7^2 + w_8^2 + w_9^2 - 2(w_4w_7 + w_5w_8 + w_6w_9) = 18\\
&\rightarrow w_4^2 + w_5^2 + w_6^2 - 28w_5 = 1309/65\\
&\rightarrow w_1^2 + w_2^2 + w_3^2 + w_4^2 + w_5^2 + w_6^2 + 2w_1w_4 + 2w_2w_5 + 2w_3w_6-\\
&24w_1 - 32w_2 - 24w_4 - 32w_5 = 20432/65\\
&\rightarrow w_1^2 + w_2^2 + w_3^2 + w_4^2 + w_5^2 + 2w_6^2 + 4w_7^2 + 4w_8^2 + 4w_9^2 + 4w_6w_9-\\
&2w_3w_6 - 4w_1w_7 + 4w_5w_8 - 2w_1w_4 - 2w_2w_5 - 4w_2w_8 - 4w_3w_9+\\
&4w_4w_7 - 32w_1 + 52w_2 + 32w_4 - 52w_5 + 64w_7 - 104w_8 = 16\\
&\rightarrow w_7^2 + w_8^2 + w_9^2 + 14w_7 + 12w_8 = 377\\
&\rightarrow 441(w_1^2 + w_2^2 + w_3^2) + 841(w_4^2 + w_5^2 + w_6^2)+\\
&676(w_7^2 + w_8^2 + w_9^2) - 1508(w_4w_7 + w_6w_9 + w_5w_8)-\\
&12184(w_2w_5 + w_1w_4 + w_3w_6) + 1092(w_1w_7 + w_3w_9 + w_2w_8)+\\
&2268w_1 + 3780w_2 - 3132w_4 - 5220w_5 + 2808w_7 + 4680w_8\\
&= 7658712/65
\end{aligned}\right\}. \tag{12.14}$$

There are several efficient methods reported in the literature to solve equations like those above; see, for instance, Tsai (2007). The improvement of such methods is not the intention of this book. Therefore, to find all the isolated solutions for Eqs. (12.14), the *Algorithm 795: PHCpack*, a mathematical tool developed by Verschelde (1999) as a general-purpose solver for polynomial systems that is based on the homotopy continuation method, was used.[1] The homotopy analysis method resorts to using the concept of the homotopy from topology to generate a convergent series solution for nonlinear systems. Usually this is enabled by utilizing a homotopy–McLaurin series to deal with the nonlinearities in the system. The efficacy of such a procedure is really impressive.

Finally, for brevity, only the resulting 40 feasible positions concerned with point O_2 are provided in Table 12.2. It is worth noting that the numerical results listed in Table 12.2 are in excellent agreement with the results published by Huang et al. (2007) even though the corresponding strategies to establish the closure equations are so different.

[1] The Algorithm 795: PHCpack can be freely used by anyone. Personal communication of Prof. Jan Verschelde with Prof. Jaime Gallardo-Alvarado, who acknowledges with thanks such a generous voluntary decision.

Table 12.2 Example 12.1. The 40 positions of point O_2

Sol.	X	Y	Z
1	−16.1829	28.9635	0.5362 I
2	−4.3926 − 0.08693 I	10.7564 − 0.5922 I	− 9.8320 + 0.3967 I
3	20.6781	13.6644	− 26.2914 I
4	−23.1687 + 56.8842 I	15.3951 + 19.0137 I	65.5912 + 14.4442 I
5	−58.8081	16.1188	0.6558 − 49.0382 I
6	9.9573 − 6.8117 I	2.9276 − 3.0687 I	−11.5819 − 10.3307 I
7	−23.1687 − 56.8842 I	15.3951 − 19.0137 I	−65.5912 + 14.4442 I
8	−16.1829	28.9635	0.4699 − 0.5362 I
9	−23.1687 + 56.8842 I	15.3951 + 19.0137 I	−65.5912 − 14.4442 I
10	−58.8081	16.1188	49.0382 I
11	20.6781 I	13.6644	26.2914 I
12	−4.3926 − 0.0869 I	10.7564 − 0.5922 I	9.8320 − 0.3967 I
13	9.9573 − 6.8117 I	2.9276 − 3.0687 I	11.5819 + 10.3307 I
14	−23.1687 − 56.8842 I	15.3951 − 19.0137 I	65.5912 − 14.4442 I
15	−24.7645 I	31.4612	37.6348 I
16	−8.0729	32.9135	30.7163 I
17	2.2138 − 3.9850 I	5.4239 − 0.4501 I	−15.9058 + 0.1680 I
18	−2.1866	10.7206	−9.2147
19	9.9573 + 6.8117 I	2.9276 + 3.06874 I	11.5819 − 10.3307 I
20	−4.3926 + 0.0869 I	10.7564 + 0.5922 I	−9.8320 − 0.3967 I
21	−3.9531	4.6728	5.8287 I
22	15.9632	−3.6670	10.3434 I
23	−8.0729	32.9135	−30.7163 I
24	−24.7645	31.4612	−37.6348 I
25	−3.9531	4.6728	−5.8287 I
26	−4.3926 + 0.0869 I	10.7564 + 0.5922 I	9.8320 + 0.3967 I
27	2.2138 − 3.9850 I	5.4239 − 0.4501 I	15.9058 − 0.1680 I
28	9.9573 + 6.8117 I	2.9276 + 3.0687 I	− 11.5819 + 10.3307 I
29	−2.1866	10.7206	9.2147
30	15.9632	−3.6670	−10.3434 I
31	17.5673	−4.8684	−5.5030 I
32	10.8048 + 0.9346 I	7.0723 −1.6629 I	−10.3332 + 0.4647 I
33	17.5673	−4.8684	5.5030 I
34	10.8048 + 0.9346 I	7.0723 − 1.6629 I	10.3332 − 0.4647 I
35	8.0000	8.9999	− 9.9997
36	10.8048 − .9346 I	7.0723+1.6629 I	− 10.3332 − .4647 I
37	2.2138 + 3.9850 I	5.4239 + 0.4501 I	− 15.9058 − 0.16809 I
38	10.8048 − 0.9346 I	7.0723 + 1.6629 I	10.3332 + 0.4647 I
39	8.0000	8.9999	9.9997
40	2.2138 + 3.9850 I	5.4239 + 0.4501 I	15.9058 + 0.1680 I

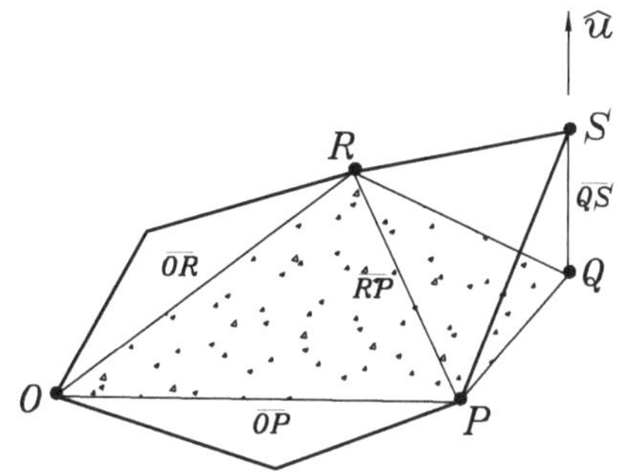

Fig. 12.5 Three-dimensional platform

12.3.1.3 Forward Displacement Analysis of the Six-Legged Parallel Manipulator with Three-Dimensional Platforms: The General Case

A generalization of the method outlined in previous paragraphs may be made by considering that both platforms of the six-legged parallel manipulator are three-dimensional bodies. The fundamental principle consists of finding the coordinates of an arbitrary point of the moving platform based on the coordinates of three noncollinear points where such a point does not belong to the plane, named the control plane, shaped by the three control points (see Fig. 12.5). In that regard, the usual cross product of three-dimensional vectorial algebra plays a central role. It is opportune to mention that although essentially the methodology suggested in this section to obtain the closure equations of the hexapod with three-dimensional platforms is the same as that developed for the hexapod with planar platforms, the corresponding kinematic constraint equations are much more complex.

Let *O*, *P*, and *R* be the control points of the manipulator, and let *S* be a corner of the moving platform, which, for generality, is out of the control plane *OPR*. Furthermore, consider that point *Q* is a projection of point *S* over the *OPRQ* plane. Then the position vector $\boldsymbol{s}$ of point *S*, expressed in the reference frame *XYZ*, may be determined as

$$\boldsymbol{s} = \boldsymbol{q} + \overline{QS}\hat{\boldsymbol{u}}, \tag{12.15}$$

where $\hat{\boldsymbol{u}}$ is a unit vector normal to the plane *OPRQ* directed from point *Q* to point *S* which is given by

$$\hat{\boldsymbol{u}} = \frac{2(\boldsymbol{p}-\boldsymbol{o}) \times (\boldsymbol{r}-\boldsymbol{o})}{\sqrt{2(\overline{OP}\,\overline{OR})^2 + 2(\overline{OP}\,\overline{RP})^2 + 2(\overline{OR}\,\overline{RP})^2 - \overline{OP}^4 - \overline{OR}^4 - \overline{RP}^4}}. \tag{12.16}$$

Finally, the closure equations associated to the forward displacement analysis are obtained based on expressions (12.11) and (12.12). The system of equations may be solved by using special software like Maple©through the library Homotopy.

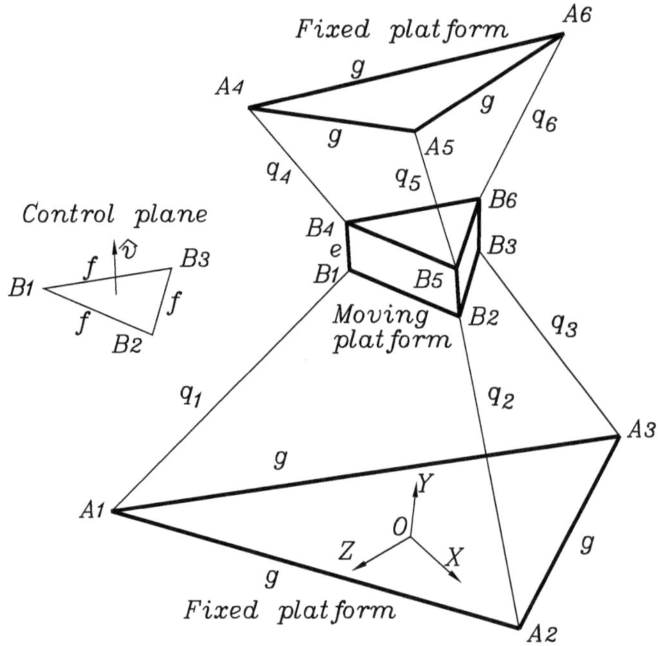

Fig. 12.6 Example 12.2. Parallel manipulator provided with a triangular-prism moving platform

However, it is unfortunately a very time-consuming task. Furthermore, not all the solutions may be obtained with such software. Thus, the efficacy of the Algorithm 795: PHCpack is noticeable.

Example 12.2. With reference to Fig. 12.6, the exercise is devoted to computing the forward displacement analysis of a six-legged parallel manipulator provided with a regular triangular-prism moving platform whose dimensions, using SI units through the numerical example, are given by $e = 1/4$ and $f = \sqrt{3}$. Furthermore, the fixed platform is also a regular triangular prism, where the side of the equilateral triangle is $g = 2\sqrt{3}$ while the height of the prism is given by $h = 3$. On the other hand, the instantaneous generalized coordinates, or limb lengths, are given by $q_1 = 2$, $q_2 = 5/2$, $q_3 = 9/4$, $q_4 = 39/20$, $q_5 = 43/20$, and $q_6 = 7/4$.

Solution. Let $B_1 = (w_1, w_2, w_3)$, $B_2 = (w_4, w_5, w_6)$, and $B_3 = (w_7, w_8, w_9)$ be the control points of the robot. As an initial step it is necessary to compute the unit vector $\hat{\boldsymbol{v}} = v_x\hat{\boldsymbol{i}} + v_y\hat{\boldsymbol{j}} + v_z\hat{\boldsymbol{k}}$ in terms of the unknowns $w_i (i = 1, 2, 3, \ldots, 9)$. Taking into account that the vector $\hat{\boldsymbol{v}}$ is normal to the plane $\Delta B_1B_2B_3$, it follows that

$$\hat{\boldsymbol{v}} = \frac{2(\boldsymbol{b}_2 - \boldsymbol{b}_1) \times (\boldsymbol{b}_3 - \boldsymbol{b}_1)}{\sqrt{3}f^2}. \tag{12.17}$$

Then the components of the vector $\hat{\boldsymbol{v}}$ are directly obtained as

$$v_x = \frac{2(w_5w_9 - w_5w_3 - w_2w_9 - w_6w_8 + w_6w_2 + w_3w_8)}{\sqrt{3}f^2},$$

$$v_y = \frac{2(w_6w_7 - w_6w_1 - w_3w_7 - w_4w_9 + w_4w_3 + w_1w_9)}{\sqrt{3}f^2},$$

$$v_z = \frac{2(w_4w_8 - w_4w_2 - w_1w_8 - w_5w_7 + w_5w_1 + w_2w_7)}{\sqrt{3}f^2}. \tag{12.18}$$

On the other hand, it is evident that the vectors $\boldsymbol{b}_i(i = 4,5,6)$, denoting the positions of points $B_i(i = 4,5,6)$, may be obtained in terms of the unknowns $w_i(i = 1,2,3,\ldots,9)$ through the following expressions:

$$\begin{aligned} \boldsymbol{b}_4 &= \boldsymbol{b}_1 + e\hat{\boldsymbol{v}} \\ \boldsymbol{b}_5 &= \boldsymbol{b}_2 + e\hat{\boldsymbol{v}} \\ \boldsymbol{b}_6 &= \boldsymbol{b}_3 + e\hat{\boldsymbol{v}}. \end{aligned} \tag{12.19}$$

Furthermore, in order to formulate the closure equations required to solve the forward displacement analysis, six equations are obtained considering that the limb lengths of the manipulator are constrained to

$$(\boldsymbol{b}_i - \boldsymbol{a}_i) \bullet (\boldsymbol{b}_i - \boldsymbol{a}_i) = q_i^2, \quad i = 1,2,\ldots,6.$$

Here vectors $\boldsymbol{a}_i$ denote the positions of the points A_i connecting the limbs to the fixed platform. The coordinates of points A_i are expressed in the fixed reference frame XYZ whose origin is located at the center of the equilateral triangle $\Delta A_1A_2A_3$. Indeed,

$$\begin{aligned} &A_1 = (2,0,0), \quad A_2 = (-1,0,-\sqrt{3}), \quad A_3 = (-1,0,\sqrt{3}), \\ &A_4 = (2,3,0), \quad A_5 = (-1,3,-\sqrt{3}), \quad A_6 = (-1,3,\sqrt{3}). \end{aligned}$$

To complete the closure equations, we may write three compatibility equations upon the coordinates of the control points as follows:

$$\begin{aligned} (\boldsymbol{b}_1 - \boldsymbol{b}_2) \bullet (\boldsymbol{b}_1 - \boldsymbol{b}_2) &= f^2, \\ (\boldsymbol{b}_2 - \boldsymbol{b}_3) \bullet (\boldsymbol{b}_2 - \boldsymbol{b}_3) &= f^2, \\ (\boldsymbol{b}_1 - \boldsymbol{b}_3) \bullet (\boldsymbol{b}_1 - \boldsymbol{b}_3) &= f^2. \end{aligned} \tag{12.20}$$

Collecting these results, we obtain a nonlinear system of nine equations in the unknowns $w_i(i = 1,2,\ldots,9)$. To illustrate the complexity of such equations, consider that one of them is given by

$$\begin{aligned} &-(1/54)w_3^2w_5w_8 - (1/54)w_2w_9^2w_5 - (1/54)w_2^2w_6w_9 - \\ &(1/54)w_2w_6^2w_8 - (1/54)w_3w_4^2w_9 - (1/54)w_3^2w_4w_7 - \end{aligned}$$

$$
\begin{aligned}
&(1/54)w_3w_7^2w_6 - (1/54)w_1w_6^2w_7 - (1/54)w_9^2w_4w_1 - \\
&(1/54)w_1^2w_6w_9 - (1/54)w_1w_5^2w_7 - (1/54)w_1^2w_5w_8 - \\
&(1/54)w_1w_8^2w_4 - (1/54)w_2w_4^2w_8 - (1/54)w_7^2w_5w_2 - \\
&(1/54)w_2^2w_4w_7 + (2/9)\sqrt{3}w_8w_6 - (1/54)w_3w_5^2w_9 - \\
&(1/54)w_8^2w_6w_3 + (2/9)\sqrt{3}w_2w_9 + (2/9)\sqrt{3}w_3w_5 - \\
&(2/9)\sqrt{3}w_9w_5 - (2/9)\sqrt{3}w_2w_6 - \\
&(2/9)\sqrt{3}w_3w_8 + w_1^2 + (1/108)w_2^2w_6^2 + \\
&(1/108)w_2^2w_9^2 + (1/108)w_8^2w_6^2 + (1/108)w_3^2w_5^2 + \\
&(1/108)w_3^2w_8^2 + (1/108)w_9^2w_5^2 + (1/108)w_3^2w_4^2 + \\
&(1/108)w_3^2w_7^2 + (1/108)w_9^2w_4^2 + (1/108)w_1^2w_6^2 + \\
&(1/108)w_1^2w_9^2 + (1/108)w_7^2w_6^2 + (1/108)w_1^2w_5^2 + \\
&(1/108)w_1^2w_8^2 + (1/108)w_7^2w_5^2 + (1/108)w_2^2w_4^2 + \\
&(1/108)w_2^2w_7^2 + (1/108)w_8^2w_4^2 - (1/3)\sqrt{3}w_7w_6 - \\
&(1/3)\sqrt{3}w_3w_4 + (1/3)\sqrt{3}w_9w_4 + \\
&(1/3)\sqrt{3}w_3w_7 + (1/3)\sqrt{3}w_1w_6 - \\
&(1/3)\sqrt{3}w_1w_9 - 4w_1 - 6w_2 - (1/54)w_3w_4w_1w_6 + \\
&(1/54)w_3w_4w_1w_9 + (1/54)w_3w_4w_7w_6 + (1/54)w_3w_7w_9w_4 + \\
&(1/54)w_3w_7w_1w_6 - (1/54)w_3w_7w_1w_9 + (1/54)w_9w_4w_1w_6 - \\
&(1/54)w_9w_4w_7w_6 + (1/54)w_1w_9w_7w_6 - (1/54)w_1w_5w_2w_4 + \\
&(1/54)w_1w_5w_2w_7 + (1/54)w_1w_5w_8w_4 + (1/54)w_1w_8w_7w_5 + \\
&(1/54)w_1w_8w_2w_4 - (1/54)w_1w_8w_2w_7 + (1/54)w_7w_5w_2w_4 - \\
&(1/54)w_7w_5w_8w_4 + (1/54)w_2w_7w_8w_4 - (1/54)w_2w_6w_3w_5 + \\
&(1/54)w_2w_6w_3w_8 + (1/54)w_2w_6w_9w_5 + (1/54)w_2w_9w_8w_6 + \\
&(1/54)w_2w_9w_3w_5 - (1/54)w_2w_9w_3w_8 + (1/54)w_8w_6w_3w_5 - \\
&(1/54)w_8w_6w_9w_5 + (1/54)w_3w_8w_9w_5 + w_2^2 + w_3^2 + \\
&(1/9)\sqrt{3}w_9w_5w_1 - (1/9)\sqrt{3}w_8w_6w_1 + \\
&(1/9)\sqrt{3}w_8w_4w_3 - (1/9)\sqrt{3}w_7w_5w_3 + \\
&(1/9)\sqrt{3}w_7w_6w_2 - (1/9)\sqrt{3}w_9w_4w_2 + 3679/400 = 0. \qquad (12.21)
\end{aligned}
$$

Later, 24 solutions for the forward displacement analysis are obtained by applying the Algorithm 795: PHCpack. The solutions, including complex and real solutions, are listed in Table 12.3. According to Table 12.3, there are only two real solutions meeting the conditions of the forward displacement analysis: solutions 6 and 13. The coordinates of the control points B_1, B_2, and B_3 corresponding to the real solutions are given in Table 12.4.

It is worth pointing out the accuracy of the proposed method; for example, for solution 13 of the example the following errors were found for the nine characteristic equations: −0.000016908, −0.000024932, 0.000180170, 0.000068922, −0.000013489, −0.000028003, −0.000025511, −0.000054521, and −0.000066191.

Example 12.3. With reference to Fig. 12.7, the exercise consists of solving the forward displacement analysis of a 6-6 fully parallel manipulator provided with three-dimensional platforms.

To approach the forward displacement analysis, let us consider points B_1, B_3, and B_5 to be the control points of the robot. Furthermore, the coordinates of points A_i are given by

$$A_1 = (2.0, 0.1, 0.01),\ A_2 = (1.1, 1.7, -0.05),\ A_3 = (-0.9, 1.8, 0.05),$$
$$A_4 = (-1.95, 0, 0),\ A_5 = (-1.05, -1.75, 0.05),\ A_6 = (0.9, -1.75, 0.05).$$

Meanwhile, relevant parameters associated to the geometry of the moving platform are as follows:

$$\overline{B_1B_2} = .9669539781,\ \overline{B_2B_3} = 1.016120072,\ \overline{B_3B_4} = 1.170469992,$$
$$\overline{B_4B_5} = .9910095850,\ \overline{B_5B_6} = 1.038761282,\ \overline{B_6B_1} = 1.077972634,$$
$$\overline{Q_2S_2} = 0.15,\ \overline{Q_4S_4} = 0.1,\ \overline{Q_6S_6} = 0.125.$$

On the other hand, in the instantaneous pose of the robot the generalized coordinates are given by $q_1 = 3.0$, $q_2 = 2.9$, $q_3 = 3.1$, $q_4 = 3.15$, $q_5 = 2.95$, and $q_6 = 2.98$

Solution. The application of the proposed method yields a higher nonlinear system of nine equations in the unknowns $w_i (i = 1, 2, \ldots, 9)$. Some of these equations contain very long terms; for instance, one equation is given by

$$\begin{aligned}
&0.105w_7w_2w_6 - 0.0055w_2w_4^2w_8 - 0.0055w_2w_9^2w_5 - 0.0055w_3^2w_4w_7 - \\
&\quad 0.0055w_1w_6^2w_7 - 0.0055w_3^2w_5w_8 + 0.4892w_7 - 0.0055w_8^2w_6w_3 + \\
&0.105w_4w_3w_8 - 0.0055w_3w_4^2w_9 + 0.0055w_9w_4w_1w_6 - 0.0055w_2w_6w_3w_5 + \\
&\quad 0.0055w_8w_6w_3w_5 + 0.0055w_2w_6w_3w_8 + 0.0055w_7w_5w_2w_4 + \\
&\quad 0.0055w_2w_6w_9w_5 + 0.0055w_1w_5w_2w_7 + 0.0055w_2w_9w_3w_5 -
\end{aligned}$$

Table 12.3 Example 12.2. Solutions of the forward displacement analysis

Sol. w_5	w_1 w_6	w_2 w_7	w_3 w_8	w_4 w_9
1	−0.12367 + 0.84764I	0.91652 + 1.0508I	−1.0517 − 0.79586I	−4.0562 − 3.6310I
1.6784 + 0.10187I	−5.6673 + 2.8634I	−1.7713 + 1.6498I	2.3067 + 1.0233I	−0.78107e−1 + 0.60098I
2	0.48900 + 1.0930I	1.6348 − 0.43858e−1I	1.3594 + 1.2677I	−1.3767 + 1.5297I
1.8274 − 0.23951e−1I	0.54465 + 0.27235I	0.28704 + 2.1682I	1.9498 − 0.95584e−2I	−0.64758 + 1.1649I
3	0.50153 − 0.98471I	1.6530 + 0.43715e−1I	−1.2429 + 1.2453I	0.51326 − 2.1280I
1.8852 + 0.82592e−2I	0.81972 + 1.2558I	−1.2866 − 1.5479I	1.9001 + 0.32175e−1I	−0.22594 + 0.25779I
4	−0.72259 + 0.98645e−1I	1.6229 − 0.28981e−2I	−0.11111 − 2.4595I	1.1046 + 0.54568I
1.9306 + 0.46453e−2I	−0.95417 − 1.4879I	0.88948 − 0.52759I	1.9755 − 0.27859e−1I	1.0750 − 1.6009I
5	−0.13625 − 0.79354I	0.90925 − 1.1117I	0.98088 − 0.69770I	−1.7634 − 1.3966I
2.3369 − 1.0280I	0.19053 + 0.69503I	−4.5173 + 4.0853I	1.7369 − 0.46840e−1I	5.9724 + 3.4080I
6	0.87781 + 0.12979e−168I	1.4111 + 0.52008e−169I	0.86568 + 0.83472e−169I	0.58093 + 0.52639e−169I
1.7058 − 0.87803e−169I	−0.81510 + 0.72585e−169I	−0.73328 − 0.21378e−169I	1.7255 − 0.86938e−168I	0.31292 + 0.20e−181I
7	−0.72259 − 0.98645e−1I	1.6229 + 0.28981e−2I	−0.11111 + 2.4595I	1.1046 − 0.54568I
1.9306 − 0.46453e−2I	−0.95417 + 1.4879I	0.88948 + 0.52759I	1.9755 + 0.27859e−1I	1.0750 + 1.6009I
8	9.1222 − 0.65804I	1.4298 + 0.87600e−1I	−0.65326 − 6.9826I	1.2524 − 1.4949I
2.4036 + 1.3861I	−1.6792 + 0.66913I	1.0222 + 0.15728e−1I	0.97786 + 1.3675I	0.11756 + 0.84797I
9	1.0498 + 7.4776I	1.4394 + 0.39530I	−7.6049 − 0.85944I	−0.79872 + 6.7677I
1.7516 + 0.35049I	−8.7424 + 0.28188I	−0.80603 + 8.1447I	1.8313 + 0.41856I	−6.5301 + 0.28398I
10	−0.13625 + 0.79354I	0.90925 + 1.1117I	0.98088 + 0.69770I	−1.7634 + 1.3966I
2.3369 + 1.0280I	0.19053 − 0.69503I	−4.5173 − 4.0853I	1.7369 + 0.46840e−1I	5.9724 − 3.4080I
11	0.48900 − 1.0930I	1.6348 + 0.43858e−1I	1.3594 − 1.2677I	−1.3767 − 1.5297I
1.8274 + 0.23951e−1I	0.54465 − 0.27235I	0.28704 − 2.1682I	1.9498 + 0.95584e−2I	−0.64758 − 1.1649I
12	8.5297 − 0.34950I	1.4334 + 0.80686e−1I	0.33969 + 6.3780I	1.1841 + 0.40386e−2I

(continued)

Table 12.3 (continued)

0.91145 + 1.2229I	−0.11187 − 0.69338I	1.0720 − 1.5088I	2.3488 + 1.3156I	1.6901 − 0.86148I
13	0.63898 + 0.50090e−152I	1.4115 + 0.14738e−152I	−0.39405 + 0.33473e−152I	−0.98800 + 0.63659e−153I
1.6411 + 0.34918e−152I	0.15385 + 0.21686e−152I	0.32772 + 0.10867e−152I	1.7579 − 0.14614e−153I	1.2742 + 0.11516e−151I
14	2.1714 + 2.1979I	1.3985 + 0.11143I	2.6266 − 0.20278I	0.36764 + 2.8100I
1.7036 + 0.15211I	1.5748 − 1.2405I	0.40827 + 1.5785I	1.7849 + 0.93213e−1I	3.6990 − 1.2147I
15	0.93356 − 0.16040I	2.2397 − 1.5308I	1.8350 + 1.7752I	−0.71572 + 0.43050I
0.80431 − 1.5277I	1.1299 + 0.38656I	−4.3551 − 7.3920I	1.7112 − 0.74643e−1I	9.2375 − 3.2874I
16	0.93356 + 0.16040I	2.2397 + 1.5308I	1.8350 − 1.7752I	−0.71572 − 0.43050I
0.80431 + 1.5277I	1.1299 − 0.38656I	−4.3551 + 7.3920I	1.7112 + 0.74643e−1I	9.2375 + 3.2874I
17	0.50153 + 0.98471I	1.6530 − 0.43715e−1I	−1.2429 − 1.2453I	0.51326 + 2.1280I
1.8852 − 0.82592e−2I	0.81972 − 1.2558I	−1.2866 + 1.5479I	1.9001 − 0.32175e−1I	−0.22594 − 0.25779I
18	0.95689 − 0.16876I	2.2818 − 1.6232I	−1.9280 − 1.8298I	−4.4026 − 7.5716I
1.6487 − 0.13281I	−9.4876 + 3.2936I	−0.68170 + 0.62581I	0.86027 − 1.6382I	−1.0051 − 0.44211I
19	8.5297 + 0.34950I	1.4334 − 0.80686e−1I	0.33969 − 6.3780I	1.1841 − 0.40386e−2I
0.91145 − 1.2229I	−0.11187 + 0.69338I	1.0720 + 1.5088I	2.3488 − 1.3156I	1.6901 + 0.86148I
20	−0.12367 − 0.84764I	0.91652 − 1.0508I	−1.0517 + 0.79586I	−4.0562 + 3.6310I
1.6784 − 0.10187I	−5.6673 − 2.8634I	−1.7713 − 1.6498I	2.3067 − 1.0233I	−0.78107e−1−0.60098I
21	9.1222 + 0.65804I	1.4298 − 0.87600e−1I	−0.65326 + 6.9826I	1.2524 + 1.4949I
2.4036 − 1.3861I	−1.6792 − 0.66913I	1.0222 − 0.15728e−1I	0.97786 − 1.3675I	0.11756 − 0.84797I
22	2.1714 − 2.1979I	1.3985 − 0.11143I	2.6266 + 0.20278I	0.36764 − 2.8100I
1.7036 − 0.15211I	1.5748 + 1.2405I	0.40827 − 1.5785I	1.7849 − 0.93213e−1I	3.6990 + 1.2147I
23	1.0498 − 7.4776I	1.4394 − 0.39530I	−7.6049 + 0.85944I	−0.79872 − 6.7677I
1.7516 − 0.35049I	−8.7424 − 0.28188I	−0.80603 − 8.1447I	1.8313 − 0.41856I	−6.5301 − 0.28398I
24	0.95689 + 0.16876I	2.2818 + 1.6232I	−1.9280 + 1.8298I	−4.4026 + 7.5716I
1.6487 + 0.13281I	−9.4876 − 3.2936I	−0.68170 − 0.62581I	0.86027 + 1.6382I	−1.0051 + 0.44211I

Table 12.4 Example 12.2. Real solutions of the forward displacement analysis

Sol.	B_1	B_2	B_3
6	(0.87781,1.4111,0.86568)	(0.58093,1.7058,-0.81510)	(−0.73328,1.7255,0.31292)
13	(0.63898,1.4115,−0.39405)	(−0.98800,1.6411,0.15385)	(0.32772,1.7579,1.2742)

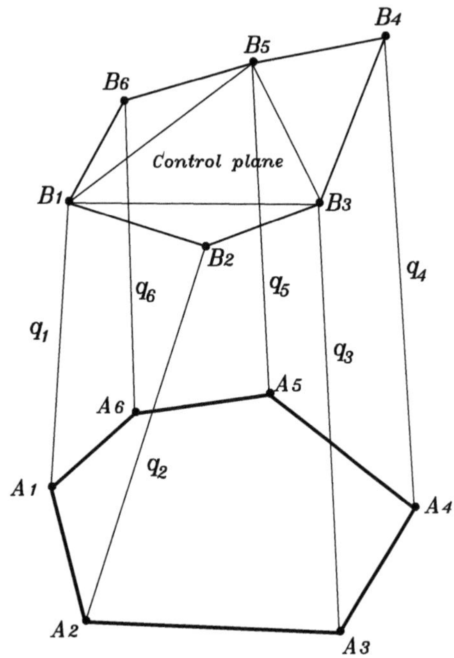

Fig. 12.7 Example 12.3. Parallel manipulator provided with three-dimensional platforms

$$0.0055w_2w_9w_3w_8 + 0.0055w_3w_8w_9w_5 + 0.0055w_3w_4w_7w_6 +$$
$$0.0027w_7^2w_5^2 - 0.2719w_8w_2 + 0.0027w_3^2w_4^2 + 0.0027w_1^2w_9^2 +$$
$$0.0027w_2^2w_7^2 - 0.2719w_9w_3 - 0.0055w_3w_7^2w_6 - 0.105w_1w_8w_6 -$$
$$0.105w_7w_3w_5 - 2.0771w_5 + 0.0611w_3 - 1.3440w_4 - 1.3451w_1 +$$
$$0.0611w_6 - 0.0222w_9 + 0.7561w_8 - 2.078935388w_2 - 0.0055w_3w_7w_1w_9 -$$
$$0.0055w_9w_4w_7w_6 + 0.0055w_2w_7w_8w_4 + 0.0055w_3w_7w_1w_6 +$$
$$0.0055w_1w_8w_2w_4 + 0.105w_1w_9w_5 - 0.0055w_3w_5^2w_9 - 0.0055w_1^2w_6w_9 -$$
$$0.0055w_2^2w_6w_9 + 0.0027w_8^2w_4^2 - .2717129654w_8w_5 - 0.2717w_7w_4 +$$
$$0.0027w_2^2w_4^2 + 0.0027w_2^2w_6^2 + 0.0027w_7^2w_6^2 - 0.0055w_1^2w_5w_8 -$$
$$0.0055w_2w_6^2w_8 - 0.0055w_9^2w_4w_1 + 0.0027w_8^2w_6^2 + 0.1155w_3w_5 -$$
$$0.1155w_9w_5 - 0.0055w_1w_5^2w_7 - 0.0055w_1w_8^2w_4 - 0.0055w_7^2w_5w_2 -$$
$$0.0055w_2^2w_4w_7 - 0.105w_4w_2w_9 + 0.0055w_3w_7w_9w_4 + 0.0055w_1w_9w_7w_6 -$$

$$\begin{aligned}
&0.0055w_7w_5w_8w_4 + 0.0055w_1w_8w_7w_5 + 0.0055w_1w_5w_8w_4 - \\
&0.0055w_1w_8w_2w_7 - 0.0055w_8w_6w_9w_5 - 0.0055w_1w_5w_2w_4 - \\
&0.0055w_3w_4w_1w_6 + 0.0055w_2w_9w_8w_6 + 0.0055w_3w_4w_1w_9 - \\
&0.2719w_7w_1 + 0.0027w_1^2w_6^2 + 0.7471w_6w_3 + 0.0027w_3^2w_7^2 + \\
&0.7471w_5w_2 + 0.0027w_3^2w_5^2 + 0.1786w_3w_7 - 0.1786w_3w_4 + \\
&0.1786w_1w_6 - 0.1786w_7w_6 - 0.0052w_1w_8 - 0.0052w_2w_4 + \\
&0.0052w_8w_4 + 0.0027w_1^2w_8^2 + 0.0027w_3^2w_8^2 + 0.0027w_9^2w_5^2 + \\
&0.0027w_1^2w_5^2 + 0.1786w_9w_4 + 0.0027w_2^2w_9^2 + 0.1155w_8w_6 + \\
&0.7471w_4w_1 + 0.0052w_1w_5 - .2717129654w_9w_6 + 0.0052w_7w_2 + \\
&0.0027w_9^2w_4^2 - 0.0052w_7w_5 - 0.1155w_2w_6 - 0.1786w_1w_9 - \\
&0.1155w_3w_8 + 0.1155w_2w_9 - 4.3075 + 0.0494w_7^2 + 0.0494w_8^2 + \\
&0.3732w_6^2 + 0.3738w_1^2 + 0.3738w_2^2 + 0.3738w_3^2 + \\
&0.3732w_4^2 + 0.3732w_5^2 + 0.0494w_9^2 = 0.
\end{aligned}$$

With the aid of the Algorithm 795: PHCpack, it was possible to find in a reasonable time the 40 solutions of the forward displacement analysis. For brevity, only the coordinates of point B_1 are provided in Table 12.5.

For solution 6 of Table 12.5, the unknowns w_i result in

$$w_1 = -0.97282,\ w_2 = 0.32074,\ w_3 = 0.34702,\ w_4 = 0.22500,\ w_5 = -1.0663,$$
$$w_6 = 0.40866,\ w_7 = 0.53619,\ w_8 = 0.30563,\ w_9 = 1.4503.$$

Next, the coordinates of points B_i meeting solution 6 can be computed. These coordinates are reported in Table 12.6.

For a quick visualization of the results of Example 12.3, the six real solutions of the forward displacement analysis are sketched in Fig. 12.8.

Finally, at this point it is worth noting how easy it is to formulate the forward displacement analysis of the general six-legged parallel manipulator, based on elementary linear algebra, when compared with conventional methods based on the handling of the rotation matrix.

12.3.2 Inverse Displacement Analysis of the Hexapod

The inverse displacement analysis of the Gough platform consists of finding the set of generalized coordinates q_i meeting a prescribed pose of the moving platform as measured from the fixed platform. With reference to Fig. 12.2, this analysis is equivalent to computing the limb lengths of the manipulator given the rotation matrix $\mathbf{R}$ and the vector $\boldsymbol{q}_1$. Toward this aim, it immediately emerges that the length of limb 1 is directly obtained as

Table 12.5 Example 12.3. The 40 reachable positions of point B_1

Sol.	X	Y	Z
1	3.7494−2.9963 I	−3.7782−0.5447 I	−1.7837−1.7445 I
2	0.0760−1.2462 I	−0.1316−2.7877 I	−3.8860+0.7811 I
3	3.7494+2.9963 I	−3.7782+.5447 I	−1.7837+1.7445 I
4	0.0760+1.2462 I	−0.1316+2.7877 I	−3.8860−0.7811 I
5	0.7288+0.5415 I	−0.2975+0.26785 I	2.7801+0.2869 I
6	−0.9728	0.3207	0.3470
7	2.1430	−0.4089	2.9631
8	−0.9590	−0.3443	0.2259
9	−0.4972	−0.7500	−1.4187
10	−0.5303+2.1960 I	3.0365+2.2008 I	1.9730−0.4614 I
11	8.1080+6.4369 I	8.0432−8.1831 I	−5.9524−4.3076 I
12	8.1080−6.4369 I	8.0432+8.1831 I	−5.9524+4.3076 I
13	−0.6229−2.0713 I	2.4868+1.0519 I	−2.9743+2.6619 I
14	1.3016−1.4189 I	−1.2405+0.5389 I	−2.9944+0.0893 I
15	−0.5311+0.4599 I	−0.8418−0.36858 I	1.5394+0.5341 I
16	−0.9211+0.1337 I	0.8305+0.0602 I	0.5799+0.6082 I
17	6.5785−8.7195 I	7.2094+6.2057 I	−7.2268+0.5798 I
18	−0.26629−1.7773 I	2.6914−1.4813 I	1.5962−0.11936 I
19	10.3080+8.6003 I	3.8477−10.3590 I	−10.7760+3.0253 I
20	17.5520−1.0330 I	−3.0869+14.0260 I	−6.4885−9.3508 I
21	−0.6229+2.0713 I	2.4868−1.0519 I	−2.9743−2.6619 I
22	−0.9211−0.1337 I	0.8305−0.0602 I	0.5799−0.6082 I
23	4.4611+8.8337 I	8.5242−1.9896 I	−3.9348+1.2622 I
24	−0.3963	1.1135	−1.4835
25	−0.5763	0.8697 I	1.3404
26	0.72882−0.5415 I	−0.2975−0.2678 I	2.7801−0.2869 I
27	−0.2662+1.7773 I	2.6914+1.4813 I	1.5962+0.1193 I
28	4.4611−8.8337 I	8.5242+1.9896 I	−3.9348−1.2622 I
29	−0.5303−2.1960 I	3.0365−2.2008 I	1.9730+0.46140 I
30	1.3016+1.4189 I	−1.2405−.5389 I	−2.9944−.0893 I
31	6.9064+3.4935 I	−4.6817+3.6641 I	0.1182+3.5098 I
32	6.9064−3.4935 I	−4.6817−3.6641 I	0.1182−3.5098 I
33	17.5520+1.0330 I	−3.0869−14.0260 I	−6.4885+9.3508 I
34	8.0782−1.4485 I	−2.0175−4.7257 I	0.4305−2.8591 I
35	4.4331+0.0525 I	0.8568−0.2645 I	1.6168+0.0450 I
36	6.5785+8.7195 I	7.2094−6.2057 I	−7.2268−0.5798 I
37	4.4331−0.0525 I	0.8568+0.2645 I	1.6168−0.0450 I
38	−0.5311−0.4599 I	−0.8418+0.3685 I	1.5394−0.5341 I
39	8.0782+1.4485 I	−2.0175+4.7257 I	0.4305+2.8591 I
40	10.3080−8.6003 I	3.8477+10.3590 I	−10.7760−3.0253 I

Table 12.6 Example 12.3. Coordinates of points B_i, solution 6 of the forward displacement analysis

i	X	Y	Z
1	−0.97282	0.32074	0.34702
2	−0.6569484103	−0.5878132292	0.2483304217
3	0.22500	−1.0663	0.40866
4	0.8607983859	−0.6154207301	1.281843897
5	0.53619	0.30563	1.4503
6	−0.3864961108	0.6928457951	1.171502498

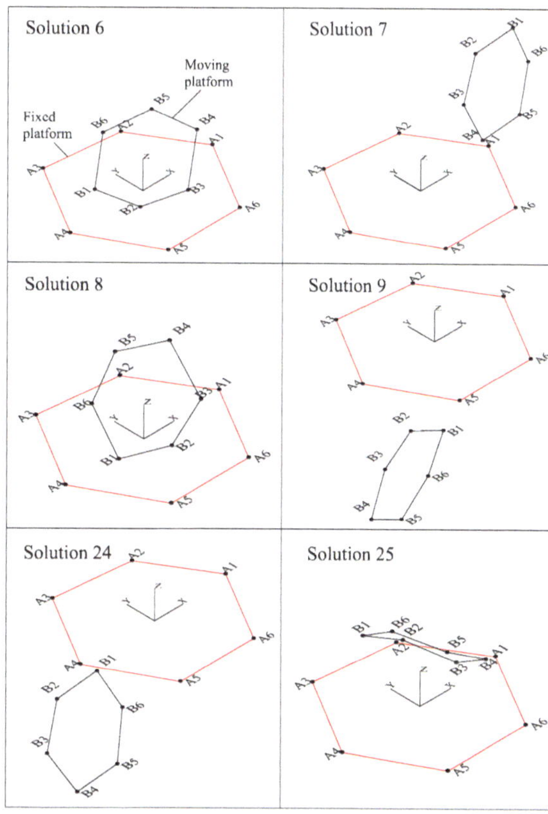

Fig. 12.8 Example 12.3. Real solutions of the forward displacement analysis

$$q_1 = \sqrt{\boldsymbol{q}_1 \bullet \boldsymbol{q}_1}. \tag{12.22}$$

On the other hand, in order to compute the length of the remaining limbs, consider that

$$\boldsymbol{b}_i = \boldsymbol{q}_1 + \mathbf{R}\boldsymbol{b}_i^*, \quad i = 2, 3, \ldots, 6,$$

where, as previously noted, $\boldsymbol{b}_i^*$ is the vector obtained through the coordinates of the ith spherical joint mounted on the moving platform expressed in the moving reference frame xyz. There are several strategies to express the rotation matrix $\mathbf{R}$; for example, a typical representation of it uses Euler angles α, β, and γ. In fact,

$$\mathbf{R} = \begin{bmatrix} \cos\beta & -\cos\gamma\sin\beta & \sin\beta\sin\gamma \\ \cos\alpha\sin\beta & \cos\alpha\cos\beta\cos\gamma - \sin\alpha\sin\gamma & -\cos\gamma\sin\alpha - \cos\alpha\cos\beta\sin\gamma \\ \sin\alpha\sin\beta & \cos\alpha\sin\gamma + \cos\beta\cos\gamma\sin\alpha & -\cos\alpha\cos\gamma - \cos\beta\sin\alpha\sin\gamma \end{bmatrix}. \tag{12.23}$$

Finally, once the vectors $\boldsymbol{b}_i (i = 2, 3, \ldots, 6)$ are computed, the lengths of the limbs are easily obtained as

$$q_i = \sqrt{(\boldsymbol{b}_i - \boldsymbol{a}_i) \bullet (\boldsymbol{b}_i - \boldsymbol{a}_i)}, \quad i = 2, 3, \ldots, 6. \tag{12.24}$$

The difference between the complexity of the forward and inverse displacement analyses of the general six-legged parallel manipulator is evident.

12.4 Infinitesimal Kinematics of the Six-Legged Parallel Manipulator

In this section we resort to reciprocal screw theory to approach the infinitesimal kinematic analyses of the 6-SPS parallel manipulator. To this end, the model of the screws is depicted in Fig. 12.9.

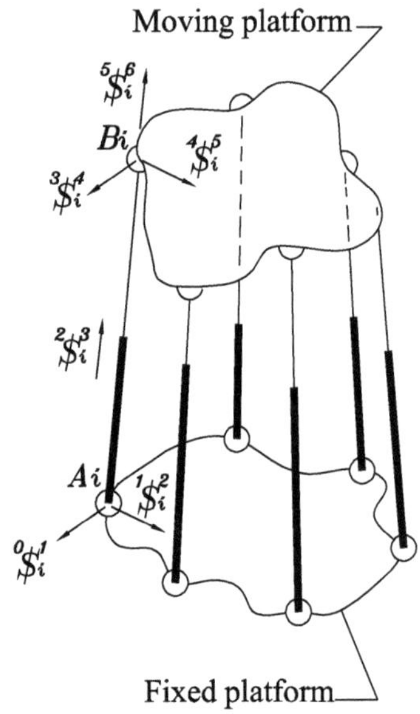

Fig. 12.9 Infinitesimal screws of the 6-SPS parallel manipulator

12.4.1 Velocity Analysis of the Six-Legged Parallel Manipulator

Let $\boldsymbol{\omega}$ be the angular velocity vector of the moving platform, body m, as observed from the fixed platform, body 0. Furthermore, let $\boldsymbol{v}_P$ be the velocity of a point P embedded in the moving platform; for example, P would be the center of the moving platform. The velocity state of the moving platform m as measured from the fixed platform 0, the six-dimensional vector

$$ {}^{0}\boldsymbol{V}_P^m = \begin{bmatrix} \boldsymbol{\omega} \\ \boldsymbol{v}_P \end{bmatrix}, $$

may be written in screw form through the connector chains as follows:

$$ {}^{0}\boldsymbol{V}_P^m = {}_0\omega_1^i\, {}^0\$_i^1 + {}_1\omega_2^i\, {}^1\$_i^2 + {}_2\omega_3^i\, {}^2\$_i^3 + {}_3\omega_4^i\, {}^3\$_i^4 + {}_4\omega_5^i\, {}^4\$_i^5 + {}_5\omega_6^i\, {}^5\$_i^6. \tag{12.25} $$

In that equation the joint-velocity rate ${}_2\omega_3^i = \dot{q}_i$ is the ith speed while ${}_5\omega_6^i$ is the passive joint-velocity rate of the ith spherical along the ith limb, which represents a redundant rotation along the same limb. It should be noted that the screw ${}^5\$_i^6$ is reciprocal to all the screws in the same limb, excluding the screw ${}^2\$_i^3$ representing the prismatic pair. Hence, applying the Klein form $\{*;*\}$ between the screw ${}^5\$_i^6$ to both sides of Eq. (12.25) and reducing terms, one obtains

$$ \left\{ {}^{0}\boldsymbol{V}_P^m ; {}^5\$_i^6 \right\} = \dot{q}_i. \tag{12.26} $$

Casting Eq. (12.26) into matrix-vector form, we get

$$ \mathbf{J}^T \boldsymbol{\Delta}\, {}^{0}\boldsymbol{V}_P^m = \dot{\mathbf{q}}, \tag{12.27} $$

where

- $\mathbf{J} = \begin{bmatrix} {}^5\$_1^6 & {}^5\$_2^6 & {}^5\$_3^6 & {}^5\$_4^6 & {}^5\$_5^6 & {}^5\$_6^6 \end{bmatrix}^T$ is the screw-coordinate Jacobian matrix of the robot,
- $\dot{\mathbf{q}} = \begin{bmatrix} \dot{q}_1 & \dot{q}_2 & \dot{q}_3 & \dot{q}_4 & \dot{q}_5 & \dot{q}_6 \end{bmatrix}^T$ is the first-order driver matrix of the parallel manipulator,
- $\boldsymbol{\Delta} = \begin{bmatrix} \mathbf{0} & \mathbf{I} \\ \mathbf{I} & \mathbf{0} \end{bmatrix}$ is an operator of polarity.

Hence, the input–output equation of velocity of the manipulator results in

$$ \mathbf{A}\, {}^{0}\boldsymbol{V}_P^m = \dot{\mathbf{q}}, \tag{12.28} $$

where $\mathbf{A} = \mathbf{J}^T \boldsymbol{\Delta}$.

12.4.2 Acceleration Analysis of the Six-Legged Parallel Manipulator

In order to obtain the input–output equation of acceleration of the hexapod, consider that $\boldsymbol{\alpha}$ is the angular acceleration vector of the moving platform as observed from the fixed platform while $\boldsymbol{a}_P$ is the acceleration vector of point P. Hence, the reduced acceleration state of the moving platform as measured from the fixed platform, the six-dimensional vector

$$ {}^0\boldsymbol{A}_P^m = \begin{bmatrix} \boldsymbol{\alpha} \\ \boldsymbol{a}_P - \boldsymbol{\omega} \times \boldsymbol{v}_P \end{bmatrix}, $$

may be written in screw form through the connector kinematic chains as follows:

$$ {}^0\boldsymbol{A}_P^m = {}_0\dot{\omega}_1^i\,{}^0\$_i^1 + {}_1\dot{\omega}_2^i\,{}^1\$_i^2 + {}_2\dot{\omega}_3^i\,{}^2\$_i^3 + {}_3\dot{\omega}_4^i\,{}^3\$_i^4 + {}_4\dot{\omega}_5^i\,{}^4\$_i^5 + {}_5\dot{\omega}_6^i\,{}^5\$_i^6 + \mathcal{A}_i, \qquad (12.29) $$

where $\mathcal{A}_i$ is the Lie screw of acceleration of the ith limb, which is computed as

$$ \begin{aligned} \mathcal{A}_i = &\left[{}_0\omega_1^i\,{}^0\$_i^1 \quad {}_1\omega_2^i\,{}^1\$_i^2 + {}_2\omega_3^i\,{}^2\$_i^3 + {}_3\omega_4^i\,{}^3\$_i^4 + {}_4\omega_5^i\,{}^4\$_i^5 + {}_5\omega_6^i\,{}^5\$_i^6\right] \\ &+ \left[{}_1\omega_2^i\,{}^1\$_i^2 \quad {}_2\omega_3^i\,{}^2\$_i^3 + {}_3\omega_4^i\,{}^3\$_i^4 + {}_4\omega_5^i\,{}^4\$_i^5 + {}_5\omega_6^i\,{}^5\$_i^6\right] \\ &+ \left[{}_2\omega_3^i\,{}^2\$_i^3 \quad {}_3\omega_4^i\,{}^3\$_i^4 + {}_4\omega_5^i\,{}^4\$_i^5 + {}_5\omega_6^i\,{}^5\$_i^6\right] + \left[{}_3\omega_4^i\,{}^3\$_i^4 \quad {}_4\omega_5^i\,{}^4\$_i^5 + {}_5\omega_6^i\,{}^5\$_i^6\right] \\ &+ \left[{}_4\omega_5^i\,{}^4\$_i^5 \quad {}_5\omega_6^i\,{}^5\$_i^6\right]. \end{aligned} $$

Following the trend of the velocity analysis, the input–output equation of acceleration of the robot manipulator results in

$$ \mathbf{A}\,{}^0\boldsymbol{A}_P^m = \ddot{\mathbf{q}} + \mathbf{C}, \qquad (12.30) $$

where $\ddot{\mathbf{q}} = \left[\ddot{q}_1\ \ddot{q}_2\ \ddot{q}_3\ \ddot{q}_4\ \ddot{q}_5\ \ddot{q}_6\right]^T$ is the second-order driver matrix of the manipulator. Meanwhile,

$$ \mathbf{C} = \begin{bmatrix} \{{}^5\$_1^6; \mathcal{A}_1\} \\ \{{}^5\$_2^6; \mathcal{A}_2\} \\ \{{}^5\$_3^6; \mathcal{A}_3\} \\ \{{}^5\$_4^6; \mathcal{A}_4\} \\ \{{}^5\$_5^6; \mathcal{A}_5\} \\ \{{}^5\$_6^6; \mathcal{A}_6\} \end{bmatrix} \qquad (12.31) $$

is the complementary matrix of acceleration.

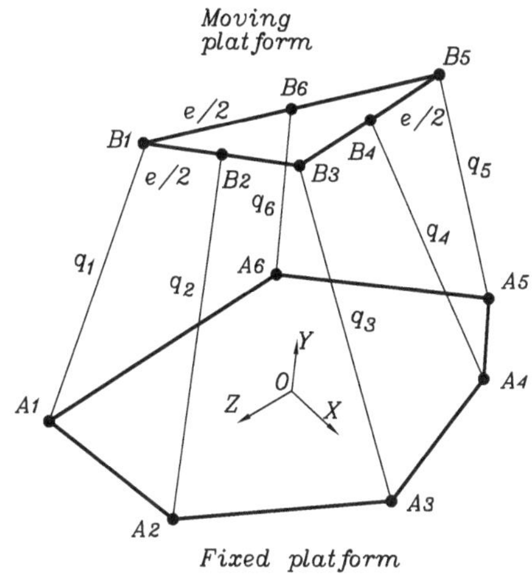

Fig. 12.10 Six-legged parallel manipulator with triangular moving platform

12.5 Exercises

1. Figure 12.10 shows the geometric scheme of a six-legged parallel manipulator whose moving platform is an equilateral triangle of side e. Assuming that B_1, B_3, and B_5 are selected as the control points of the robot, determine the closure equations required to approach the forward displacement analysis of the parallel manipulator based on the data of the control points. *Hint:* Point B_2 is located in the middle of points B_1 and B_3. Dealing with points B_4 and B_6, consider similar geometric conditions.
2. Determine two singular configurations of Gough's tyre testing machine. *Hint:* The singularity analysis may be performed based on the input–output equation of velocity of the parallel manipulator [see Eq. (12.28)].

References

Bonev, I. A. (2011). The true origins of parallel robots. Available at Parallel Mic Center, ParalleMIC online review, http://www.parallemic.org/Reviews/Review007.html

Bottema, O., & Roth, B. (1979). *Theoretical kinematics*. New York: Dover Publications.

Dai, J. S., Huang, Z., & Lipkin, H. (2006). Mobility of overconstrained parallel mechanisms. *ASME Journal of Mechanical Design, 128*(1), 220–229.

Faugère, J. C., & Lazard, D. (1995). Combinatorial classes of parallel manipulators. *Mechanism and Machine Theory, 30*(6), 765–776.

Fichter, E. F., Kerr, D. R., & Rees-Jones, J. (2009). The Gough-Stewart platform parallel manipulator: a retrospective appreciation. *Proceedings Institution of Mechanical Engineers Part C: Journal of Mechanical Engineering Science (JMES), 223*(1) 50th Anniversary Issue Paper, 243–281.

Gallardo-Alvarado, J., Orozco-Mendoza, H., & Rico-Martinez, J. M. (2010). A novel five degrees of freedom decoupled robot. *Robotica, 28*(6), 909–917.

Gough, V. E. (1957). Contribution to discussion of papers on research in automobile stability. *Proceedings of the Automobile Division of the Institution of Mechanical Engineers, 171*, 392–394.

Gough, V. E., & Whitehall, S. G. (1962). Universal tyre testing machine. In: G. Eley (Ed.), *Proceedings 9th International Automobile Technical Congress*, discussion pp. 250ff. Fédèration Internationale des Sociétés d'Ingénieurs des Techniques de lÁutomobile (FISITA), Institution of Mechanical Engineers, London (Vol. 1, pp. 117–137).

Huang, X., Liao, Q., Wei, S., Qiang, X., & Huang, S. (2007). Forward kinematics of the 6-6 Stewart platform with planar base and platform using algebraic elimination. *Proceedings IEEE International Conference on Automation and Logistics, August 18–21, Jinan, China* (pp. 2655–2659).

Husty, M. L. (1996). An algorithm for solving the direct kinematic of general Stewart-Gough platforms. *Mechanism and Machine Theory, 31*(4), 365–380.

Innocenti, C., & Wenger, P. (2004). Position analysis of the RRP-3(SS) multiloop spatial structure. *Proceedings DETC'04 ASME 2004 Design Engineering Technical Conferences and Computers and Information in Engineering Conference, Salt Lake City, UT*. Paper DETC2004-57382 CD-ROM format.

Lazard, D. (1993). On the representation of rigid-body motions and its application to generalized platform manipulators. In J. Angeles, P. Kovac, & G. Hommel (Eds.), *Computational kinematics93* (pp. 175–181). Dordrecht: Kluwer.

Merlet, J.-P. (2008). *Parallel robots* (2nd ed.). New York: Springer.

Mourrain, B. (1993). The 40 generic positions of a parallel robot. In M. Bronstein (Ed.), *Proceedings ISSSAC'93* (pp. 173–182). Kiev: ACM.

Raghavan, M. (1993). The Stewart platform of general geometry has 40 configurations. *ASME Journal of Mechanical Design, 115*(2), 277–282.

Rolland, L. (2005). Certified solving of the forward kinematics problem with an exact algebraic method for the general parallel manipulator. *Advanced Robotics, 19*(9), 995–1025.

Stewart, D. (1965–1966). A platform with six degrees of freedom. *Proceedings Institution of Mechanical Engineers, 180*(1), 371–386.

Stewart, D. (1966). A platform with six degrees of freedom: a new form of mechanical linkage which enables a platform to move simultaneously in all six degrees of freedom developed by Elliott-automation. *Aircraft Engineering and Aerospace Technology, 38*(4), 30–35.

Tsai, L.-W. (1999). *Robot analysis*. New York: Wiley.

Verschelde, J. (1999). Algorithm 795: PHCpack: a general-purpose solver for polynomial systems by homotopy continuation. *ACM Transactions on Mathematical Software, 25*(2), 251–276.

Wampler, C. W. (1995). Forward displacement analysis of general six-in-parallel SPS (Stewart) platform manipulators using soma coordinates. *Mechanism and Machine Theory, 31*(3), 331–337.

Chapter 13
The Original Stewart Platform

13.1 Introduction

At the beginning of the 1900s, a two-degree-of-freedom (DOF) parallel manipulator intended to operate like a flight simulator named the *"Antoinette Barrel Trainer"* was designed by a French company to help pilots control wheel motion before an aircraft was flown. To do this, the orientation of the Antoinette flight simulator was controlled by two wheels mounted left and right of the pilot, one for pitch and the other for roll. Despite the simplicity of this first *virtual plane*, at that time the benefits of flight simulators were immediately recognized mainly by the military industry; for instance, during World Wars I and II, flight simulators were designed for purposes like teaching pilots how to operate flight controls, improving pilots' skills in air gunner operations, instructing pilots how to navigate at night guided by the stars, and so forth.

In the mid-1960s, when working for the Space and Weapons Research Establishment on aviation simulation at the Elliott Automation Company, Stewart published his celebrated theoretical paper entitled "A platform with six degrees of freedom" (Stewart 1965–1966), where flight simulators were one of its possible applications (Fig. 13.1). Interesting topics like mobility and singularity analyses were included in this contribution, attracting the attention of engineers and scientists. However, an unusual number of ardent communications followed Stewart's theoretical paper (Fichter et al. 2009). Meir offered positive and exhaustive comments about potential applications of the Stewart platform; Meir also appreciated the elegant mathematical treatment Stewart used to approach the kinematic analysis of the new spatial mechanism. On the other hand, Meir also considered that the inclusion of six servo motors could be a serious obstacle to developing the new parallel manipulator due to the *obvious* high cost of such a mechatronic system. It is difficult to imagine how a new machine would be cheaper than commercially available options. For example, one of the main problems today with introducing the general hexapod as a multiaxis machine tool has been the use of spherical joints. Fortunately, this has not been a

J. Gallardo-Alvarado, *Kinematic Analysis of Parallel Manipulators by Algebraic Screw Theory*, DOI 10.1007/978-3-319-31126-5_13

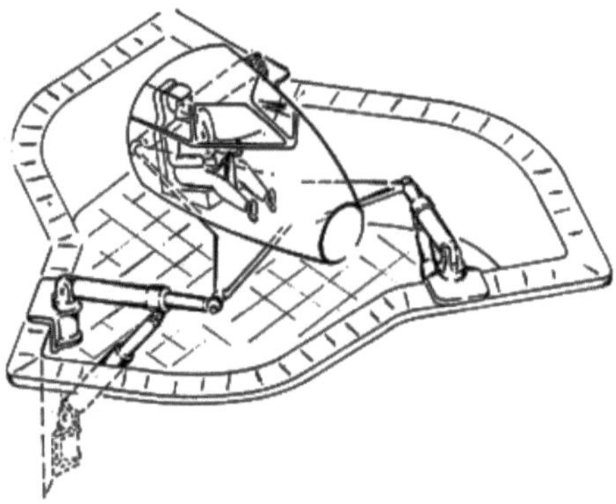

Fig. 13.1 The original Stewart platform proposed as a flight simulator

reason to stop the publication of remarkable contributions devoted to the development of parallel manipulators. The mobility analysis of the Stewart platform was commended by Hunt, who encouraged Stewart to explore other criteria concerned with the effective computation of the degrees of freedom of the proposed spatial mechanism; Hunt believed all the unstable (singular) configurations would be reconsidered. It is worth noting that many mobility formulas have been set in the last 160 years, and none of them is infallible. Thus, the procedure Stewart used to explain the mobility of his invention is acceptable and generous if one considers that at that time such computations were usually ignored by most designers. The mobility of the Stewart platform was also questioned by Thomas, who innocently asked, "Are six degrees of freedom really necessary?" With a certain sarcasm, Stewart simply replied that perhaps Mr. Thomas wanted to know if flight simulations really were worthwhile. Limitations concerning the centrifuge motions of the Stewart platform were pointed out by Nicholls. More communications followed Stewart's theoretical paper, undoubtedly the most abrasive the review by Gough 1966. The originality of the Stewart platform was bitterly disappointing to Gough, who claimed that he had designed a machine with similar characteristics, called the *tyre test machine* (Gough 1957; Gough and Whitehall 1962) (see Fig. 12.1), many years before the Stewart platform. Curiously, it seems that Gough was one of the reviewers of Stewart's paper. It is a little surprising that Stewart himself accepted Gough's commentary and simply justified the originality of his contribution based on the argument that he had been unaware of Gough's invention although clearly both mechanisms are quite different. Ironically, six-legged in-parallel manipulators are—incorrectly—known today as Gough–Stewart platforms. In fact, it is not absolutely true when some kinematicians consider that *the Gough–Stewart platform is a 6-DOF 6-UPS parallel mechanism* (where U, P, and S stand for universal, prismatic, and spherical joints, respectively). There are hundreds of papers describing the Stewart platform as a six-legged 6-DOF in-parallel manipulator; see, for instance, Merkle (1997), Zuo et al. (1995), Lee et al. (2003), Korobeynikov and Turlapov (2005), and Fichter (1986). Furthermore, the situation became more confused when Cappel noted that in 1961 he invented a flight simulator based on a platform system (Cappel 2002)

Fig. 13.2 The flight simulator of Klaus Cappel

(see Fig. 13.2). Mechanically, Cappel's flight simulator is practically identical to Gough's tyre testing machine, but none of these inventions looks like Stewart's flight simulator.

In a paper commemorating the *Journal of Mechanical Engineering Science*'s 50th anniversary, Fichter et al. (2009) made a brief review, including reprints, of the seminal contributions of Gough and Stewart. Prior to Fichter's paper, the differences between the original flight simulator introduced by Stewart and the tyre testing machine built by Gough were carefully examined by Bonev (2011).

In this chapter, as a tribute to Stewart's extraordinary theoretical paper, we use screw theory to address the infinitesimal kinematics of the Stewart platform. For the sake of completeness, the displacement analysis of the Stewart platform is also investigated.

13.2 Description of the Stewart Platform

The Stewart platform (Fig. 13.1), as explained by his inventor, is a 6-DOF mechanism controlled in any pose by six motors, each having a ground abutment, which may form an elegant design for simulating flight conditions in the training of pilots (Stewart 1966). Other applications proposed by Stewart include the study of the behavior of bodies subject to spatial general forces or random motions, helicopter simulators, and new topologies for machine tools. As Figs. 12.1 and 13.1 show, evidently the Stewart and Gough platforms are mechanically different. Any kinematician who has tried to solve the forward displacement analysis of the tyre testing machine credited to Gough can perfectly understand that both robot manipulators are definitively different. In what follows, a technical description of the Stewart platform is given.

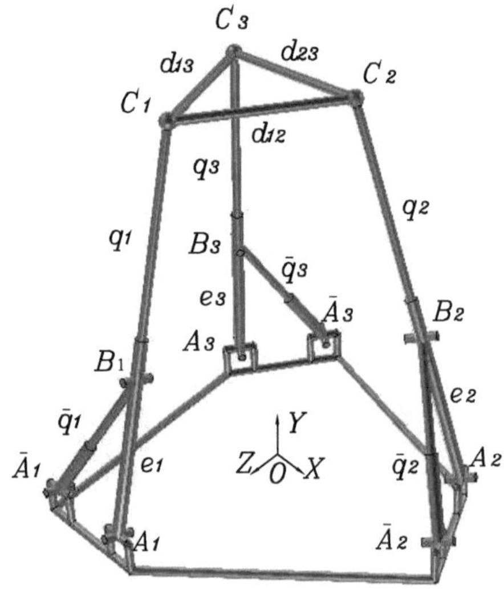

Fig. 13.3 Geometric scheme of the Stewart platform

With reference to Fig. 13.3, the Stewart platform is a parallel manipulator composed of a triangular moving platform and a fixed platform connected to each other by three adjustable-length limbs. Each limb, called *a double leg* here, is built with a hydraulic jack (primary limb) whose orientation is constrained by a second jack (secondary limb) attached to the distal end of the first cylinder. In order to describe the geometry of the robot, the reference coordinate system XYZ (see Fig. 13.3) is selected in such a way that its origin is located at the center O of the fixed platform with the Y-axis normal to the plane of the fixed platform. The geometric centers of the spherical joints, connecting the primary limbs to the moving platform, are denoted by C_i (unless otherwise specified in this chapter, consider $i = 1, 2, 3$). These points are located by vectors $\boldsymbol{c}_i$. The nominal positions of the universal joints, connecting the primary limbs to the fixed platform, are denoted by points $A_i = (ax_i, 0, az_i)$ located by vectors $\boldsymbol{a}_i$. Furthermore, in a double leg the primary and secondary limbs are linked to each other through a revolute joint whose nominal position $B_i = (X_i, Y_i, Z_i)$ is located by vector $\boldsymbol{b}_i$. On the other hand, the secondary limbs are connected to the fixed platform by universal joints whose nominal positions are denoted by points $\bar{A}_i = (\bar{ax}_i, 0, \bar{az}_i)$ located by vectors $\bar{\boldsymbol{a}}_i$. The distance between the centers of adjacent spherical joints are denoted by d_{ij}, while e_i indicates the distance between points A_i and B_i. In a double leg, the axes of the ith universal joint are denoted by unit vectors $\boldsymbol{u}_i$ and $\boldsymbol{v}_i$, where the axis $\boldsymbol{u}_i$ is along the line $A_i\bar{A}_i$, while the vector $\boldsymbol{v}_i$, which is also parallel to the axis of the revolute joint in the same double leg, is normal to the plane $A_i\bar{A}_iB_i$. Finally, q_i denotes the generalized coordinate associated to the prismatic joint of the ith primary limb, whereas $\bar{q}_i$ denotes the generalized coordinate representing the prismatic joint of the secondary limb.

13.3 Finite Kinematics of the Stewart Platform

In this section the displacement analysis of the Stewart platform is reported. To this end, the geometric scheme of the robot is sketched in Fig. 13.3.

13.3.1 Forward Displacement Analysis of the Stewart Platform

The forward displacement analysis of the Stewart platform consists of finding the pose of the triangular moving platform with respect to the fixed platform for a given set $\{q_i, \bar{q}_i\}$ of generalized coordinates. Evidently this analysis is equivalent to computing the coordinates of points $C_i = (X_i, Y_i, Z_i)$ denoting the centers of the spherical joints attached to the moving platform. To this end, according to the parameters e_i and $\bar{q}_i$, six closure equations may be written as follows:

$$(\boldsymbol{b}_i - \boldsymbol{a}_i) \bullet (\boldsymbol{b}_i - \boldsymbol{a}_i) = {e_i}^2 \tag{13.1}$$

and

$$(\boldsymbol{b}_i - \bar{\boldsymbol{a}}_i) \bullet (\boldsymbol{b}_i - \bar{\boldsymbol{a}}_i) = \bar{q}_i^2. \tag{13.2}$$

Solving Eqs. (13.1) and (13.2), it follows that

$$X_i = \tilde{A}_i Z_i + \tilde{B}_i, \tag{13.3}$$

where the coefficients from Eq. (13.3) are given by

$$\begin{aligned} \tilde{A}_i &= (az_i - \bar{a}z_i)/(\bar{a}x_i - ax_i), \\ \tilde{B}_i &= (\bar{q}_i^2 - \bar{a}x_i^2 - \bar{a}z_i^2 + {az_i}^2 - {e_i}^2 + {ax_i}^2)/2(ax_i - \bar{a}x_i). \end{aligned} \tag{13.4}$$

Furthermore, from Eq. (13.1) we obtain

$$Y_i = \sqrt{\tilde{C}_i Z_i^2 + \tilde{D}_i Z_i + \tilde{E}_i}, \tag{13.5}$$

where the coefficients $\tilde{C}_i$, $\tilde{D}_i$, and $\tilde{E}_i$ are given by

$$\begin{aligned} \tilde{C}_i &= -\tilde{A}_i - 1, \\ \tilde{D}_i &= 2(-\tilde{A}_i\tilde{B}_i + \tilde{A}_i ax_i + az_i), \\ \tilde{E}_i &= -\tilde{B}_i^2 + 2\tilde{B}_i ax_i - ax_i^2 - az_i^2 + e_i^2. \end{aligned} \tag{13.6}$$

Later, the vector $\boldsymbol{c}_i$ may be computed as

$$\boldsymbol{c}_i = \boldsymbol{a}_i + q_i(\boldsymbol{b}_i - \boldsymbol{a}_i)/e_i. \tag{13.7}$$

On the other hand, according to the parameters d_{ij} it is possible to obtain three closure equations as follows:

$$(\boldsymbol{c}_i - \boldsymbol{c}_j) \bullet (\boldsymbol{c}_i - \boldsymbol{c}_j) = d_{ij}^2, \quad i,j = 1,2,3 \text{ mod}(3). \tag{13.8}$$

After a few computations, substituting Eq. (13.7) into expressions (13.8) leads to a nonlinear system of three equations in the unknowns Z_1, Z_2, and Z_3 given by

$$\begin{aligned} &A_{ij}Z_i^2Z_j^2 + B_{ij}Z_i^2Z_j + C_{ij}Z_iZ_j^2 + D_{ij}Z_i^2 + E_{ij}Z_j^2 + \\ &F_{ij}Z_iZ_j + G_{ij}Z_i + H_{ij}Z_j + I_{ij} = 0, \quad i,j = 1,2,3 \text{ mod}(3). \end{aligned} \tag{13.9}$$

Finally, the nonlinear system of Eq. (13.9) is solved using Sylvester's dialytic method of elimination.

Example 13.1. The numerical example consists of solving the forward displacement analysis of a Stewart-type platform. To do this, consider that the coordinates of points connecting the limbs to the fixed platform, the points A_i and $\bar{A}_i$, are chosen as $A_1 = (0.866, 0, 0.5)$, $\bar{A}_1 = (0.75, 0, 0.1)$, $A_2 = (0, 0, -1)$, $\bar{A}_2 = (-0.1, 0, -0.5)$, $A_3 = (-0.866, 0, 0.5)$, and $\bar{A}_3 = (-0.5, 0, 0.25)$. Meanwhile, the remaining parameters of the mechanism are given by $d_{12} = d_{23} = d_{31} = 0.75$ and $e_1 = e_2 = e_3 = 1$. On the other hand, considering that the instantaneous generalized coordinates of the robot manipulator are given by $q_1 = 1.9613$, $\bar{q}_1 = 1.0427$, $q_2 = 2.3278$, $\bar{q}_2 = 0.9343$, $q_3 = 2.0042$, and $\bar{q}_3 = 1.03059$, we are required to compute all the poses that the moving platform can reach with respect to the fixed platform.

Solution. Let $C_i = (X_i, Y_i, Z_i), i = 1, 2, 3$, be the unknown coordinates of the centers of the spherical joints mounted on the triangular moving platform. According to the formulas reported in Sect. 13.3.1, in terms of the coordinates Z_i the coordinates X_i and Y_i are obtained as

$$\left.\begin{aligned} X_1 &= 2.218445216 - 3.448275862Z_1 \\ X_2 &= 3.064582450 + 5.0Z_2 \\ X_3 &= -1.024005120 + .6830601093Z_3 \\ Y_1 &= \sqrt{-1.079108062 + 10.32720839Z_1 - 12.89060642Z_1^2} \\ Y_2 &= \sqrt{-9.391665592 - 32.64582450Z_2 - 26.0Z_2^2} \\ Y_3 &= \sqrt{0.7250343821 + 1.215853989Z_3 - 1.466571113Z_3^2} \end{aligned}\right\}. \tag{13.10}$$

Furthermore, three equations in the unknowns $Z_i (i = 1, 2, 3)$ of the manipulator are given by

$$\left.\begin{array}{l}
696.6516250 + 1684.131570Z_1 + 3684.441990Z_2 + \\
2713.74085Z_1Z_2 - 3446.421040Z_1^2 - \\
10904.18670Z_1^2Z_2 + 4771.064149Z_2^2 - \\
2623.21160Z_1Z_2^2 - 5950.80266Z_1^2Z_2^2 = 0 \\
\\
1167.437592 - 214.9868327Z_3 + 3269.441648Z_2 + \\
208.876191Z_3Z_2 - 563.0540435Z_3^2 - \\
2090.215770Z_3^2Z_2 + 2276.872647Z_2^2 + \\
674.858694Z_3Z_2^2 - 1622.502736Z_3^2Z_2^2 = 0 \\
\\
460.4789265 - 1428.281002Z_1 - 246.7242776Z_3 + \\
40.5710168Z_1Z_3 + 1143.132433Z_1^2 + \\
461.9131814Z_1^2Z_3 - 32.62443132Z_3^2 + \\
764.0218494Z_1Z_3^2 - 1054.899758Z_1^2Z_3^2 = 0
\end{array}\right\} . \tag{13.11}$$

Equation (13.11) are systematically reduced into a 16th-order univariate polynomial equation by applying Sylvester's dialytic method of elimination. The 16 solutions of Eq. (13.11) are listed in Table 13.1.

Furthermore, the imaginary solutions for the forward displacement analysis of the mechanism yield the coordinates of points C_i listed in Table 13.2.

Table 13.1 Example 13.1. Solutions of the characteristic equations

Sol.	Z_1	Z_2	Z_3
1	0.1696258517e−1	−0.4424184551	1.574188580
2	0.1023998348	−0.4463388730	1.893015072
3	0.1032053603	−0.4209969765	4.890000963
4	0.1234252301	−0.4414360108	2.491264413
5	0.4849732583	−0.6227557932	0.4968729622
6	0.5192112739	−0.6153372672	0.3834696346
7	0.6776660162	−0.9700963909	4.043356765
8	0.6994512257	−9.112924538	−0.9655678705
9	0.7045255523	−0.8240778330	−0.8302050320
10	0.7589078276	−0.8093462018	−0.4514713312
11	0.8278256605	−0.8188300025	−0.4021805390
12	0.9221839938	−0.2993994213	1.231487056
13	1.060066824	−0.8725236485	−0.4593693428
14	0.6342106049+ 0.3434751217e−1I	−0.6349783898− 0.2050544099I	0.9514005181− 0.2301986390I
15	−0.6854559084	1.298328510	1.343688867
16	0.6342106049− 0.3434751217e−1I	−0.6349783898+ 0.2050544099I	0.9514005181+ 0.2301986390I

Table 13.2 Example 13.1. Imaginary solutions of the forward displacement analysis of the Stewart platform

Sol.	C_1	C_2	C_3
1	(8.154,7.394I,−1.825)	(22.244,22.760I,4.350)	(0.656,1.077I,2.190)
2	(−2.718,3.112I,1.328)	(3.649,3.249I,0.630)	(0.503,0.085I,1.966)
3	(2.820,0.761I,−0.278)	(2.233,1.177I,0.347)	(5.511,10.680I,9.298)
4	(2.683,0.056I,−0.238)	(1.995,0.505I,0.300)	(2.227,4.634I,4.490)
5	(3.403,1.868I,−0.447)	(1.984,0.451I,0.297)	(0.972,1.999I,2.652)
6	(2.826,0.776I,−0.279)	(1.938,0.036I,0.288)	(1.408,2.992I,3.291)
7	(−1.614,1.600I,1.007)	(−2.286,0.072I,−0.556)	(−1.800,0.702I,−1.406)
8	(−2.080,2.290I,1.142)	(−2.396,0.709I,−0.578)	(−1.733,0.068I,−1.308)
9	(−1.2462,0.880I,0.901)	(−2.457,0.888I,−0.590)	(−2.319,2.280I,−2.165)
10	(−3.650,4.214I,1.598)	(−3.021,1.949I,−0.703)	(−1.811,0.757I,−1.422)
11	(−1.064,0.042I,0.848)	(−4.157,3.445I,−0.930)	(4.352,8.581I,7.601)
12	(−1.211,0.789I,0.891)	(−98.931,100.691I,−19.885)	(−2.504,2.701I,−2.437)

On the other hand, the method yields two complex solutions for the forward displacement analysis where the coordinates of points C_i are obtained as

$$\begin{aligned}
C_1 &= (-0.7706 \mp 0.2322I, 1.13172 \mp .3516I, .7632 \pm 0.6736e - 1I),\\
C_2 &= (-0.2567 \mp 2.3862I, 3.24940 \mp 0.6378e - 1I, -0.1503 \mp 0.4773I),\\
C_3 &= (0.11978 \mp 0.3151I, 1.65309 \pm .4404I, 1.4046 \mp 0.4613I).
\end{aligned}$$

Finally, there are only two real solutions generated for the forward displacement analysis. The first solution yields the coordinates for points C_i as follows:

$$\begin{aligned}
C_1 &= (0.238626487, 1.858018182, 0.4705280515),\\
C_2 &= (-0.1145196499, 2.152763876, -0.121850935),\\
C_3 &= (-.502460206, 1.970943221, 0.4937327908).
\end{aligned}$$

Meanwhile, the second solution yields the following coordinates for points C_i:

$$\begin{aligned}
C_1 &= 0.70712454e - 2, 1.762815752, 0.537679072),\\
C_2 &= (-0.2817542583e - 1, 2.148507796, -0.104582091),\\
C_3 &= (-0.657708123, 1.979617755, 0.2664498417).
\end{aligned}$$

At this point it is worth noting how different the forward displacement analyses of the Stewart and Gough platforms are. Certainly, the mathematical procedure to obtain a 40th-order univariate polynomial equation based on the handling of a rotation matrix and complicated strategies for reducing variables is a formidable task when compared with the semiclosed-form solutions generated for the Stewart platform. Furthermore, it is straightforward to demonstrate that the cost to determine the actual configuration of any parallel manipulator depends on the real available postures of the robot; for instance, robots having multiple solutions associated to their displacement analyses demand more sensors than the number required for robots with closed- or semiclosed-form solutions.

13.3.2 Inverse Displacement Analysis of the Stewart Platform

The inverse displacement analysis of the Stewart platform consists of finding the generalized coordinates of the robot given the pose of the moving platform with respect to the fixed platform. Indeed, given the coordinates of the centers of the spherical joints we are required to compute the leg lengths of the robot manipulator. The inverse displacement analysis of the Stewart platform is straightforward and is included here only for the sake of completeness. In fact, given the coordinates of points C_i, the generalized coordinates q_i are directly computed as

$$q_i = \sqrt{(\boldsymbol{c}_i - \boldsymbol{a}_i) \bullet (\boldsymbol{c}_i - \boldsymbol{a}_i)}. \tag{13.12}$$

Furthermore, in order to compute the generalized coordinates $\bar{q}_i$, consider that, according to Eq. (13.7), points B_i may be obtained considering that

$$\boldsymbol{b}_i = e_i \boldsymbol{c}_i / q_i + \boldsymbol{a}_i. \tag{13.13}$$

Afterward, the generalized coordinates $\bar{q}_i$ result in

$$\bar{q}_i = \sqrt{(\boldsymbol{b}_i - \bar{\boldsymbol{a}}_i) \bullet (\boldsymbol{b}_i - \bar{\boldsymbol{a}}_i)}. \tag{13.14}$$

As a closure note to the finite kinematic analysis of the Stewart platform, it is worth noting how easy it is to formulate the forward displacement analysis of the Stewart platform given the introduction of Eq. (13.7). Consider, for instance, that the forward displacement analysis of the Gough platform is governed by a set of seven quadratic equations (Husty 1996). Certainly, while the forward displacement analysis of the general Gough platform is a formidable task and does not allow for closed-form solutions (see, for instance, Raghavan 1993; Wampler 1995; Husty 1996; Innocenti 2001; Rolland 2005; Tarokh 2007; Wang 2007), the forward displacement analysis of the Stewart platform requires only a few paragraphs, yielding nearly closed-form solutions because of the decoupled topology of the Stewart flight simulator (Gallardo-Alvarado 2012).

13.4 Infinitesimal Kinematics of the Stewart Platform

In this section the input–output equations of velocity and acceleration of the Stewart platform are obtained by resorting to reciprocal screw theory. To approach this, the model of the screws is outlined in Fig. 13.4.

13.4.1 *Velocity Analysis of the Stewart Platform*

Let $\boldsymbol{\omega}$ be the angular velocity vector of the moving platform as observed from the fixed platform. Furthermore, let $\boldsymbol{v}_O$ be the velocity vector of a point embedded in the moving platform that is instantaneously coincident with the origin O of the reference coordinate system *XYZ*. The velocity state of the moving platform with respect to the fixed platform, the six-dimensional vector

$$ {}^0\boldsymbol{V}_O^6 = \begin{bmatrix} \boldsymbol{\omega} \\ \boldsymbol{v}_O \end{bmatrix}, $$

may be expressed in two ways through any of the double legs of the Stewart platform as follows:

$$ {}_0\omega_1^i\,{}^0\$_i^1 + {}_1\omega_2^i\,{}^1\$_i^2 + \dot{q}_i\,{}^2\$_i^3 + {}_3\omega_4^i\,{}^3\$_i^4 + {}_4\omega_5^i\,{}^4\$_i^5 + {}_5\omega_6^i\,{}^5\$_i^6 = {}^0\boldsymbol{V}_O^6 \tag{13.15} $$

and

$$ {}_0\bar{\omega}_1^i\,{}^0\bar{\$}_i^1 + {}_1\bar{\omega}_2^i\,{}^1\bar{\$}_i^2 + \dot{\bar{q}}_i\,{}^2\bar{\$}_i^3 + {}_3\bar{\omega}_4^i\,{}^3\bar{\$}_i^4 + \dot{q}_i\,{}^2\$_i^3 + {}_3\omega_4^i\,{}^3\$_i^4 + {}_4\omega_5^i\,{}^4\$_i^5 + {}_5\omega_6^i\,{}^5\$_i^6 = {}^0\boldsymbol{V}_O^6, \tag{13.16} $$

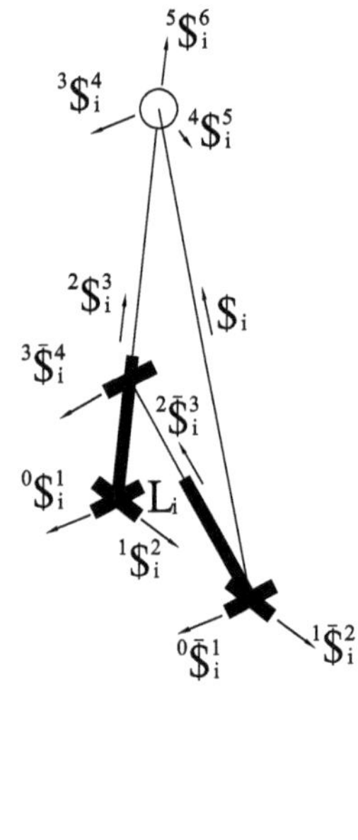

Fig. 13.4 Infinitesimal screws of the Stewart platform

where the screws are computed taking point O as the reference pole. Note that because of the selected orientation for the universal joints, it is evident that ${}^{1}\bar{\$}_i^2 = {}^{1}\$_i^2$. On the other hand, by using college kinematics, we may use the vectors $\boldsymbol{\omega}$ and $\boldsymbol{v}_O$ of the velocity state ${}^{0}\boldsymbol{V}_O^6$ to determine the velocity of any point P of the moving platform. In fact,

$$\boldsymbol{v}_P = \boldsymbol{v}_O + \boldsymbol{\omega} \times \boldsymbol{r}_{P/O}, \tag{13.17}$$

where $\boldsymbol{r}_{P/O}$ is a vector pointed from O to P.

Equations (13.15) and (13.16) are the fundamental expressions to obtain the input–output equation of velocity of the robot. As an intermediate step, it is necessary to first compute the joint-velocity rate ${}_3\bar{\omega}_4^i$. To achieve this, in agreement with the closed kinematic chain L_i formed with the lower cylinders of the double leg, it is possible to write the following closure equations:

$${}_0\omega_1^i\, {}^{0}\$_i^1 + {}_1\omega_2^i\, {}^{1}\$_i^2 - {}_0\bar{\omega}_1^i\, {}^{0}\bar{\$}_i^1 - {}_1\bar{\omega}_2^i\, {}^{1}\bar{\$}_i^2 - \dot{\bar{q}}_i\, {}^{2}\bar{\$}_i^3 - {}_3\bar{\omega}_4^i\, {}^{3}\bar{\$}_i^4 = \mathbf{0}. \tag{13.18}$$

Hence, by applying the Klein form of the screw ${}^{1}\$_i^2$ to both sides of Eq. (13.18) and reducing terms, it follows that

$${}_3\bar{\omega}_4^i = \{{}^{1}\$_i^2; -\dot{\bar{q}}_i{}^{2}\bar{\$}_i^3\}/\{{}^{1}\$_i^2; {}^{3}\bar{\$}_i^4\}. \tag{13.19}$$

Furthermore, taking into account that the screw ${}^{5}\$_i^6$ is reciprocal to all the screws, excluding the screw associated to the prismatic joint, in the same primary limb, applying the Klein form of the screw ${}^{5}\$_i^6$ to both sides of Eq. (13.15) and canceling terms yields

$$\dot{q}_i = \{{}^{5}\$_i^6; {}^{0}\boldsymbol{V}_O^6\}. \tag{13.20}$$

Following a similar procedure, from Eq. (13.16) one obtains

$$\{\dot{\bar{q}}_i \$_i^P + \dot{q}_i{}^{2}\$_i^3; \$_i\} = \{\$_i; {}^{0}\boldsymbol{V}_O^6\}, \tag{13.21}$$

where $\$_i^P = {}^{2}\bar{\$}_i^3 - \frac{\{{}^{1}\$_i^2; {}^{2}\bar{\$}_i^3\}}{\{{}^{1}\$_i^2; {}^{3}\bar{\$}_i^4\}}{}^{3}\bar{\$}_i^4$ and $\$_i$ denotes a line in Plücker coordinates directed from point $\bar{A}_i$ to point C_i.

Casting expressions (13.20) and (13.21) into matrix-vector form, we find that the input–output equation of velocity of the robot results in

$$\mathbf{A}\, {}^{0}\boldsymbol{V}_O^6 = \mathbf{B}\, \dot{\mathbf{q}}, \tag{13.22}$$

where $\mathbf{A} = \mathbf{J}^{\mathrm{T}}\ \boldsymbol{\Delta}$, so that $\mathbf{J} = \left[{}^{5}\$_1^6\ {}^{5}\$_2^6\ {}^{5}\$_3^6\ \$_1\ \$_2\ \$_3\right]$ is the screw-coordinate Jacobian matrix of the robot, and $\boldsymbol{\Delta}$ is an operator of polarity. Meanwhile, $\dot{\mathbf{q}} = \left[\dot{q}_1\ \dot{q}_2\ \dot{q}_3\ \dot{\bar{q}}_1\ \dot{\bar{q}}_2\ \dot{\bar{q}}_3\right]^T$ is the first-order driver matrix of the robot. Furthermore,

$$\mathbf{B} = \begin{bmatrix} 1 & 0 & 0 & 0 & 0 & 0 \\ 0 & 1 & 0 & 0 & 0 & 0 \\ 0 & 0 & 1 & 0 & 0 & 0 \\ \{^2\$_1^3; \$_1\} & 0 & 0 & \{\$_1^P; \$_1\} & 0 & 0 \\ 0 & \{^2\$_2^3; \$_2\} & 0 & 0 & \{\$_2^P; \$_2\} & 0 \\ 0 & 0 & \{^2\$_3^3; \$_3\} & 0 & 0 & \{\$_3^P; \$_3\} \end{bmatrix} \tag{13.23}$$

is called the first-order coefficient matrix of the robot.

13.4.2 Acceleration Analysis of the Stewart Platform

Let $\boldsymbol{\alpha}$ be the angular acceleration vector of the moving platform with respect to the fixed platform. Furthermore, let $\boldsymbol{a}_O$ be the acceleration vector of point O. The reduced acceleration state, or accelerator for brevity, of the moving platform with respect to the fixed platform, the six-dimensional vector

$$^0\mathbf{A}_O^6 = \begin{bmatrix} \boldsymbol{\alpha} \\ \boldsymbol{a}_O - \boldsymbol{\omega} \times \boldsymbol{v}_O \end{bmatrix},$$

may be expressed through any of the double legs of the Stewart platform as

$${}_0\alpha_1^i\,{}^0\$_i^1 + {}_1\alpha_2^i\,{}^1\$_i^2 + \ddot{q}_i\,{}^2\$_i^3 + {}_3\alpha_4^i\,{}^3\$_i^4 + {}_4\alpha_5^i\,{}^4\$_i^5 + {}_5\alpha_6^i\,{}^5\$_i^6 + {}^0\mathcal{A}_{Oi}^6 = {}^0\mathbf{A}_O^6 \tag{13.24}$$

and

$${}_0\bar{\alpha}_1^i\,{}^0\bar{\$}_i^1 + {}_1\bar{\alpha}_2^i\,{}^1\bar{\$}_i^2 + \ddot{\bar{q}}_i\,{}^2\bar{\$}_i^3 + {}_3\bar{\alpha}_4^i\,{}^3\bar{\$}_i^4 + \ddot{\bar{q}}_i\,{}^2\$_i^3 + {}_3\alpha_4^i\,{}^3\$_i^4 + {}_4\alpha_5^i\,{}^4\$_i^5 + \\ {}_5\alpha_6^i\,{}^5\$_i^6 + {}^0\bar{\mathcal{A}}_{Oi}^6 = {}^0\mathbf{A}_O^6. \tag{13.25}$$

Here the Lie screws of acceleration ${}^0\mathcal{A}_{Oi}^6$ and ${}^0\bar{\mathcal{A}}_{Oi}^6$ are computed, respectively, as

$$\begin{aligned} {}^0\mathcal{A}_{Oi}^6 = \big[{}_0\omega_1^i\,{}^0\$_i^1 \quad {}_1\omega_2^i\,{}^1\$_i^2 + \dot{q}_i\,{}^2\$_i^3 + {}_3\omega_4^i\,{}^3\$_i^4 + {}_4\omega_5^i\,{}^4\$_i^5 + {}_5\omega_6^i\,{}^5\$_i^6\big] + \\ \big[{}_1\omega_2^i\,{}^1\$_i^2 \quad \dot{q}_i\,{}^2\$_i^3 + {}_3\omega_4^i\,{}^3\$_i^4 + {}_4\omega_5^i\,{}^4\$_i^5 + {}_5\omega_6^i\,{}^5\$_i^6\big] + \\ \big[\dot{q}_i\,{}^2\$_i^3 \quad {}_3\omega_4^i\,{}^3\$_i^4 + {}_4\omega_5^i\,{}^4\$_i^5 + {}_5\omega_6^i\,{}^5\$_i^6\big] + \\ \big[{}_3\omega_4^i\,{}^3\$_i^4 \quad {}_4\omega_5^i\,{}^4\$_i^5 + {}_5\omega_6^i\,{}^5\$_i^6\big] + \big[{}_4\omega_5^i\,{}^4\$_i^5 \quad {}_5\omega_6^i\,{}^5\$_i^6\big] \end{aligned}$$

and

$$
\begin{aligned}
&{}^{0}\bar{\mathcal{A}}^{6}_{Oi} = \\
&\left[{}_{0}\bar{\omega}^{i}_{1}\,{}^{0}\bar{\$}^{1}_{i} \quad {}_{1}\bar{\omega}^{i}_{2}\,{}^{1}\bar{\$}^{2}_{i} + \dot{\bar{q}}_{i}\,{}^{2}\bar{\$}^{3}_{i} + {}_{3}\bar{\omega}^{i}_{4}\,{}^{3}\bar{\$}^{4}_{i} + \dot{\bar{q}}_{i}\,{}^{2}\$^{3}_{i} + {}_{3}\omega^{i}_{4}\,{}^{3}\$^{4}_{i} + {}_{4}\omega^{i}_{5}\,{}^{4}\$^{5}_{i} + {}_{5}\omega^{i}_{6}\,{}^{5}\$^{6}_{i}\right] \\
&+\left[{}_{1}\bar{\omega}^{i}_{2}\,{}^{1}\bar{\$}^{2}_{i} \quad \dot{\bar{q}}_{i}\,{}^{2}\bar{\$}^{3}_{i} + {}_{3}\bar{\omega}^{i}_{4}\,{}^{3}\bar{\$}^{4}_{i} + \dot{\bar{q}}_{i}\,{}^{2}\$^{3}_{i} + {}_{3}\omega^{i}_{4}\,{}^{3}\$^{4}_{i} + {}_{4}\omega^{i}_{5}\,{}^{4}\$^{5}_{i} + {}_{5}\omega^{i}_{6}\,{}^{5}\$^{6}_{i}\right] + \\
&\left[\dot{\bar{q}}_{i}\,{}^{2}\bar{\$}^{3}_{i} \quad {}_{3}\bar{\omega}^{i}_{4}\,{}^{3}\bar{\$}^{4}_{i} + \dot{\bar{q}}_{i}\,{}^{2}\$^{3}_{i} + {}_{3}\omega^{i}_{4}\,{}^{3}\$^{4}_{i} + {}_{4}\omega^{i}_{5}\,{}^{4}\$^{5}_{i} + {}_{5}\omega^{i}_{6}\,{}^{5}\$^{6}_{i}\right] + \\
&\left[{}_{3}\bar{\omega}^{i}_{4}\,{}^{3}\bar{\$}^{4}_{i} \quad \dot{\bar{q}}_{i}\,{}^{2}\$^{3}_{i} + {}_{3}\omega^{i}_{4}\,{}^{3}\$^{4}_{i} + {}_{4}\omega^{i}_{5}\,{}^{4}\$^{5}_{i} + {}_{5}\omega^{i}_{6}\,{}^{5}\$^{6}_{i}\right] + \\
&\left[\dot{\bar{q}}_{i}\,{}^{2}\$^{3}_{i} \quad {}_{3}\omega^{i}_{4}\,{}^{3}\$^{4}_{i} + {}_{4}\omega^{i}_{5}\,{}^{4}\$^{5}_{i} + {}_{5}\omega^{i}_{6}\,{}^{5}\$^{6}_{i}\right] + \\
&\left[{}_{3}\omega^{i}_{4}\,{}^{3}\$^{4}_{i} \quad {}_{4}\omega^{i}_{5}\,{}^{4}\$^{5}_{i} + {}_{5}\omega^{i}_{6}\,{}^{5}\$^{6}_{i}\right] + \left[{}_{4}\omega^{i}_{5}\,{}^{4}\$^{5}_{i} + {}_{5}\omega^{i}_{6}\,{}^{5}\$^{6}_{i}\right].
\end{aligned}
$$

On the other hand, to compute the joint-acceleration rate ${}_{3}\bar{\alpha}^{i}_{4}$, according to the closed loop L_i let us consider that

$$
{}_{0}\alpha^{i}_{1}\,{}^{0}\$^{1}_{i} + {}_{1}\alpha^{i}_{2}\,{}^{1}\$^{2}_{i} - {}_{0}\bar{\alpha}^{i}_{1}\,{}^{0}\bar{\$}^{1}_{i} - {}_{1}\bar{\alpha}^{i}_{2}\,{}^{1}\bar{\$}^{2}_{i} - \ddot{\bar{q}}_{i}\,{}^{2}\bar{\$}^{3}_{i} - {}_{3}\bar{\alpha}^{i}_{4}\,{}^{3}\bar{\$}^{4}_{i} + \mathcal{A}_i = \mathbf{0}, \tag{13.26}
$$

where the Lie screw of acceleration $\mathcal{A}_i$ is given by

$$
\begin{aligned}
\mathcal{A}_i = &\left[{}_{0}\omega^{i}_{1}\,{}^{0}\$^{1}_{i} \quad {}_{1}\omega^{i}_{2}\,{}^{1}\$^{2}_{i} - {}_{0}\bar{\omega}^{i}_{1}\,{}^{0}\bar{\$}^{1}_{i} - {}_{1}\bar{\omega}^{i}_{2}\,{}^{1}\bar{\$}^{2}_{i} - \dot{\bar{q}}_{i}\,{}^{2}\bar{\$}^{3}_{i} - {}_{3}\bar{\omega}^{i}_{4}\,{}^{3}\bar{\$}^{4}_{i}\right] - \\
&\left[{}_{1}\omega^{i}_{2}\,{}^{1}\$^{2}_{i} \quad {}_{0}\bar{\omega}^{i}_{1}\,{}^{0}\bar{\$}^{1}_{i} + {}_{1}\bar{\omega}^{i}_{2}\,{}^{1}\bar{\$}^{2}_{i} + \dot{\bar{q}}_{i}\,{}^{2}\bar{\$}^{3}_{i} + {}_{3}\bar{\omega}^{i}_{4}\,{}^{3}\bar{\$}^{4}_{i}\right] + \\
&\left[{}_{0}\bar{\omega}^{i}_{1}\,{}^{0}\bar{\$}^{1}_{i} \quad {}_{1}\bar{\omega}^{i}_{2}\,{}^{1}\bar{\$}^{2}_{i} + \dot{\bar{q}}_{i}\,{}^{2}\bar{\$}^{3}_{i} + {}_{3}\bar{\omega}^{i}_{4}\,{}^{3}\bar{\$}^{4}_{i}\right] + \left[{}_{1}\bar{\omega}^{i}_{2}\,{}^{1}\bar{\$}^{2}_{i} \quad \dot{\bar{q}}_{i}\,{}^{2}\bar{\$}^{3}_{i} + {}_{3}\bar{\omega}^{i}_{4}\,{}^{3}\bar{\$}^{4}_{i}\right] + \\
&\left[\dot{\bar{q}}_{i}\,{}^{2}\bar{\$}^{3}_{i} \quad {}_{3}\bar{\omega}^{i}_{4}\,{}^{3}\bar{\$}^{4}_{i}\right].
\end{aligned}
$$

Following the trend of the velocity analysis, the joint-acceleration rate ${}_{3}\bar{\alpha}^{i}_{4}$ of the manipulator is computed based on Eq. (13.26) as

$$
{}_{3}\bar{\alpha}^{i}_{4} = \{{}^{1}\$^{2}_{i}; -\ddot{\bar{q}}_{i}{}^{2}\bar{\$}^{3}_{i}\}/\{{}^{1}\$^{2}_{i}; {}^{3}\bar{\$}^{4}_{i}\} + \{{}^{1}\$^{2}_{i}; \mathcal{A}_i\}/\{{}^{1}\$^{2}_{i}; {}^{3}\bar{\$}^{4}_{i}\}. \tag{13.27}
$$

Finally, the input–output equation of acceleration of the Stewart platform results in

$$
\mathbf{A}\,{}^{0}\mathbf{A}^{6}_{O} = \mathbf{B}\,\ddot{\mathbf{q}} + \mathbf{C}, \tag{13.28}
$$

where $\ddot{\mathbf{q}} = \left[\ddot{q}_1\ \ddot{q}_2\ \ddot{q}_3\ \ddot{\bar{q}}_1\ \ddot{\bar{q}}_2\ \ddot{\bar{q}}_3\right]^T$ is the second-order driver matrix of the robot. Meanwhile,

$$
\mathbf{C} = \begin{bmatrix} \left\{ {}^{5}\$_{1}^{6}; {}^{0}\mathcal{A}_{O1}^{6} \right\} \\ \left\{ {}^{5}\$_{2}^{6}; {}^{0}\mathcal{A}_{O2}^{6} \right\} \\ \left\{ {}^{5}\$_{3}^{6}; {}^{0}\mathcal{A}_{O3}^{6} \right\} \\ \left\{ \$_{1}; {}^{0}\bar{\mathcal{A}}_{O1} + \frac{\{ {}^{1}\$_{1}^{2}; \mathcal{A}_{1} \}}{\{ {}^{1}\$_{1}^{2}; {}^{3}\bar{\$}_{1}^{4} \}} {}^{3}\bar{\$}_{1}^{4} \right\} \\ \left\{ \$_{2}; {}^{0}\bar{\mathcal{A}}_{O2} + \frac{\{ {}^{1}\$_{2}^{2}; \mathcal{A}_{2} \}}{\{ {}^{1}\$_{2}^{2}; {}^{3}\bar{\$}_{2}^{4} \}} {}^{3}\bar{\$}_{2}^{4} \right\} \\ \left\{ \$_{3}; {}^{0}\bar{\mathcal{A}}_{O3} + \frac{\{ {}^{1}\$_{3}^{2}; \mathcal{A}_{3} \}}{\{ {}^{1}\$_{3}^{2}; {}^{3}\bar{\$}_{3}^{4} \}} {}^{3}\bar{\$}_{3}^{4} \right\} \end{bmatrix} \tag{13.29}
$$

is called the complementary matrix of acceleration of the Stewart platform.

Example 13.2. The example consists of computing the time history of the velocity and acceleration of the center of the moving platform of a Stewart-type platform given the generalized coordinates, and their time derivatives, of the parallel manipulator. To this end, using SI units through the numerical example, consider that the corners of the fixed platform are given by the coordinates $A_1 = (0.25, 0, 0.433)$, $A_2 = (0.25, 0, -0.433)$, $A_3 = (-0.5, 0, 0)$, $\bar{A}_1 = (0.03349, 0, 0.558)$, $\bar{A}_2 = (0.4665, 0, -0.308)$, and $\bar{A}_3 = (-0.5, 0, -0.25)$ while the remaining parameters of the mechanism are chosen as $e_1 = e_2 = e_3 = 0.4$, $d_{12} = d_{23} = d_{13} = 0.41773$. On the other hand, consider that in the reference configuration of the flight simulator also known as the home position (see Fig. 13.3), it follows that $B_1 = (0.198, 0.386, 0.343)$, $B_2 = (0.198, 0.386, -0.343)$, and $B_3 = (-0.396, 0.386, 0)$, whereas the coordinates of the centers of the spherical joints mounted on the moving platform are given by $C_1 = (0.120, 0.965, 0.208)$, $C_2 = (0.120, 0.965, -0.208)$, and $C_3 = (-0.241, 0.965, 0)$. Furthermore, the generalized coordinates of the mechanism, upon the reference configuration of the parallel manipulator, are commanded to follow periodic functions given by $q_1 = -0.1\sin(t)\cos(t)$, $\bar{q}_1 = 0.05\sin(t)\cos(t)$, $q_2 = 0.1\sin(t)$, $\bar{q}_2 = -0.075\sin(t)$, $q_3 = 0.15\sin(t)$, and $\bar{q}_3 = 0.05\sin(t)\cos(t)$, where the time t is confined to the interval $0 < t < 2\pi$.

Solution. Taking into account that only one instantaneous solution of the forward displacement analysis is required in the example, we used the Newton–Raphson method to solve the finite kinematics of the robot manipulator. After this consideration, the resulting time history of the center of the moving platform by applying the theory of screws is given in Fig. 13.5.

Finally, it is worth closing this example by noting that although there are no physical prototypes following the topology of the Stewart platform, the kinematic analysis of such a robot is an interesting task that would help us to understand the big differences between the Stewart and Gough platforms.

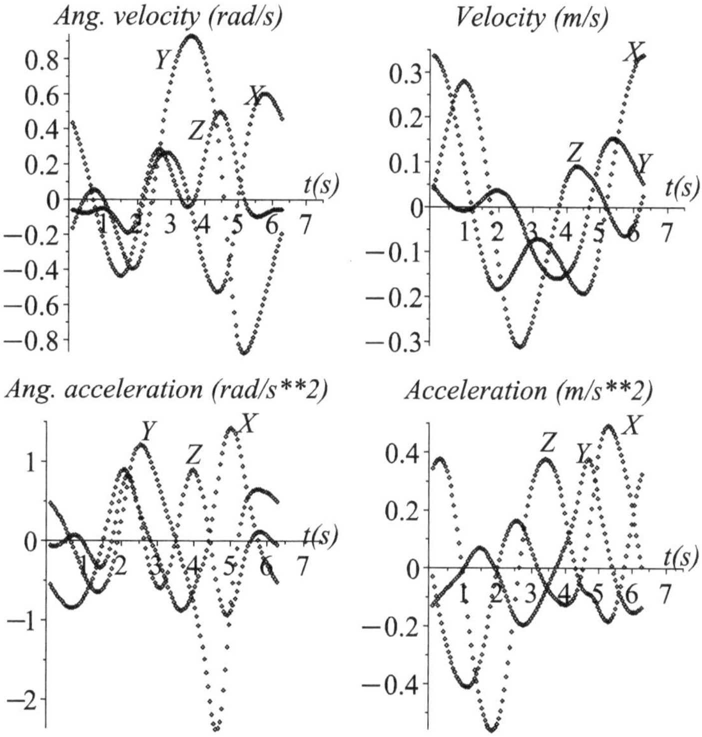

Fig. 13.5 Example 13.2. Temporal behavior of the velocity and acceleration of the center of the moving platform of a Stewart-type platform

13.4.3 Singularity Analysis of the Stewart Platform

The infinitesimal kinematics of the Stewart platform is completed by studying its singularities. The input–output equation of velocity is the means to elucidate the singularities of the Stewart platform. Hence, matrices **A** and **B** [see expression (13.22)] are studied in order to approach the singularity analysis of the robot under study. The Stewart platform is at a singular configuration when (1) matrix **A** is singular, known as direct singularity; (2) matrix **B** is singular, known as inverse singularity; (3) matrices **A** and **B** are both singular, known as combined singularity.

13.4.3.1 Direct Singularity of the Stewart Platform

Taking into account that $\mathbf{A} = \mathbf{J}^{\mathrm{T}} \boldsymbol{\Delta}$, then matrix **A** becomes singular when the elements of matrix $\mathbf{J} = [{}^5\$_1^6 \quad {}^5\$_2^6 \quad {}^5\$_3^6 \quad \$_1 \quad \$_2 \quad \$_3]$ are linearly dependent, for example, when the directions of the screws are coplanar or, in other words, when the double legs are coplanar. In general, at a direct singularity the manipulator loses degrees of freedom; more specifically, if the double legs are coplanar, then the Stewart platform loses 3 of its 6 DOF. Hence, there are two main possibilities to be considered: (1) The actuators $\bar{q}_i$ assume the role of passive kinematic pairs and depend on the actuators q_i; (2) the actuators q_i assume the role of passive kinematic pairs and depend on the actuators $\bar{q}_i$. The first possibility implies that the Stewart platform with coplanar double legs is kinematically equivalent to the 3-RPR (where R represents revolute joints) 3-DOF planar parallel manipulator. The singularity analysis of nonredundant planar parallel manipulators was successfully achieved by Bonev et al. (2003), a fine piece of work, by applying the theory of screws, and therefore this possibility is not included in the chapter. On the other hand, to determine the direct singularity loci of the second possibility, consider a submatrix $\bar{\mathbf{J}} = [{}^5\$_1^6 \, {}^5\$_2^6 \, {}^5\$_3^6]$. The direct loci singularity is expressed as $\det(\bar{\mathbf{J}}) = 0$, where the screws of $\bar{\mathbf{J}}$ are computed considering that $B_i = (X_i, 0, Z_i)$. Hence, solving Eqs. (13.2) and (13.3), it follows that

$$X_i = \tilde{a}_i + \tilde{b}_i\bar{q}_i^2 + \tilde{c}_i Z_i, \quad Z_i = \tilde{d}_i + \tilde{e}_i\bar{q}_i^2 + \sqrt{\tilde{f}_i + \tilde{g}_i\bar{q}_i^2 + \tilde{h}_i\bar{q}_i^4}. \tag{13.30}$$

Furthermore, taking into account that in Plücker coordinates the screw ${}^5\$_i^6$ is given by

$${}^5\$_i^6 = \begin{bmatrix} {}^5\hat{\boldsymbol{s}}_i^6 \\ \boldsymbol{a}_i \times {}^5\hat{\boldsymbol{s}}_i^6 \end{bmatrix}, \tag{13.31}$$

where the primal part ${}^5\hat{\boldsymbol{s}}_i^6$ is computed as

$${}^5\hat{\boldsymbol{s}}_i^6 = (\boldsymbol{b}_i - \boldsymbol{a}_i)/e_i, \tag{13.32}$$

then, with the removal of rows with zero values, we see that the screws of matrix $\bar{\mathbf{J}}$ are given by

$${}^5\$_i^6 = \begin{bmatrix} \tilde{i}_i + \tilde{j}_i\bar{q}_i^2 + \sqrt{\tilde{k}_i + \tilde{l}_i\bar{q}_i^2 + \tilde{m}_i\bar{q}_i^4} \\ \tilde{n}_i + \tilde{p}_i\bar{q}_i^2 + \sqrt{\tilde{q}_i + \tilde{r}_i\bar{q}_i^2 + \tilde{s}_i\bar{q}_i^4} \\ \tilde{t}_i + \tilde{u}_i\bar{q}_i^2 + \sqrt{\tilde{v}_i + \tilde{w}_i\bar{q}_i^2 + \tilde{x}_i\bar{q}_i^4} \end{bmatrix}. \tag{13.33}$$

Then the computation of the determinant of $\bar{\mathbf{J}}$ is straightforward. Furthermore, as proved by Stewart, another direct singularity occurs when the primal parts of the screws of $\bar{\mathbf{J}}$ are along the Y-axis, which implies that ${}^5\$_i^6 = [1 \quad -az_i \quad ax_i]^T$. Computing the direct singularity loci in accordance with this possibility is a trivial endeavor.

13.4.3.2 Inverse Singularity of the Stewart Platform

Clearly, matrix $\mathbf{B}$ is singular if and only if $\det(\mathbf{B}) = 0$. Hence, the condition of inverse singularity of the Stewart platform is given by

$$\{\$_1^P; \$_1\}\,\{\$_2^P; \$_2\}\,\{\$_3^P; \$_3\} = 0. \tag{13.34}$$

Consider that computing any of the Klein forms of Eq. (13.34) leads to

$$\{{}^{2}\bar{\$}_i^3 - \frac{\{{}^{1}\$_i^2; {}^{2}\bar{\$}_i^3\}}{\{{}^{1}\$_i^2; {}^{3}\bar{\$}_i^4\}} {}^{3}\bar{\$}_i^4; \$_i\} = 0. \tag{13.35}$$

Therefore, an inverse singularity occurs when in a double leg the corresponding points A_i, $\bar{A}_i$, and B_i are collinear; in other words, an inverse singularity occurs only when a double leg of the Stewart platform is fully extended or folded.

13.4.3.3 Combined Singularity of the Stewart Platform

If the lines ${}^{5}\$_1^6$, ${}^{5}\$_2^6$, and ${}^{5}\$_3^6$ are coplanar, then $\det(\mathbf{A}) = 0$. Furthermore, if in a double leg the points A_i, $\bar{A}_i$, and B_i are collinear, then $\det(\mathbf{B}) = 0$. This is known as a combined singularity, for example, $\det(\mathbf{A}) = \det(\mathbf{B}) = 0$.

Although the singularity analysis of the Stewart platform developed in this chapter is not exhaustive, it provides some insight about the *illness regions* of the robot at hand. The right interpretation and comprehension of such results must be taken into proper account at the design stage of the robot in order to avoid damage to physical prototypes. Finally, in terms of singular configurations, it is worth noting that compared with the Gough platform, the Stewart platform has fewer singularities because of its decoupled topology; see, for instance, St-Onge and Gosselin (2000), Bandyopadhyay and Ghosal (2006), and Li et al. (2006, 2007). The fact that not all the active limbs of the Stewart platform connect the moving and fixed platforms reduces the presence of singular regions into the workspace of the robot. This is another relevant argument that ensures that the Gough and Stewart platforms are robots possessing vastly different kinematic characteristics.

13.5 Exercises

1. In your own words, what are the main differences between the Gough and Stewart platforms?
2. Calculating the infinitesimal kinematics of the Stewart platform by resorting to screw theory appears to be a very involved procedure. The introduction of virtual kinematic chains is a viable option to simplify the kinematic analysis of parallel manipulators. Explore this possibility in order to simplify the velocity analysis of the Stewart platform.
3. Investigate possible applications of the Stewart platform.

References

Bandyopadhyay, S., & Ghosal, A. (2006). Geometric characterization and parametric representation of the singularity manifold of a 6-6 Stewart platform manipulator. *Mechanism and Machine Theory, 41*(11), 1377–1400.

Bonev, I. A. (2011). *The true origins of parallel robots*. Available at Parallel Mic Center, ParalleMIC online review. http://www.parallemic.org/Reviews/Review007.html.

Bonev, I. A., Zlatanov, D., & Gosselin, C. M. (2003). Singularity analysis of 3-DOF planar parallel mechanisms via screw theory. *ASME Journal of Mechanical Design, 125*(3), 573–581.

Cappel, K. L. (2002). Letter to the editor: six degrees of freedom. In *Mechanical engineering*. New York: ASME.

Fichter, E. F. (1986). A Stewart platform-based manipulator: general theory and practical construction. *International Journal of Robotics Research, 5*(2), 157–182.

Fichter, E. F., Kerr, D. R., & Rees-Jones, J. (2009). The Gough–Stewart platform parallel manipulator: a retrospective appreciation. *Proceedings Institution of Mechanical Engineers Part C: Journal of Mechanical Engineering Science, 223*(1), 243–281. JMES's 50th Anniversary Issue Paper.

Gallardo-Alvarado, J. (2012). Enhancing the originality of the Stewart platform through its kinematic analyses. *Proceedings of the Institution of Mechanical Engineers, Part C: Journal of Mechanical Engineering Science, 226*(12), 3013–3025.

Gough, V. E. (1957). Contribution to discussion of papers on research in automobile stability. *Proceedings of the Automobile Division of the Institution of Mechanical Engineers, 171*, 392–394.

Gough, V.E. (1966) Communications to Stewart, D. (1965–1966). A platform with six degrees of freedom. *Proceedings Institution of Mechanical Engineers, 180*(1), 371–386, pp. 379–380.

Gough, V. E., & Whitehall, S. G. (1962). Universal tyre testing machine. In G. Eley (Ed.), *Proceedings of the 9th International Automobile Technical Congress*, discussion pp. 250ff, Fédération Internationale des Sociétés d'Ingénieurs des Techniques de lÁutomobile (FISITA) (pp. 117–137). London: IMechE 1.

Husty, M. L. (1996). An algorithm for solving the direct kinematic of general Stewart–Gough platforms. *Mechanism and Machine Theory, 31*(4), 365–380.

Innocenti, C. (2001). Forward kinematics in polynomial form of the general Stewart platform. *ASME Journal of Mechanical Design, 123*(2), 254–260.

Korobeynikov, A. V., & Turlapov, V. E. (2005). Modeling and evaluating of the Stewart platform. In *International Conference Graphicon 2005*, Novosibirsk: Akademgorodok.

Lee, S.-H., Song, J.-B., Choi, W. C., & Hong, D. (2003). Position control of a Stewart platform using inverse dynamics control with approximate dynamics. *Mechatronics, 13*(6), 605–619.

Li, H. D., Gosselin, C. M., Richard, M. J., & St-Onge, B. M. (2006). Analytic form of the six-dimensional singularity locus of the general Gough-Stewart platform. *ASME Journal of Mechanical Design, 128*(1), 279–287.

Li, H., Gosselin, C. M., & Richard, M. J. (2007). Determination of the maximal singularity-free zones in the six-dimensional workspace of the general Gough-Stewart platform. *Mechanism and Machine Theory, 42*(4), 497–511.

Merkle, R. C. (1997). A new family of six degree of freedom positional devices. *Nanotechnology, 8*(2), 47–52.

Raghavan, M. (1993). The Stewart platform of general geometry has 40 configurations. *ASME Journal of Mechanical Design, 115*(2), 277–282.

Rolland, L. (2005). Certified solving of the forward kinematics problem with an exact algebraic method for the general parallel manipulator. *Advanced Robotics, 19*(9), 995–1025.

Stewart, D. (1965–1966). A platform with six degrees of freedom. *Proceedings Institution of Mechanical Engineers, 180*(1), 371–386.

Stewart, D. (1966). A platform with six degrees of freedom: a new form of mechanical linkage which enables a platform to move simultaneously in all six degrees of freedom developed by Elliott-automation. *Aircraft Engineering and Aerospace Technology, 38*(4), 30–35.

St-Onge, B. M., & Gosselin, C. (2000). Singularity analysis and representation of the general Gough–Stewart platform. *International Journal of Robotics Research, 19*(3), 271–288.

Tarokh, M. (2007). Real time forward kinematics solutions for general Stewart platforms. In *Proceedings IEEE International Conference on Robotics and Automation, Rome* (pp. 901–906).

Wampler, C. W. (1995). Forward displacement analysis of general six-in-parallel SPS (Stewart) platform manipulators using SOMA coordinates. *Mechanism and Machine Theory, 31*(3), 331–337.

Wang, Y. (2007). A direct numerical solution to forward kinematics of general Stewart–Gough platforms. *Robotica, 25*(1), 121–128.

Zuo, A., Wu, Q. M. J., & Gruver, W. A. (2002). Stereo vision guided control of a Stewart platform. In *Proceedings IEEE International Symposium on Intelligent Control, Vancouver* (pp. 27–30).

Chapter 14
Delta Robot

14.1 Introduction

The Delta robot (see Fig. 2.4), a limited-degree-of-freedom (DOF) parallel manipulator invented by Clavel (1990, 1991), is one of the most emblematic and celebrated machines with parallel kinematic structures successfully introduced in both academia and industry. With the reflection of Prof. Clavel—"... why use robots with multi-kilowatt motors to handle products of only a few grams?"—began an intensive and prolific research field of modern robotics. The Delta robot was designed to serve in the electronic, food, and pharmaceutical industries, which require a high level of hygiene and reliable standards for products. Today there are robots based on parallelogram topologies showing very impressive kinematic abilities. Consider, for instance, that the Quattro robot, released by Adept Technology©, can make 240 pick-and-place cycles per minute because of its ability to generate sophisticated translational motions (Bonev 2011; Pierrot et al. 2009). Certainly, the core advantage of Delta-type robots is speed. As explained by Clavel in his thesis (Clavel 1990), the Delta robot, named DELTA4, was introduced as a 4-DOF parallel manipulator whose main characteristic is the inclusion of three parallelogram-type closed-loop chains to perform translational motions of the moving platform as observed from the fixed platform. The bars of the parallelograms are connected to each other by ball joints (*rotule*). Furthermore, spherical joints are used in the physical prototypes reported by Clavel in his thesis. However, curiously, in a patent granted to Clavel (Clavel 1991; Bonev 2011), in 1990, it is possible to distinguish the use of revolute joints instead of spherical joints. In 1991, Clavel presented his doctoral thesis, "Conception d'un robot parallèle rapide à 4 degrés de liberté," and was granted the Golden Robot award in 1999 for his work. Several translational parallel manipulators follow the Delta-type architecture introduced by Clavel; one of the most popular is the Maryland robot (Tsai 1997; Stamper 1997). In his thesis, Stamper (1997) described a robot proposed by Tsai (1997) as a new translational parallel manipulator free of spherical joints. Elucidating who

J. Gallardo-Alvarado, *Kinematic Analysis of Parallel Manipulators by Algebraic Screw Theory*, DOI 10.1007/978-3-319-31126-5_14

was the first person to develop a 3-DOF translational parallel manipulator based on parallelogram architectures free of spherical joints is not the intention of this chapter; the goal here is to introduce a simple method to approach the kinematics of a Delta-type robot. In that way, it is noticeable that despite its simple topology, the Delta robot is covered by a family of more than 30 patents around the world. Without a doubt, the most important are the WIPO patent granted to Clavel on June 18, 1987 (WO 87/03528), the U.S. Patent 4976582 issued in 1990, and the European Patent EP 0250470 issued in 1991. At the time the potential applications of the Delta robot were immediately recognized. In 1987, brothers Marc-Olivier and Pascal Demaurex purchased a license for the Delta robot to commercialize the new parallel manipulator. A few years later, a Japanese company named Hitach Seiki was licensed by the same Demaurex brothers to manufacture Delta robots of small dimensions. In 1999 emerged the most famous Delta robot, the IRB 340 FlexPicker robot, launched by ABB Flexible Automation and devoted to the food, pharmaceutical, and electronics industries. In reality, the list of successful applications of Clavel's ingenious robot is long, and special citations go to the Delta robots M-1iA and M-2iA developed by FANUC.

On the other hand, screw theory has proved to be an efficient and trusted mathematical resource in the analysis of Delta-type robots. So far the optimization of workspace, singularities, parametrical synthesis, structural analysis, and reciprocal screw systems (mobility analysis) are issues of Delta-type robots that have been successfully investigated via screw theory; see, for instance, Glazunov et al. (1990, 2007), Briot et al. (2008), and Zhao et al. (2009). A long list of references related to the first-order kinematic analysis of Delta-type robots using screw theory could be included here. However, the acceleration analysis via screw theory of Delta-type robots has been overlooked by most kinematicians, perhaps because of the unjustified lack of credibility of the expressions in screw form of kinematic chains developed by Rico and Duffy (1996b).

In this chapter, the kinematics of the translational Delta parallel manipulator is investigated mainly by screw theory. The analysis does not include the rotational freedom since it could be considered a trivial task from a kinematical point of view, in this case, because of the decoupled rotational and translational DOF of the robot manipulator. The inclusion of the rotational freedom is left as an exercise.

14.2 Description of the Delta Robot

In this section the topology of the robot under analysis is outlined.

The robot under study (see Fig. 14.1) is a variant of the original 4-DOF parallel manipulator introduced by Clavel (1990); that is, the rotational freedom of the moving platform is disregarded from the analysis and only the translational freedoms of the robot manipulator are considered in the analysis. It consists of a triangular moving platform, labeled as body m, and a triangular fixed platform, labeled body 0, connected to each other by three identical parallel chains RR-(4R)-R,

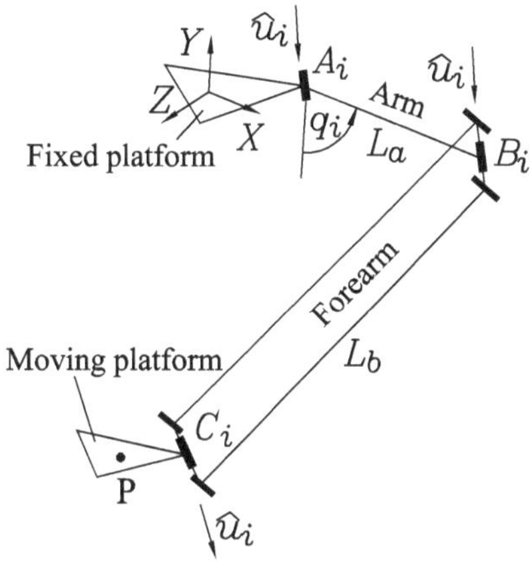

Fig. 14.1 Geometric scheme of the Delta robot

where (4R) denotes a planar four-bar linkage articulated with revolute joints. The dimensions of each parallelogram allow curvilinear translational motions of the coupler link, as observed from the moving platform, constraining the motions of the robot in such a way that the moving platform cannot rotate with respect to the fixed platform. Substituting spherical joints by revolute joints is assumed to be a simplification of the mechanical assembly of the parallelograms originally designed for the Delta robot. The actuators are active revolute joints mounted conveniently near the fixed platform; hence, we have a minimum of moving bodies. To explain the geometry of the mechanism, let XYZ be a reference frame attached to the center of the fixed platform with associated unit vectors $\boldsymbol{ijk}$. The active revolute joints shape an equilateral triangle $\Delta A_1A_2A_3$ where each corner is notated as $A_i = (A_{Xi}, A_{Yi}, A_{Zi})$, located by vector $\boldsymbol{a}_i$, called the nominal position of the ith revolute joint. Unless otherwise specified, in the remainder of this chapter consider $i = 1, 2, 3$. Furthermore, the axes of the active revolute joints are denoted as $\hat{\boldsymbol{u}}_i = u_{Xi}\boldsymbol{i} + u_{Yi}\boldsymbol{j} + u_{Zi}\boldsymbol{k}$ and are tangential to the circle circumscribed to the triangle $\Delta A_1A_2A_3$. Similarly, $B_i = (B_{Xi}, B_{Yi}, B_{Zi})$ denotes the coordinates of the nominal position of the revolute joint connecting the arm to the forearm (parallelogram) in the same complex extremity. Point B_i is located by vector $\boldsymbol{b}_i$. On the other hand, the moving platform is connected to the forearms through revolute joints whose nominal positions $C_i = (C_{Xi}, C_{Yi}, C_{Zi})$ are located by vectors $\boldsymbol{c}_i$. Points C_i form an equilateral triangle $\Delta C_1C_2C_3$ of side r. Because a constant orientation of the moving platform is generated, the axes of the revolute joints C_i are also $\hat{\boldsymbol{u}}_i$. Finally, point $P = (p_X, p_Y, p_Z)$, located by vector $\boldsymbol{p}$, denotes the center of the triangle $\Delta C_1C_2C_3$, which is known as the robot's control point.

14.3 Finite Kinematics of the Delta Robot

The kinematics of the Delta robot is a well-known issue that was successfully approached and yielded closed-form solutions; see, for instance, Pierrot et al. (1990). The formulation reported in this section differs from classical approaches because it is based on the concept that the coordinates of any point of a rigid body may be expressed as a linear combination of the coordinates of key points embedded in the same body (Merlet 2004; Gallardo-Alvarado 2014).

14.3.1 Forward Displacement Analysis of the Delta Robot

The forward displacement analysis of the Delta robot consists of finding the coordinates of the control point P for a given set of generalized coordinates q_i.

In computing the coordinates of points $B_i = (B_{Xi}, B_{Yi}, B_{Zi})$ (see Fig. 14.1), we immediately see that

$$B_{Yi} = A_{Yi} - L_a \cos(q_i). \tag{14.1}$$

Furthermore, to determine the coordinates B_{Xi} and B_{Zi}, let us consider that the active revolute joint A_i constrains the motion of the corresponding arm in such a way that

$$(\boldsymbol{b}_i - \boldsymbol{a}_i) \bullet \hat{\boldsymbol{u}}_i = 0. \tag{14.2}$$

Meanwhile, according to the vectors $\boldsymbol{a}_i$ and $\boldsymbol{b}_i$, three closure equations may be written as follows:

$$(\boldsymbol{b}_i - \boldsymbol{a}_i) \bullet (\boldsymbol{b}_i - \boldsymbol{a}_i) = L_a^2. \tag{14.3}$$

After a few computations, solving Eqs. (14.2) and (14.3), the coordinates of point B_i result in

$$B_i = \left[A_{Xi} - L_a u_{Zi} \sin(q_i), A_{Yi} - L_a \cos(q_i), A_{Zi} + L_a\, u_{Xi} \sin(q_i)\right]. \tag{14.4}$$

Once the coordinates of points B_i are computed, it is possible to determine the coordinates of points C_i. In that way, taking into account that $\Delta C_1 C_2 C_3$ is an equilateral triangle of side r where point C_1 is composed of unknown coordinates X, Y, and Z [e.g., $C_1 = (X, Y, Z)$], the coordinates of points C_2 and C_3 may be written according to the coordinates of point C_1 as follows:

$$C_2 = (X - \sqrt{3}r/2, Y, Z - r/2), \quad C_3 = (X - \sqrt{3}r/2, Y, Z + r/2). \tag{14.5}$$

Furthermore, three closure equations may be written as

$$(\boldsymbol{c}_i - \boldsymbol{b}_i) \bullet (\boldsymbol{c}_i - \boldsymbol{b}_i) = L_b^2. \tag{14.6}$$

Next, the systematic substitution of $C_1 = (X, Y, Z)$ and Eq. (14.5) into Eq. (14.6) yields three quadratic equations as follows:

$$X^2 + Y^2 + Z^2 + K_{i1}X + K_{i2}Y + K_{i3}Z + K_{i4} = 0, \tag{14.7}$$

where the coefficients $K_{ij}(i = 1,2,3; j = 1,2,3,4)$ are computed according to the parameters and generalized coordinates of the mechanism.

Solving Eq. (14.7) for X and Y, we get

$$X = W_1Z + W_2, \quad Y = W_3Z + W_4, \tag{14.8}$$

where the coefficients $W_j(j = 1,2,3,4)$ are computed according to the coefficients K_{ij}; for instance, the coefficient W_1 is given by

$$W_1 = \frac{K_{22}K_{33} + K_{12}K_{23} - K_{13}K_{22} - K_{23}K_{32} - K_{12}K_{33} + K_{13}K_{32}}{K_{21}K_{32} - K_{21}K_{12} - K_{11}K_{32} - K_{31}K_{22} + K_{31}K_{12} + K_{11}K_{22}}. \tag{14.9}$$

Finally, substituting Eq. (14.8) into any of the Eq. (14.7) yields a quadratic equation in the unknown Z as follows:

$$\begin{aligned}\left(1 + {W_1}^2 + {W_3}^2\right) Z^2 + (K_{11}W_1 + 2\,W_1W_2 + K_{13} + 2\,W_3W_4 + K_{12}W_3)\,Z \\ +K_{11}W_2 + {W_4}^2 + {W_2}^2 + K_{14} + K_{12}W_4 = 0.\end{aligned} \tag{14.10}$$

Hence, given the generalized coordinates q_i, the moving platform can reach at most two possible positions with respect to the fixed platform.

Once the coordinates X, Y, and Z are computed, the coordinates of the control point P are obtained as

$$P = (X - \sqrt{3}r/3, Y, Z). \tag{14.11}$$

Example 14.1. The first example of this chapter is devoted to computing the reachable positions of the center of the translational moving platform given the rotational generalized coordinates q_i of a Delta-type parallel manipulator. To this end consider that the nominal positions of the revolute joints, using SI units through the example, connecting the arms to the fixed platform are given by $A_1 = (0.15, -0.05, 0.0)$, $A_2 = (-0.075, -0.05, -0.1299)$, and $A_3 = (-0.075, -0.05, 0.1299)$, where the unit vectors along the axes of the revolute joints are given by $\hat{\boldsymbol{u}}_1 = \hat{\boldsymbol{k}}$, $\hat{\boldsymbol{u}}_2 = 0.866\hat{\boldsymbol{\imath}} - 0.5\hat{\boldsymbol{k}}$, and $\hat{\boldsymbol{u}}_3 = -0.866\hat{\boldsymbol{\imath}} - 0.5\hat{\boldsymbol{k}}$. Furthermore, the lengths of the arm and forearm are selected, respectively, as $L_a = 0.2$ and $L_b = 0.4$. On the other hand, the moving platform is an equilateral triangle of side $r = 0.0866$. Finally, consider that the generalized coordinates are given by $q_1 = \frac{\pi}{3}$, $q_2 = \frac{\pi}{6}$, and $q_3 = \frac{\pi}{4}$.

Solution. The unique coordinates of the nominal positions of the kinematic pairs connecting the arms to the forearms of the robot result in

$$B_1 = (-0.023205, -0.15, 0),$$
$$B_2 = (-0.025, -0.223205, -0.2165),$$
$$B_3 = (-0.0042893, -0.1914213, 0.2523708).$$

Furthermore, the proposed method yields three quadratic equations in the unknowns X, Y, and Z as follows:

$$\left.\begin{array}{l} X^2 + Y^2 + Z^2 + 0.046410X + 0.3Y - 0.136961 = 0 \\ X^2 + Y^2 + Z^2 - 0.1X + 0.446410Y + 0.3464Z - 0.077681 = 0 \\ X^2 + Y^2 + Z^2 - 0.141416X + 0.38284Y - 0.4181Z - 0.0746 = 0 \end{array}\right\}. \tag{14.12}$$

Solving Eq. (14.12), we obtain X and Y in terms of Z as follows:

$$X = 0.2740511381 - 5.849720592Z, \quad Y = -0.1308473848 - 8.215500783Z,$$

where the unknown Z is computed according to the quadratic equation

$$102.7136841Z^2 - 3.792428300Z - 0.07127191760 = 0. \tag{14.13}$$

Furthermore, the coordinates of points C_i are written in terms of the unknown Z as

$$\begin{aligned} C_1 &= (0.2740511 - 5.849720592Z, -0.1308473 - 8.215500783Z, Z), \\ C_2 &= (0.1990533 - 5.849720592Z, -0.1308473 - 8.215500783Z, Z - 0.0433), \\ C_3 &= (0.1990533 - 5.849720592Z, -0.1308473 - 8.215500783Z, Z + 0.0433). \end{aligned} \tag{14.14}$$

After, the coordinates of the control point P are computed as

$$\begin{aligned} P &= (C_1 + C_2 + C_3)/3 \\ &= (0.2240526047 - 5.849720592Z, -0.1308473848 - 8.215500783Z, Z). \end{aligned} \tag{14.15}$$

Solving Eq. (14.13) and substituting the values of Z into Eq. (14.15), the two solutions of the forward displacement analysis result in

$$P = \left\{\begin{array}{l} (-0.07210690702, -0.5467815585, 0.05062797567) \\ (0.3042268163, -0.01824862280, -0.01370564802) \end{array}\right. .$$

Example 14.2. Performing the forward displacement analysis of a parallel manipulator is mandatory to determine its workspace. However, the lack of closed-form solutions for most parallel manipulators makes this work a rather complicated task demanding the implementation of numerical techniques. In this area, only specific regions of the workspace may be evaluated. In this example, the workspace of a Delta-type robot is determined based on the forward displacement analysis. To this end, let us consider that the intervals of the generalized coordinates are given by $1.658 \leq q_i \leq 3.054$ with increments $\Delta q_i = 0.174$. Furthermore, the following parameters of Example 14.1 are reconsidered here:

$$
\begin{aligned}
&A_1 = (0.15, -0.05, 0.0),\\
&A_2 = (-0.075, -0.05, -0.1299),\\
&A_3 = (-0.075, -0.05, 0.1299),\\
&\hat{\boldsymbol{u}}_1 = \hat{\boldsymbol{k}}, \quad \hat{\boldsymbol{u}}_2 = 0.866\hat{\boldsymbol{\imath}} - 0.5\hat{\boldsymbol{k}}, \quad \hat{\boldsymbol{u}}_3 = -0.866\hat{\boldsymbol{\imath}} - 0.5\hat{\boldsymbol{k}},\\
&L_a = 0.2, L_b = 0.4, r = 0.0866.
\end{aligned}
$$

Solution. The formulas developed for the forward displacement analysis were translated into a simple algorithm. Next, the reachable points of point $P = (p_X, p_Y, p_Z)$ computed with such an algorithm are shown in Fig. 14.2.

To show how the chosen interval affects the resulting workspace, we performed another *run* of the algorithm. Besides the same parameters of Example 14.1, consider the following interval: $0.087 \leq q_i \leq 1.483$ with $\Delta q_i = 0.174$. The corresponding evaluated workspace is shown in Fig. 14.3.

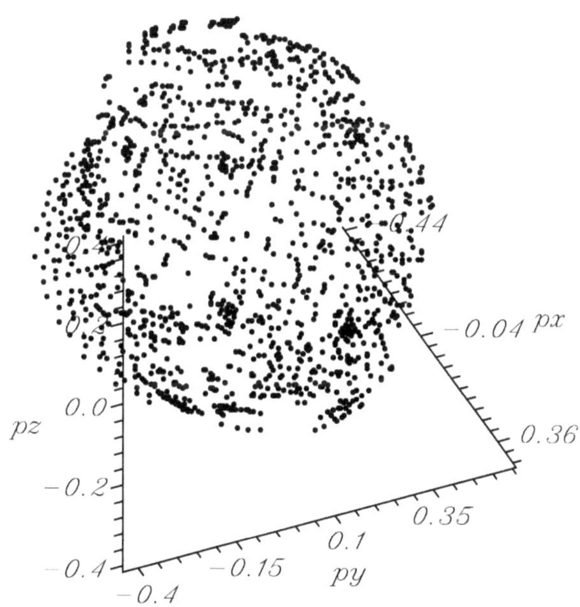

Fig. 14.2 Example 14.2. Workspace 1 (volume) of the Delta robot

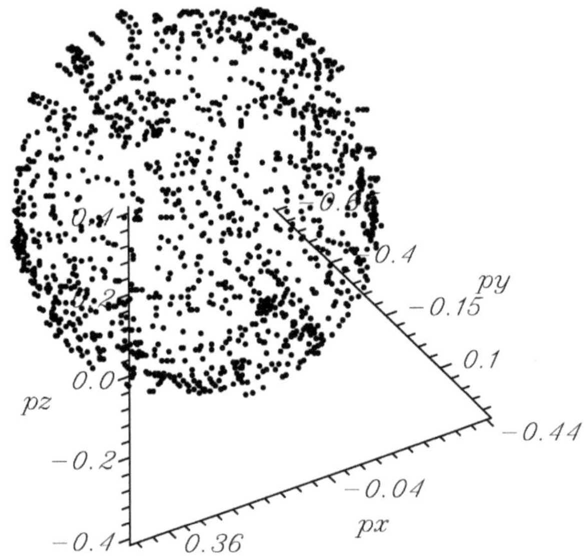

Fig. 14.3 Example 14.2. Workspace 2 (volume) of the Delta robot

14.3.2 *Inverse Displacement Analysis of the Delta Robot*

The inverse displacement analysis of the Delta robot consists of finding the generalized coordinates q_i given the coordinates of the control point P. To this end, let $P = (p_X, p_Y, p_Z)$ be the coordinates of point P. As an initial step, it is evident that the coordinates of points C_i may be written in terms of the coordinates p_X, p_Y, and p_Z as follows:

$$\left.\begin{aligned} C_1 &= (p_X + \sqrt{3}r/3, p_Y, p_Z) \\ C_2 &= (p_X - \sqrt{3}r/6, p_Y, p_Z - r/2) \\ C_3 &= (p_X - \sqrt{3}r/6, p_Y, p_Z + r/2) \end{aligned}\right\}. \tag{14.16}$$

To compute the coordinates B_{Yi}, a necessary step to compute q_i, three closure equations for each point B_i may be written as

$$\left.\begin{aligned} (\boldsymbol{b}_i - \boldsymbol{a}_i) \bullet (\boldsymbol{b}_i - \boldsymbol{a}_i) &= L_a^2 \\ (\boldsymbol{b}_i - \boldsymbol{c}_i) \bullet (\boldsymbol{b}_i - \boldsymbol{c}_i) &= L_b^2 \\ (\boldsymbol{b}_i - \boldsymbol{a}_i) \bullet \hat{\boldsymbol{u}}_i &= 0 \end{aligned}\right\}. \tag{14.17}$$

After a few computations (see the procedure outlined for the forward displacement analysis), a quadratic equation in the unknown B_{Yi} is obtained. Finally, the generalized coordinates q_i are computed [see Eq. (14.1)] as

$$q_i = \arccos\left(\frac{A_{Yi} - B_{Yi}}{L_a}\right). \tag{14.18}$$

Note that according to the inverse displacement analysis, the robot admits at most eight different configurations given the coordinates of the center P of the moving platform.

Example 14.3. The example consists of finding the generalized coordinates q_i given the location of the moving platform. To this end, the parameters of the mechanism are given by $A_1 = (0.15, -0.05, 0.0)$, $A_2 = (-0.075, -0.05, -0.1299)$, $A_3 = (-0.075, -0.05, 0.1299)$, $La = 0.2$, $Lb = 0.4$, $r = 0.0866$, $\hat{\boldsymbol{u}}_1 = \hat{\boldsymbol{k}}$, $\hat{\boldsymbol{u}}_2 = 0.866\hat{\boldsymbol{\imath}} - 0.5\hat{\boldsymbol{k}}$, $\hat{\boldsymbol{u}}_3 = -0.866\hat{\boldsymbol{\imath}} - 0.5\hat{\boldsymbol{k}}$. In other words, most of the data of Example 14.1 is repeated here. Furthermore, the coordinates of the center of the moving platform are given by $P = (0.1, -0.5, -0.25)$.

Solution. The coordinates of the corners of the moving platform, the points C_i, are immediately obtained as $C_1 = (0.15, -0.5, -0.25)$, $C_2 = (0.075, -0.5, -0.2933)$, and $C_3 = (0.075, -0.5, -0.2067)$. To compute the feasible angles q_1, three equations in the unknown coordinates of point $B_1 = (B_{X1}, B_{Y1}, B_{Z1})$ are obtained as

$$\left.\begin{array}{l}(B_{X1} - 0.15)^2 + (B_{Y1} + 0.05)^2 + B_{Z1}^2 - 0.04 = 0 \\ (B_{X1} - 0.15)^2 + (B_{Y1} + 0.5)^2 + (B_{Z1} + 0.25)^2 - 0.16 = 0 \\ B_{Z1} = 0\end{array}\right\}. \tag{14.19}$$

Solving Eq. (14.19), we obtain the reachable positions of point B_1 as

$$B_1 = \begin{cases}(0.0314954, -0.21111, 0) \\ (0.268503, -0.211111, 0)\end{cases}$$

Hence, we get two solutions for q_1: 0.6341870999 and 0.6341805811.

It is straightforward to show that q_2 and q_3 are computed using similar procedures. Table 14.1 lists the complete results of the inverse displacement analysis.

Certainly, for the inverse displacement analysis, depending on the coordinates B_{Yi}, there are two feasible values for each generalized coordinate q_i; that is, there are eight feasible solutions for the inverse displacement analysis of the Delta robot.

It is worth noting how easy it is to approach the displacement analysis of Delta-type robots with the method used in this chapter. For instance, Stamper (1997) obtained two equations in the tangent of the half-angle to solve the forward displacement analysis. These two equations are then solved by the dialytic method of elimination. Tsai (2007) carried out the displacement analysis of the Delta robot by handling the equations of spheres. Pierrot et al. (2009) reported the displacement analysis of a 4-DOF parallel manipulator built with parallelograms

Table 14.1 Example 14.3. Inverse displacement analysis

Sol.	q_1	q_2	q_3
1	0.6341870999	0.9082253901	1.030621301
2	0.6341805811	0.6147696441	0.3363628323

and suggests solving the forward displacement analysis with a numerical iterative scheme instead of searching for all solutions analytically. None of the contributions cited in this paragraph considers combining Eqs. (14.2) and (14.5) as an approach to the displacement analysis of the Delta robot. Of course, such a combination can be implemented in the robot developed by Pierrot et al. (2009).

14.4 Infinitesimal Kinematics of the Delta Robot

In this section the velocity and acceleration analyses of the robot at hand are addressed by screw theory. It is important to note here that the Delta robot is kinematically equivalent to the 3-RUU translational parallel manipulator if the orientations of the universal joints ("U") are properly established (see Fig. 14.4).

To avoid handling Jacobian matrices with deficient ranks, pseudo-kinematic pairs are introduced in the analysis of the mechanism; for instance, the Delta robot is modeled as a 3-R*RUU translational parallel manipulator, where R* denotes an imaginary revolute joint. With these assumptions in mind, the model of the screws is depicted in Fig. 14.4. The Plücker coordinates of the screws contained in the ith complex limb are computed under the following considerations: (1) ${}^{0}\$_{i}^{1}$ denotes the screw associated to the virtual revolute joint R*; (2) ${}^{1}\$_{i}^{2}$ is the screw associated to the active revolute joint; (3) ${}^{2}\$_{i}^{3}$ and ${}^{3}\$_{i}^{4}$ are the screws representing the universal joint connecting the arm to the forearm whose primal parts intersect at point B_i. Meanwhile, the universal joint connecting the forearm to the moving platform is simulated by the screws ${}^{4}\$_{i}^{5}$ and ${}^{5}\$_{i}^{6}$, whose primal parts intersect at point C_i. It is important to note that the primal parts of the screws ${}^{3}\$_{i}^{4}$ and ${}^{5}\$_{i}^{6}$ are normal to the plane of the ith parallelogram.

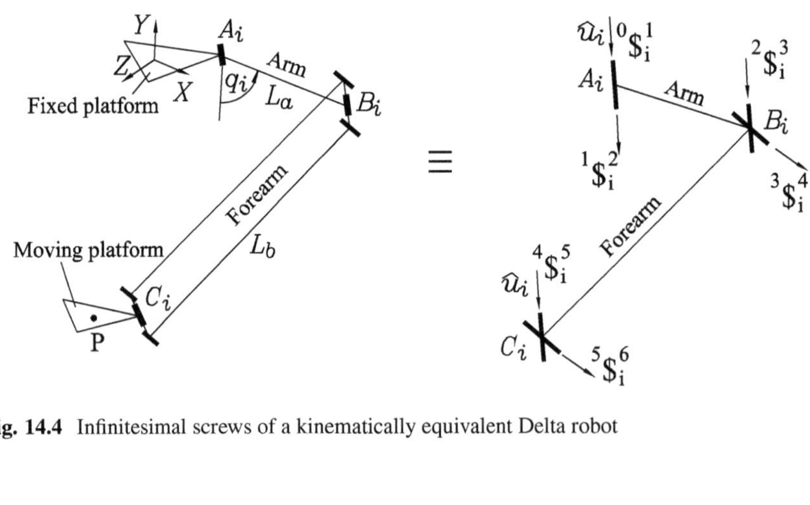

Fig. 14.4 Infinitesimal screws of a kinematically equivalent Delta robot

14.4.1 Velocity Analysis of the Delta Robot

Let

$$V = \begin{bmatrix} \mathbf{0} \\ \boldsymbol{v} \end{bmatrix}$$

be the velocity state of the moving platform as observed from the fixed platform where $\boldsymbol{v}$ is the velocity of the moving platform. Evidently, the primal part of the velocity state of any translational manipulator is the vector $\mathbf{0}$. On the other hand, the velocity state $\boldsymbol{V}$ may be expressed in screw form through the connector limbs of the robot as follows:

$$_0\omega_1^i\,{}^0\$_i^1 + {}_1\omega_2^i\,{}^1\$_i^2 + {}_2\omega_3^i\,{}^2\$_i^3 + {}_3\omega_4^i\,{}^3\$_i^4 + {}_4\omega_5^i\,{}^4\$_i^5 + {}_5\omega_6^i\,{}^5\$_i^6 = \boldsymbol{V}, \tag{14.20}$$

where $_0\omega_1^i = 0$ is the joint-velocity rate associated to the pseudo-revolute joint. Meanwhile, $_1\omega_2^i = \dot{q}_i$ is the ith generalized speed of the robot. To obtain the input–output equation of velocity of the robot, let us consider that

$$L_i = \begin{bmatrix} \hat{\boldsymbol{p}}_i \\ \boldsymbol{d}_i \end{bmatrix} \tag{14.21}$$

is a line in Plücker coordinates passing through points C_i and B_i. It is evident that L_i is reciprocal to all the screws in the same limb except the screw associated to the actuated revolute joint. Hence, systematically applying the Klein form $\{*;*\}$ of the line L_i to both sides of Eq. (14.20) and reducing terms yields the input–output equation of velocity as

$$\mathbf{A}\boldsymbol{v} = \mathbf{B}\dot{\mathbf{q}}, \tag{14.22}$$

where

$$\mathbf{A} = \left[\hat{\boldsymbol{p}}_1\ \hat{\boldsymbol{p}}_2\ \hat{\boldsymbol{p}}_3\right]^T \tag{14.23}$$

is called the *reduced active matrix* of the manipulator. Meanwhile,

$$\mathbf{B} = \text{diag}\left[\{{}^1\$_1^2;\boldsymbol{L}_1\}\ \{{}^1\$_2^2;\boldsymbol{L}_2\}\ \{{}^1\$_3^2;\boldsymbol{L}_3\}\right] \tag{14.24}$$

is the first-order influence coefficient matrix of the generalized velocities, whereas $\dot{\mathbf{q}} = \left[\dot{q}_1\ \dot{q}_2\ \dot{q}_3\right]^T$ is the first-order driver matrix of the robot.

Expression (14.22) allows us to explore some interesting kinematic properties of the Delta robot by studying the matrices $\mathbf{A}$ and $\mathbf{B}$; see, for instance, Stamper (1997), Joshi and Tsai (2002), López et al. (2006), Kelaiaia et al. (2012), and Zhao (2013). In particular, the singularity analysis of the Delta robot deserves special attention.

14.4.1.1 Singularity Analysis of the Delta Robot

According to the trend introduced by Gosselin and Angeles (1990), it is possible to detect three types of singularities in closed-loop kinematic chains. Singularities for the Delta robot occur when (1) **A** is singular, (2) **B** is singular, and (3) both **A** and **B** are singular matrices. It is evident that matrix **A** is singular when $\det(\mathbf{A}) = 0$, namely, direct singularity. Usually, obtaining the direct singular regions of parallel manipulators in loci form is a rather complicated task because of the nonlinearity of the equations associated to the forward displacement analysis. A simple way to approach this issue is using elementary linear algebra. In fact, matrix **A** is clearly singular when the vectors $\hat{\boldsymbol{p}}_i$ are linearly dependent, such as when these vectors are collinear or parallel. If the robot passes through these configurations, the moving platform gains freedoms; that is, Cartesian velocities are available for the moving platform even when the joint rates associated to the motors are null. On the other hand, it is evident that **B** is singular when $\det(\mathbf{B}) = 0$, which occurs when any of the elements of the diagonal of **B** vanishes: For example, when a limb is folded/unfolded, the screw ${}^1\$_i^2$ and the line $\boldsymbol{L}_i$ in the same limb are reciprocal. That is, their primal parts intersect at a common point B_i, causing $\left\{{}^1\$_i^2; \boldsymbol{L}_i\right\} = 0$. In such a configuration, the moving platform loses DOF. In fact, assuming that $F < 3$, where $F = \dim(\mathbf{B})$, the manipulator loses $3 - F$ DOF. Finally, a combined singularity emerges when $\det(\mathbf{A}) = \det(\mathbf{B}) = 0$, for instance, when for the triangular platforms we have $A_1A_2A_3 = C_1C_2C_3$. In other words, the limbs are vertical. Singularities must be taken into proper account at the design stage of the manipulator; otherwise, uncontrollable conditions could damage the robot.

The approach reported in this subsection requires only three reciprocal lines to solve the velocity analysis of the Delta robot and to be applied to the corresponding singular analysis. On that topic, it is worth noting that other strategies also based on screw theory reported in the literature require handling at least six reciprocal wrenches; see, for instance, Glazunov et al. (1990, 2007), Briot et al. (2008).

14.4.2 Acceleration Analysis of the Delta Robot

Unlike the velocity analysis, the validity of the equations of the acceleration analysis in screw form of mechanisms introduced by Rico and Duffy (1996b) has been bitterly discussed by some researchers even though such a methodology has been widely verified by the author of this book through the publication of several papers in well-known Journal Citation Reports (JCR) journals. Following the trend of the methodology introduced by Rico and Duffy (1996a,b) in this section, we address the acceleration analysis of the Delta robot using screw theory, a little proof of the correctness of the pioneering method introduced by Rico and Duffy (1996b) and its application to the acceleration analysis of parallel manipulators (Rico and Duffy 1996a).

Let

$$\boldsymbol{A} = \begin{bmatrix} \mathbf{0} \\ \boldsymbol{a} \end{bmatrix}$$

be the reduced acceleration state of the moving platform as observed from the fixed platform where $\boldsymbol{a}$ is the acceleration of any point embedded in the moving platform. The accelerator $\boldsymbol{A}$ may be expressed in screw form through the limbs of the robot as follows:

$${}_0\dot{\omega}_1^i\,{}^0\$_i^1 + {}_1\dot{\omega}_2^i\,{}^1\$_i^2 + {}_2\dot{\omega}_3^i\,{}^2\$_i^3 + {}_3\dot{\omega}_4^i\,{}^3\$_i^4 + {}_4\dot{\omega}_5^i\,{}^4\$_i^5 + {}_5\dot{\omega}_6^i\,{}^5\$_i^6 + \mathcal{L}_i = \boldsymbol{A}, \qquad (14.25)$$

where ${}_0\dot{\omega}_1^i = 0$, ${}_1\dot{\omega}_2^i = \ddot{q}_i$. Meanwhile, $\mathcal{L}_i$ is the Lie screw of acceleration, which is computed as

$$\begin{aligned}\mathcal{L}_i = {} & \left[{}_0\omega_1^i\,{}^0\$_i^1 \quad {}_1\omega_2^i\,{}^1\$_i^2 + {}_2\omega_3^i\,{}^2\$_i^3 + {}_3\omega_4^i\,{}^3\$_i^4 + {}_4\omega_5^i\,{}^4\$_i^5 + {}_5\omega_6^i\,{}^5\$_i^6\right] + \\ & \left[{}_1\omega_2^i\,{}^1\$_i^2 \quad {}_2\omega_3^i\,{}^2\$_i^3 + {}_3\omega_4^i\,{}^3\$_i^4 + {}_4\omega_5^i\,{}^4\$_i^5 + {}_5\omega_6^i\,{}^5\$_i^6\right] + \\ & \left[{}_2\omega_3^i\,{}^2\$_i^3 \quad {}_3\omega_4^i\,{}^3\$_i^4 + {}_4\omega_5^i\,{}^4\$_i^5 + {}_5\omega_6^i\,{}^5\$_i^6\right] + \\ & \left[{}_3\omega_4^i\,{}^3\$_i^4 \quad {}_4\omega_5^i\,{}^4\$_i^5 + {}_5\omega_6^i\,{}^5\$_i^6\right] + \left[{}_4\omega_5^i\,{}^4\$_i^5 \quad {}_5\omega_6^i\,{}^5\$_i^6\right].\end{aligned} \qquad (14.26)$$

Furthermore, the joint-velocity rates are computed by rewriting Eq. (14.20) as follows:

$$\mathbf{J}_i\boldsymbol{\Omega}_i = \boldsymbol{V}, \qquad (14.27)$$

where

$$\boldsymbol{\Omega}_i = \left[{}_0\omega_1^i \; {}_1\omega_2^i \; {}_2\omega_3^i \; {}_3\omega_4^i \; {}_4\omega_5^i \; {}_5\omega_6^i\right]^T \qquad (14.28)$$

is a matrix containing the joint-velocity rates of the manipulator in the same limb, whereas

$$\mathbf{J}_i = \left[{}^0\$_i^1 \; {}^1\$_i^2 \; {}^2\$_i^3 \; {}^3\$_i^4 \; {}^4\$_i^5 \; {}^5\$_i^6\right] \qquad (14.29)$$

is the Jacobian of the ith planar leg. Thus, the joint-velocity rates of the ith planar limb are computed as follows:

$$\boldsymbol{\Omega}_i = \mathbf{J}_i^{-1}\boldsymbol{V}. \qquad (14.30)$$

At this point it is worth noting the importance of the introduction of pseudo-kinematic joints to simplify the inverse velocity analysis of the Delta robot.

Finally, following the trend of the velocity analysis, the input–output equation of acceleration of the Delta robot results in

$$\mathbf{A}\boldsymbol{a} = \mathbf{B}\ddot{\mathbf{q}} + \mathbf{C}, \tag{14.31}$$

where

$$\mathbf{C} = \left[\{\mathcal{L}_1; \boldsymbol{L}_1\} \; \{\mathcal{L}_2; \boldsymbol{L}_2\} \; \{\mathcal{L}_3; \boldsymbol{L}_3\}\right]^T \tag{14.32}$$

is the complementary matrix of acceleration. Meanwhile,

$$\ddot{\mathbf{q}} = \left[\ddot{q}_1 \; \ddot{q}_2 \; \ddot{q}_3\right]^T \tag{14.33}$$

is the second-order driver matrix of the parallel manipulator.

Example 14.4. The example is devoted to the forward kinematics of the Delta robot. To do so, the reference configuration of the parallel manipulator is specified according to the following data:

$$\begin{aligned}
A_1 &= (0.15, -0.05, 0.0), A_2 = (-0.075, -0.05, -0.1299),\\
A_3 &= (-0.075, -0.05, 0.1299), B_1 = (0.35, -0.05, 0.0),\\
B_2 &= (-0.175, -0.05, -0.303), B_3 = (-0.175, -0.05, 0.303),\\
C_1 &= (0.05, -0.314, 0.0), C_2 = (-0.025, -0.314, -0.043),\\
C_3 &= (-0.025, -0.314, 0.043), La = 0.2, Lb = 0.4, r = 0.0866,\\
\hat{\boldsymbol{u}}_1 &= \hat{\boldsymbol{k}}, \hat{\boldsymbol{u}}_2 = 0.866\hat{\boldsymbol{i}} - 0.5\hat{\boldsymbol{k}}, \hat{\boldsymbol{u}}_3 = -0.866\hat{\boldsymbol{i}} - 0.5\hat{\boldsymbol{k}}.
\end{aligned}$$

Furthermore, the generalized coordinates q_i are commanded to follow periodic functions given by

$$\begin{aligned}
q_1 &= 0.5\pi + 1.5\sin(t)\cos(t),\\
q_2 &= 0.5\pi + 0.75\sin(t)\cos(t),\\
q_3 &= 0.5\pi + 1.75\sin(t)\cos(t),
\end{aligned}$$

where the time t is confined to the interval $0 \le t \le 2\pi$.

The example consists of finding the temporal behavior of the displacement, velocity, and acceleration of the center of the moving platform as observed from the fixed platform.

Solution. The equations of the forward displacement analysis of the example are solved using the Newton–Raphson method. The temporal behavior of the displacement of the center of the moving platform is provided in Fig. 14.5.

On the other hand, the resulting temporal behaviors of the velocity and acceleration of the translational moving platform computed by screw theory are given, respectively, in Figs. 14.6 and 14.7.

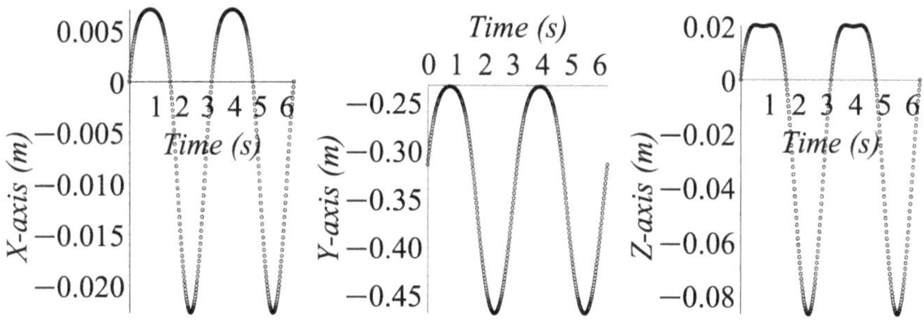

Fig. 14.5 Example 14.4. Time history of the displacement of the center of the moving platform

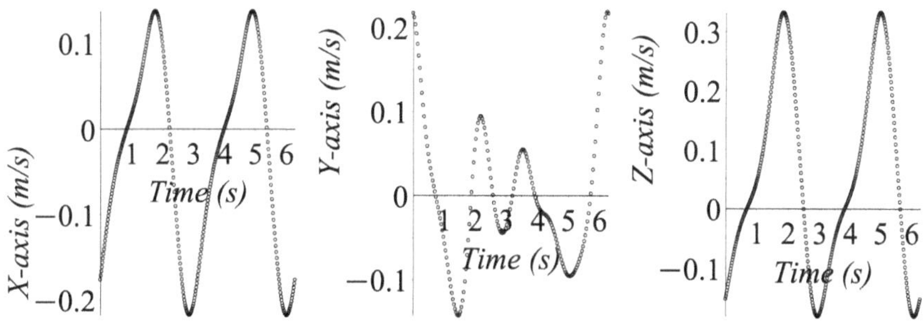

Fig. 14.6 Example 14.4. Time history of the velocity of the moving platform

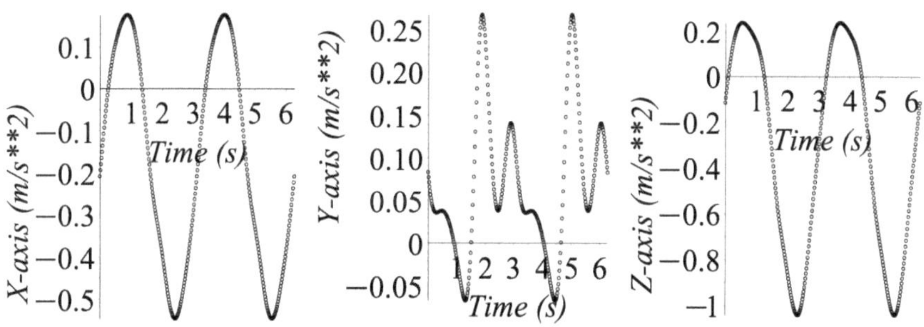

Fig. 14.7 Example 14.4. Time history of the acceleration of the moving platform

14.5 Exercises

1. The Delta robot is based on parallelogram-type mechanisms that belong to the pantograph family. Conduct brief research about the conventional pantograph.
2. Solve Example 14.1 considering that the generalized coordinates are given by $q_1 = \frac{\pi}{4}$, $q_2 = \frac{\pi}{3}$, and $q_3 = \frac{\pi}{4}$.
3. Solve Example 14.3 considering that the coordinates of point P are given by $P = (-0.125, -0.45, -0.25)$.
4. Perform the kinematic analysis of the original 4-DOF Delta parallel manipulator.

References

Bonev, I. A. (2011). *The true origins of parallel robots*. Available at Parallel Mic Center, ParalleMIC online review. http://www.parallemic.org/Reviews/Review007.html.

Briot, S., Arakelian, V., & Glazunov, V. (2008). Design and analysis of the properties of the Delta inverse robot. In *Proceedings X International Conference on the Theory of Machines and Mechanisms, Liberec* (pp. 113–118).

Clavel, R. (1990). Device for the movement and positioning of an element in space, U.S. Patent No. 4,976,582, December 11.

Clavel, R. (1991). *Conception d'un robot parallèle rapide à 4 degrés de liberté*. Ph.D. thesis, Ecole Polytechique Fédérale de Lausanne (EPFL).

Gallardo-Alvarado, J. (2014). A simple method to solve the forward displacement analysis of the general six-legged parallel manipulator. *Robotics and Computer-Integrated Manufacturing*, 30(1), 55–61.

Glazunov, V., Gruntovich, R., Lastochkin, A., & Than, N. M. (2007). Representations of constraints imposed by kinematic chains of parallel mechanisms. In *Proceedings 12th IFToMM World Congress, Besancon*, June 18–21.

Glazunov, V. A., Koliskor, A. S., Krainev, A. F., & Model, B. I. (1990). Classification principles and analysis methods for parallel-structure spatial mechanisms. *Journal of Machinery Manufacture and Reliability, N1*, 30–37.

Gosselin, C., & Angeles, J. (1990). Singularity analysis of closed-loop kinematic chains. *IEEE Journal of Robotics and Automation, 6*(3), 281–290.

Joshi, S. A., & Tsai, L.W. (2002). Jacobian analysis of limited-DOF parallel manipulators. *ASME Journal of Mechanical Design, 124*(2), 254–258.

Kelaiaia, R., Company, O., & Zaatri, A. (2012). Multiobjective optimization of a linear Delta parallel robot. *Mechanism and Machine Theory, 50*, 159–178.

López, M., Castillo, E., García, G., & Bashir, A. (2006). Delta robot: inverse, direct and intermediate Jacobians. *Proceedings Institution of Mechanical Engineers Part C: Journal of Mechanical Engineering Science, 220*(1), 103–109.

Merlet, J.-P. (2004). Solving the forward kinematics of a Gough-type parallel manipulator with interval analysis. *International Journal of Robotics Research, 23*(3), 221–235.

Pierrot, F., Nabat, V., Company, O., Krut, S., & Poignet, P. (2009). Optimal design of a 4-DOF parallel manipulator: from academia to industry. *IEEE Transactions on Robotics, 25*(2), 213–224.

Pierrot, F., Reynaud, C., & Fournier, A. (1990). Delta: a simple and efficient parallel robot. *Robotica, 8*(2), 105–109.

Rico, J. M., & Duffy, J. (1996a). An efficient inverse acceleration analysis of in-parallel manipulators. In *Proceedings 1996 ASME Design Engineering Technical Conference and Computers in Engineering Conference*.

Rico, J. M., & Duffy, J. (1996b). An application of screw algebra to the acceleration analysis of serial chains. *Mechanism and Machine Theory, 31*(4), 445–457.

Stamper, R. E. (1997). *A three degree of freedom parallel manipulator with only translational degrees of freedom*. Ph.D. dissertation, University of Maryland, Ph.D. 97-4.

Tsai, L.-W. (1997). Multi-degree-of-freedom mechanisms for machine tools and the like. U.S. Patent No. 5656905.

Tsai, L.-W. (1999). *Robot analysis*. New York: Wiley.

Zhao, Y. (2013). Dimensional synthesis of a three translational degrees of freedom parallel robot while considering kinematic anisotropic property. *Robotics and Computer-Integrated Manufacturing, 29*(1), 169–179.

Zhao, J., Li, B., Yang, X., & Yu, H. (2009). Geometrical method to determine the reciprocal screws and applications to parallel manipulators. *Robotica, 27*(6), 929–940.

Part VI
Solved Exercises

Chapter 15
Full Answers to Selected Exercises

15.1 Chapter 3: Mathematical Background

1. *Complete the proof of Proposition 3.2.* Transitive: If ${}^A\boldsymbol{V}^B_O \sim {}^A\boldsymbol{V}^B_{O*}$ and ${}^A\boldsymbol{V}^B_{O*} \sim {}^A\boldsymbol{V}^B_{O\#}$, where

$$
{}^A\boldsymbol{V}^B_O = \begin{bmatrix} {}^A\boldsymbol{\omega}^B_O \\ {}^A\boldsymbol{v}^B_O \end{bmatrix}, \quad {}^A\boldsymbol{V}^B_{O*} = \begin{bmatrix} {}^A\boldsymbol{\omega}^B_{O*} \\ {}^A\boldsymbol{v}^B_{O*} \end{bmatrix}, \quad {}^A\boldsymbol{V}^B_{O\#} = \begin{bmatrix} {}^A\boldsymbol{\omega}^B_{O\#} \\ {}^A\boldsymbol{v}^B_{O\#} \end{bmatrix},
$$

then it follows that ${}^A\boldsymbol{V}^B_{O*} \sim {}^A\boldsymbol{V}^B_{O\#}$.

Solution. If we consider that the angular velocity vector is a property of a rigid body, it is evident that

$$
{}^A\boldsymbol{\omega}^B_O = {}^A\boldsymbol{\omega}^B_{O*} = {}^A\boldsymbol{\omega}^B_{O\#} = {}^A\boldsymbol{\omega}^B.
$$

On the other hand, using elementary kinematics we have

$$
{}^A\boldsymbol{v}^B_{O*} = {}^A\boldsymbol{v}^B_O + {}^A\boldsymbol{\omega}^B \times \boldsymbol{r}_{O*/O}
$$

and

$$
{}^A\boldsymbol{v}^B_{O\#} = {}^A\boldsymbol{v}^B_{O*} + {}^A\boldsymbol{\omega}^B \times \boldsymbol{r}_{O\#/O*}.
$$

Therefore,

$$
\begin{aligned}
{}^A\boldsymbol{v}^B_{O\#} &= \left({}^A\boldsymbol{v}^B_O + {}^A\boldsymbol{\omega}^B \times \boldsymbol{r}_{O*/O}\right) + {}^A\boldsymbol{\omega}^B \times \boldsymbol{r}_{O\#/O*} \\
&= {}^A\boldsymbol{v}^B_O + {}^A\boldsymbol{\omega}^B \times \left(\boldsymbol{r}_{O*/O} + \boldsymbol{r}_{O\#/O*}\right) \\
&= {}^A\boldsymbol{v}^B_O + {}^A\boldsymbol{\omega}^B \times \boldsymbol{r}_{O\#/O},
\end{aligned}
$$

which finally leads us to conclude that in fact ${}^A\boldsymbol{V}^B_O \sim {}^A\boldsymbol{V}^B_{O\#}$. ◇

J. Gallardo-Alvarado, *Kinematic Analysis of Parallel Manipulators by Algebraic Screw Theory*, DOI 10.1007/978-3-319-31126-5_15

2. *Prove Proposition 3.4. Let us consider that P is a point of body B such that the component of the linear velocity ${}^A\boldsymbol{v}_P^B$ perpendicular to the angular velocity ${}^A\boldsymbol{\omega}^B$ vanishes. Then there is a line passing through point P whose direction is parallel to ${}^A\boldsymbol{\omega}^B$. Furthermore, the points along this line have the same linear velocity of P; in other words, the component of the linear velocity perpendicular to ${}^A\boldsymbol{\omega}^B$ is also null.*

Solution. Let O be an arbitrary point embedded in the line described in Proposition 3.4. Hence, it is evident that

$$ {}^A\boldsymbol{v}_P^B = {}^A\boldsymbol{v}_O^B + {}^A\boldsymbol{\omega}^B \times \boldsymbol{r}_{P/O}, $$

where ${}^A\boldsymbol{v}_P^B$ is the velocity of point P fixed in body B as observed from body A. Meanwhile, ${}^A\boldsymbol{v}_O^B$ is the velocity of point O fixed in body B as measured from body A, whereas $\boldsymbol{r}_{P/O}$ is the position vector of P with respect to O. Thus, it is sufficient to prove that

$$ {}^A\boldsymbol{\omega}^B \times \boldsymbol{r}_{P/O} = \mathbf{0}. \tag{15.1} $$

Premultiplying both sides of Eq. (15.1) by the angular velocity vector ${}^A\boldsymbol{\omega}^B$, we have

$$ {}^A\boldsymbol{\omega}^B \times ({}^A\boldsymbol{\omega}^B \times \boldsymbol{r}_{P/O}) = \mathbf{0}. \tag{15.2} $$

Furthermore, according to Wylie and Barret (1982), the triple vectorial product may be decomposed as

$$ {}^A\boldsymbol{\omega}^B \times ({}^A\boldsymbol{\omega}^B \times \boldsymbol{r}_{P/O}) = ({}^A\boldsymbol{\omega}^B \bullet \boldsymbol{r}_{P/O}){}^A\boldsymbol{\omega}^B - {}_A\omega_B^2\boldsymbol{r}_{P/O}, \tag{15.3} $$

where ${}_A\omega_B$ denotes the magnitude of the angular velocity vector ${}^A\boldsymbol{\omega}^B$. On the other hand, if we take into account that the vector $\boldsymbol{r}_{P/O}$ is alone to the vector ${}^A\boldsymbol{\omega}^B$, then it follows that

$$ \boldsymbol{r}_{P/O} = \lambda\, {}^A\boldsymbol{\omega}^B, \quad \text{where} \quad \lambda \in \Re. \tag{15.4} $$

Substituting Eq. (15.4) into Eq. (15.3) leads to

$$ {}^A\boldsymbol{\omega}^B \times ({}^A\boldsymbol{\omega}^B \times \boldsymbol{r}_{P/O}) = ({}^A\boldsymbol{\omega}^B \bullet \lambda{}^A\boldsymbol{\omega}^B){}^A\boldsymbol{\omega}^B - {}_A\omega_B^2\boldsymbol{r}_{P/O} $$

or

$$ \begin{aligned} {}^A\boldsymbol{\omega}^B \times ({}^A\boldsymbol{\omega}^B \times \boldsymbol{r}_{P/O}) &= \lambda_A\omega_B^{2A}\boldsymbol{\omega}^B - {}_A\omega_B^2\boldsymbol{r}_{P/O} \\ &= \lambda_A\omega_B^{2A}\boldsymbol{\omega}^B - \lambda_A\omega_B^{2A}\boldsymbol{\omega}^B = \mathbf{0}. \end{aligned} $$

Hence,

$$^{A}\boldsymbol{\omega}^{B} \times (^{A}\boldsymbol{\omega}^{B} \times \boldsymbol{r}_{P/O}). = \mathbf{0}.$$

Furthermore, considering that $^{A}\boldsymbol{\omega}^{B}$ is an arbitrary vector, the only possibility to satisfy Eq. (15.2) is that $^{A}\boldsymbol{\omega}^{B} \times \boldsymbol{r}_{P/O} = \mathbf{0}$. ◇

On the other hand, note that the proof of Proposition 3.4 is immediate according to Eqs. (15.1) and (15.4). In fact, if we take into account that $^{A}\boldsymbol{\omega}^{B} \times {}^{A}\boldsymbol{\omega}^{B} = \mathbf{0}$, it follows that

$$^{A}\boldsymbol{\omega}^{B} \times \boldsymbol{r}_{P/O} = {}^{A}\boldsymbol{\omega}^{B} \times (\lambda\, {}^{A}\boldsymbol{\omega}^{B}) = \mathbf{0}.\diamond$$

3. *Prove that the screw representing a prismatic joint, in Eq. (3.60), is well established.*

   Solution. Let

$$\$ = \begin{bmatrix} \hat{s} \\ h\hat{s} + \hat{s} \times \boldsymbol{r} \end{bmatrix}$$

   be the general representation of a screw in Plücker coordinates. Furthermore, let us consider that

$$\$/h = \begin{bmatrix} \hat{s}/h \\ \hat{s} + \hat{s} \times \boldsymbol{r}/h \end{bmatrix}.$$

   Hence, since in a prismatic joint the pitch goes to infinity, it follows that

$$\lim_{h\to\infty} \$/h = \begin{bmatrix} \lim_{h\to\infty} \hat{s}/h \\ \hat{s} + \lim_{h\to\infty} \hat{s} \times \boldsymbol{r}/h \end{bmatrix} = \begin{bmatrix} \mathbf{0} \\ \hat{s} \end{bmatrix}.\diamond$$

4. *Which condition must satisfy revolute joints in order to model spherical or universal joints?*

   Solution. The primal parts (directions of the lines) must be concurrent.

5. *The plane joint is another type of lower kinematic pair. How is it modeled with infinitesimal screws?*

   Solution. A plane joint can be modeled as the combination of two prismatic joints and one revolute joint.

6. *Complete the proof of Proposition 3.9. Consider the classes of equivalence $\overline{\boldsymbol{V}_1} = (\boldsymbol{\omega}_1; \boldsymbol{v}_{O1})$, $\overline{\boldsymbol{V}_2} = (\boldsymbol{\omega}_1; \boldsymbol{v}_{O2})$, $\overline{\boldsymbol{V}_3} = (\boldsymbol{\omega}_1; \boldsymbol{v}_{O3})$, and $\lambda_1, \lambda_2, \lambda_3 \in \Re$. Then the Lie product, as it is given in Definition 3.3, of the Lie algebra se(3) satisfies the Jacobi identity.*

Solution. It is necessary to prove that

$$[\boldsymbol{V}_1 \quad [\boldsymbol{V}_2 \quad \boldsymbol{V}_3]] + [\boldsymbol{V}_3 \quad [\boldsymbol{V}_1 \quad \boldsymbol{V}_2]] + [\boldsymbol{V}_2 \quad [\boldsymbol{V}_3 \quad \boldsymbol{V}_1]] = \begin{bmatrix} \mathbf{0} \\ \mathbf{0} \end{bmatrix}.$$

To do so, consider that

$$\begin{aligned}
&[(\boldsymbol{\omega}_1; \boldsymbol{v}_{O1}) \quad [(\boldsymbol{\omega}_1; \boldsymbol{v}_{O2}) \quad (\boldsymbol{\omega}_1; \boldsymbol{v}_{O3})]] + \\
&[(\boldsymbol{\omega}_1; \boldsymbol{v}_{O3}) \quad [(\boldsymbol{\omega}_1; \boldsymbol{v}_{O1}) \quad (\boldsymbol{\omega}_1; \boldsymbol{v}_{O2})]] + \\
&[(\boldsymbol{\omega}_1; \boldsymbol{v}_{O2}) \quad [(\boldsymbol{\omega}_1; \boldsymbol{v}_{O3}) \quad (\boldsymbol{\omega}_1; \boldsymbol{v}_{O1})]] \\
= &\begin{bmatrix} \mathbf{0} \\ \boldsymbol{\omega}_1 \times (\boldsymbol{\omega}_1 \times \boldsymbol{v}_{O3} - \boldsymbol{\omega}_1 \times \boldsymbol{v}_{O2}) \end{bmatrix} + \\
&\begin{bmatrix} \mathbf{0} \\ \boldsymbol{\omega}_1 \times (\boldsymbol{\omega}_1 \times \boldsymbol{v}_{O2} - \boldsymbol{\omega}_1 \times \boldsymbol{v}_{O1}) \end{bmatrix} + \\
&\begin{bmatrix} \mathbf{0} \\ \boldsymbol{\omega}_1 \times (\boldsymbol{\omega}_1 \times \boldsymbol{v}_{O1} - \boldsymbol{\omega}_1 \times \boldsymbol{v}_{O3}) \end{bmatrix} = \begin{bmatrix} \mathbf{0} \\ \mathbf{0} \end{bmatrix}. \diamond
\end{aligned}$$

7. *Give two examples of reciprocal screws.*

   Solution.

   - Let $\$_1 = (\hat{\boldsymbol{s}}_1; \boldsymbol{s}_{O1})$ and $\$_2 = (\hat{\boldsymbol{s}}_2; \boldsymbol{s}_{O2})$ be two infinitesimal screws representing revolute joints whose primal parts, the unit vectors $\hat{\boldsymbol{s}}_1$ and $\hat{\boldsymbol{s}}_2$, intersect a point O. If point O is chosen as the reference pole to compute the Plücker coordinates of $\$_1$ and $\$_2$, then because the dual parts of the screws $\$_1$ and $\$_2$ vanish, it follows that the Klein form between both screws is obtained as

   $$\{\$_1; \$_2\} = \hat{\boldsymbol{s}}_1 \bullet \mathbf{0} + \hat{\boldsymbol{s}}_2 \bullet \mathbf{0} = 0.$$

   - Let $\$_1 = (\hat{\boldsymbol{s}}_1; \boldsymbol{s}_{O1})$ be a screw representing a revolute joint and let $\$_2 = (\mathbf{0}; \hat{\boldsymbol{s}}_2)$ be a screw representing a prismatic joint. If the unit vectors $\hat{\boldsymbol{s}}_1$ and $\hat{\boldsymbol{s}}_2$ are orthogonal, then applying the Klein form between both screws leads to

   $$\{\$_1; \$_2\} = \hat{\boldsymbol{s}}_1 \bullet \hat{\boldsymbol{s}}_2 + \mathbf{0} \bullet \boldsymbol{s}_{O1} = 0.$$

8. *Prove that the Lie product is nonassociative. To do this, consider the equivalence classes* $\overline{\boldsymbol{V}_1} = (\boldsymbol{\omega}_1; \boldsymbol{v}_{O1})$, $\overline{\boldsymbol{V}_2} = (\boldsymbol{\omega}_1; \boldsymbol{v}_{O2})$, $\overline{\boldsymbol{V}_3} = (\boldsymbol{\omega}_1; \boldsymbol{v}_{O3})$, *and* λ_1, λ_2, $\lambda_3 \in \Re$. *The Lie product is nonassociative; for example,*

   $$[\boldsymbol{V}_1 \quad [\boldsymbol{V}_2 \quad \boldsymbol{V}_3]] \neq [[\boldsymbol{V}_1 \quad \boldsymbol{V}_2] \quad \boldsymbol{V}_3].$$

Solution. It is evident that

$$[\boldsymbol{V}_1 \quad [\boldsymbol{V}_2 \quad \boldsymbol{V}_3]] = \begin{bmatrix} \mathbf{0} \\ \boldsymbol{\omega}_1 \times (\boldsymbol{\omega}_1 \times (\boldsymbol{v}_{O3} - \boldsymbol{v}_{O2})) \end{bmatrix}.$$

Meanwhile,

$$[[\boldsymbol{V}_1 \quad \boldsymbol{V}_2] \quad \boldsymbol{V}_3] = \begin{bmatrix} \mathbf{0} \\ -\boldsymbol{\omega}_1 \times (\boldsymbol{\omega}_1 \times (\boldsymbol{v}_{O2} - \boldsymbol{v}_{O1})) \end{bmatrix}.$$

Hence, if we take into account that $\boldsymbol{v}_{O3} \neq \boldsymbol{v}_{O1}$, it follows that

$$[\boldsymbol{V}_1 \quad [\boldsymbol{V}_2 \quad \boldsymbol{V}_3]] \neq [[\boldsymbol{V}_1 \quad \boldsymbol{V}_2] \quad \boldsymbol{V}_3]\,.\diamond$$

9. *Prove that any point of the instantaneous screw axis (ISA) of a screw can be used to compute the dual part of the screw.*

   Solution. Let P and Q be two points of the ISA, and let $\hat{s}$ be a unit vector along the ISA. Furthermore, let O be the reference pole. The dual part of the screw may be computed based on P and Q, respectively, as follows:

$$\boldsymbol{s}_O = h\hat{s} + \hat{s} \times \boldsymbol{r}_{O/P}$$

and

$$\boldsymbol{s}_O^* = h\hat{s} + \hat{s} \times \boldsymbol{r}_{O/Q}.$$

Hence,

$$\boldsymbol{s}_O = \boldsymbol{s}_O^* - \hat{s} \times \boldsymbol{r}_{O/Q} + \hat{s} \times \boldsymbol{r}_{O/P} = \boldsymbol{s}_O^* + \hat{s} \times \boldsymbol{r}_{Q/P}.$$

However, the vector $\boldsymbol{r}_{Q/P}$ is along the ISA, and therefore it is possible to write $\boldsymbol{r}_{Q/P} = \lambda\hat{s}\ \lambda\Re$, which implies that

$$\boldsymbol{s}_O = \boldsymbol{s}_O^* + \hat{s} \times \lambda\hat{s} = \boldsymbol{s}_O^*.$$

10. *Identify the infinitesimal screws associated to the kinematic pairs of Gough's tyre testing machine.*

    Solution. Gough's tyre testing machine is a 6-UPS parallel manipulator where U, P, and S stand for universal, prismatic, and spherical joints, respectively. The explanation of the screws is as follows. The screws associated to the universal joint are notated as ${}^0\$^1$ and ${}^1\$^2$, which represent two revolute joints whose primal parts are concurrent, while ${}^2\$^3$ denotes the screw associated to the prismatic joint. Finally, the screws ${}^3\$^4$, ${}^4\$^5$, and ${}^5\$^6$ represent the screws of the spherical joint connecting the moving platform to the limb where the corresponding primal parts intersect at the center of the spherical joint.

Infinitesimal screws of a limb of the Gough platform

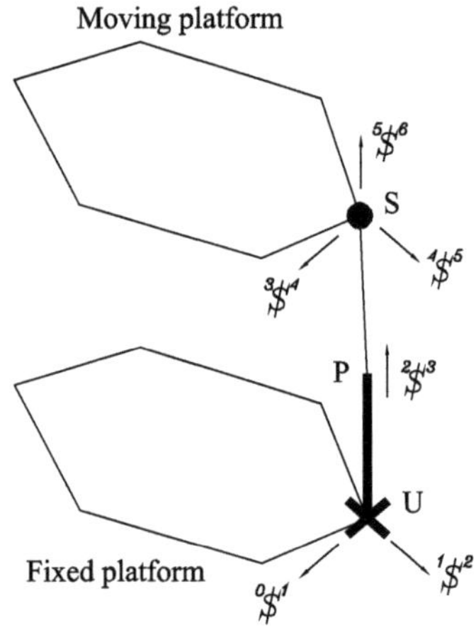

15.2 Chapter 4: Velocity Analysis

1. *Explore the possibility, if any, to obtain the angular velocity as the time derivative of a vector.*

   Solution. The angular velocity is defined as the rate of change of angular displacement. Since an angular displacement cannot be expressed as a vector, then it is impossible to obtain the angular velocity vector as the time derivative of another vector.

2. *In your own words, give an explanation of Eq. (4.11).*

   Solution. For clarity, Eq. (4.11) is repeated here:

$$^{j}\boldsymbol{\omega}^{m} \equiv \boldsymbol{b}_1 \frac{^{j}d\boldsymbol{b}_2}{dt} \bullet \boldsymbol{b}_3 + \boldsymbol{b}_2 \frac{^{j}d\boldsymbol{b}_3}{dt} \bullet \boldsymbol{b}_1 + \boldsymbol{b}_3 \frac{^{j}d\boldsymbol{b}_1}{dt} \bullet \boldsymbol{b}_2.$$

Orthonormal basis $\{\boldsymbol{b}_1, \boldsymbol{b}_2 . \boldsymbol{b}_2\}$. Time derivatives

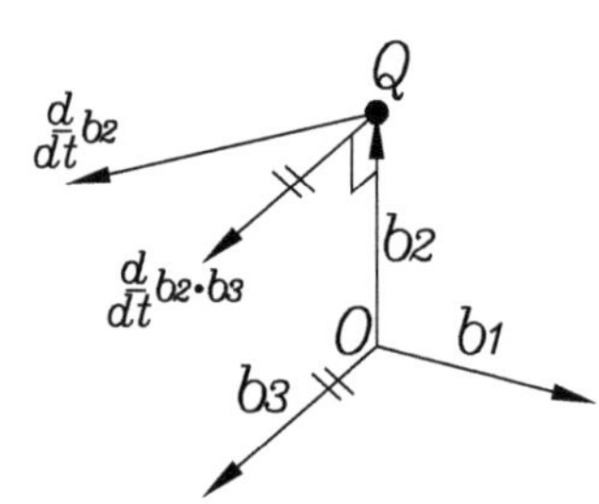

With reference to the figure given above, let Q be the point located by vector $\boldsymbol{b}_2$. Then the velocity of Q is obtained as $\boldsymbol{v}_Q = \frac{d}{dt}\boldsymbol{b}_2$. Furthermore, the projection of $\boldsymbol{v}_Q$ into the b_3-axis is obtained as $\frac{d}{dt}\boldsymbol{b}_2 \bullet \boldsymbol{b}_3$. If we take into account that $\boldsymbol{b}_2$ is a unit vector, then it is possible to ensure that because the velocity $\frac{d}{dt}\boldsymbol{b}_2 \bullet \boldsymbol{b}_3$ is located in the $b_2 b_3$ plane and orthogonal to the b_2-axis, then $\frac{d}{dt}\boldsymbol{b}_2 \bullet \boldsymbol{b}_3$ is precisely the angular velocity of body m as observed from body j along the $\boldsymbol{b}_1$ vector. The conclusion about the components of ${}^j\boldsymbol{\omega}^m$ along the remaining unit vectors is straightforward.

3. *Let m and j be two rigid bodies, and let ${}^j\boldsymbol{\omega}^m$ be the angular velocity vector of m with respect to j. If $\boldsymbol{\beta}$ is a bound vector in body m, then prove, without using Proposition 4.2, that*

$$\frac{{}^j d}{dt}\boldsymbol{\beta} = {}^j\boldsymbol{\omega}^m \times \boldsymbol{\beta}.$$

Solution. This equation is the classical definition of velocity of a point embedded in a rigid body undergoing only rotational motions.

4. *Although this book is devoted mainly to parallel manipulators, the material provided in this chapter is available for serial manipulators. On that topic, determine the velocity state of the end-effector as measured from the base link of the SCARA robot.*

Solution. The acronym **SCARA** stands for **selective compliance assembly robot arm** or **selective compliance articulated robot arm.**

SCARA robot. Infinitesimal screws

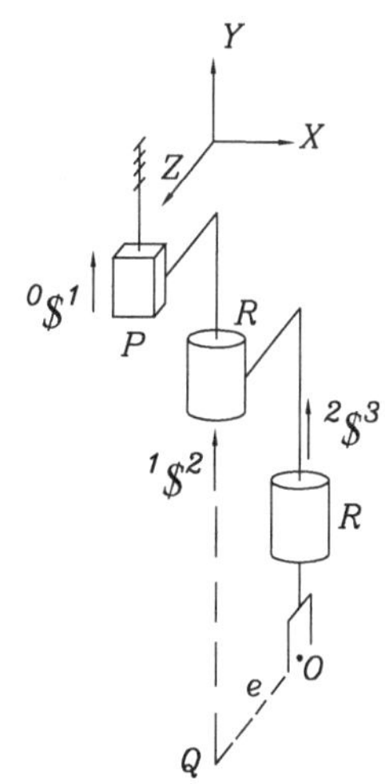

Let O be the reference point chosen for the velocity analysis and let $\boldsymbol{e} = e_X\hat{\boldsymbol{i}} + e_Z\hat{\boldsymbol{k}}$ be the eccentricity vector denoting the normal separation between the revolute joints; that is, vector $\boldsymbol{e}$ is pointed from O to point Q. Thus, the velocity state of the end-effector as observed from the base link, the vector ${}^0\boldsymbol{V}_O^3$, is given by

$$ {}^0\boldsymbol{V}_O^3 = {}_0\omega_1{}^0\$^1 + {}_1\omega_2{}^1\$^2 + {}_2\omega_3{}^2\$^3, $$

where

$$ {}^0\$^1 = \begin{bmatrix} 0 \\ 0 \\ 0 \\ 0 \\ 1 \\ 0 \end{bmatrix}; \quad {}^1\$^2 = \begin{bmatrix} 0 \\ 1 \\ 0 \\ e_Z \\ 0 \\ -e_X \end{bmatrix}; \quad {}^2\$^3 = \begin{bmatrix} 0 \\ 1 \\ 0 \\ 0 \\ 0 \\ 0 \end{bmatrix}. $$

Hence,

$$ {}^0\boldsymbol{V}_O^3 = \begin{bmatrix} 0 \\ {}_1\omega_2 + {}_2\omega_3 \\ 0 \\ {}_1\omega_2 e_Z \\ {}_0\omega_1 \\ -{}_1\omega_2 e_X \end{bmatrix}. $$

5. *Carry out the singularity analysis of the four-bar planar mechanism.*

Solution. The input–output equation of velocity of the four-bar linkage is given by

$$ \mathbf{J}^T \boldsymbol{\Delta}{}^0\boldsymbol{V}_O^2 = \mathbf{q}, $$

where the Jacobian of the closed kinematic chain $\mathbf{J}$ is given by

$$ \mathbf{J} = \begin{bmatrix} 1 & 0 & -a_1 \sin(q) \\ 0 & 1 & a_1 \cos(q) \\ -\cos(\gamma) & \sin(\gamma) & a_0 \sin(\gamma) \end{bmatrix}. $$

The input–output equation of velocity of the mechanism may be rewritten as

$$ \mathbf{J}^T \boldsymbol{\Delta}{}^0\boldsymbol{V}_O^2 = \mathbf{I}_3\mathbf{q}, $$

where $\mathbf{I}_3$ is the 3×3 identity matrix. Hence, the manipulator is free of inverse singularities since $\det(\mathbf{I}_3) = 1$. On the other hand, the closed kinematic chain is at a singular configuration concerned with the forward velocity analysis when $\det(\mathbf{J}) = 0$, in other words, when

$$ a_0 \sin(\gamma) - a_1 \sin(\gamma + q) = 0. $$

6. *Based on screw theory, determine the velocity of the middle of the coupler platform of the four-bar planar mechanism.*

Solution. The velocity state ${}^0V_O^2$ of the coupler platform is computed based on Eq. (15.5). Hence, if Q is the center of the middle platform, by applying the theory of helicoidal vector fields, we obtain the velocity state of the coupler platform with Q chosen as the reference as

$$ {}^0\mathbf{V}_Q^2 = \begin{bmatrix} {}^0\boldsymbol{\omega}^2 \\ \boldsymbol{v}_O + {}^0\boldsymbol{\omega}^2 \times \boldsymbol{r}_{Q/O} \end{bmatrix}. $$

Next, the velocity of point Q is determined as the dual part of the velocity state ${}^0\boldsymbol{V}_Q^2$, that is, $\boldsymbol{v}_Q = \boldsymbol{v}_O + {}^0\boldsymbol{\omega}^2 \times \boldsymbol{r}_{Q/O}$, where $\boldsymbol{r}_{Q/O}$ is the position vector of Q with respect to O.

7. *Determine in loci form the singular regions of the crank-slider linkage mechanism. In other words, find an analytic expression to compute the singularities of the mechanism.*

Solution. The input–output equation of velocity of the crank-slider mechanism [see Eq. (4.100)] is given by

$$ \mathbf{A}\boldsymbol{\Omega} = \mathbf{I}\dot{\mathbf{q}}, $$

where

$$ \mathbf{A} = \begin{bmatrix} -1 & -1 & 0 \\ a_1 \sin q & 0 & -1 \\ a_1 \cos q & d & 0 \end{bmatrix} $$

is the first-order coefficient matrix of the mechanism, while $\mathbf{I}$ is the 3×3 identity matrix. Thus, since $\det(\mathbf{I}) = 1$, the mechanism is free of singularities concerned with the inverse velocity analysis. On the other hand, in loci form the equation of singularities dealing with the forward velocity analysis is given by $\det(\mathbf{A}) = a_1 \cos(q) - d = 0$.

8. *Using classical methods, Eq. (4.107) leads to a linear system of two equations in the unknowns ω_2 and v_P''; determine such a system.*

Solution. With reference to Fig. 4.13, the equation of velocity was derived as follows:

$$ \boldsymbol{\omega}_1 \times \boldsymbol{r}_{P/O_1} = \boldsymbol{\omega}_2 \times \boldsymbol{r}_{P/O_2} + v_P''\hat{\boldsymbol{u}}_{P/O_2}. $$

Taking into account that the fixed reference frame $O - XYZ$ has associated unit vectors $\hat{\boldsymbol{i}}\hat{\boldsymbol{j}}\hat{\boldsymbol{k}}$, we have $\boldsymbol{\omega}_1 = -\omega_1\hat{\boldsymbol{k}}$ and $\boldsymbol{\omega}_2 = \omega_2\hat{\boldsymbol{k}}$, while the position vectors $\boldsymbol{r}_{P/O_1}$ and $\boldsymbol{r}_{P/O_2}$ may be written as $\boldsymbol{r}_{P/O_1} = r_1\cos(\theta)\hat{\boldsymbol{i}} + r_1\sin(\theta)\hat{\boldsymbol{i}}$ and $\boldsymbol{r}_{P/O_2} = -r_2\cos(\beta)\hat{\boldsymbol{i}} + r_2\sin(\beta)\hat{\boldsymbol{i}}$. Furthermore, the unit vector $\hat{\boldsymbol{u}}_{P/O_2}$ may be written as

$\hat{\boldsymbol{u}}_{P/O_2} = \boldsymbol{r}_{P/O_2}/r_2$. Equating $\hat{\boldsymbol{\imath}}$ with $\hat{\boldsymbol{\imath}}$ and $\hat{\boldsymbol{\jmath}}$ with $\hat{\boldsymbol{\jmath}}$ in the aforementioned equation of velocity and reducing terms, one obtains a linear system of two equations given by

$$\omega_2 r_2 - \omega_1 r_1 \cos(\theta + \beta) = 0, \quad v_P'' - \omega_1 r_1 \sin(\theta + \beta).$$

9. *Apply another classical method to solve the velocity analysis of the Geneva wheel, Example 4.7. Hint: Obtain two closure equations from the triangle O_1O_2P indicated in Fig. 4.13. Finding the corresponding time derivatives of such equations allows us to solve the velocity analysis.*

   Solution. From the closed loop O_1O_2P it is possible to write two closure equations as follows:

$$r_1 \cos(\theta) + r_2 \cos(\beta) - \overline{O_1O_2} = 0, \quad r_1 \sin(\theta) - r_2 \sin(\beta) = 0,$$

   where $\overline{O_1O_2}$ denotes the signed distance between points O_2 and O_1. Then the time derivatives of the aforementioned equations lead to the following input–output equation of velocity:

$$\begin{bmatrix} -r_2 \sin(\beta) & \cos(\beta) \\ -r_2 \cos(\beta) & -\sin(\beta) \end{bmatrix} \begin{bmatrix} \dot{\beta} \\ \dot{r}_2 \end{bmatrix} = \begin{bmatrix} r_1 \dot{\theta} \sin(\theta) \\ -r_1 \dot{\theta} \cos(\theta) \end{bmatrix},$$

   where $\dot{\theta} = -\omega_1$, $\dot{\beta} = -\omega_2$. Meanwhile, $\dot{r}_2 = \boldsymbol{v}_P''$.

10. *In Example 4.7 we considered that $v_4 = -\omega_2$. Give a justification, if any, for this assumption.*

    Solution. The joint rate v_4 denotes the angular velocity of Earth as observed from body D while ω_2 denotes the angular velocity of body D as observed from Earth. Hence, $v_4 = -\omega_2$.

11. *The Geneva wheel of Example 4.7 is a little case that illustrates how easy it is to formulate the velocity analysis of parallel manipulators when reciprocal screw theory is used. Determine the input–output equation of velocity of the planar parallel manipulators shown in Fig. 4.16 by resorting to reciprocal screw theory.*

    Solution.

    3-RPR parallel manipulator. The velocity state of the moving platform m, body labeled 3, as observed from the fixed platform f, body labeled 0, may be written in screw form through any of the connector limbs as follows:

$${}_0\omega_1^i \, {}^0\$_i^1 + {}_1\omega_2^i \, {}^1\$_i^2 + {}_2\omega_3^i \, {}^2\$_i^3 = {}^0\boldsymbol{V}^3, \quad i = 1, 2, 3,$$

    where ${}_1\omega_2^i = \dot{q}_i$ is the active joint rate of the robot manipulator. Furthermore, consider that $\$_i$ is a line in Plücker coordinates pointed from point A_i to point B_i. Then systematically applying the Klein form of the ith line to both sides of the equation of velocity in screw form and reducing terms leads to

$$\left\{\$_i;{}^0\mathbf{V}^3\right\} = \dot{q}_i.$$

In other words, the ith line $\$_i$ is reciprocal to the screws representing revolute joints in the same limb. If we cast the preceding equations into matrix-vector form, the input–output equation of velocity of the 3-RPR parallel manipulator results in

$$\mathbf{J}^T \,\Delta\, {}^0\mathbf{V}^3 = \dot{\mathbf{q}},$$

where $\mathbf{J} = \left[\$_1\ \$_2\ \$_3\right]$ is the screw-coordinate Jacobian matrix of the robot, Δ is an operator of polarity, while $\dot{\mathbf{q}} = \left[\dot{q}_1\ \dot{q}_2\ \dot{q}_3\right]^T$ is the first-order driver matrix of the parallel manipulator.

3-RRR parallel manipulator. The velocity state of the moving platform m, body labeled 3, as observed from the fixed platform f, body labeled 0, may be written in screw form through any of the connector limbs as follows:

$${}_0\omega_1^i\,{}^0\$_i^1 + {}_1\omega_2^i\,{}^1\$_i^2 + {}_2\omega_3^i\,{}^2\$_i^3 = {}^0\mathbf{V}^3, \quad i = 1, 2, 3,$$

where ${}_0\omega_1^i = \dot{q}_i$ is the active joint rate of the robot manipulator. Furthermore, consider that $\$_i$ is a line in Plücker coordinates pointed from point B_i to point C_i. Then systematically applying the Klein form of the ith line to both sides of the equation of velocity in screw form and reducing terms leads to

$$\left\{\$_i;{}^0\mathbf{V}^3\right\} = \dot{q}_i\left\{{}^0\$_i^1;\$_i\right\}.$$

In other words, the ith line $\$_i$ is reciprocal to the screws ${}^1\$_i^2$ and ${}^2\$_i^3$ in the same limb. Casting the preceding equations into matrix form, the input–output equation of velocity of the 3-RRR parallel manipulator results in

$$\mathbf{J}^T \,\Delta\, {}^0\mathbf{V}^3 = \dot{\mathbf{q}},$$

where $\mathbf{J} = \left[\$_1\ \$_2\ \$_3\right]$ is the screw-coordinate Jacobian matrix of the robot, Δ is an operator of polarity, while
$\dot{\mathbf{q}} = \left[\dot{q}_1\left\{{}^0\$_1^1;\$_1\right\}\ \dot{q}_2\left\{{}^0\$_2^1;\$_2\right\}\ \dot{q}_3\left\{{}^0\$_3^1;\$_3\right\}\right]^T$ is the first-order driver matrix of the 3-RRR parallel manipulator.

2-RPR+RRR parallel manipulator. Following the trend of the velocity analysis of the 3-RPR and 3-RRR parallel manipulators, the input–output equation of velocity of the robot at hand results in

$$\mathbf{J}^T \,\Delta\, {}^0\mathbf{V}^3 = \dot{\mathbf{q}}.$$

In this equation, $\mathbf{J} = [\$_1\ \$_2\ \$_3]$ is the screw-coordinate Jacobian matrix of the robot where $\$_1$ and $\$_3$ are lines in Plücker coordinates pointed from their respective points A_i to points B_i. Meanwhile, $\$_2$ is a line in Plücker coordinates pointed from B_2 to the intermediate revolute joint. Furthermore, Δ is an operator of polarity, while $\dot{\mathbf{q}} = \left[\dot{q}_1\ \dot{q}_2 \left\{{}^0\$_2^1; \$_2\right\} \dot{q}_3\right]^T$ is the first-order driver matrix of the considered parallel manipulator.

2-RPR+RRR parallel manipulator. The input–output equation of velocity of the robot at hand results in

$$\mathbf{J}^T\ \Delta\ {}^0\mathbf{V}^3 = \dot{\mathbf{q}},$$

where $\mathbf{J} = [\$_1\ \$_2\ \$_3]$ is the screw-coordinate Jacobian matrix of the robot where $\$_1$ and $\$_2$ are lines in Plücker coordinates pointed from their respective points B_i to points C_i, while $\$_3$ is a line in Plücker coordinates pointed from point A_3 to point C_3. Furthermore, Δ is an operator of polarity, while $\dot{\mathbf{q}} = \left[\dot{q}_1 \left\{{}^0\$_1^1; \$_1\right\} \dot{q}_2 \left\{{}^0\$_2^1; \$_2\right\} \dot{q}_3\right]^T$ is the first-order driver matrix of the parallel manipulator under consideration.

12. *The link c of the six-bar dwell mechanism (see Fig. 4.17) rotates with a generalized velocity $\dot{q}$. The topology of the mechanism is such that the angular velocity $\dot{q}$ produces an oscillatory linear velocity $\dot{x}$. Determine the input–output equation of velocity of the planar mechanism.*

Six-bar dwell mechanism. Infinitesimal screws

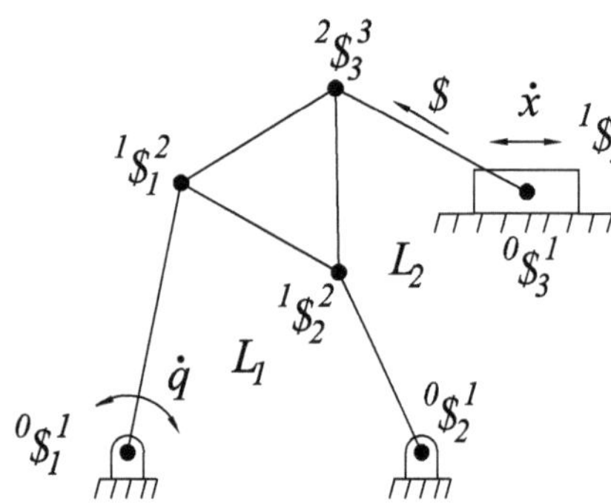

Solution. Referring to the above figure, we can obtain the velocity state of the moving platform m as observed from the fixed platform 0 by solving the kinematics of the four-bar linkage indicated by the loop L_1. Then the velocity state may be written in screw form through the third limb as follows:

$$\dot{x}\,{}^0\$_3^1 + {}_1\omega_2^3\ {}^1\$_3^2 + {}_2\omega_3^3\ {}^2\$_3^3 = {}^0\mathbf{V}^m.$$

Finally, applying the Klein form of the line $ to both sides of such equation yields

$$\left\{\$; {}^0\mathbf{V}^m\right\} = \dot{x}\left\{\$; {}^0\$_3^1\right\}.$$

13. *The mechanism shown in Fig. 4.18 is used to produce high torque in the shaft at B and is a good example to which to apply screw theory. The gear unit is pivoted at A and produces a constant angular velocity of 5 rad/s counterclockwise as observed from A to C over the right-handed screw. If the pitch of the screw is six single threads per inch of length, determine the joint-velocity rate of link BC.*

High torque mechanism. Infinitesimal screws

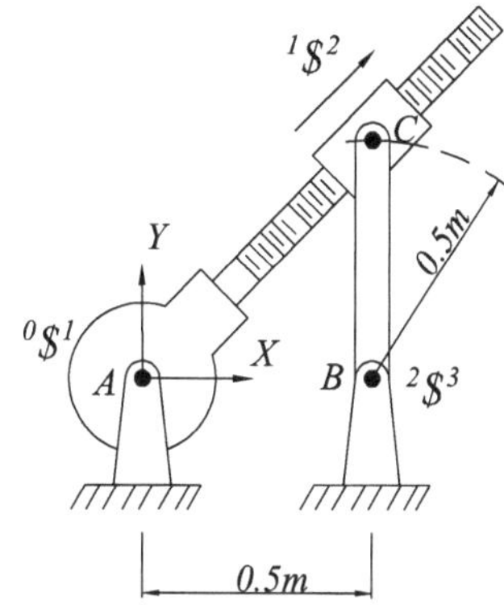

Solution. The equation of velocity of the mechanism may be written in screw form as follows:

$$_0\omega_1{}^0\$^1 + {}_1\omega_2{}^1\$^2 + {}_2\omega_3{}^2\$^3 + {}_3\omega_4{}^3\$^4 = \mathbf{0},$$

where $_1\omega_2 = 5/6$ in./s. Furthermore, according to the fixed reference frame XYZ, whose origin is located at point A, the infinitesimal screws are computed as

$$^0\$^1 = \begin{bmatrix} 0 \\ 0 \\ 1 \\ 0 \\ 0 \\ 0 \end{bmatrix}, \quad {}^1\$^2 = \begin{bmatrix} 0 \\ 0 \\ 0 \\ \sqrt{2}/2 \\ \sqrt{2}/2 \\ 0 \end{bmatrix}, \quad {}^2\$^3 = \begin{bmatrix} 0 \\ 0 \\ 1 \\ 1/2 \\ -1/2 \\ 0 \end{bmatrix}, \quad {}^3\$^4 = \begin{bmatrix} 0 \\ 0 \\ 1 \\ 0 \\ -1/2 \\ 0 \end{bmatrix}.$$

Hence, one obtains the following linear equations:

$$_0\omega_1 + {}_2\omega_3 + {}_3\omega_4 = 0, \; \sqrt{2}{}_1\omega_2 + {}_2\omega_3/2 = 0, \; \sqrt{2}{}_1\omega_2 - {}_2\omega_3/2 - {}_3\omega_4/2 = 0$$

whose solution yields $_3\omega_4 = 2\sqrt{2}{}_1\omega_2 = 2.357$ rad/s. Hence, the link BC is undergoing an angular velocity of 2.357 rad/s clockwise.

15.3 Chapter 5: Acceleration Analysis

1. *It was demonstrated that* ${}^j\boldsymbol{v}_O^m = {}^j\boldsymbol{v}_O^k + {}^k\boldsymbol{v}_O^m$. *Prove that even though such an expression is correct, the acceleration vector* ${}^j\boldsymbol{a}_O^m$, *where* ${}^j\boldsymbol{a}_O^m \equiv \frac{{}^jd}{dt}({}^j\boldsymbol{v}_O^m)$, *cannot be determined correctly based on such an expression.*

   Solution. Using the involved reference frames properly, we have

$$
\begin{aligned}
{}^j\boldsymbol{a}_O^m &\equiv \frac{{}^jd}{dt}({}^j\boldsymbol{v}_O^k + {}^k\boldsymbol{v}_O^m)\\
&= \frac{{}^jd}{dt}({}^j\boldsymbol{v}_O^k) + \frac{{}^jd}{dt}({}^k\boldsymbol{v}_O^m)\\
&= {}^j\boldsymbol{a}_O^k + {}^j\boldsymbol{\omega}^k \times {}^k\boldsymbol{v}_O^m + \frac{{}^kd}{dt}({}^k\boldsymbol{v}_O^m)\\
&= {}^j\boldsymbol{a}_O^k + {}^k\boldsymbol{a}_O^m + {}^j\boldsymbol{\omega}^k \times {}^k\boldsymbol{v}_O^m.
\end{aligned}
$$

   Therefore, because the last term must be $2{}^j\boldsymbol{\omega}^k \times {}^k\boldsymbol{v}_O^m$, the equation obtained to compute ${}^j\boldsymbol{a}_O^m$ with this procedure yields an incomplete result.

2. *Determine the input–output equation of acceleration of the four-bar planar mechanism by resorting to screw theory. Afterward, determine the angular acceleration of the driven link given the angular acceleration of the driver link. Hint: Write the equation of acceleration in screw form and apply the appropriate Klein form to both sides of such an equation.*

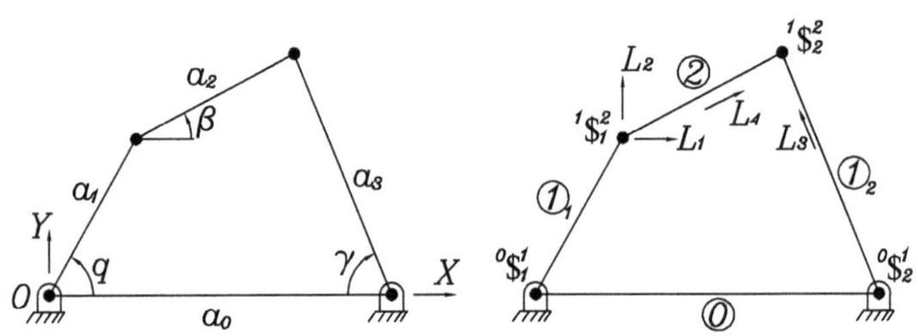

Four-bar linkage mechanism

   Solution. Let $\boldsymbol{\alpha} = \alpha_X\hat{\boldsymbol{i}} + \alpha_Y\hat{\boldsymbol{j}} + \alpha_Z\hat{\boldsymbol{k}}$ be the angular acceleration vector of the moving platform with respect to the fixed platform, where $\hat{\boldsymbol{i}}, \hat{\boldsymbol{j}}$, and $\hat{\boldsymbol{k}}$ are unit vectors associated to the X-, Y-, and Z-axes, respectively. Furthermore, let $\boldsymbol{a}_O = a_{OX}\hat{\boldsymbol{i}} + a_{OY}\hat{\boldsymbol{j}} + a_{OZ}\hat{\boldsymbol{k}}$ be the acceleration vector of the point embedded in the moving platform that is instantaneously coincident with the origin O of the reference frame XYZ attached to the fixed platform. Then the reduced acceleration state of the moving platform with respect to the fixed platform, the vector ${}^0\boldsymbol{A}_O^2$, may be expressed through any of the two connector chains as

$$ {}^0\boldsymbol{A}_O^2 = {}_0\alpha_1^i\,{}^0\$_i^1 + {}_1\alpha_2^i\,{}^1\$_i^2 + \mathcal{L}_i, \quad i = 1, 2. $$

Here ${}_*\alpha_*^* = \frac{d}{dt}({}_*\omega_*^*)$, so that ${}_0\alpha_1^1 = \ddot{q}$ and $\mathcal{L}_i = \left[{}_0\omega_1^i\,{}^0\$_i^1 \quad {}_1\omega_2^i\,{}^1\$_i^2\right]$. Meanwhile, the screws are given in Plücker coordinates by

$$ {}^0\$_1^1 = \begin{bmatrix} 0 \\ 0 \\ 1 \\ 0 \\ 0 \\ 0 \end{bmatrix}, {}^1\$_1^2 = \begin{bmatrix} 0 \\ 0 \\ 1 \\ a_1 \sin q \\ -a_1 \cos q \\ 0 \end{bmatrix}, {}^0\$_2^1 = \begin{bmatrix} 0 \\ 0 \\ 1 \\ 0 \\ -a_0 \\ 0 \end{bmatrix}, {}^1\$_2^2 = \begin{bmatrix} 0 \\ 0 \\ 1 \\ a_3 \sin \gamma \\ -a_1 \cos q - a_2 \cos \beta \\ 0 \end{bmatrix}. $$

On the other hand, consider that there are three lines $\boldsymbol{L}_1$, $\boldsymbol{L}_2$, and $\boldsymbol{L}_3$ whose Plücker coordinates are given by

$$ \boldsymbol{L}_1 = \begin{bmatrix} 1 \\ 0 \\ 0 \\ 0 \\ 0 \\ -a_1 \sin q \end{bmatrix}, \boldsymbol{L}_2 = \begin{bmatrix} 0 \\ 1 \\ 0 \\ 0 \\ 0 \\ a_1 \cos q \end{bmatrix}, \boldsymbol{L}_3 = \begin{bmatrix} -\cos \gamma \\ \sin \gamma \\ 0 \\ 0 \\ 0 \\ a_0 \sin \gamma \end{bmatrix}. $$

Then, by applying the Klein form of $\boldsymbol{L}_1$ to both sides of the reduced acceleration state in screw form computed through the first limb and reducing terms, one obtains

$$ \left\{\boldsymbol{L}_1; {}^0\boldsymbol{A}_O^2\right\} = -a_1\ddot{q}\sin q + \{\boldsymbol{L}_1; \mathcal{L}_1\}. $$

Similarly, we have

$$ \left\{\boldsymbol{L}_2; {}^0\boldsymbol{A}_O^2\right\} = a_1\ddot{q}\cos q + \{\boldsymbol{L}_2; \mathcal{L}_2\} $$

and

$$ \left\{\boldsymbol{L}_3; {}^0\boldsymbol{A}_O^2\right\} = \{\boldsymbol{L}_3; \mathcal{L}_2\}. $$

Finally, gathering these results, we see that the input–output equation of acceleration of the four-bar linkage results in

$$ \mathbf{J}^T \boldsymbol{\Delta}\, {}^0\boldsymbol{A}_O^2 = \mathbf{q}_a, \tag{15.5} $$

where $\mathbf{J} = [\boldsymbol{L}_1 \ \boldsymbol{L}_2 \ \boldsymbol{L}_3]$ is the 3×3 reduced active screw-coordinate Jacobian matrix of the manipulator where the third, fourth, and fifth rows were removed from the six-dimensional vectors $\boldsymbol{L}_i (i = 1, 2, 3)$ because of the planar arrangement of the links of the mechanism at hand. Similarly, the reduced acceleration state ${}^0\boldsymbol{A}_O^2$ is reduced into a three-dimensional vector considering that $\alpha_X = \alpha_Y = a_Z = 0$. Meanwhile, $\mathbf{q}_a = [-a_1\ddot{q}\sin q + \{\boldsymbol{L}_1; \mathcal{L}_1\} \quad a_1\dot{q}\cos q + \{\boldsymbol{L}_2; \mathcal{L}_2\} \quad \{\boldsymbol{L}_3; \mathcal{L}_2\}]^T$ is the second-order driver matrix of the robot. Finally,

$$\boldsymbol{\Delta} = \begin{bmatrix} 0 & 0 & 1 \\ 0 & 1 & 0 \\ 1 & 0 & 0 \end{bmatrix}$$

is an operator of polarity. Once the reduced acceleration state ${}^0\boldsymbol{A}_O^2$ is calculated upon the input–output equation of acceleration, the angular acceleration of the driven link is obtained as the primal part of the vector ${}^0\boldsymbol{A}_O^2$.

3. *Using a standard method, determine the symbolic equation of the angular acceleration of the driven wheel of the Geneva mechanism considering that the driver wheel rotates at a constant angular velocity ω_1 counterclockwise.*

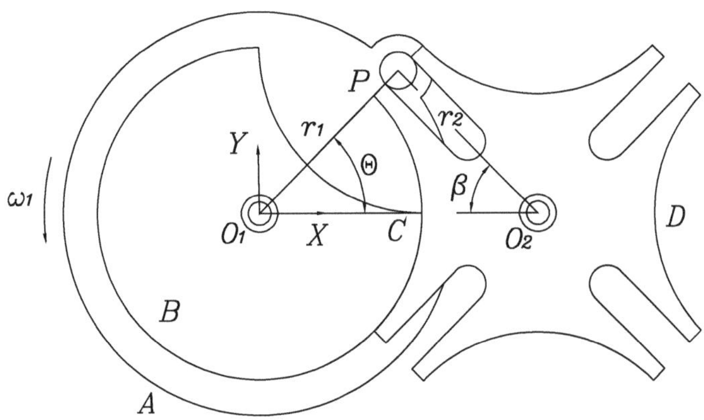

Geneva wheel

Solution. As an intermediate step it is necessary to first solve the velocity analysis of the mechanism. Using elementary kinematics, the velocity of pin P can be computed as

$$\boldsymbol{v}_P = \boldsymbol{\omega}_1 \times \boldsymbol{r}_{P/O_1},$$

where $\boldsymbol{r}_{P/O_1}$ is a vector pointed from O_1 to point P; that is, $\boldsymbol{r}_{P/O_1} = r_1\cos(\theta)\hat{\boldsymbol{i}} + r_1\sin(\theta)\hat{\boldsymbol{j}}$ and $\boldsymbol{\omega}_1 = \omega_1\hat{\boldsymbol{k}}$. Furthermore, the same velocity vector $\boldsymbol{v}_P$ may be obtained upon the driven wheel D as follows:

$$\boldsymbol{v}_P = \boldsymbol{v}'_P + \boldsymbol{v}''_P,$$

where $\boldsymbol{v}'_P$ is the velocity of the point of body D, which is instantaneously coincident with point P, while $\boldsymbol{v}''_P$ is the velocity of point P relative to the wheel D. Indeed,

$$\boldsymbol{v}'_P = \boldsymbol{\omega}_2 \times \boldsymbol{r}_{P/O_2}, \quad \boldsymbol{v}''_P = v''_P \hat{\boldsymbol{u}}_{P/O_2},$$

where $\boldsymbol{r}_{P/O_2}$ is the vector pointed from O_2 to point P, that is, $\boldsymbol{r}_{P/O_2} = -r_2 \cos(\beta)\hat{\boldsymbol{i}} + r_2 \sin(\beta)\hat{\boldsymbol{i}}$, while $\hat{\boldsymbol{u}}_{P/O_2}$ is a unit vector pointed from O_2 to P, namely, $\hat{\boldsymbol{u}}_{P/O_2} = -\cos(\beta)\hat{\boldsymbol{i}} + \sin(\beta)\hat{\boldsymbol{j}}$. Hence, it follows that

$$\boldsymbol{\omega}_1 \times \boldsymbol{r}_{P/O_1} = \boldsymbol{\omega}_2 \times \boldsymbol{r}_{P/O_2} + v''_P \hat{\boldsymbol{u}}_{P/O_2},$$

which yields a linear system of two equations in the unknowns ω_2 and v''_P whose solution is given by

$$\omega_2 = -\cos(\beta + \theta)\omega_1 r_1 / r_2, \quad v''_P = -\omega_1 r_1 \sin(\beta + \theta).$$

In what follows the acceleration analysis is presented. Since the angular velocity ω_1 of the driver wheel is constant, then the acceleration of pin P is given by

$$\boldsymbol{a}_P = \boldsymbol{\omega}_1 \times (\boldsymbol{\omega}_1 \times \boldsymbol{r}_{P/O_1}).$$

Furthermore, the same acceleration vector $\boldsymbol{a}_P$ may be obtained upon the driven wheel as

$$\boldsymbol{a}_P = \boldsymbol{a}'_P + \boldsymbol{a}''_P + 2\boldsymbol{\omega}_2 \times \boldsymbol{v}''_P, \tag{15.6}$$

where $\boldsymbol{a}'_P$ is the acceleration of the point of body D, which is instantaneously coincident with point P, while $\boldsymbol{a}''_P$ is the acceleration of point P relative to the wheel D, and $2\boldsymbol{\omega}_2 \times \boldsymbol{v}''_P$ is the Coriolis acceleration. Indeed,

$$\boldsymbol{a}'_P = \boldsymbol{\alpha}_2 \times \boldsymbol{r}_{P/O_2} + \boldsymbol{\omega}_2 \times (\boldsymbol{\omega}_2 \times \boldsymbol{r}_{P/O_2}), \quad \boldsymbol{a}''_P = a''_P \hat{\boldsymbol{u}}_{P/O_2}.$$

Hence, it follows that

$$\boldsymbol{\omega}_1 \times (\boldsymbol{\omega}_1 \times \boldsymbol{r}_{P/O_1}) = \boldsymbol{\alpha}_2 \times \boldsymbol{r}_{P/O_2} + \boldsymbol{\omega}_2 \times (\boldsymbol{\omega}_2 \times \boldsymbol{r}_{P/O_2}) + a''_P \hat{\boldsymbol{u}}_{P/O_2} + 2\boldsymbol{\omega}_2 \times \boldsymbol{v}''_P,$$

which leads to a linear system of two equations in the unknowns $\boldsymbol{\alpha}_2$ and a''_P, whose solution for the angular acceleration of the driven wheel D results in

$$\alpha_2 = \left[\omega_1^2 r_1 \sin(\theta + \beta) + 2v''_P \omega_2\right] / r_2.$$

4. *Carry out the singularity analysis of the 2RRR+PPR parallel manipulator based on the input–output equation of velocity.*

Solution. The input–output equation of velocity of the manipulator was obtained as

$$\tilde{\mathbf{A}} \boldsymbol{V}_P = \tilde{\mathbf{B}} \dot{\mathbf{q}},$$

2RRR+PPR parallel manipulator

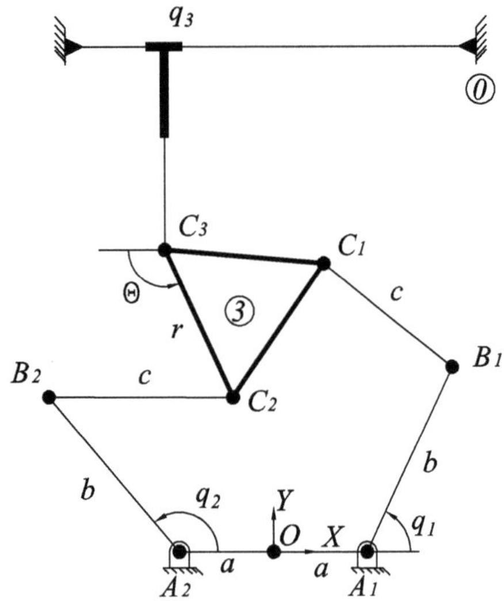

where $\tilde{\mathbf{A}} = \mathbf{J}^T \boldsymbol{\Delta}$ is the active matrix of the robot where the 3×3 Jacobian matrix $\mathbf{J}$ is given by $\mathbf{J} = \begin{bmatrix} \boldsymbol{L}_1 \, \boldsymbol{L}_2 \, \boldsymbol{L}_3 \end{bmatrix}$ and $\boldsymbol{\Delta} = \begin{bmatrix} 0 & 1 & 0 \\ 0 & 0 & 1 \\ 1 & 0 & 0 \end{bmatrix}$ is an operator of polarity. Furthermore, $\dot{\mathbf{q}} = \begin{bmatrix} \dot{q}_1 \, \dot{q}_2 \, \dot{q}_3 \end{bmatrix}^T$ is the first-order driver matrix of the manipulator, while matrix $\tilde{\mathbf{B}}$ is given by

$$\tilde{\mathbf{B}} = \begin{bmatrix} \{{}^1\$^2_1; \boldsymbol{L}_1\} & 0 & 0 \\ 0 & \{{}^1\$^2_2; \boldsymbol{L}_2\} & 0 \\ 0 & 0 & \{{}^0\$^1_3; \boldsymbol{L}_3\} \end{bmatrix}.$$

Thus, the parallel manipulator is at a singular configuration when (i) matrix $\mathbf{J}$ is singular, that is, while $det(\begin{bmatrix} \boldsymbol{L}_1 \, \boldsymbol{L}_2 \, \boldsymbol{L}_3 \end{bmatrix}) = 0$; (ii) matrix $\tilde{\mathbf{B}}$ is singular, which occurs when $\{{}^1\$^2_1; \boldsymbol{L}_1\} \{{}^1\$^2_2; \boldsymbol{L}_2\} \{{}^0\$^1_3; \boldsymbol{L}_3\} = 0$; (iii) both matrices $\mathbf{J}$ and $\tilde{\mathbf{B}}$ are singular.

5. *The figure shows a planar timing mechanism. The motion of pin B is controlled by the linear slot A, which is being elevated with a constant velocity upward of $1.5\, m/s$ as measured from the fixed reference frame. If $r = 0.3\, m$ and $\theta = 30°$, compute (a) the acceleration of pin B as measured from the Earth, (b) the relative acceleration of pin B as observed from the circular slot. Hint: Subdivide the timing mechanism into two different planar mechanisms and identify the corresponding j, k, and m bodies for each mechanism.*

Solution. The velocity of B as observed from the semidisk is given by

$$\boldsymbol{v}_B = v_B(\sin\theta \hat{\boldsymbol{i}} + \cos\theta \hat{\boldsymbol{j}}).$$

Planar timing mechanism

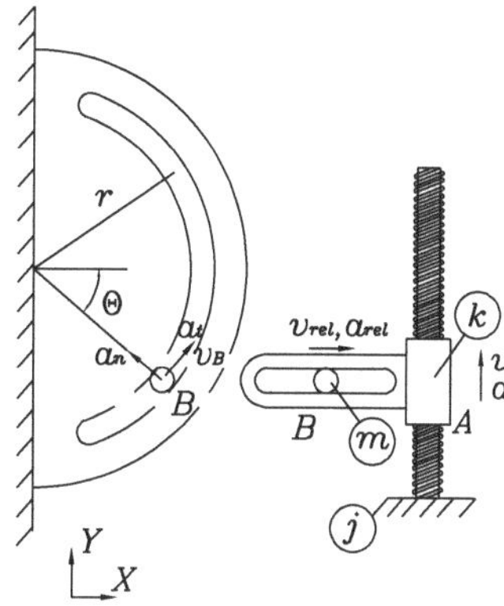

Furthermore, the velocity of B may be computed upon the lifting mechanism as

$$\boldsymbol{v}_B = {}^j\boldsymbol{v}_B^k + {}^k\boldsymbol{v}_B^m = v_A\hat{\boldsymbol{j}} + v_{rel}\hat{\boldsymbol{i}}.$$

Hence, it follows that $v_{rel} = v_A \tan\theta$ and $v_B = v_A/\cos\theta$. After, the acceleration of pin B can be computed upon the semidisk as

$$\boldsymbol{a}_B = \boldsymbol{a}_t + \boldsymbol{a}_n = a_t(\sin\theta\hat{\boldsymbol{i}} + \cos\theta\hat{\boldsymbol{j}}) + v_B^2/r(-\cos\theta\hat{\boldsymbol{i}} + \sin\theta\hat{\boldsymbol{j}}).$$

Meanwhile, the acceleration vector $\boldsymbol{a}_B$ may be computed according to the lifting mechanism as follows:

$$\boldsymbol{a}_B = {}^j\boldsymbol{a}_B^k + {}^k\boldsymbol{a}_B^m + 2{}^j\boldsymbol{\omega}^k \times {}^k\boldsymbol{v}_B^m = a_A\hat{\boldsymbol{j}} + a_{rel}\hat{\boldsymbol{i}} = a_A\hat{\boldsymbol{j}} + a_{rel}\hat{\boldsymbol{i}}.$$

Hence, one obtains

$$a_t\sin\theta - v_B^2\cos\theta/r = a_{rel}, a_t\cos\theta + v_B^2\sin\theta/r = a_A.$$

Furthermore, the acceleration of B results in

$$a_B^2 = \left[(ra_A\sin\theta\cos^2\theta - v_A^2)/(r\cos^3\theta)\right]^2 + a_A^2.$$

Meanwhile, the relative acceleration of the pin results in

$$a_{rel} = (ra_A\sin\theta\cos^2\theta + v_A^2)/(r\cos^3\theta).$$

Therefore, by substituting data we have $a_B = a_{rel} = 11.547\,\text{m/s}^2$.

6. *Determine the general input–output equations of velocity and acceleration of the timing mechanism considering that the active kinematic pair is the lead screw whose pitch is h. Hint: The timing mechanism may be modeled as a PPR closed*

kinematic chain where the first prismatic joint is related to the active kinematic joint through the pitch h of the lead screw.

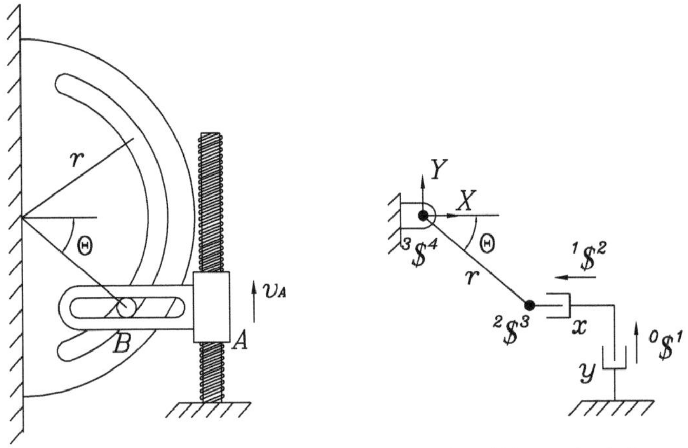

Planar timing mechanism. Equivalent PPR closed kinematic chain

The equation of velocity in screw form of the closed kinematic chain is given by

$$v_A {}^0\$^1 + {}_1\omega_2 {}^1\$^2 + {}_2\omega_3 {}^2\$^3 + {}_3\omega_4 {}^3\$^4 =$$

$$v_A \begin{bmatrix} 0 \\ 0 \\ 0 \\ 0 \\ 1 \\ 0 \end{bmatrix} + {}_1\omega_2 \begin{bmatrix} 0 \\ 0 \\ 0 \\ -1 \\ 0 \\ 0 \end{bmatrix} + {}_2\omega_3 \begin{bmatrix} 0 \\ 0 \\ 1 \\ -r\sin\theta \\ -r\cos\theta \\ 0 \end{bmatrix} + {}_3\omega_4 \begin{bmatrix} 0 \\ 0 \\ 1 \\ 0 \\ 0 \\ 0 \end{bmatrix} = \begin{bmatrix} 0 \\ 0 \\ 0 \\ 0 \\ 0 \\ 0 \end{bmatrix}.$$

Hence, the input–output equation of velocity of the manipulator results in

$$\mathbf{A}\Omega = \mathbf{q}_v,$$

where $\mathbf{A} = \begin{bmatrix} 0 & 1 & 1 \\ -1 & -r\sin\theta & 0 \\ 0 & -r\cos\theta & 0 \end{bmatrix}$, $\Omega = \begin{bmatrix} {}_1\omega_2 & {}_2\omega_3 & {}_3\omega_4 \end{bmatrix}^T$, and $\mathbf{q}_v = \begin{bmatrix} 0 & 0 & -v_A \end{bmatrix}^T$.

On the other hand, the equation of acceleration of the closed kinematic chain may be written in screw form as

$$a_A{}^0\$^1 + {}_1\alpha_2{}^1\$^2 + {}_2\alpha_3{}^2\$^3 + {}_3\alpha_4{}^3\$^4 + \mathcal{L} = \begin{bmatrix} 0 \\ 0 \\ 0 \\ 0 \\ 0 \\ 0 \end{bmatrix},$$

where the Lie screw of acceleration $\mathcal{L}$ is given by

$$\mathcal{L} = \left[v_A{}^0\$^1 \quad {}_1\omega_2{}^1\$^2 + {}_2\omega_3{}^2\$^3 + {}_3\omega_4{}^3\$^4\right] +$$
$$\left[{}_1\omega_2{}^1\$^2 \quad {}_2\omega_3{}^2\$^3 + {}_3\omega_4{}^3\$^4\right] + \left[{}_2\omega_3{}^2\$^3 \quad {}_3\omega_4{}^3\$^4\right].$$

Hence, the input–output equation of acceleration of the robot manipulator results in

$$\mathbf{A}\dot{\Omega} = \mathbf{q}_a - \mathcal{L},$$

where $\mathbf{q}_a = \left[0\ 0\ -a_A\right]^T$. Meanwhile, $\Omega = \left[{}_1\alpha_2\ {}_2\alpha_3\ {}_3\alpha_4\right]^T$.

15.4 Chapter 6: Jerk Analysis

1. *The nozzle shown in the figure rotates with constant angular velocity ω about a fixed horizontal axis passing through point O. The velocity of the water relative to the nozzle is v and is constant. Determine the velocity, acceleration, and jerk of a water particle as it passes point A.*

Nozzle. Jerk analysis

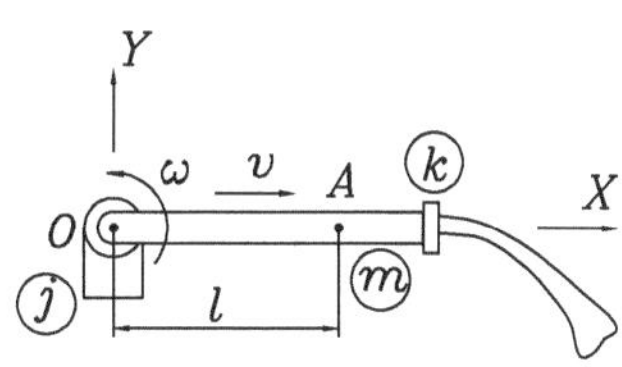

Solution. Let j, k, and m be three rigid bodies in relative motion where j plays the role of fixed reference frame. The velocity of a water particle passing through point A, the vector $\boldsymbol{v}_A$, is given by

$$\boldsymbol{v}_A = {}^j\boldsymbol{v}_A^k + {}^k\boldsymbol{v}_A^m,$$

where ${}^j\boldsymbol{v}_A^k = \boldsymbol{\omega} \times \boldsymbol{r}_{A/O} = \omega\hat{\boldsymbol{k}} \times l\hat{\boldsymbol{i}} = \omega l\hat{\boldsymbol{j}}$. Meanwhile, ${}^k\boldsymbol{v}_A^m = \boldsymbol{v} = v\hat{\boldsymbol{i}}$. Hence, one obtains

$$\boldsymbol{v}_A = v\hat{\boldsymbol{i}} + \omega l\hat{\boldsymbol{j}}.$$

On the other hand, the acceleration of A, the vector $\boldsymbol{a}_A$, is computed as

$$\boldsymbol{a}_A = {}^j\boldsymbol{a}_A^k + {}^k\boldsymbol{a}_A^m + 2{}^j\boldsymbol{\omega}^k \times {}^k\boldsymbol{v}_A^m,$$

where ${}^j\boldsymbol{a}_A^k = -\omega^2 \boldsymbol{r}_{A/O} = -\omega^2 l\hat{\boldsymbol{i}}$. Meanwhile, ${}^k\boldsymbol{a}_A^m = \boldsymbol{0}$ whereas $2{}^j\boldsymbol{\omega}^k \times {}^k\boldsymbol{v}_A^m = 2\omega\hat{\boldsymbol{k}} \times v\hat{\boldsymbol{i}} = 2\omega v\hat{\boldsymbol{j}}$. Therefore,

$$\boldsymbol{a}_A = -\omega^2 l\hat{\boldsymbol{i}} + 2\omega v\hat{\boldsymbol{j}}.$$

Finally, the jerk of the particle A, the vector ${}^j\boldsymbol{\rho}_A^m$, is given by

$${}^j\boldsymbol{\rho}_A^m = {}^j\boldsymbol{\rho}_A^k + {}^k\boldsymbol{\rho}_A^m + 3{}^j\boldsymbol{\alpha}^k \times {}^k\boldsymbol{v}_A^m + 3{}^j\boldsymbol{\omega}^k \times {}^k\boldsymbol{a}_A^m + 3{}^j\boldsymbol{\omega}^k \times \left({}^j\boldsymbol{\omega}^k \times {}^k\boldsymbol{v}_A^m\right),$$

where

$$\begin{aligned}
{}^j\boldsymbol{\rho}_A^k &= {}^j\boldsymbol{\rho}^k \times \boldsymbol{r}_{A/O} + 2{}^j\boldsymbol{\alpha}^k \times \left({}^j\boldsymbol{\omega}^k \times \boldsymbol{r}_{A/O}\right) + \\
&\quad {}^j\boldsymbol{\omega}^k \times \left({}^j\boldsymbol{\alpha}^k \times \boldsymbol{r}_{A/O}\right) + {}^j\boldsymbol{\omega}^k \times \left[{}^j\boldsymbol{\omega}^k \times \left({}^j\boldsymbol{\omega}^k \times \boldsymbol{r}_{A/O}\right)\right] \\
&= {}^j\boldsymbol{\omega}^k \times \left[{}^j\boldsymbol{\omega}^k \times \left({}^j\boldsymbol{\omega}^k \times \boldsymbol{r}_{A/O}\right)\right] \\
&= \omega\hat{\boldsymbol{k}} \times \left[\omega\hat{\boldsymbol{k}} \times \left(\omega\hat{\boldsymbol{k}} \times l\hat{\boldsymbol{i}}\right)\right] = -\omega^3 l\hat{\boldsymbol{j}}.
\end{aligned}$$

Furthermore, considering that ${}^k\boldsymbol{\rho}_A^m = 3{}^j\boldsymbol{\alpha}^k \times {}^k\boldsymbol{v}_A^m = 3{}^j\boldsymbol{\omega}^k \times {}^k\boldsymbol{a}_A^m = \boldsymbol{0}$, one obtains

$${}^j\boldsymbol{\rho}_A^m = -\omega^3 l\hat{\boldsymbol{j}} + 3{}^j\boldsymbol{\omega}^k \times \left({}^j\boldsymbol{\omega}^k \times {}^k\boldsymbol{v}_A^m\right) = -3\omega^2 v\hat{\boldsymbol{i}} - \omega^3 l\hat{\boldsymbol{j}}.$$

2. *The slider P can be moved by the string S while the bar OA rotates around the pivot O. The orientation of the bar is given according to the expression $\theta = 0.25 + 0.1t + 0.05t^2 + 0.075t^3$, where the angle θ is given in radians and the time t is in seconds. The position of the slider is commanded to follow the expression $r = 0.6 - 0.25t - 0.025t^2 + 0.001t^3$, where r is given in meters and t in seconds. Determine the velocity, acceleration, and jerk of the slider at time $t = 3$ s.*

Slider-spring mechanism

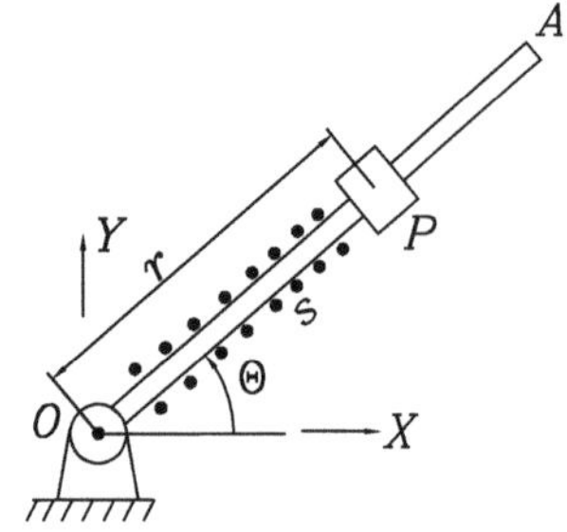

Solution. Consider that ${}^j\boldsymbol{\omega}^k = \dot{\theta}\hat{\boldsymbol{k}}$, ${}^j\boldsymbol{\alpha}^k = \ddot{\theta}\hat{\boldsymbol{k}}$, ${}^j\boldsymbol{\rho}^k = \dddot{\theta}\hat{\boldsymbol{k}}$, ${}^k\boldsymbol{v}_P^m = \dot{r}\hat{\boldsymbol{r}}_{P/O}$, ${}^k\boldsymbol{a}_P^m = \ddot{r}\hat{\boldsymbol{r}}_{P/O}$, ${}^k\boldsymbol{\rho}_P^m = \dddot{r}\hat{\boldsymbol{r}}_{P/O}$, where $\dot{\theta} = 0.1+0.1t+0.225t^2$, $\ddot{\theta} = 0.1+0.45t$, $\dddot{\theta} = 0.45$, $\dot{r} = -0.25-0.05t+0.003t^2$, $\ddot{r} = -0.05+0.006t$, and $\dddot{r} = 0.006$. Thus, at time $t = 3$ we have $r = -0.348$, $\dot{r} = -0.373$, $\ddot{r} = -0.032$, $\dddot{r} = 0.006$, $\theta = 3.025$, $\dot{\theta} = 2.425$, $\ddot{\theta} = 1.45$, and $\dddot{\theta} = 0.45$. After, the velocity of point P, the vector $\boldsymbol{v}_P$, may be calculated as follows:

$$\boldsymbol{v}_P = {}^j\boldsymbol{v}_P^k + {}^k\boldsymbol{v}_P^m,$$

where ${}^j\boldsymbol{v}_P^k = {}^j\boldsymbol{\omega}^k \times \boldsymbol{r}_{P/O}$ while ${}^k\boldsymbol{v}_P^m = \dot{r}\hat{\boldsymbol{r}}_{P/O}$. Hence, after a few computations one obtains $v_P = 0.922$ m/s. On the other hand, since the acceleration of point A, the vector $\boldsymbol{a}_P$ is given by

$$\begin{aligned}\boldsymbol{a}_P &= {}^j\boldsymbol{a}_P^k + {}^k\boldsymbol{a}_P^m + 2{}^j\boldsymbol{\omega}^k \times {}^k\boldsymbol{v}_P^m \\ &= {}^j\boldsymbol{\alpha}^k \times \boldsymbol{r}_{P/O} + {}^j\boldsymbol{\omega}^k \times ({}^j\boldsymbol{\omega}^k \times \boldsymbol{r}_{P/O}) + \ddot{r}\hat{\boldsymbol{r}}_{P/O} + 2{}^j\boldsymbol{\omega}^k \times \dot{r}\hat{\boldsymbol{r}}_{P/O}.\end{aligned}$$

Then, substituting the corresponding data, one obtains $\boldsymbol{a}_P = -1.731\hat{\boldsymbol{\imath}} + 2.532\hat{\boldsymbol{\jmath}}$ or $a_P = 3.067$ m/s^2.

Finally, it was proved that the jerk may be computed as

$$\boldsymbol{\rho}_P = {}^j\boldsymbol{\rho}_P^k + {}^k\boldsymbol{\rho}_P^m + 3{}^j\boldsymbol{\alpha}^k \times {}^k\boldsymbol{v}_P^m + 3{}^j\boldsymbol{\omega}^k \times {}^k\boldsymbol{a}_P^m + 3{}^j\boldsymbol{\omega}^k \times \left({}^j\boldsymbol{\omega}^k \times {}^k\boldsymbol{v}_P^m\right),$$

where

$$\begin{aligned}{}^j\boldsymbol{\rho}_P^k = {}^j\boldsymbol{\rho}^k \times \boldsymbol{r}_{P/O} + 2{}^j\boldsymbol{\alpha}^k \times \left({}^j\boldsymbol{\omega}^k \times \boldsymbol{r}_{P/O}\right) + \\ {}^j\boldsymbol{\omega}^k \times \left({}^j\boldsymbol{\alpha}^k \times \boldsymbol{r}_{P/O}\right) + {}^j\boldsymbol{\omega}^k \times \left[{}^j\boldsymbol{\omega}^k \times \left({}^j\boldsymbol{\omega}^k \times \boldsymbol{r}_{P/O}\right)\right] \\ = -4.205\hat{\boldsymbol{\imath}} - 4.346\hat{\boldsymbol{\jmath}}.\end{aligned}$$

Meanwhile,

$${}^k\boldsymbol{\rho}_P^m = \dddot{r}(\cos\theta\hat{\boldsymbol{\imath}} + \sin\theta\hat{\boldsymbol{\jmath}}) = -0.005\hat{\boldsymbol{\imath}} - 0.043\hat{\boldsymbol{\imath}}.$$

Therefore, after a few computations it follows that $\boldsymbol{\rho}_P = -10.530\hat{\boldsymbol{\imath}} - 1.781\hat{\boldsymbol{\jmath}}$ or $\rho_P = 10.68$ m/s^3.

3. *In the development of cam profiles due to tribological implications and the ability of the actuated body to follow the cam profile without chatter, the jerk must be taken into proper account. The cam shown in the figure is designed in such a way that the position of the center of the roller A must follow the function $r = b - c\cos(\theta)$ where $b > c$. Considering that the cam does not rotate, determine the magnitude of the jerk of A in terms of θ if the slotted arm revolves with a constant counterclockwise angular rate $\dot{\theta} = \omega$.*

Solution. Consider that ${}^j\boldsymbol{\omega}^k = \dot{\theta}\hat{\boldsymbol{k}}$, ${}^j\boldsymbol{\alpha}^k = \boldsymbol{0}$, ${}^j\boldsymbol{\rho}^k = \boldsymbol{0}$, ${}^k\boldsymbol{v}_P^m = \dot{r}\hat{\boldsymbol{r}}_{A/O}$, ${}^k\boldsymbol{a}_P^m = \ddot{r}\hat{\boldsymbol{r}}_{A/O}$, ${}^k\boldsymbol{\rho}_P^m = \dddot{r}\hat{\boldsymbol{r}}_{A/O}$, where $\hat{\boldsymbol{r}}_{A/O} = \cos\theta\hat{\boldsymbol{\imath}} + \sin\theta\hat{\boldsymbol{\jmath}}$, $\dot{r} = c\dot{\theta}\sin\theta$, $\ddot{r} = c\dot{\theta}^2\cos\theta$, and $\dddot{r} = -c\dot{\theta}^3\sin\theta + 3c\dot{\theta}\ddot{\theta}\cos\theta + c\dddot{\theta}\sin\theta$. Take into account that the jerk of point A is given by

Cam mechanism

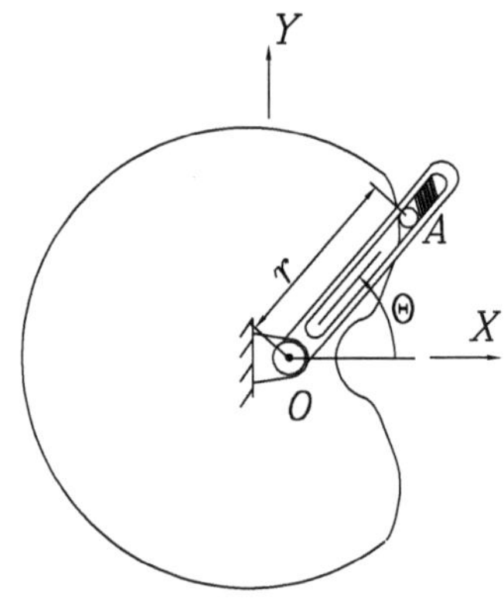

$$\boldsymbol{\rho}_A = {}^j\boldsymbol{\rho}_A^k + {}^k\boldsymbol{\rho}_A^m + 3{}^j\boldsymbol{\alpha}^k \times {}^k\boldsymbol{v}_A^m + 3{}^j\boldsymbol{\omega}^k \times {}^k\boldsymbol{a}_A^m + 3{}^j\boldsymbol{\omega}^k \times \left({}^j\boldsymbol{\omega}^k \times {}^k\boldsymbol{v}_A^m\right),$$

where

$$\begin{aligned} {}^j\boldsymbol{\rho}_A^k &= {}^j\boldsymbol{\rho}^k \times \boldsymbol{r}_{A/O} + 2{}^j\boldsymbol{\alpha}^k \times \left({}^j\boldsymbol{\omega}^k \times \boldsymbol{r}_{A/O}\right) + \\ &{}^j\boldsymbol{\omega}^k \times \left({}^j\boldsymbol{\alpha}^k \times \boldsymbol{r}_{A/O}\right) + {}^j\boldsymbol{\omega}^k \times \left[{}^j\boldsymbol{\omega}^k \times \left({}^j\boldsymbol{\omega}^k \times \boldsymbol{r}_{A/O}\right)\right]. \end{aligned}$$

Meanwhile,

$${}^k\boldsymbol{\rho}_A^m = \dddot{r}(\cos\theta\hat{\boldsymbol{i}} + \sin\theta\hat{\boldsymbol{j}}).$$

Gathering these results, we see that the magnitude of the vector jerk $\boldsymbol{\rho}_A$ results in

$$\rho_A = \dot{\theta}\sqrt{K_1\dot{\theta}^4 + K_2\dot{\theta}^2 + K_3}, \tag{15.7}$$

where the coefficients from Eq. (15.7) are given by

$$\begin{aligned} K_1 &= 40c^2\sin^2(\theta)\cos^2(\theta) + 16c^2\cos^4(\theta) - 8bc\cos^3(\theta) - \\ &\quad 10bc\sin(\theta)\cos(\theta) + 9c^2\sin^4(\theta) + b^2, \\ K_2 &= -2bc\sin^2(\theta)\cos(\theta) + 8c^2\sin^2(\theta)\cos^2(\theta) - 6c^2\sin^2(\theta), \\ K_3 &= c^2\sin^4(\theta). \end{aligned}$$

4. *The disk shown in the figure rotates about a fixed axis passing through point O with angular velocity $\omega = 5$ rad/s, angular acceleration $\alpha = 2\,rad/s^2$, and angular jerk $\rho = 3\,rad/s^3$. The small sphere A moves in the circular slot in such a way that in the same instant we have $\beta = 30°$, $\dot{\beta} = 2\,rad/s$, $\ddot{\beta} = -4\,rad/s^2$, and $\dddot{\beta} = 4\,rad/s^3$. Determine the velocity, acceleration, and jerk of A.*

Solution. Consider the body j to be the fixed reference frame while m is the proper body A and k is the body relating both entities. Hence, according to the aforementioned data of the problem we have ${}^j\boldsymbol{\omega}^k = 5\hat{\boldsymbol{k}}$, ${}^j\boldsymbol{\alpha}^k = 2\hat{\boldsymbol{k}}$, and ${}^j\boldsymbol{\rho}^k = 3\hat{\boldsymbol{k}}$.

Rotating disk with circular slot

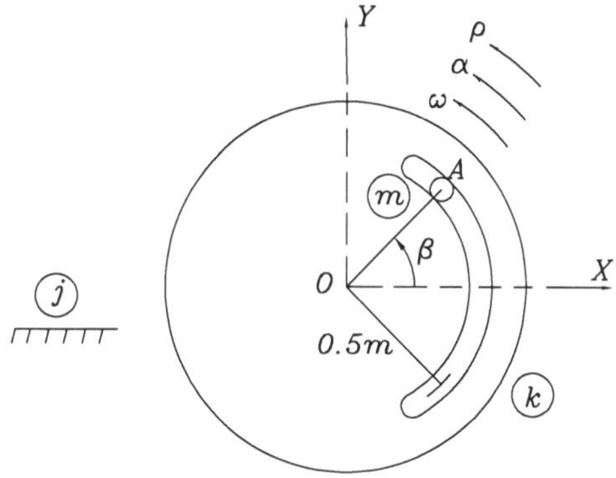

Meanwhile, ${}^k\boldsymbol{\omega}^m = \dot{\boldsymbol{\beta}} = 2\hat{\boldsymbol{k}}$, ${}^k\boldsymbol{\alpha}^m = \ddot{\boldsymbol{\beta}} - 4\hat{\boldsymbol{k}}$, and ${}^k\boldsymbol{\rho}^m = \dddot{\boldsymbol{\beta}} = 4\hat{\boldsymbol{k}}$. Next, the velocity of point A, the vector $\boldsymbol{v}_A$, may be calculated as follows:

$$\boldsymbol{v}_A = {}^j\boldsymbol{v}_A^k + {}^k\boldsymbol{v}_A^m.$$

Here ${}^j\boldsymbol{v}_A^k = {}^j\boldsymbol{\omega}^k \times \boldsymbol{r}_{A/O}$ while ${}^k\boldsymbol{v}_A^m = {}^k\boldsymbol{\omega}^m \times \boldsymbol{r}_{A/O}$, where $\boldsymbol{r}_{A/O} = r\big[\cos(\beta)\hat{\boldsymbol{i}} + \sin(\beta)\hat{\boldsymbol{j}}\big]$. Hence, ${}^j\boldsymbol{v}_A^k = -1.25\hat{\boldsymbol{i}} + 2.165\hat{\boldsymbol{j}}$, whereas ${}^k\boldsymbol{v}_A^m = -0.5\hat{\boldsymbol{i}} + 0.866\hat{\boldsymbol{j}}$. Thus, $\boldsymbol{v}_A = -1.75\hat{\boldsymbol{i}} + 3.031\hat{\boldsymbol{j}}$ or $v_A = 3.5\,\text{m/s}$. On the other hand, the acceleration of point A, the vector $\boldsymbol{a}_A$, is given by

$$\boldsymbol{a}_A = {}^j\boldsymbol{a}_A^k + {}^k\boldsymbol{a}_A^m + 2{}^j\boldsymbol{\omega}^k \times {}^k\boldsymbol{v}_A^m,$$

where ${}^j\boldsymbol{a}_A^k = {}^j\boldsymbol{\alpha}^k \times \boldsymbol{r}_{A/O} + {}^j\boldsymbol{\omega}^k \times ({}^j\boldsymbol{\omega}^k \times \boldsymbol{r}_{A/O}) = -11.325\hat{\boldsymbol{i}} - 5.383\hat{\boldsymbol{j}}$ and ${}^k\boldsymbol{a}_A^m = {}^k\boldsymbol{\alpha}^m \times \boldsymbol{r}_{A/O} + {}^k\boldsymbol{\omega}^m \times ({}^k\boldsymbol{\omega}^m \times \boldsymbol{r}_{A/O}) = -0.732\hat{\boldsymbol{i}} - 2.732\hat{\boldsymbol{j}}$. Therefore, after adding the Coriolis acceleration, one obtains $\boldsymbol{a}_A = -20.717\hat{\boldsymbol{i}} - 13.116\hat{\boldsymbol{j}}$, or $a_A = 24.52\,\text{m/s}^2$.

Finally, the jerk of point A may be computed as

$$\boldsymbol{\rho}_A = {}^j\boldsymbol{\rho}_A^k + {}^k\boldsymbol{\rho}_A^m + 3{}^j\boldsymbol{\alpha}^k \times {}^k\boldsymbol{v}_A^m + 3{}^j\boldsymbol{\omega}^k \times {}^k\boldsymbol{a}_A^m + 3{}^j\boldsymbol{\omega}^k \times \left({}^j\boldsymbol{\omega}^k \times {}^k\boldsymbol{v}_A^m\right),$$

where

$$\begin{aligned} {}^j\boldsymbol{\rho}_A^k &= {}^j\boldsymbol{\rho}^k \times \boldsymbol{r}_{A/O} + 2{}^j\boldsymbol{\alpha}^k \times \left({}^j\boldsymbol{\omega}^k \times \boldsymbol{r}_{A/O}\right) + \\ &{}^j\boldsymbol{\omega}^k \times \left({}^j\boldsymbol{\alpha}^k \times \boldsymbol{r}_{A/O}\right) + {}^j\boldsymbol{\omega}^k \times \left[{}^j\boldsymbol{\omega}^k \times \left({}^j\boldsymbol{\omega}^k \times \boldsymbol{r}_{A/O}\right)\right] \\ &= 17.509\hat{\boldsymbol{i}} - 60.327\hat{\boldsymbol{j}}. \end{aligned}$$

Similarly,

$$\begin{aligned}{}^k\boldsymbol{\rho}_A^m &= {}^k\boldsymbol{\rho}^m \times \boldsymbol{r}_{A/O} + 2\,{}^k\boldsymbol{\alpha}^m \times \left({}^k\boldsymbol{\omega}^m \times \boldsymbol{r}_{A/O}\right) + \\ &\quad {}^k\boldsymbol{\omega}^m \times \left({}^k\boldsymbol{\alpha}^m \times \boldsymbol{r}_{A/O}\right) + {}^k\boldsymbol{\omega}^m \times \left[{}^k\boldsymbol{\omega}^m \times \left({}^k\boldsymbol{\omega}^m \times \boldsymbol{r}_{A/O}\right)\right] \\ &= 11.392\hat{\boldsymbol{i}} + 4.267\hat{\boldsymbol{j}}.\end{aligned}$$

Therefore, after a few computations it follows that $\boldsymbol{\rho}_A = 102.186\hat{\boldsymbol{i}} - 134.992\hat{\boldsymbol{j}}$ or $\rho_A = 169.307\,\mathrm{m/s^3}$.

5. *Determine the input–output equation of jerk of the 3-RPR parallel manipulator shown in the figure by resorting to reciprocal screw theory.*

3RPR planar parallel manipulator

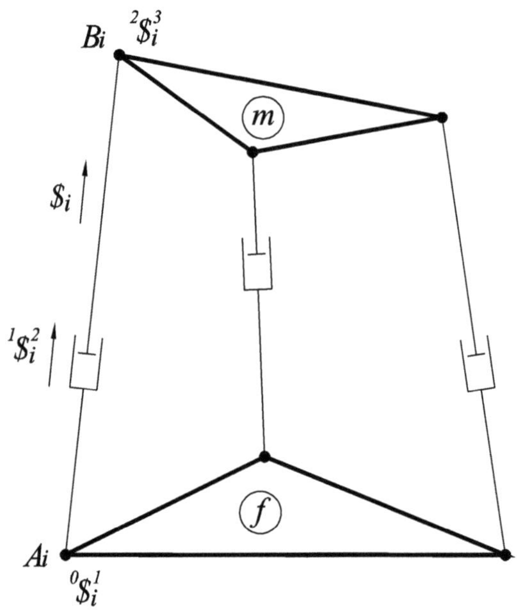

Solution. The jerk of the moving platform m as measured from the fixed platform f may be obtained through any of the connector chains as follows:

$${}^j\boldsymbol{J}^m = {}_0\rho_1^i\,{}^0\$_i^1 + \dddot{q}_i\,{}^1\$_i^2 + {}_2\rho_3^i\,{}^2\$_i^3 + \mathcal{J}_i, \quad i = 1, 2, 3,$$

where the ith screw of jerk $\mathcal{J}_i$ is given by

$$\begin{aligned}\mathcal{J}_i &= 2\left[{}^0\mathbf{V}_i^1 \;\; {}^1\mathbf{A}_i^3\right] + 2\left[{}^1\mathbf{V}_i^2 \;\; {}^2\mathbf{A}_i^3\right] + \left[{}^0\mathbf{A}_i^1 \;\; {}^1\mathbf{V}_i^3\right] + \\ &\quad \left[{}^1\mathbf{A}_i^2 \;\; {}^2\mathbf{V}_i^3\right] + \left[{}^0\mathbf{V}_i^1 \left[{}^0\mathbf{V}_i^1 \;\; {}^1\mathbf{V}_i^3\right]\right] + \left[{}^1\mathbf{V}_i^2 \left[{}^1\mathbf{V}_i^2 \;\; {}^2\mathbf{V}_i^3\right]\right].\end{aligned}$$

Furthermore, consider that $\$_i$ is a line in Plücker coordinates pointed from A_i to B_i. It is evident that the line $\$_i$ is reciprocal to the screws representing revolute joints in the same limb. Then, the application of the Klein form leads to

$$\{\$_i; {}^j\boldsymbol{J}^m\} = \dddot{q}_i.$$

Therefore, the input–output equation of jerk of the planar parallel manipulator results in

$$\mathbf{J}^T \, \Delta \, {}^j\boldsymbol{J}^m = \dddot{\mathbf{q}}^T + \mathcal{H}^T,$$

where $\mathbf{J} = \left[\$_1 \ \$_2 \ \$_3\right]$ is the screw-coordinate Jacobian matrix of the robot while $\dddot{\mathbf{q}} = \left[\dddot{q}_1 \ \dddot{q}_2 \ \dddot{q}_3\right]$ is the third-order driver matrix of the parallel manipulator and Δ is an operator of polarity of order 3. Finally, $\mathcal{H} = \left[\{\$_1; \mathcal{J}_1\} \ \{\$_2; \mathcal{J}_2\} \ \{\$_3; \mathcal{J}_3\}\right]$ is the complementary matrix of jerk.

Naturally, to apply the input–output equation of jerk, we must properly account for the three-dimensional nature of the involved screws and kinematic states.

15.5 Chapter 7: Hyper-Jerk Analysis

1. *The disk shown in the figure rotates about a fixed axis passing through point O with angular velocity $\omega = 5\,rad/s$, angular acceleration $\alpha = 2\,rad/s^2$, angular jerk $\rho = 3\,rad/s^3$, and angular hyper-jerk $\sigma = 2\,rad/s^4$ in the indicated directions. On the other hand, the small sphere A moves in the slot undergoing relative motions with respect to the slot as follows: $\dot{u} = 2\,m/s$, $\ddot{u} = 1.5\,m/s^2$, $\dddot{u} = 0.5\,m/s^3$, and $\ddddot{u} = 3\,m/s^4$ in the indicated directions. It is required to compute the instantaneous velocity, acceleration, jerk, and hyper-jerk of sphere A considering $x = 0.75\,m$ and $y = 0.5\,m$.*

Rotating disk with right slot

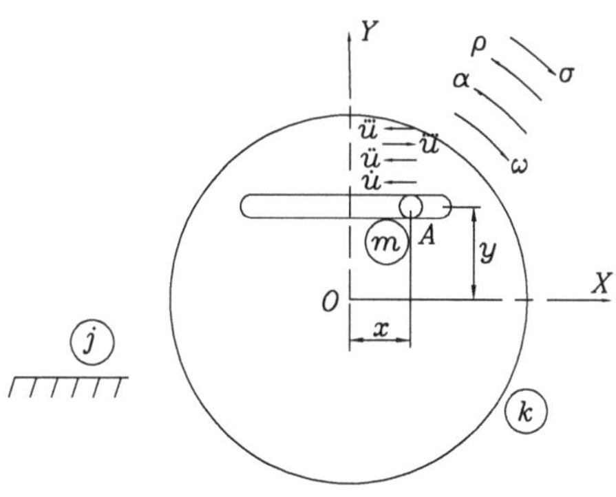

Solution. Consider that body j is the fixed reference frame while bodies m and k are, respectively, the proper sphere A and the disk. Hence, according to the XYZ reference frame whose origin O is attached to the center of the disk, we have ${}^{j}\boldsymbol{\omega}^{k} = -5\hat{\boldsymbol{k}}$, ${}^{j}\boldsymbol{\alpha}^{k} = 2\hat{\boldsymbol{k}}$, ${}^{j}\boldsymbol{\rho}^{k} = 3\hat{\boldsymbol{k}}$, ${}^{j}\boldsymbol{\sigma}^{k} = -2\hat{\boldsymbol{k}}$, ${}^{k}\boldsymbol{v}_{A}^{m} = -2\hat{\boldsymbol{\imath}}$, ${}^{k}\boldsymbol{a}_{A}^{m} = -1.5\hat{\boldsymbol{\imath}}$, ${}^{k}\boldsymbol{\rho}_{A}^{m} = 0.5\hat{\boldsymbol{\imath}}$, and ${}^{k}\boldsymbol{\sigma}_{A}^{m} = -3\hat{\boldsymbol{\imath}}$.

The velocity of the sphere A may be computed as

$$\boldsymbol{v}_A = {}^{j}\boldsymbol{v}_{A}^{k} + {}^{k}\boldsymbol{v}_{A}^{m},$$

where ${}^{j}\boldsymbol{v}_{A}^{k} = {}^{j}\boldsymbol{\omega}^{k} \times \boldsymbol{r}_{A/O}$ where $\boldsymbol{r}_{A/O} = 0.75\hat{\boldsymbol{\imath}} + 0.5\hat{\boldsymbol{\jmath}}$. Therefore, ${}^{j}\boldsymbol{v}_{A}^{k} = 2.5\hat{\boldsymbol{\imath}} - 3.75\hat{\boldsymbol{\jmath}}$. Next, the velocity of the sphere results in $\boldsymbol{v}_A = 0.5\hat{\boldsymbol{\imath}} - 3.75\hat{\boldsymbol{\jmath}}$ or $v_A = 3.783\,\text{m/s}$. On the other hand, the acceleration of the sphere may be calculated as

$$\boldsymbol{a}_A = {}^{j}\boldsymbol{a}_{A}^{k} + {}^{k}\boldsymbol{a}_{A}^{m} + 2\,{}^{j}\boldsymbol{\omega}^{k} \times {}^{k}\boldsymbol{v}_{A}^{m},$$

where ${}^{j}\boldsymbol{a}_{A}^{k} = {}^{j}\boldsymbol{\alpha}^{k} \times \boldsymbol{r}_{A/O} + {}^{j}\boldsymbol{\omega}^{k} \times ({}^{j}\boldsymbol{\omega}^{k} \times \boldsymbol{r}_{A/O}) = -19.75\hat{\boldsymbol{\imath}} - 11\hat{\boldsymbol{\jmath}}$. Hence, one obtains $\boldsymbol{a}_A = -21.25\hat{\boldsymbol{\imath}} + 9\hat{\boldsymbol{\jmath}}$ or $a_A = 23.077\,\text{m/s}^2$. Furthermore, the jerk of the sphere may be computed as

$$\begin{aligned}\boldsymbol{\rho}_A = {}^{j}\boldsymbol{\rho}_{A}^{k} + {}^{k}\boldsymbol{\rho}_{A}^{m} + 3\,{}^{j}\boldsymbol{\alpha}^{k} \times {}^{k}\boldsymbol{v}_{A}^{m} + 3\,{}^{j}\boldsymbol{\omega}^{k} \times {}^{k}\boldsymbol{a}_{A}^{m} + \\ 3\,{}^{j}\boldsymbol{\omega}^{k} \times \left({}^{j}\boldsymbol{\omega}^{k} \times {}^{k}\boldsymbol{v}_{A}^{m}\right),\end{aligned}$$

where

$$\begin{aligned}{}^{j}\boldsymbol{\rho}_{A}^{k} &= {}^{j}\boldsymbol{\rho}^{k} \times \boldsymbol{r}_{A/O} + 2\,{}^{j}\boldsymbol{\alpha}^{k} \times \left({}^{j}\boldsymbol{\omega}^{k} \times \boldsymbol{r}_{A/O}\right) + \\ &\quad {}^{j}\boldsymbol{\omega}^{k} \times \left({}^{j}\boldsymbol{\alpha}^{k} \times \boldsymbol{r}_{A/O}\right) + {}^{j}\boldsymbol{\omega}^{k} \times \left[{}^{j}\boldsymbol{\omega}^{k} \times \left({}^{j}\boldsymbol{\omega}^{k} \times \boldsymbol{r}_{A/O}\right)\right] \\ &= -41.5\hat{\boldsymbol{\imath}} + 111\hat{\boldsymbol{\jmath}}.\end{aligned}$$

Therefore, the jerk vector of A is obtained as $\boldsymbol{\rho}_A = 109\hat{\boldsymbol{\imath}} + 121.5\hat{\boldsymbol{\jmath}}$ or $\rho_A = 163.227\,\text{m/s}^3$. Finally, the hyper-jerk of the sphere A may be computed as

$$\begin{aligned}\boldsymbol{\sigma}_A = {}^{j}\boldsymbol{\sigma}_{A}^{k} + {}^{k}\boldsymbol{\sigma}_{A}^{m} + 4\,{}^{j}\boldsymbol{\omega}^{k} \times {}^{k}\boldsymbol{\rho}_{A}^{m} + 4\,{}^{j}\boldsymbol{\rho}^{k} \times {}^{k}\boldsymbol{v}_{A}^{m} + 8\,{}^{j}\boldsymbol{\alpha}^{k} \times \left({}^{j}\boldsymbol{\omega}^{k} \times {}^{k}\boldsymbol{v}_{A}^{m}\right) + \\ 6\,{}^{j}\boldsymbol{\alpha}^{k} \times {}^{k}\boldsymbol{a}_{A}^{m} + 6\,{}^{j}\boldsymbol{\omega}^{k} \times \left({}^{j}\boldsymbol{\omega}^{k} \times {}^{k}\boldsymbol{a}_{A}^{m}\right) + 4\,{}^{j}\boldsymbol{\omega}^{k} \times \left({}^{j}\boldsymbol{\alpha}^{k} \times {}^{k}\boldsymbol{v}_{A}^{m}\right) + \\ 4\,{}^{j}\boldsymbol{\omega}^{k} \times \left[{}^{j}\boldsymbol{\omega}^{k} \times \left({}^{j}\boldsymbol{\omega}^{k} \times {}^{k}\boldsymbol{v}_{A}^{m}\right)\right],\end{aligned}$$

where

$$\begin{aligned}{}^{j}\boldsymbol{\sigma}_{A}^{k} = {}^{j}\boldsymbol{\sigma}^{k} \times \boldsymbol{r}_{A/O} + 3\,{}^{j}\boldsymbol{\sigma}^{k} \times \left({}^{j}\boldsymbol{\omega}^{k} \times \boldsymbol{r}_{A/O}\right) + \\ 3\,{}^{j}\boldsymbol{\alpha}^{k} \times \left({}^{j}\boldsymbol{\alpha}^{k} \times \boldsymbol{r}_{A/O}\right) + 3\,{}^{j}\boldsymbol{\alpha}^{k} \times \left[{}^{j}\boldsymbol{\omega}^{k} \times \left({}^{j}\boldsymbol{\omega}^{k} \times \boldsymbol{r}_{A/O}\right)\right] +\end{aligned}$$

$$2\,{}^{j}\boldsymbol{\omega}^{k}\times\left({}^{j}\boldsymbol{\rho}^{k}\times\boldsymbol{r}_{A/O}\right)+{}^{j}\boldsymbol{\omega}^{k}\times\left({}^{j}\boldsymbol{\rho}^{k}\times\boldsymbol{r}_{A/O}\right)+$$
$${}^{j}\boldsymbol{\omega}^{k}\times\left[{}^{j}\boldsymbol{\omega}^{k}\times\left({}^{j}\boldsymbol{\alpha}^{k}\times\boldsymbol{r}_{A/O}\right)\right]+{}^{j}\boldsymbol{\omega}^{k}\times\left\{{}^{j}\boldsymbol{\omega}^{k}\times\left[{}^{j}\boldsymbol{\omega}^{k}\times\left({}^{j}\boldsymbol{\omega}^{k}\times\boldsymbol{r}_{A/O}\right)\right]\right\}$$
$$=628.25\hat{\boldsymbol{i}}+200\hat{\boldsymbol{j}}.$$

Therefore, the hyper-jerk of the sphere is obtained as $\boldsymbol{\sigma}_A = 610.25\hat{\boldsymbol{i}} - 602\hat{\boldsymbol{j}}$ or $\sigma_A = 857.21\ \text{m/s}^4$.

2. *The figure shows an actuating mechanism for a telescoping antenna on a spacecraft. The supporting shaft j rotates about the fixed Y-axis while the orientation of the telescoping antenna is determined according to the angles* $\theta(t) = \frac{\pi}{9} + \frac{\pi}{6}\sin(t)$ *and* $\phi(t) = \frac{\pi}{6} + \frac{\pi}{9}\sin(t)$*. At the same time, the antenna extends according to the variable length* $l(t) = 1 + \frac{\pi}{18}\sin(t)$ *while body m rotates freely with respect to body k along the unit axis* $\hat{\boldsymbol{u}}_{P/O}$ *according to the angle* $\beta(t) = \frac{\pi}{4} + \frac{\pi}{12}\sin(t)$*. Determine the velocity, acceleration, jerk, and hyper-jerk of P as measured from the fixed body at time* $t = 4$*.*

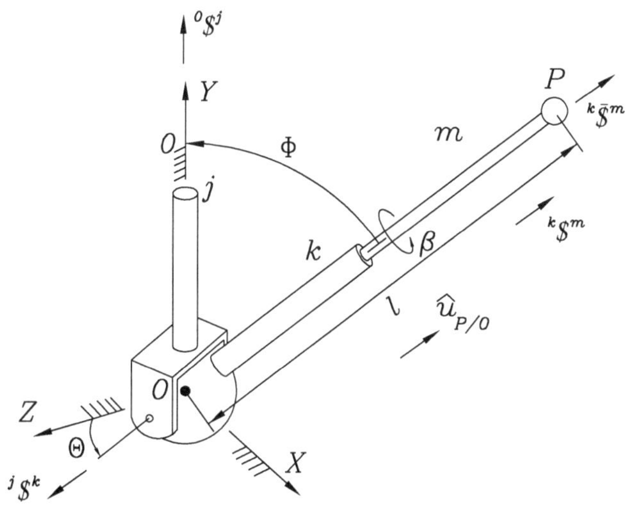

Actuating mechanism for a telescoping antenna on a spacecraft

Solution. The velocity state of body m as measured from the fixed reference 0 taking the origin O of the reference frame XYZ as the reference pole may be computed as

$${}^{0}\boldsymbol{V}_{O}^{m}=\dot{\theta}\,{}^{0}\$^{j}+\dot{\phi}\,{}^{j}\$^{k}+\dot{\beta}\,{}^{k}\$^{m}+\dot{l}\,{}^{k}\overline{\$}^{m},$$

where the screws are given by

$$
{}^{0}\$^{j} = \begin{bmatrix} 0 \\ 1 \\ 0 \\ 0 \\ 0 \\ 0 \end{bmatrix}, {}^{j}\$^{k} = \begin{bmatrix} \sin\theta \\ 0 \\ \cos\theta \\ 0 \\ 0 \\ 0 \end{bmatrix}, {}^{k}\$^{m} = \begin{bmatrix} \sin\phi\cos\theta \\ \cos\phi \\ -\sin\phi\sin\theta \\ 0 \\ 0 \\ 0 \end{bmatrix}, {}^{k}\overline{\$}^{m} = \begin{bmatrix} 0 \\ 0 \\ 0 \\ \sin\phi\cos\theta \\ \cos\theta \\ -\sin\phi\sin\theta \end{bmatrix}.
$$

Thus,

$$
{}^{0}\boldsymbol{V}_{O}^{m} = \begin{bmatrix} \dot{\phi}\sin\theta + \dot{\beta}\sin\phi\cos\theta \\ \dot{\theta} + \dot{\beta}\cos\phi \\ \dot{\phi}\cos\theta - \dot{\beta}\sin\phi\sin\theta \\ \dot{l}\sin\phi\cos\theta \\ \dot{l}\cos\phi \\ -\dot{l}\sin\phi\sin\theta \end{bmatrix},
$$

which yields

$$
{}^{0}\boldsymbol{\omega}^{m} = (\dot{\phi}\sin\theta + \dot{\beta}\sin\phi\cos\theta)\hat{\boldsymbol{i}} + (\dot{\theta} + \dot{\beta}\cos\phi)\hat{\boldsymbol{j}} + (\dot{\phi}\cos\theta - \dot{\beta}\sin\phi\sin\theta)\hat{\boldsymbol{k}}
$$

and

$$
{}^{0}\boldsymbol{v}_{O}^{m} = \dot{l}\sin\phi\cos\theta\hat{\boldsymbol{i}} + \dot{l}\cos\phi\hat{\boldsymbol{j}} - \dot{l}\sin\phi\sin\theta\hat{\boldsymbol{k}}.
$$

Next, taking into account that the position vector of point P with respect to O is given by

$$
\boldsymbol{r}_{P/O} = l\sin\phi\cos\theta\hat{\boldsymbol{i}} + l\cos\phi\hat{\boldsymbol{j}} - l\sin\phi\sin\theta\hat{\boldsymbol{k}},
$$

we can obtain the velocity of point P as

$$
\begin{aligned}
\boldsymbol{v}_P &= {}^{0}\boldsymbol{v}_{O}^{m} + {}^{0}\boldsymbol{\omega}^{m} \times \boldsymbol{r}_{P/O} \\
\rightarrow v_P &= \sqrt{\dot{l}^2 + l^2\dot{\phi}^2 + \dot{\theta}^2 l^2 \sin^2\phi} = 0.24\,\text{m/s}.
\end{aligned}
$$

In what follows the computation of the acceleration of point P is presented. The reduced acceleration state of body m, as observed from the fixed reference frame 0, may be calculated as

$$
\begin{aligned}
{}^{0}\boldsymbol{A}_{O}^{m} &= \ddot{\theta}\,{}^{0}\$^{j} + \ddot{\phi}\,{}^{j}\$^{k} + \ddot{\beta}\,{}^{k}\$^{m} + \ddot{l}\,{}^{k}\overline{\$}^{m} + \left[\dot{\theta}\,{}^{0}\$^{j} \quad \dot{\phi}\,{}^{j}\$^{k} + \dot{\beta}\,{}^{k}\$^{m} + \dot{l}\,{}^{k}\overline{\$}^{m}\right] \\
&+ \left[\dot{\phi}\,{}^{j}\$^{k} \quad \dot{\beta}\,{}^{k}\$^{m} + \dot{l}\,{}^{k}\overline{\$}^{m}\right] + \left[\dot{\beta}\,{}^{k}\$^{m} \quad \dot{l}\,{}^{k}\overline{\$}^{m}\right].
\end{aligned}
$$

Furthermore, if we take into account that the reduced acceleration state of body m considering point P as the reference pole can be computed upon the reduced acceleration state ${}^0\boldsymbol{A}_O^m$ as

$$
{}^0\boldsymbol{A}_P^m = \begin{bmatrix} P({}^0\boldsymbol{A}_O^m) \\ D({}^0\boldsymbol{A}_O^m) + P({}^0\boldsymbol{A}_O^m) \times \boldsymbol{r}_{P/O} \end{bmatrix},
$$

then the acceleration vector of point P may be computed as $\boldsymbol{a}_P = D({}^0\boldsymbol{A}_P^m) + {}^0\boldsymbol{\omega}^m \times \boldsymbol{v}_P$. After a few computations, the magnitude of the acceleration vector is obtained as $\boldsymbol{a}_P = 0.317\,\mathrm{m/s^2}$.

In what follows the computation of the jerk of point P is presented. The reduced jerk state of body m, as observed from the fixed reference frame 0, may be calculated as

$$
\begin{aligned}
{}^0\boldsymbol{J}_O^m = \dddot{\theta}\,{}^0\$^j + \dddot{\phi}\,{}^j\$^k + \dddot{\beta}\,{}^k\$^m + \dddot{l}\,{}^k\overline{\$}^m + {} \\
2\big[{}^0\boldsymbol{V}_O^j \quad {}^j\boldsymbol{A}_O^{\overline{m}}\big] + 2\big[{}^j\boldsymbol{V}_O^k \quad {}^k\boldsymbol{A}_O^{\overline{m}}\big] + 2\big[{}^k\boldsymbol{V}_O^m \quad {}^m\boldsymbol{A}_O^{\overline{m}}\big] + {} \\
\big[{}^0\boldsymbol{A}_O^j \quad {}^j\boldsymbol{V}_O^{\overline{m}}\big] + \big[{}^j\boldsymbol{A}_O^k \quad {}^k\boldsymbol{V}_O^{\overline{m}}\big] + \big[{}^k\boldsymbol{A}_O^m \quad {}^m\boldsymbol{V}_O^{\overline{m}}\big] + {} \\
\big[{}^0\boldsymbol{V}_O^j\big[{}^0\boldsymbol{V}_O^j \quad {}^j\boldsymbol{V}_O^{\overline{m}}\big]\big] + \big[{}^j\boldsymbol{V}_O^k\big[{}^j\boldsymbol{V}_O^k \quad {}^k\boldsymbol{V}_O^{\overline{m}}\big]\big] + \big[{}^k\boldsymbol{V}_O^m\big[{}^k\boldsymbol{V}_O^m \quad {}^m\boldsymbol{V}_O^{\overline{m}}\big]\big].
\end{aligned}
$$

Furthermore, taking into account that the reduced jerk state of body m considering point P as the reference pole can be computed upon the reduced acceleration state ${}^0\boldsymbol{J}_O^m$ as follows:

$$
{}^0\boldsymbol{J}_P^m = \begin{bmatrix} P({}^0\boldsymbol{J}_O^m) \\ D({}^0\boldsymbol{J}_O^m) + P({}^0\boldsymbol{J}_O^m) \times \boldsymbol{r}_{P/O} \end{bmatrix}.
$$

then we can compute the jerk vector of point P as

$$
\boldsymbol{\rho}_P = D({}^0\boldsymbol{J}_P^m) + 2\,{}^0\boldsymbol{\alpha}^m \times \boldsymbol{v}_P + {}^0\boldsymbol{\omega}^m \times \boldsymbol{a}_P.
$$

After a few computations, the magnitude of the jerk vector is obtained as $\boldsymbol{\rho}_P = 0.546\,\mathrm{m/s^3}$.

The exercise finishes with the computation of the hyper-jerk of point P.

The reduced hyper-jerk state of body m, as observed from the fixed reference frame 0, may be calculated as

$$
\begin{aligned}
{}^0\boldsymbol{H}_O^m = \ddddot{\theta}\,{}^0\$^j + \ddddot{\phi}\,{}^j\$^k + \ddddot{\beta}\,{}^k\$^m + \ddddot{l}\,{}^k\overline{\$}^m + {} \\
3\big[{}^0\boldsymbol{V}_O^j \quad {}^j\boldsymbol{J}_O^{\overline{m}}\big] + 3\big[{}^j\boldsymbol{V}_O^k \quad {}^k\boldsymbol{J}_O^{\overline{m}}\big] + 3\big[{}^k\boldsymbol{V}_O^m \quad {}^m\boldsymbol{J}_O^{\overline{m}}\big] + {} \\
3\big[{}^0\boldsymbol{A}_O^j \quad {}^j\boldsymbol{A}_O^{\overline{m}}\big] + 3\big[{}^j\boldsymbol{A}_O^k \quad {}^k\boldsymbol{A}_O^{\overline{m}}\big] + 3\big[{}^k\boldsymbol{A}_O^m \quad {}^m\boldsymbol{A}_O^{\overline{m}}\big] + {} \\
3\big[{}^0\boldsymbol{V}_O^j\big[{}^0\boldsymbol{V}_O^j \quad {}^j\boldsymbol{A}_O^{\overline{m}}\big]\big] + 3\big[{}^j\boldsymbol{V}_O^k\big[{}^j\boldsymbol{V}_O^k \quad {}^k\boldsymbol{A}_O^{\overline{m}}\big]\big] + 3\big[{}^k\boldsymbol{V}_O^m\big[{}^k\boldsymbol{V}_O^m \quad {}^m\boldsymbol{A}_O^{\overline{m}}\big]\big] + {} \\
2\big[{}^0\boldsymbol{A}_O^j\big[{}^0\boldsymbol{V}_O^j \quad {}^j\boldsymbol{V}_O^{\overline{m}}\big]\big] + 2\big[{}^j\boldsymbol{A}_O^k\big[{}^j\boldsymbol{V}_O^k \quad {}^k\boldsymbol{V}_O^{\overline{m}}\big]\big] + 2\big[{}^k\boldsymbol{A}_O^m\big[{}^k\boldsymbol{V}_O^m \quad {}^m\boldsymbol{V}_O^{\overline{m}}\big]\big] + {}
\end{aligned}
$$

$$+\left[{}^{0}\boldsymbol{J}^{j}_{O}\quad {}^{j}\boldsymbol{V}\tfrac{m}{O}\right]+\left[{}^{j}\boldsymbol{J}^{k}_{O}\quad {}^{k}\boldsymbol{V}\tfrac{m}{O}\right]+\left[{}^{k}\boldsymbol{J}^{m}_{O}\quad {}^{m}\boldsymbol{V}\tfrac{m}{O}\right]+$$
$$\left[{}^{0}\boldsymbol{V}^{j}_{O}\left[{}^{0}\boldsymbol{V}^{j}_{O}\left[{}^{0}\boldsymbol{V}^{j}_{O}\quad {}^{j}\boldsymbol{V}\tfrac{m}{O}\right]\right]\right]+\left[{}^{j}\boldsymbol{V}^{k}_{O}\left[{}^{j}\boldsymbol{V}^{k}_{O}\left[{}^{j}\boldsymbol{V}^{k}_{O}\quad {}^{k}\boldsymbol{V}\tfrac{m}{O}\right]\right]\right]+$$
$$\left[{}^{k}\boldsymbol{V}^{m}_{O}\left[{}^{k}\boldsymbol{V}^{m}_{O}\left[{}^{k}\boldsymbol{V}^{m}_{O}\quad {}^{m}\boldsymbol{V}\tfrac{m}{O}\right]\right]\right].$$

Furthermore, taking into account that the reduced hyper-jerk state of body m considering point P as the reference pole can be computed upon the reduced hyper-jerk state ${}^{0}\boldsymbol{J}^{m}_{O}$ as follows:

$${}^{0}\boldsymbol{H}^{m}_{P}=\begin{bmatrix} P({}^{0}\boldsymbol{H}^{m}_{O}) \\ D({}^{0}\boldsymbol{H}^{m}_{O})+P({}^{0}\boldsymbol{H}^{m}_{O})\times \boldsymbol{r}_{P/O}\end{bmatrix},$$

then we can compute the hyper-jerk vector of point P as $\boldsymbol{\sigma}_P = D({}^{0}\boldsymbol{H}^{m}_{P}) + 3{}^{0}\boldsymbol{\sigma}^{m}\times \boldsymbol{v}_P + 3{}^{0}\boldsymbol{\alpha}^{m}\times \boldsymbol{a}_P + {}^{0}\boldsymbol{\omega}^{m}\times \boldsymbol{\rho}_P$. After a lengthy computation, the magnitude of the hyper-jerk vector is obtained as $\boldsymbol{\sigma}_P = 1.017\,\mathrm{m/s^4}$.

3. *Determine the input–output equation of hyper-jerk of the two-degree-of-freedom parallel manipulator shown in the figure that was introduced by Nam and Park (2006).*

   Solution. The velocity state of body m as measured from the fixed reference frame 0, the vector ${}^{0}\boldsymbol{V}^{m}_{P}$, taking point P as the reference pole, may be computed through any of the outer limbs of the planar parallel manipulator as follows:

$${}^{0}\boldsymbol{V}^{m}_{P}=\dot{q}_i\,{}^{0}\$^{1}_{i}+{}_{1}\omega^{i}_{2}\,{}^{1}\$^{2}_{i}+{}_{2}\omega^{i}_{3}\,{}^{2}\$^{3}_{i},\quad i=1,2,$$

Two-degree-of-freedom parallel manipulator

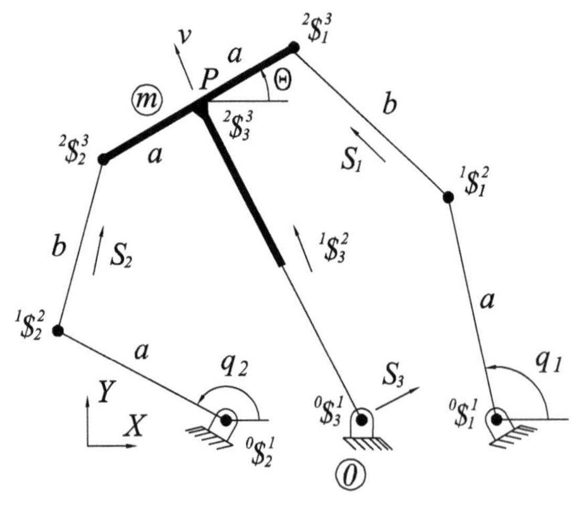

where, according to the architecture of the robot, the motion of the moving platform is such that ${}^0\boldsymbol{V}_P^m = \left[-\dot{v}\cos(\theta)\ \dot{v}\sin(\theta)\ 0\ 0\ 0\ \dot{\theta}\right]^T$. Furthermore, consider that $\boldsymbol{S}_i(i = 1, 2, 3)$ is a line in Plücker coordinates introduced to simplify the velocity analysis. For example, applying the Klein form of such lines to both sides of the equation of velocity and reducing terms, we have

$$\left\{\boldsymbol{S}_i; {}^0\boldsymbol{V}_P^m\right\} = \dot{q}_i \left\{\boldsymbol{S}_i; {}^0\$_i^1\right\}, \quad i = 1, 2.$$

On the other hand, in order to satisfy an algebraic requirement, consider that the velocity state ${}^0\boldsymbol{V}_P^m$ may be written in screw form through the central kinematic chain as follows:

$${}^0\boldsymbol{V}_P^m = {}_0\omega_1^3\ {}^0\$_3^1 + {}_1\omega_2^{3\,1}\$_3^2 + {}_2\omega_3^{3\,2}\$_3^3,$$

where ${}^2\$_3^3$ is a screw representing a virtual kinematic joint, such as ${}_2\omega_3^3 = 0$. Next, if we take into account that the line $\boldsymbol{S}_3$ is reciprocal to all the screws in the same limb, it follows that

$$\left\{\boldsymbol{S}_3; {}^0\boldsymbol{V}_P^m\right\} = \boldsymbol{0}.$$

Hence, gathering these results, we see that the input–output equation of velocity of the robot gives us

$$\mathbf{J}^T \Delta {}^0\boldsymbol{V}_P^m = \left[\dot{q}_1 \left\{\boldsymbol{S}_1; {}^0\$_1^1\right\}\ \dot{q}_2 \left\{\boldsymbol{S}_2; {}^0\$_2^1\right\}\ 0\right]^T,$$

where $\mathbf{J} = \left[\boldsymbol{S}_1\ \boldsymbol{S}_2\ \boldsymbol{S}_3\right]$ is the Jacobian of the parallel manipulator. Meanwhile, Δ is an operator of polarity. Naturally, in order to apply the input–output equation of velocity, we must achieve the corresponding reduction of the involved matrices.

Following a similar procedure, by resorting to reciprocal screw theory, we find that the input–output equation of acceleration of the robot results in

$$\mathbf{J}^T \Delta {}^0\boldsymbol{A}_P^m = \left[\ddot{q}_1 \left\{\boldsymbol{S}_1; {}^0\$_1^1\right\}\ \dot{q}_2 \left\{\boldsymbol{S}_2; {}^0\$_2^1\right\}\ 0\right]^T + \left[\left\{\boldsymbol{S}_1; \mathcal{L}_1\right\}\ \left\{\boldsymbol{S}_2; \mathcal{L}_2\right\}\ \left\{\boldsymbol{S}_3; \mathcal{L}_3\right\}\right]^T,$$

where the Lie screw of acceleration $\mathcal{L}_i$ is given by

$$\mathcal{L}_i = \left[\dot{q}_i\ {}^0\$_i^1 \quad {}_1\omega_2^{i\,1}\$_i^2 + {}_2\omega_3^{i\,2}\$_i^3\right] + \left[{}_1\omega_2^{i\,1}\$_i^2 \quad {}_2\omega_3^{i\,2}\$_i^3\right], i = 1, 2.$$

Meanwhile,

$$\mathcal{L}_3 = \left[{}_0\omega_1^3\ {}^0\$_3^1 \quad {}_1\omega_2^{3\,1}\$_3^2 + {}_2\omega_3^{3\,2}\$_3^3\right] + \left[{}_1\omega_2^{3\,1}\$_3^2 \quad {}_2\omega_3^{3\,2}\$_3^3\right].$$

On the other hand, following in this fashion, we see that the input–output equation of jerk of the robot manipulator results in

$$\mathbf{J}^T \Delta {}^0\boldsymbol{J}_P^m = \left[\dddot{q}_1 \left\{\boldsymbol{S}_1; {}^0\$_1^1\right\} \dddot{q}_2 \left\{\boldsymbol{S}_2; {}^0\$_2^1\right\} 0\right]^T + \left[\{\boldsymbol{S}_1; \mathcal{J}_1\} \{\boldsymbol{S}_2; \mathcal{J}_2\} \{\boldsymbol{S}_3; \mathcal{J}_3\}\right]^T,$$

where for each limb the Lie screw of jerk $\mathcal{J}_*$ is computed as

$$\mathcal{J}_* = 2\left[{}^0\boldsymbol{V}_P^1 \quad {}^1\boldsymbol{A}_P^3\right] + 2\left[{}^1\boldsymbol{V}_P^2 \quad {}^2\boldsymbol{A}_P^3\right] + \left[{}^0\boldsymbol{A}_P^1 \quad {}^1\boldsymbol{V}_P^3\right] + \left[{}^1\boldsymbol{A}_P^2 \quad {}^2\boldsymbol{V}_P^3\right] + \left[{}^0\boldsymbol{V}_P^1\left[{}^1\boldsymbol{V}_P^2 \quad {}^2\boldsymbol{V}_P^3\right]\right].$$

Finally, the input–output equation of hyper-jerk results in

$$\mathbf{J}^T \Delta {}^0\boldsymbol{H}_P^m = \left[\ddddot{q}_1 \left\{\boldsymbol{S}_1; {}^0\$_1^1\right\} \ddddot{q}_2 \left\{\boldsymbol{S}_2; {}^0\$_2^1\right\} 0\right]^T + \left[\{\boldsymbol{S}_1; \mathcal{H}_1\} \{\boldsymbol{S}_2; \mathcal{H}_2\} \{\boldsymbol{S}_3; \mathcal{H}_3\}\right]^T,$$

where for each limb the Lie screw of hyper-jerk $\mathcal{H}_*$ is calculated according to

$$\begin{aligned}\mathcal{H}_* = {} & 3\left[{}^0\boldsymbol{V}_P^1 \quad {}^1\boldsymbol{J}_P^3\right] + 3\left[{}^1\boldsymbol{V}_P^2 \quad {}^2\boldsymbol{J}_P^3\right] + 3\left[{}^0\boldsymbol{A}_P^1 \quad {}^1\boldsymbol{A}_P^3\right] + \\ & 3\left[{}^1\boldsymbol{A}_P^2 \quad {}^2\boldsymbol{A}_P^3\right] + 3\left[{}^0\boldsymbol{V}_P^1\left[{}^1\boldsymbol{V}_P^2 \quad {}^2\boldsymbol{A}_P^3\right]\right] + \\ & 2\left[{}^0\boldsymbol{A}_P^1\left[{}^1\boldsymbol{V}_P^2 \quad {}^2\boldsymbol{V}_P^3\right]\right] + \\ & \left[{}^0\boldsymbol{J}_P^1 \quad {}^1\boldsymbol{V}_P^3\right] + \left[{}^1\boldsymbol{J}_P^2 \quad {}^2\boldsymbol{V}_P^3\right] + \\ & \left[{}^0\boldsymbol{V}_P^1\left[{}^1\boldsymbol{A}_P^2 \quad {}^2\boldsymbol{V}_P^3\right]\right] + \\ & \left[{}^0\boldsymbol{V}_P^1\left[{}^0\boldsymbol{V}_P^1\left[{}^1\boldsymbol{V}_P^2 \quad {}^2\boldsymbol{V}_P^3\right]\right]\right]\end{aligned}$$

15.6 Chapter 12: Gough's Tyre Testing Machine

1. *The figure shows the geometric scheme of a six-legged parallel manipulator whose moving platform is an equilateral triangle of side e. Assuming that B_1, B_3, and B_5 are selected as the control points of the robot, determine the closure equations required to approach the forward displacement analysis of the parallel manipulator based on the data of the control points. Hint: Point B_2 is located at the middle of points B_1 and B_3. Dealing with points B_4 and B_6, consider similar geometric conditions.*

   Solution. Let $B_1 = (w_1, w_2, w_3)$, $B_3 = (w_4, w_5, w_6)$, and $B_5 = (w_7, w_8, w_9)$ be the control points of the moving platform where $w_i (i = 1, 2, 3, \ldots, 9)$ are unknown coordinates. Then the vectors associated to the points B_2, B_4, and B_6 may be expressed upon the control points as follows:

$$\boldsymbol{B}_2 = (\boldsymbol{B}_1 + \boldsymbol{B}_3)/2, \quad \boldsymbol{B}_4 = (\boldsymbol{B}_3 + \boldsymbol{B}_5)/2, \quad \boldsymbol{B}_6 = (\boldsymbol{B}_1 + \boldsymbol{B}_5)/2.$$

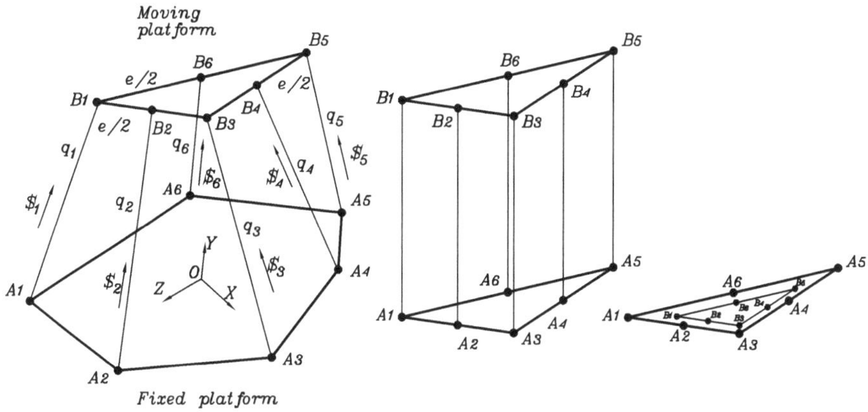

Six-legged parallel manipulator with triangular moving platform

Next, in order to compute the unknown coordinates, nine closure equations may be written as follows:

$$(\boldsymbol{B}_i - \boldsymbol{A}_i) \bullet (\boldsymbol{B}_i - \boldsymbol{A}_i) = q_i^2, \quad i = 1, 2, , 3, \ldots, 6,$$
$$(\boldsymbol{B}_1 - \boldsymbol{B}_3) \bullet (\boldsymbol{B}_1 - \boldsymbol{B}_3) = e^2,$$
$$(\boldsymbol{B}_5 - \boldsymbol{B}_3) \bullet (\boldsymbol{B}_5 - \boldsymbol{B}_3) = e^2,$$
$$(\boldsymbol{B}_1 - \boldsymbol{B}_5) \bullet (\boldsymbol{B}_1 - \boldsymbol{B}_5) = e^2.$$

2. *Determine two singular configurations of Gough's tyre testing machine. Hint: The singularity analysis may be performed based on the input–output equation of velocity of the parallel manipulator.*

Solution. The input–output equation of velocity of the hexapod is given by

$$\boldsymbol{J}^T \Delta \boldsymbol{V} = \mathbf{Q}_v,$$

where $\mathbf{J} = \begin{bmatrix} \$_1 \ \$_2 \ \$_3 \ \$_4 \ \$_5 \ \$_6 \end{bmatrix}$ is the screw-coordinate Jacobian matrix of the manipulator, $\Delta = \begin{bmatrix} \mathbf{0} \ \mathbf{I} \\ \mathbf{I} \ \mathbf{0} \end{bmatrix}$ is a 6×6 operator of polarity, and $\$_i (i = 1, 2, 3, \ldots, 6)$ are lines in Plücker coordinates directed from points A_i to points B_i. Meanwhile, $\mathbf{Q}_v = \begin{bmatrix} \dot{q}_1 \ \dot{q}_2 \ \dot{q}_3 \ \dot{q}_4 \ \dot{q}_5 \ \dot{q}_6 \end{bmatrix}$ is the first-order driver matrix of the robot. Hence, the hexapod is at a singular configuration when matrix $\mathbf{J}$ is singular, that is, when the lines $\$_i (i = 1, 2, 3, \ldots, 6)$ are parallel or coplanar.

References

Nam, Y.-J., & Park, M.-K. (2006). Kinematics and optimization of 2-dof parallel manipulator with revolute actuators and a passive leg. *Journal of Mechanical Science and Technology*, *20*(6), 828–839.

Wylie, C. R., & Barret, L. C. (1982). *Advanced engineering mathematics* (5th ed.). New York: McGraw-Hill.

Chapter 16
Appendix 1: A Simple Method to Compute the Rotation Matrix

There are several well-known methods to compute the rotation matrix. However, such methods require a little experience dealing with the handling of two related reference frames through the introduction of a wide class of orientation angles, such as Euler angles, Tait–Bryan angles (roll, pitch, and yaw), XYZ angles, and so on. This appendix provides a simple method for computing the rotation matrix. The method is based on the fact that the coordinates of three points of a rigid body are sufficient to determine the pose of the rigid body as observed from another body or reference frame.

With reference to Fig. 16.1, let us consider two identical triangles $\Delta A_1A_2A_3$, body labeled 0, and $\Delta a_1a_2a_3$, body labeled 1, where the coordinates of the vertices of both triangles are expressed in accordance with the reference frame XYZ attached to body 0. The rotation matrix $\mathbf{R}$ of body 1 with respect to body 0 may be expressed as

$$\mathbf{R} = [\hat{\boldsymbol{u}}_X \quad \hat{\boldsymbol{u}}_Y \quad \hat{\boldsymbol{u}}_Z], \tag{16.1}$$

where the unit vector $\hat{\boldsymbol{u}}_X$ is given by

$$\hat{\boldsymbol{u}}_X = \frac{(\boldsymbol{\tau} - \boldsymbol{a}_2) + \lambda(\boldsymbol{a}_3 - \boldsymbol{a}_2)}{\mid (\boldsymbol{\tau} - \boldsymbol{a}_2) + \lambda(\boldsymbol{a}_3 - \boldsymbol{a}_2) \mid}. \tag{16.2}$$

Here $\boldsymbol{\tau}$ is the position vector of point o with respect to point O. Meanwhile, taking into account that $d = r\sin\theta/\sin(\pi - \beta - \theta)$, we compute the scalar λ as

$$\lambda = d/(\mid \boldsymbol{a}_3 - \boldsymbol{a}_2 \mid). \tag{16.3}$$

Furthermore, the unit vector $\hat{\boldsymbol{u}}_Y$ is obtained as

$$\hat{\boldsymbol{u}}_Y = \frac{(\boldsymbol{a}_1 - \boldsymbol{a}_2) \times (\boldsymbol{a}_3 - \boldsymbol{a}_2)}{(\boldsymbol{a}_1 - \boldsymbol{a}_2) \times (\boldsymbol{a}_3 - \boldsymbol{a}_2)}. \tag{16.4}$$

J. Gallardo-Alvarado, *Kinematic Analysis of Parallel Manipulators by Algebraic Screw Theory*, DOI 10.1007/978-3-319-31126-5_16

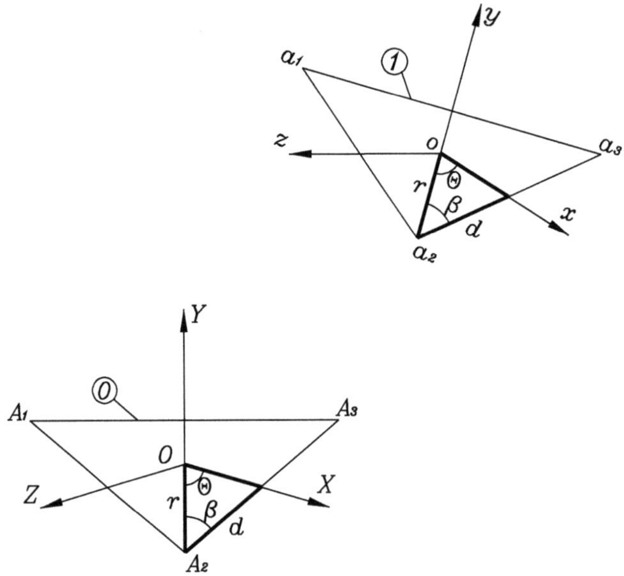

Fig. 16.1 Parameters of the rotation matrix

Finally, the unit vector $\hat{\boldsymbol{u}}_Z$ is determined according to the usual definition of cross product as follows:

$$\hat{\boldsymbol{u}}_Z = \hat{\boldsymbol{u}}_X \times \hat{\boldsymbol{u}}_Y. \tag{16.5}$$

Chapter 17
Appendix 2: Computer Codes

In this book, Maple was chosen as the software package to develop a few computer codes because of its affordability and higher characteristics to implement symbolic computation.

17.1 Maple Procedures

- Procedure *Lie*. The Lie product is computed based on the command *CrossProduct*, which performs the usual cross product of two three-dimensional vectors. The primal part of the resulting Lie product is obtained through the command *Primal*, while the dual part is computed through the command *Dual*.
- Procedure *Killing*. The Killing form is computed based on the command *DotProduct*, which performs the usual inner product of two three-dimensional vectors. The scalar is obtained directly by applying the command *DotProduct* to the primal parts of the involved six-dimensional vectors.
- Procedure *Klein*. Similar to the Killing form, the Klein form is computed based on the command *DotProduct*. The scalar is computed by a systematic application of the command *DotProduct* to the involved six-dimensional vectors.
- Procedure *Jacobi*. The identity of Jacobi of three vectors V1, V2, and V3 are computed by means of the procedure *Jacobi*, which resorts recursively to the procedure *Lie*. Evidently, the final result depends on the order of the involved six-dimensional vectors.
- Procedure *jVm:* Compute the velocity state of body *m* as observed from body *j*.
- Procedure *jLieAm:* Compute the Lie screw of acceleration of body *m* with respect to body *j*.
- Procedure *jAm:* Compute the reduced acceleration state of body *m* as measured from body *j*.

J. Gallardo-Alvarado, *Kinematic Analysis of Parallel Manipulators by Algebraic Screw Theory*, DOI 10.1007/978-3-319-31126-5_17

- Procedure *JLieJm:* Compute the Lie screw of jerk of body *m* as observed from body *j*.
- Procedure *jJm:* Compute the reduced jerk state of body *m* as measured from body *j*.
- Procedure *JLieHJm:* Compute the Lie screw of hyper-jerk of body *m* with respect to body *j*.
- Procedure *jHJm:* Compute the reduced hyper-jerk state of body *m* as observed from body *j*.

In what follows the Maple sheets are provided.

```
> # Lie algebra se(3) of the Euclidean group SE(3). Basic
  operations.
>
> restart: with(LinearAlgebra):
>
> # Procedure to compute the Lie product between two six-
  dimensional vectors V1 and V2.
> Lie:=proc(V1,V2) local Primal,Dual:
> Primal:=CrossProduct(<V1[1],V1[2],V1[3]>,<V2[1],V2[2],V2[3]>):
> Dual:=CrossProduct(<V1[1],V1[2],V1[3]>,<V2[4],V2[5],V2[6]>)-
  CrossProduct(<V2[1],V2[2],V2[3]>,<V1[4],V1[5],V1[6]>):
> <Primal[1],Primal[2],Primal[3],Dual[1],Dual[2],Dual[3]>:
> end:
>
> # Procedure to compute the Killing form between two six-
  dimensional vectors V1 and V2.
> Killing:=proc(V1,V2):
> DotProduct(<V1[1],V1[2],V1[3]>,<V2[1],V2[2],V2[3]>):
> end:
>
> # Procedure to compute the Klein form between two six-dimensional
  vectors V1 and V2.
> Klein:=proc(V1,V2):
> DotProduct(<V1[1],V1[2],V1[3]>,<V2[4],V2[5],V2[6]>)+DotProduct
  (<V1[4],V1[5],V1[6]>,<V2[1],V2[2],V2[3]>):
> end:

> # Procedure to compute the Jacobi identity upon three six-
  dimensional vectors V1, V2 and V3.
> Jacobi:=proc(V1,V2,V3):
> Lie(V1,Lie(V2,V3))+Lie(V3,Lie(V1,V2))+Lie(V2,Lie(V3,V1)):
> end:
```

Maple sheet C-1

```
>
> # Symbolic example.
> Vector1:=<a1,b1,c1,d1,e1,f1>:Vector2:=<a2,b2,c2,d2,e2,
  f2>:Vector3:=<a3,b3,c3,d3,e3,f3>:print(Vector1,Vector2,Vector3);
```

$$\begin{bmatrix} a1 \\ b1 \\ c1 \\ d1 \\ e1 \\ f1 \end{bmatrix}, \begin{bmatrix} a2 \\ b2 \\ c2 \\ d2 \\ e2 \\ f2 \end{bmatrix}, \begin{bmatrix} a3 \\ b3 \\ c3 \\ d3 \\ e3 \\ f3 \end{bmatrix} \tag{1}$$

```
> Lie(Vector1,Vector2);
```

$$\begin{bmatrix} b1\,c2 - c1\,b2 \\ c1\,a2 - a1\,c2 \\ a1\,b2 - b1\,a2 \\ b1\,f2 - c1\,e2 - b2\,f1 + c2\,e1 \\ c1\,d2 - a1\,f2 - c2\,d1 + a2\,f1 \\ a1\,e2 - b1\,d2 - a2\,e1 + b2\,d1 \end{bmatrix} \tag{2}$$

```
> Killing(Vector1,Vector2);
```

$$\overline{a1}\,a2 + \overline{b1}\,b2 + \overline{c1}\,c2 \tag{3}$$

```
> Klein(Vector1,Vector2);
```

$$\overline{a1}\,d2 + \overline{b1}\,e2 + \overline{c1}\,f2 + \overline{d1}\,a2 + \overline{e1}\,b2 + \overline{f1}\,c2 \tag{4}$$

```
> simplify(Jacobi(Vector1,Vector2,Vector3));
```

$$\begin{bmatrix} 0 \\ 0 \\ 0 \\ 0 \\ 0 \\ 0 \end{bmatrix} \tag{5}$$

Maple sheet C-1, continuation

```
>
> # Numerical example.
> Vector1:=<0.25,-0.1,0.27,4,-8,12>:Vector2:=<0.5,0.3,0.58,0.0,15,
  -8>:Vector3:=<0.15,0.1,0.18,1.0,1.5,3.5>:print(Vector1,Vector2,
  Vector3);
```

$$\begin{bmatrix} 0.25 \\ -0.1 \\ 0.27 \\ 4 \\ -8 \\ 12 \end{bmatrix}, \begin{bmatrix} 0.5 \\ 0.3 \\ 0.58 \\ 0. \\ 15 \\ -8 \end{bmatrix}, \begin{bmatrix} 0.15 \\ 0.1 \\ 0.18 \\ 1.0 \\ 1.5 \\ 3.5 \end{bmatrix} \qquad (1)$$

```
> Lie(Vector1,Vector2);
```

$$\begin{bmatrix} -0.139 \\ -0.0100 \\ 0.125 \\ -11.4900000000000 \\ 5.68000000000000 \\ 8.95000000000000 \end{bmatrix} \qquad (2)$$

```
> Killing(Vector1,Vector2);
```

$$0.251599999999999990 \qquad (3)$$

```
> Klein(Vector1,Vector2);
```

$$2.900000000 \qquad (4)$$

```
> Jacobi(Vector1,Vector2,Vector3);
```

$$\begin{bmatrix} 0. \\ 0. \\ 0. \\ 1.11022302462516\ 10^{-16} \\ 4.44089209850063\ 10^{-16} \\ 0. \end{bmatrix} \qquad (5)$$

```
>
```

Maple sheet C-1, continuation

> *# Four−bar linkage mechanism. Symbolic computation of the velocity state of the coupler platform as measured from the fixed platform*

> *restart* : *with*(*linalg*) :

> *# Plücker coordinates of the reciprocal lines*

> $L_1 := vector(3, [1, 0, -a_1 \cdot \sin(q)]); L_2 := vector(3, [0, 1, a_1 \cdot \cos(q)]); L_3 := vector(3, [-\cos(\text{gamma}), \sin(\text{gamma}), a_0 \cdot \sin(\text{gamma})]);$

$$L_1 := \begin{bmatrix} 1 & 0 & -a_1 \sin(q) \end{bmatrix}$$

$$L_2 := \begin{bmatrix} 0 & 1 & a_1 \cos(q) \end{bmatrix}$$

$$L_3 := \begin{bmatrix} -\cos(\gamma) & \sin(\gamma) & a_0 \sin(\gamma) \end{bmatrix} \quad \textbf{(1)}$$

> *#Jacobian matrix of the robot*

> $J := augment(L_1, L_2, L_3);$

$$J := \begin{bmatrix} 1 & 0 & -\cos(\gamma) \\ 0 & 1 & \sin(\gamma) \\ -a_1 \sin(q) & a_1 \cos(q) & a_0 \sin(\gamma) \end{bmatrix} \quad \textbf{(2)}$$

> *# Operator of polarity*

> Delta := *matrix*(3, 3, [0, 1, 0, 0, 0, 1, 1, 0, 0]);

$$\Delta := \begin{bmatrix} 0 & 1 & 0 \\ 0 & 0 & 1 \\ 1 & 0 & 0 \end{bmatrix} \quad \textbf{(3)}$$

> *#Vector of generalized speeds,* $q_v = \frac{d(q)}{dt}$

> $Q := vector(3, [-a_1 \cdot q_v \cdot \sin(q), a_1 \cdot q_v \cdot \cos(q), 0]);$

$$Q := \begin{bmatrix} -a_1 q_v \sin(q) & a_1 q_v \cos(q) & 0 \end{bmatrix} \quad \textbf{(4)}$$

> *#Symbolic computation of the velocity state V*

> $V := simplify(multiply(inverse(multiply(transpose(J), \text{Delta})), Q));$

$$V := \left[\frac{a_1 q_v (\cos(\gamma) \sin(q) + \sin(\gamma) \cos(q))}{a_1 \sin(q) \cos(\gamma) + a_1 \cos(q) \sin(\gamma) - a_0 \sin(\gamma)}, \frac{\sin(\gamma) a_1 q_v \sin(q) a_0}{a_1 \sin(q) \cos(\gamma) + a_1 \cos(q) \sin(\gamma) - a_0 \sin(\gamma)}, -\frac{a_1 \cos(q) q_v a_0 \sin(\gamma)}{a_1 \sin(q) \cos(\gamma) + a_1 \cos(q) \sin(\gamma) - a_0 \sin(\gamma)} \right] \quad \textbf{(5)}$$

>

Maple sheet C-2

```
> # Symbolic computation of the velocity state, the accelerator, the jerkor and the reduced hyper-
    jerk state between two links, labeled as body j and body m, of a serial kinematic chain.
>
> restart : with(linalg) :
>
> # jVm: Procedure to compute the velocity state of body m with respect to body j.
> jVm :=proc(j, m) add(omega[i, i + 1]·S[i, i + 1], i=j..m − 1) :end:
> jVm(0, 3);
```

$$\omega_{0,1} S_{0,1} + \omega_{1,2} S_{1,2} + \omega_{2,3} S_{2,3} \tag{1}$$

```
>
> # jLieAm: Procedure to compute the Lie screw of acceleration of body m with respect to body j.
> jLieAm :=proc(j, m) add(Lie(omega[i, i + 1]·S[i, i + 1], jVm(i + 1, m)), i=j .. m − 2) :end:
> jLieAm(0, 3);
```

$$Lie(\omega_{0,1} S_{0,1}, \omega_{1,2} S_{1,2} + \omega_{2,3} S_{2,3}) + Lie(\omega_{1,2} S_{1,2}, \omega_{2,3} S_{2,3}) \tag{2}$$

```
>
> # jAm: Procedure to compute the reduced acceleration state of body m with respec to body j.
> jAm :=proc(j, m) add(alpha[i, i + 1]·S[i, i + 1], i=j..m − 1) + jLieAm(j, m) :end:
> jAm(0, 3);
```

$$\alpha_{0,1} S_{0,1} + \alpha_{1,2} S_{1,2} + \alpha_{2,3} S_{2,3} + Lie(\omega_{0,1} S_{0,1}, \omega_{1,2} S_{1,2} + \omega_{2,3} S_{2,3}) + Lie(\omega_{1,2} S_{1,2}, \omega_{2,3} S_{2,3}) \tag{3}$$

```
> # jLieJm: Procedure to compute the Lie screw of jerk of body m with respect to body j
> jLieJm :=proc(j, m) 2.· add(Lie(omega[i, i + 1]·S[i, i + 1], jAm(i + 1, m)), i=j .. m − 2)
    + add(Lie(alpha[i, i + 1]·S[i, i + 1], jVm(i + 1, m)), i=j .. m-2) + add(Lie(omega[i, i
    + 1]·S[i, i + 1], Lie(omega[i, i + 1]·S[i, i + 1], jVm(i + 1, m))), i=j .. m-2) :end:
> jLieJm(0, 3);
```

$$2.\, Lie(\omega_{0,1} S_{0,1}, \alpha_{1,2} S_{1,2} + \alpha_{2,3} S_{2,3} + Lie(\omega_{1,2} S_{1,2}, \omega_{2,3} S_{2,3})) + 2.\, Lie(\omega_{1,2} S_{1,2}, \alpha_{2,3} S_{2,3}) + Lie(\alpha_{0,1} S_{0,1}, \omega_{1,2} S_{1,2} + \omega_{2,3} S_{2,3}) + Lie(\alpha_{1,2} S_{1,2}, \omega_{2,3} S_{2,3}) + Lie(\omega_{0,1} S_{0,1}, Lie(\omega_{0,1} S_{0,1}, \omega_{1,2} S_{1,2} + \omega_{2,3} S_{2,3})) + Lie(\omega_{1,2} S_{1,2}, Lie(\omega_{1,2} S_{1,2}, \omega_{2,3} S_{2,3})) \tag{4}$$

```
>
> # jJm: Procedure to compute the reduced jerk state of body m with respect to body j.
> jJm :=proc(j, m) add(rho[i, i + 1]·S[i, i + 1], i=j..m − 1) + jLieJm(j, m) :end:
> jJm(0, 3);
```

$$\rho_{0,1} S_{0,1} + \rho_{1,2} S_{1,2} + \rho_{2,3} S_{2,3} + 2.\, Lie(\omega_{0,1} S_{0,1}, \alpha_{1,2} S_{1,2} + \alpha_{2,3} S_{2,3} + Lie(\omega_{1,2} S_{1,2}, \tag{5}$$

Maple sheet C-3

$$\omega_{2,3} S_{2,3})) + 2.\, Lie(\omega_{1,2} S_{1,2}, \alpha_{2,3} S_{2,3}) + Lie(\alpha_{0,1} S_{0,1}, \omega_{1,2} S_{1,2} + \omega_{2,3} S_{2,3})$$
$$+ Lie(\alpha_{1,2} S_{1,2}, \omega_{2,3} S_{2,3}) + Lie(\omega_{0,1} S_{0,1}, Lie(\omega_{0,1} S_{0,1}, \omega_{1,2} S_{1,2} + \omega_{2,3} S_{2,3}))$$
$$+ Lie(\omega_{1,2} S_{1,2}, Lie(\omega_{1,2} S_{1,2}, \omega_{2,3} S_{2,3}))$$

```
>
> # jLieHJm: Procedure to compute the Lie screw of hyper-jerk of body m with respect to body j
> jLieHJm :=proc( j, m)  3· add(Lie(omega[i, i + 1]·S[i, i + 1], jJm(i + 1, m)), i = j .. m-2)
      + 3.· add(Lie(alpha[i, i + 1]·S[i, i + 1], jAm(i + 1, m)), i = j .. m-2) + 3.
      · add(Lie(omega[i, i + 1]·S[i, i + 1], Lie(omega[i, i + 1]·S[i, i + 1], jAm(i + 1, m))), i
      = j .. m-2) + 2.· add(Lie(alpha[i, i + 1]·S[i, i + 1], Lie(omega[i, i + 1]·S[i, i + 1], jVm(i
      + 1, m))), i = j .. m-2) + add(Lie(rho[i, i + 1]·S[i, i + 1], jVm(i + 1, m)), i = j .. m-2)
      + add(Lie(omega[i, i + 1]·S[i, i + 1], Lie(alpha[i, i + 1]·S[i, i + 1], jVm(i + 1, m))), i
      = j .. m-2) + add(Lie(omega[i, i + 1]·S[i, i + 1], Lie(omega[i, i + 1]·S[i, i + 1],
      Lie(omega[i, i + 1]·S[i, i + 1], jVm(i + 1, m)))), i = j .. m − 2) :end:
> jLieHJm(0, 3);
```

$$3\, Lie(\omega_{0,1} S_{0,1}, \rho_{1,2} S_{1,2} + \rho_{2,3} S_{2,3} + 2.\, Lie(\omega_{1,2} S_{1,2}, \alpha_{2,3} S_{2,3}) + Lie(\alpha_{1,2} S_{1,2}, \omega_{2,3} S_{2,3}) \tag{6}$$
$$+ Lie(\omega_{1,2} S_{1,2}, Lie(\omega_{1,2} S_{1,2}, \omega_{2,3} S_{2,3}))) + 3\, Lie(\omega_{1,2} S_{1,2}, \rho_{2,3} S_{2,3})$$
$$+ 3.\, Lie(\alpha_{0,1} S_{0,1}, \alpha_{1,2} S_{1,2} + \alpha_{2,3} S_{2,3} + Lie(\omega_{1,2} S_{1,2}, \omega_{2,3} S_{2,3})) + 3.\, Lie(\alpha_{1,2} S_{1,2},$$
$$\alpha_{2,3} S_{2,3}) + 3.\, Lie(\omega_{0,1} S_{0,1}, Lie(\omega_{0,1} S_{0,1}, \alpha_{1,2} S_{1,2} + \alpha_{2,3} S_{2,3} + Lie(\omega_{1,2} S_{1,2},$$
$$\omega_{2,3} S_{2,3}))) + 3.\, Lie(\omega_{1,2} S_{1,2}, Lie(\omega_{1,2} S_{1,2}, \alpha_{2,3} S_{2,3})) + 2.\, Lie(\alpha_{0,1} S_{0,1},$$
$$Lie(\omega_{0,1} S_{0,1}, \omega_{1,2} S_{1,2} + \omega_{2,3} S_{2,3})) + 2.\, Lie(\alpha_{1,2} S_{1,2}, Lie(\omega_{1,2} S_{1,2}, \omega_{2,3} S_{2,3}))$$
$$+ Lie(\rho_{0,1} S_{0,1}, \omega_{1,2} S_{1,2} + \omega_{2,3} S_{2,3}) + Lie(\rho_{1,2} S_{1,2}, \omega_{2,3} S_{2,3}) + Lie(\omega_{0,1} S_{0,1},$$
$$Lie(\alpha_{0,1} S_{0,1}, \omega_{1,2} S_{1,2} + \omega_{2,3} S_{2,3})) + Lie(\omega_{1,2} S_{1,2}, Lie(\alpha_{1,2} S_{1,2}, \omega_{2,3} S_{2,3}))$$
$$+ Lie(\omega_{0,1} S_{0,1}, Lie(\omega_{0,1} S_{0,1}, Lie(\omega_{0,1} S_{0,1}, \omega_{1,2} S_{1,2} + \omega_{2,3} S_{2,3}))) + Lie(\omega_{1,2} S_{1,2},$$
$$Lie(\omega_{1,2} S_{1,2}, Lie(\omega_{1,2} S_{1,2}, \omega_{2,3} S_{2,3})))$$

```
>
> # jHJm: Procedure to compute the reduced hyper-jerk state of body m with respect to body j.
> jHJm :=proc( j, m) add(sigma[i, i + 1]·S[i, i + 1], i = j ..m − 1) + jLieHJm( j, m) :end:
> jHJm(0, 3);
```

$$\sigma_{0,1} S_{0,1} + \sigma_{1,2} S_{1,2} + \sigma_{2,3} S_{2,3} + 3\, Lie(\omega_{0,1} S_{0,1}, \rho_{1,2} S_{1,2} + \rho_{2,3} S_{2,3} + 2.\, Lie(\omega_{1,2} S_{1,2}, \tag{7}$$
$$\alpha_{2,3} S_{2,3}) + Lie(\alpha_{1,2} S_{1,2}, \omega_{2,3} S_{2,3}) + Lie(\omega_{1,2} S_{1,2}, Lie(\omega_{1,2} S_{1,2}, \omega_{2,3} S_{2,3})))$$
$$+ 3\, Lie(\omega_{1,2} S_{1,2}, \rho_{2,3} S_{2,3}) + 3.\, Lie(\alpha_{0,1} S_{0,1}, \alpha_{1,2} S_{1,2} + \alpha_{2,3} S_{2,3} + Lie(\omega_{1,2} S_{1,2},$$
$$\omega_{2,3} S_{2,3})) + 3.\, Lie(\alpha_{1,2} S_{1,2}, \alpha_{2,3} S_{2,3}) + 3.\, Lie(\omega_{0,1} S_{0,1}, Lie(\omega_{0,1} S_{0,1}, \alpha_{1,2} S_{1,2}$$

Maple sheet C-3, continuation

$$+ \alpha_{2,3} S_{2,3} + Lie(\omega_{1,2} S_{1,2}, \omega_{2,3} S_{2,3}))) + 3.\, Lie(\omega_{1,2} S_{1,2}, Lie(\omega_{1,2} S_{1,2}, \alpha_{2,3} S_{2,3}))$$
$$+ 2.\, Lie(\alpha_{0,1} S_{0,1}, Lie(\omega_{0,1} S_{0,1}, \omega_{1,2} S_{1,2} + \omega_{2,3} S_{2,3})) + 2.\, Lie(\alpha_{1,2} S_{1,2},$$
$$Lie(\omega_{1,2} S_{1,2}, \omega_{2,3} S_{2,3})) + Lie(\rho_{0,1} S_{0,1}, \omega_{1,2} S_{1,2} + \omega_{2,3} S_{2,3}) + Lie(\rho_{1,2} S_{1,2},$$
$$\omega_{2,3} S_{2,3}) + Lie(\omega_{0,1} S_{0,1}, Lie(\alpha_{0,1} S_{0,1}, \omega_{1,2} S_{1,2} + \omega_{2,3} S_{2,3})) + Lie(\omega_{1,2} S_{1,2},$$
$$Lie(\alpha_{1,2} S_{1,2}, \omega_{2,3} S_{2,3})) + Lie(\omega_{0,1} S_{0,1}, Lie(\omega_{0,1} S_{0,1}, Lie(\omega_{0,1} S_{0,1}, \omega_{1,2} S_{1,2}$$
$$+ \omega_{2,3} S_{2,3}))) + Lie(\omega_{1,2} S_{1,2}, Lie(\omega_{1,2} S_{1,2}, Lie(\omega_{1,2} S_{1,2}, \omega_{2,3} S_{2,3})))$$

```
>
```

Maple sheet C-3, continuation

```
> # Application of the Algorithm 795: PHCpack, a mathematical  tool developed by Verschelde
    (1999), to the forward displacement analysis of a Gough platform with planar platforms, Eqs.
    (12.11) of Example 12.1. The example was originally proposed by Huang et al (2007).

> # Library PHCpack
> restart : with(linalg) : with(phc) : setPHCloc("C:\\PHC") :
>
>
> # Higher non-linear system of nine equations in the nine unknows wi(i=1,2,3,...,9) associated to
    the control points of the moving platform. Eqs. (12-11) of Example 12.1
> Eq1 := w1^2 - 18. w1 - 124.2307694 + w2^2 - 6. w2 + w3^2 : Eq2 := 0.25 w4^2 + 0.5 w4 w1
    - 6 w4 + 0.25 w1^2 - 6. w1 - 78.5846153 + 0.25 w5^2 + 0.5 w5 w2 - 8 w5 + 0.25 w2^2
    - 8. w2 + 0.25 w6^2 + 0.5 w6 w3 + 0.25 w3^2 : Eq3 := w4^2 + w5^2 - 28. w5 - 20.1384616
    + w6^2 : Eq4 := -4. + 8 w4 - 13 w5 + 16. w7 - 26 w8 - 8 w1 + 13 w2 + w8^2 + w9^2 + w7^2
    + 0.25 w4^2 + 0.25 w5^2 + 0.25 w6^2 + 0.25 w3^2 + w8 w5 + w7 w4 - w1 w7 - w3 w9
    - w2 w8 + w9 w6 + 0.25 w1^2 + 0.25 w2^2 - 0.5 w4 w1 - 0.5 w5 w2 - 0.5 w6 w3 : Eq5
    := w7^2 + 14. w7 - 377. + w8^2 + 12. w8 + w9^2 : Eq6 := -3132.000000 w4
    - 5220.000000 w5 + 2807.999997 w7 + 4679.999995 w8 + 2268.000003 w1
    + 3780.000004 w2 - 1.178263385 10^5 + 675.9999987 w8^2 + 675.9999987 w9^2
    + 675.9999987 w7^2 + 840.9999999 w4^2 + 840.9999999 w5^2 + 840.9999999 w6^2
    + 441.0000009 w3^2 - 1507.999998 w8 w5 - 1507.999998 w7 w4 + 1092.000000 w1 w7
    + 1092.000000 w3 w9 + 1092.000000 w2 w8 - 1507.999998 w9 w6 + 441.0000009 w1^2
    + 441.0000009 w2^2 - 1218.000001 w4 w1 - 1218.000001 w5 w2 - 1218.000001 w6 w3 :
    Eq7 := w1^2 - 2 w4 w1 + w4^2 + w2^2 - 2 w5 w2 + w5^2 + w3^2 - 2 w6 w3 + w6^2 - 20 : Eq8
    := w7^2 - 2 w1 w7 + w1^2 + w8^2 - 2 w2 w8 + w2^2 + w9^2 - 2 w3 w9 + w3^2 - 26 : Eq9
    := w7^2 - 2 w4 w7 + w4^2 + w8^2 - 2 w5 w8 + w5^2 + w9^2 - 2 w6 w9 + w6^2 - 18 :
>
> # Assembly of the nine equations
> T := makeSystem([w1, w2, w3, w4, w5, w6, w7, w8, w9], [ ], [Eq1, Eq2, Eq3, Eq4, Eq5, Eq6,
    Eq7, Eq8, Eq9]) :
>
> # Solutions of the system T
> sols := solve(T) :
>
> # Refining the forty solutions of T.
> refsols := refine(sols, T, digits = 32, residual_tol = 1. 10^-30, error_tol = 1. 10^-30, singular_tol
    = 1. 10^-20, max_iterations = 10) :
>
> # Printing the forty solutions, including complex solutions.
> printSolutions(T, refsols);
```

Maple sheet C-4

```
(1) [w1 = -1.9357+1.0126*I, w2 = -6.7454-3.2539*I, w3 = 5.1995
-3.9691*I, w4 = 1.6092+3.5948*I, w5 = -1.1441+.37696*I, w6 =
-.46138e-1+1.6530*I, w7 = 15.381-7.5718*I, w8 = 8.8921-.94107*I,
w9 = 10.408+17.629*I]
(2) [w1 = 13.060-.24296*I, w2 = 9.1043-2.9878*I, w3 =
13.101+1.4675*I, w4 = 8.1355+.86950*I, w5 = 11.092-1.4988*I, w6
= 12.058-.94805*I, w7 = 7.3465+1.9068*I, w8 = 7.6276-.43359*I,
w9 = 8.9465-2.3972*I]
(3) [w1 = 61.324-33.536*I, w2 = 57.833-42.473*I, w3 =
54.635+74.744*I, w4 = 37.168-1.0400*I, w5 = 56.600-19.666*I, w6
= 16.386+53.486*I, w7 = -61.113-44.986*I, w8 = 2.5655+.72147*I,
w9 = 46.865-52.074*I]
(4) [w1 = 13.060+.24296*I, w2 = 9.1043+2.9878*I, w3 =
-13.101+1.4675*I, w4 = 8.1355-.86950*I, w5 = 11.092+1.4988*I, w6
= -12.058-.94805*I, w7 = 7.3465-1.9068*I, w8 = 7.6276+.43359*I,
w9 = -8.9465-2.3972*I]
(5) [w1 = 20.757+.15961e-286*I, w2 = -6.5123+.16067e-286*I, w3 =
.91496e-287-3.8067*I, w4 = 20.337+.13312e-286*I, w5 =
6.4315+.16134e-286*I, w6 = .93120e-287-15.961*I, w7 =
16.416+.33278e-301*I, w8 = 2.9977+.16141e-301*I, w9 =
.71481e-301-12.933*I]
(6) [w1 = -1.9357+1.0126*I, w2 = -6.7454-3.2539*I, w3 =
-5.1995+3.9691*I, w4 = 1.6092+3.5948*I, w5 = -1.1441+.37696*I,
w6 = .46138e-1-1.6530*I, w7 = 15.381-7.5718*I, w8 = 8.8921
-.94107*I, w9 = -10.408-17.629*I]
(7) [w1 = 61.324-33.536*I, w2 = 57.833-42.473*I, w3 = -54.635
-74.744*I, w4 = 37.168-1.0400*I, w5 = 56.600-19.666*I, w6 =
-16.386-53.486*I, w7 = -61.113-44.986*I, w8 = 2.5655+.72147*I,
w9 = -46.865+52.074*I]
(8) [w1 = 13.060+.24296*I, w2 = 9.1043+2.9878*I, w3 = 13.101
-1.4675*I, w4 = 8.1355-.86950*I, w5 = 11.092+1.4988*I, w6 =
12.058+.94805*I, w7 = 7.3465-1.9068*I, w8 = 7.6276+.43359*I, w9
= 8.9465+2.3972*I]
(9) [w1 = 13.060-.24296*I, w2 = 9.1043-2.9878*I, w3 = -13.101
-1.4675*I, w4 = 8.1355+.86950*I, w5 = 11.092-1.4988*I, w6 =
-12.058+.94805*I, w7 = 7.3465+1.9068*I, w8 = 7.6276-.43359*I, w9
= -8.9465+2.3972*I]
(10) [w1 = 20.757-.14521e-263*I, w2 = -6.5123-.42248e-263*I, w3
= -.60720e-263+3.8067*I, w4 = 20.337-.37215e-263*I, w5 = 6.4315
-.10e-262*I, w6 = -.93078e-278+15.961*I, w7 =
16.416+.53701e-278*I, w8 = 2.9977+.39854e-279*I, w9 =
-.10000e-277+12.933*I]
```

Maple sheet C-4, continuation

```
(11) [w1 = -21.163-.96547e-297*I, w2 = 33.004+.70623e-296*I, w3
= -.60333e-296+39.948*I, w4 = -25.754-.23388e-296*I, w5 =
14.231+.67e-310*I, w6 = -.15158e-309+21.146*I, w7 = -28.417
-.19192e-310*I, w8 = 20.293+.38652e-310*I, w9 =
-.54415e-310+26.230*I]
(12) [w1 = -6.5526-.299e-233*I, w2 = 41.747+.31e-233*I, w3 =
-.30e-233+39.103*I, w4 = -19.607-.38e-234*I, w5 = 20.359
-.60e-234*I, w6 = .20e-234+14.447*I, w7 = -15.967+.63081e-234*I,
w8 = 18.168+.14870e-233*I, w9 = -.21279e-233+14.231*I]
(13) [w1 = -5.0912-.56191e-265*I, w2 = 6.9611-.18234e-264*I, w3
= -.47264e-264+.14715*I, w4 = -9.0843-.22104e-265*I, w5 = 1.1622
-.10760e-264*I, w6 = -.28327e-264+5.5851*I, w7 =
25.759+.55760e-264*I, w8 = 12.266-.10e-263*I, w9 = .15822e-294
-30.738*I]
(14) [w1 = -18.637+.227e-250*I, w2 = -12.053+.24e-250*I, w3 =
.33e-250+27.860*I, w4 = -8.3980+.135e-250*I, w5 =
-7.2242+.141e-250*I, w6 = .22e-250+17.460*I, w7 =
6.0931+.46041e-251*I, w8 = 12.211-.17531e-251*I, w9 =
.44246e-251-6.4088*I]
(15) [w1 = 8.9273-.10000e-254*I, w2 = -11.638+.11051e-261*I, w3
= -.57723e-262-.19249*I, w4 = 12.241-.20306e-261*I, w5 =
-.20936e-1-.23216e-262*I, w6 = -.18925e-261-11.414*I, w7 =
19.867+.45715e-262*I, w8 = 5.2487+.99801e-262*I, w9 =
.11960e-261-19.656*I]
(16) [w1 = -2.0900-.44285e-282*I, w2 = 7.6880+.10325e-282*I, w3
= -8.3226+.64826e-282*I, w4 = -2.3113-.21416e-281*I, w5 =
7.0404+.27666e-295*I, w6 = -12.742-.52806e-295*I, w7 = -2.3407
-.19323e-294*I, w8 = 11.217+.13171e-295*I, w9 = -11.995
-.56153e-295*I]
(17) [w1 = -21.163+.28e-309*I, w2 = 33.004-.24e-309*I, w3 =
-.44e-309-39.948*I, w4 = -25.754+.47759e-309*I, w5 = 14.231
-.17075e-309*I, w6 = -.58351e-309-21.146*I, w7 =
-28.417+.46947e-309*I, w8 = 20.293-.40444e-310*I, w9 =
-.42387e-309-26.230*I]
(18) [w1 = -5.0912-.56191e-265*I, w2 = 6.9611-.18234e-264*I, w3
= .47264e-264-.14715*I, w4 = -9.0843-.22104e-265*I, w5 = 1.1622
-.10760e-264*I, w6 = .28327e-264-5.5851*I, w7 =
25.759+.55760e-264*I, w8 = 12.266-.10e-263*I, w9 =
-.15822e-294+30.738*I]
(19) [w1 = -6.5526-.71488e-209*I, w2 = 41.747+.11031e-208*I, w3
= .13774e-208-39.103*I, w4 = -19.607-.44553e-209*I, w5 =
20.359+.97259e-210*I, w6 = .64745e-209-14.447*I, w7 = -15.967
```

Maple sheet C-4, continuation

```
-.21972e-209*I, w8 = 18.168+.50729e-209*I, w9 = .10e-208-14.231*
I]
(20) [w1 = -18.637-.18e-309*I, w2 = -12.053-.17e-309*I, w3 =
.278e-309-27.860*I, w4 = -8.3980-.1094e-309*I, w5 = -7.2242
-.112e-309*I, w6 = .19e-309-17.460*I, w7 = 6.0931-.366e-310*I,
w8 = 12.211+.101e-310*I, w9 = .47180e-310+6.4088*I]
(21) [w1 = 8.9273-.31443e-311*I, w2 = -11.638+.43374e-313*I, w3
= .21112e-311+.19249*I, w4 = 12.241-.30678e-311*I, w5 =
-.20936e-1-.51088e-312*I, w6 = .26624e-311+11.414*I, w7 = 19.867
-.25266e-312*I, w8 = 5.2487+.60346e-312*I, w9 = .10e-324+19.656*
I]
(22) [w1 = -2.0900+.10e-281*I, w2 = 7.6880+.78485e-296*I, w3 =
8.3226+.41749e-296*I, w4 = -2.3113-.24368e-296*I, w5 =
7.0404+.23427e-296*I, w6 = 12.742+.83754e-297*I, w7 =
-2.3407+.30632e-296*I, w8 = 11.217+.16063e-296*I, w9 = 11.995
-.34955e-296*I]
(23) [w1 = -1.8459+1.5583*I, w2 = 8.2472+1.9084*I, w3 = -8.7041
-.79126*I, w4 = -1.9058+.56762*I, w5 = 7.7898+1.3788*I, w6 =
-13.292-.72554*I, w7 = -5.1545-.89956*I, w8 = 11.212-.86484e-1*
I, w9 = -12.776-.24644*I]
(24) [w1 = 6.3558-5.2538*I, w2 = 3.6532-3.3474*I, w3 =
-15.690+.74604*I, w4 = 6.1043-2.5121*I, w5 = 7.8707-2.6747*I, w6
= -12.441+.85151e-1*I, w7 = .88675-2.0559*I, w8 = 8.3399+.76204*
I, w9 = -14.111-.37468*I]
(25) [w1 = -1.9357-1.0126*I, w2 = -6.7454+3.2539*I, w3 =
5.1995+3.9691*I, w4 = 1.6092-3.5948*I, w5 = -1.1441-.37696*I, w6
= -.46138e-1-1.6530*I, w7 = 15.381+7.5718*I, w8 = 8.8921+.94107*
I, w9 = 10.408-17.629*I]
(26) [w1 = -1.9357-1.0126*I, w2 = -6.7454+3.2539*I, w3 = -5.1995
-3.9691*I, w4 = 1.6092-3.5948*I, w5 = -1.1441-.37696*I, w6 =
.46138e-1+1.6530*I, w7 = 15.381+7.5718*I, w8 = 8.8921+.94107*I,
w9 = -10.408+17.629*I]
(27) [w1 = -1.8459+1.5583*I, w2 = 8.2472+1.9084*I, w3 =
8.7041+.79126*I, w4 = -1.9058+.56762*I, w5 = 7.7898+1.3788*I, w6
= 13.292+.72554*I, w7 = -5.1545-.89956*I, w8 = 11.212-.86484e-1*
I, w9 = 12.776+.24644*I]
(28) [w1 = 6.3558-5.2538*I, w2 = 3.6532-3.3474*I, w3 = 15.690
-.74604*I, w4 = 6.1043-2.5121*I, w5 = 7.8707-2.6747*I, w6 =
12.441-.85151e-1*I, w7 = .88675-2.0559*I, w8 = 8.3399+.76204*I,
w9 = 14.111+.37468*I]
(29) [w1 = 61.324+33.536*I, w2 = 57.833+42.473*I, w3 =
-54.635+74.744*I, w4 = 37.168+1.0400*I, w5 = 56.600+19.666*I, w6
```

Maple sheet C-4, continuation

```
= -16.386+53.486*I, w7 = -61.113+44.986*I, w8 = 2.5655-.72147*I,
w9 = -46.865-52.074*I]
(30) [w1 = 54.225+.11981e-296*I, w2 = 74.270+.53353e-296*I, w3 =
.52260e-296-83.130*I, w4 = 22.509-.31464e-296*I, w5 =
59.189+.11716e-296*I, w6 = -.37027e-297-48.296*I, w7 =
-54.423+.95814e-296*I, w8 = 7.4011+.81511e-297*I, w9 =
.10e-295+44.346*I]
(31) [w1 = -1.8459-1.5583*I, w2 = 8.2472-1.9084*I, w3 =
-8.7041+.79126*I, w4 = -1.9058-.56762*I, w5 = 7.7898-1.3788*I,
w6 = -13.292+.72554*I, w7 = -5.1545+.89956*I, w8 =
11.212+.86484e-1*I, w9 = -12.776+.24644*I]
(32) [w1 = 36.346+.21049e-290*I, w2 = 49.200+.15214e-290*I, w3 =
-.24751e-290+51.653*I, w4 = 25.349+.17307e-290*I, w5 =
50.076+.10885e-290*I, w6 = -.20001e-290+41.568*I, w7 =
-92.279+.158e-294*I, w8 = 7.1677-.349e-296*I, w9 = -.164e-294
-83.570*I]
(33) [w1 = 6.3558+5.2538*I, w2 = 3.6532+3.3474*I, w3 = -15.690
-.74604*I, w4 = 6.1043+2.5121*I, w5 = 7.8707+2.6747*I, w6 =
-12.441-.85151e-1*I, w7 = .88675+2.0559*I, w8 = 8.3399-.76204*I,
w9 = -14.111+.37468*I]
(34) [w1 = 9.0000+.1075e-293*I, w2 = 10.154-.73e-294*I, w3 =
-12.769-.34e-294*I, w4 = 4.6000+.1037e-293*I, w5 = 10.462
-.70e-295*I, w6 = -13.508+.42e-294*I, w7 = 5.2000+.11e-293*I, w8
= 8.8462-.306e-294*I, w9 = -9.6308+.70e-294*I]
(35) [w1 = 61.324+33.536*I, w2 = 57.833+42.473*I, w3 = 54.635
-74.744*I, w4 = 37.168+1.0400*I, w5 = 56.600+19.666*I, w6 =
16.386-53.486*I, w7 = -61.113+44.986*I, w8 = 2.5655-.72147*I, w9
= 46.865+52.074*I]
(36) [w1 = -1.8459-1.5583*I, w2 = 8.2472-1.9084*I, w3 = 8.7041
-.79126*I, w4 = -1.9058-.56762*I, w5 = 7.7898-1.3788*I, w6 =
13.292-.72554*I, w7 = -5.1545+.89956*I, w8 = 11.212+.86484e-1*I,
w9 = 12.776-.24644*I]
(37) [w1 = 54.225+.30401e-290*I, w2 = 74.270+.28163e-290*I, w3 =
-.40685e-290+83.130*I, w4 = 22.509-.66867e-291*I, w5 =
59.189+.99334e-291*I, w6 = -.22502e-304+48.296*I, w7 = -54.423
-.91094e-304*I, w8 = 7.4011-.21400e-304*I, w9 = .90948e-304
-44.346*I]
(38) [w1 = 9.0000-.79615e-293*I, w2 = 10.154+.58957e-292*I, w3 =
12.769-.33030e-292*I, w4 = 4.6000-.40800e-293*I, w5 =
10.462+.19558e-292*I, w6 = 13.508+.65129e-293*I, w7 = 5.2000
-.20848e-292*I, w8 = 8.8462+.17132e-292*I, w9 = 9.6308
-.48009e-304*I]
```

Maple sheet C-4, continuation

```
(39) [w1 = 36.346+.11562e-306*I, w2 = 49.200+.83567e-307*I, w3 =
.13596e-306-51.653*I, w4 = 25.349-.10e-307*I, w5 =
50.076+.54e-328*I, w6 = .345e-328-41.568*I, w7 =
-92.279+.335e-327*I, w8 = 7.1677-.761e-329*I, w9 =
.344e-327+83.570*I]
(40) [w1 = 6.3558+5.2538*I, w2 = 3.6532+3.3474*I, w3 =
15.690+.74604*I, w4 = 6.1043+2.5121*I, w5 = 7.8707+2.6747*I, w6
= 12.441+.85151e-1*I, w7 = .88675+2.0559*I, w8 = 8.3399-.76204*
I, w9 = 14.111-.37468*I]
```

Maple sheet C-4, continuation

```
> # Workspace of a Robot Delta-type robot. Example 14.2, Chapter 14
>
> restart : with(linalg) : with(plots) :
>
> # Procedure to compute the inner product of two three-dimensional vectors V1 and V2
> dotP :=proc(V1, V2) : V1[1]·V2[1] + V1[2]·V2[2] + V1[3]·V2[3] : end:
>
> # Parameters of the parallel manipulator
> A[1] := vector(3, [0.15, -0.05, 0.0]) : A[2] := vector(3, [-0.075, -0.05, -0.1299]) : A[3]
     := vector(3, [-0.075, -0.05, 0.1299]) : u[1] := vector(3, [0.0, 0.0, 1.0]) : u[2]
     := vector(3, [0.866, 0.0, -0.5]) : u[3] := vector(3, [-0.866, 0.0, -0.5]) : La := 0.2 : Lb
     := 0.4 : r := 0.0866 :
>
> # Procedure to compute the coordinates of the center P of the moving platform
> PointP :=proc( )
     local i, B, C, Eq, Eq12, Eq23, solEq, XX, YY, ZZ, X, Y, Z, EqZ, solZ, P1, P2;
     global P, n;
     for i to 3 do
         B[i] := linalg:-vector(3, [A[i][1] − La * u[i][3] * sin(q[i]), A[i][2] − La
           * cos(q[i]), A[i][3] + La * u[i][1] * sin(q[i])])
     end do;
     C[1] := linalg:-vector(3, [X, Y, Z]);
     C[2] := linalg:-vector(3, [X + (−1) * 0.866 * r, Y, Z + (−1) * 0.5 * r]);
     C[3] := linalg:-vector(3, [X + (−1) * 0.866 * r, Y, Z + 0.5 * r]);
     for i to 3 do
         Eq[i] := expand(dotP(evalm(C[i] − B[i]), evalm(C[i] − B[i])) + (−1) * Lb
           * Lb)
     end do;
     Eq12 := simplify(Eq[1] − Eq[2]);
     Eq23 := simplify(Eq[2] − Eq[3]);
     solEq := solve({Eq12, Eq23}, {X, Y});
     XX := subs(solEq, X);
     YY := subs(solEq, Y);
     C[1] := linalg:-vector(3, [XX, YY, Z]);
     EqZ := expand(dotP(evalm(C[1] − B[1]), evalm(C[1] − B[1])) − Lb * Lb);
     solZ := solve(EqZ);
     P1 := linalg:-vector(3, [subs(Z = solZ[1], XX) − 0.577 * r, subs(Z = solZ[1], YY),
     solZ[1]]);
     P2 := linalg:-vector(3, [subs(Z = solZ[2], XX) − 0.577 * r, subs(Z = solZ[2], YY),
     solZ[2]]);
     if ℑ(P1[1] + P1[2] + P1[3]) = 0 then n := n + 1; P[n] := P1 end if;
     if ℑ(P2[1] + P2[2] + P3[3]) = 0 then
         n := n + 1; P[n] := P2
     end if
   end proc:
>
> # Computation of the coordinates of point P according to the generalized coordinates qi
```

Maple sheet C-5

```
> n := 0 :for q[1] from 0.087 by 0.174 to 1.483 do for q[2] from 0.088 by 0.174 to 1.483
    do for q[3] from 0.089 by 0.174 to 1.483 do PointP( ) :end:end:end:
>
> # The workspace of the Delta-type robot. Coordinates of the center of the moving platform
> polyhedraplot( {seq(P[i], i = 1 ..n)}, polytype = dodecahedron, polyscale = 0.5, style
    = WIREFRAME, scaling = CONSTRAINED, labels = [px, py, pz], labelfont = [ITALIC,
    ITALIC, 10], font = [ITALIC, ITALIC, 10], axes = framed, gridstyle = rectangular);
```

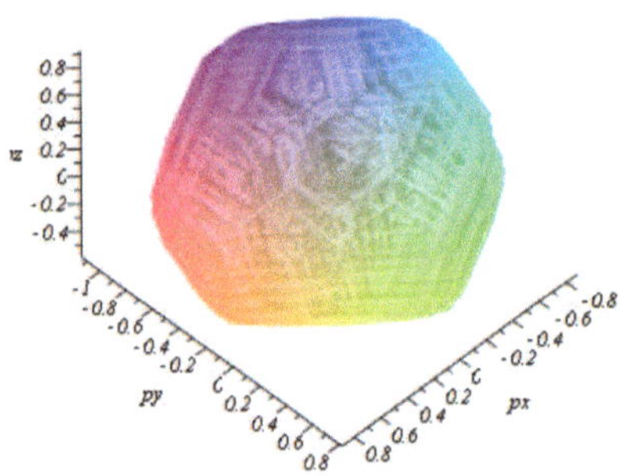


```
>
```

Maple sheet C-5, continuation

Index

J. Gallardo-Alvarado, *Kinematic Analysis of Parallel Manipulators by Algebraic Screw Theory*, DOI 10.1007/978-3-319-31126-5

MIX
Papier aus verantwortungsvollen Quellen
Paper from responsible sources
FSC® C105338

If you have any concerns about our products,
you can contact us on
ProductSafety@springernature.com

In case Publisher is established outside the EU,
the EU authorized representative is:
Springer Nature Customer Service Center GmbH
Europaplatz 3, 69115 Heidelberg, Germany

Printed by Libri Plureos GmbH
in Hamburg, Germany